BASIC MATHEMATICS
ARITHMETIC AND ALGEBRA

BASIC MATHEMATICS
ARITHMETIC AND ALGEBRA

Richard Williams
City Colleges of Chicago

SCOTT, FORESMAN AND COMPANY
Glenview, Illinois
Dallas, Texas
Oakland, New Jersey
Palo Alto, California
Tucker, Georgia
London, England

This book is dedicated to Thomas Rolle and Guerin Fischer who have always provided the support and encouragement that I have needed to facilitate my academic, professional, and personal development.

Library of Congress Cataloging in Publication Data

Williams, Richard W.
Basic mathematics.

Includes index.
1. Arithmetic—1961- . 2. Algebra. I. Title.
QA107.W57 1984 513'.122 83-19620

ISBN 0-673-15482-3

Printed in the United States of America.
2 3 4 5 6-DBH-88 87 86 85 84

PREFACE

This text is designed for a course in arithmetic and algebra or a course in elementary algebra. Students who successfully complete this textbook will be prepared for a subsequent algebra course, a technical mathematics course, an allied health mathematics course, or a liberal arts mathematics course.

All the basic arithmetic topics, except whole numbers, and all the topics required for a course in elementary algebra are covered in this text. This combination of topics was chosen because many college freshmen have forgotten most of their computational skills with fractions, decimals, and percents, but few of them need a complete course in arithmetic. Many of the students in an elementary algebra class need to review fractions, decimals, or percents in order to successfully complete the course.

In an informal, nonrigorous, and nonthreatening style the text guides students through the instructional material like an effective tutor. Moreover, the text can be utilized in a traditional lecture-discussion format, a mastery-learning or competency-based program, or in a mathematics laboratory. For ideas on using this text, see "To the Student."

SPECIAL FEATURES

Based upon the results of a three-year investigation to improve achievement scores and attitudes toward mathematics of college students, this book is designed with the following special features:

Basic **computational skills** with fractions, decimals, and signed numbers are continually **reinforced** once introduced. See, for example, Chapter 6, Example 8 (page 152), Example 10 (pages 154–55); Chapter 7, Examples 5b and 5c (page 177); Chapter 11, Examples 4e and 4f (pages 303–4); Chapter 13, Examples 10a and 10b (page 373).

Each chapter begins with a list of **OBJECTIVES** and a **PRETEST** keyed to the appropriate section and **Idea**.

Each new Idea is indicated by a boxed number in the left-hand margin such as 8 or 14. These Ideas are keyed to the problems in the Pretest.

Each chapter ends with a summary of **Important Terms** and **Important Skills** referenced by section.

Rules and procedures are written in a step-by-step format followed by several examples. Before examples are solved the rule is restated, and a color annotation or other pedagogical technique is used in guiding students through critical steps.

EXERCISES An average of 361 exercises per chapter.

An average of 48 ***Practice Problems*** per chapter. These problems allow the students to practice each new skill before proceeding to the next skill or concept.

Review Exercises are randomly arranged so that students must select the proper solution strategy.

Fill in the blank exercises. This type of problem helps develop the student's vocabulary and serves as an excellent review of the rules and procedures presented in the chapter.

Verbal statements of problems. These problems will sharpen the skills the students need to solve word problems. They will also help to develop the vocabulary needed for most mathematics courses. There are **over 4700 exercises**, **620 Practice Problems**, and **770 worked examples** in the text.

Word problems have been integrated into practically every section of the book and an entire chapter is devoted to the algebraic solution of word problems. Moreover, these word problems include applications to business, consumer mathematics, science, and technical mathematics.

FLEXIBLE ORGANIZATION Integers are discussed in Chapter 1 and reinforced in the last section of Chapters 2 and 3 on fractions and decimals. This approach was chosen to stimulate students with a new concept while reinforcing basic skills with whole numbers. However, if preferred, Chapter 1 along with Section 2.8 (Signed Fractions) and 3.7 (Signed Decimals) can be taught directly after Chapter 4 (Percents). Then Chapters 5–13 or the remaining material can be selected to meet the abilities and needs of the students.

INSTRUCTOR'S GUIDE To assist instructors in preparing to teach a course using *Basic Mathematics,* an Instructor's Guide is available. It contains the answers to the even-numbered exercises, teaching suggestions, quizzes, chapter tests, and final examinations.

ACKNOWLEDGMENTS The author wishes to extend his gratitude to the individuals whose expertise and generous assistance made this book possible. First, I am indebted to the students, staff, tutors, and faculty at Malcolm X College who offered helpful comments and suggestions in the development of the text. I am especially indebted to Drs. Wayne Watson and Rocco Caponigri for their assistance in making copies of the first draft of this text available for class testing. Secondly, I wish to thank the reviewers, Professors Wayne Andrepont and Charles Blanchard, University of Southwestern Louisiana; Harriet Bogue, West Georgia College; Terry Czerwinski, University of Illinois Circle Campus; Paul Pontius, Pan American University; Joan Richardson, University of Northern Colorado; Douglas F. Robertson, University of Minnesota; Richard Spangler, Tacoma Community College; Faye Thames, Lamar University; and staff of Scott, Foresman and Company whose comments and constructive criticism influenced the final format and contents of this text. In particular, I wish to thank Pamela Carlson, Barbara Maring, Lynn Friedman, and Susan Moss. A special thanks and sincere appreciation are extended to Dan Davis, Ernie Berman, Princess Coleman, Rochelle Colbert, and Vanessa Griffith for their detailed reviewing of the manuscript and their support throughout its preparation.

A final and loving thanks to my parents Bernis and Willie Ruth Williams whose sacrifices provided me with opportunities to develop the expertise and whose guidance provided me the tenacity required to initiate and complete this project.

Richard Williams

To the Student

If your instructor wishes, you can proceed at your own pace. Each unit begins with a list of **OBJECTIVES** and a **PRETEST**. Quickly look them over.

If you feel you are not ready to take the Pretest, we suggest you skip it. Instead, turn the page to the first section of the unit and work through the entire unit.

If you feel you can work the problems on the Pretest, do so. Then check your answers with those at the back of the book.

If you missed any problems and feel you need just a *quick review*, read the Ideas that explain the problems you missed.

If you missed any problems and feel you need a *more thorough review*, read the sections containing the problems you missed and work the *Practice Problems*. Then work the Exercises at the end of each section.

Answers to the Practice Problems are at the end of the section; answers to odd-numbered exercises and to all Pretest and Review problems are at the back of the book. We encourage you, however, to work each problem before looking at the answer. Only through working math will you learn to do math.

CONTENTS

1 INTEGERS

2 FRACTIONS AND MIXED NUMBERS

3 DECIMALS

4 PERCENTS

5 EXPONENTS

6 POLYNOMIALS

11 SPECIAL PRODUCTS AND FACTORING

12 SQUARE ROOTS AND QUADRATIC EQUATIONS

13 GEOMETRY

BASIC MATHEMATICS
ARITHMETIC AND ALGEBRA

INTEGERS

OBJECTIVES

1. Find the absolute value of an integer.
2. Add two or more integers.
3. Find the additive inverse of an integer.
4. Subtract one integer from another.
5. Find the value of an expression containing the operations of addition and subtraction.
6. Multiply two or more integers.
7. Divide one integer by another.
8. Simplify an expression containing more than one operation.
9. Evaluate a formula.

PRETEST	EXPLANATION
1. Find the following absolute values. **a.** $\vert -10\vert$ **b.** $\vert 30\vert$ **c.** $\vert +15\vert$	Section 1.1 Idea 5
2. Add. **a.** $-30 + (-15)$ **b.** $30 + (-60)$ **c.** $-5 + 10 + (-6) + (-30) + 40$	Section 1.2 Idea 7 Idea 9 Idea 13
3. Find the additive inverse. **a.** -10 **b.** 30 **c.** $+15$	Section 1.3 Idea 16
4. Subtract. **a.** $-30 - 15$ **b.** $-10 - (-45)$	Section 1.3 Idea 17
5. Find the value. **a.** $-16 + 10 - 25 + 2$ **b.** $15 - 25 + (-8) - (-8)$	Section 1.4 Idea 21 Idea 21
6. Multiply. **a.** $(5)(-10)$ **b.** $(-2)(3)(-5)$	Section 1.5 Idea 28
7. Divide. **a.** $\frac{-60}{10}$ **b.** $\frac{-30}{-5}$	Section 1.6 Idea 31
8. Simplify. **a.** $5 - 6 \cdot 2 + 12 \div (-4) + 8$ **b.** $3[-6 + 8(-13 + 2 \cdot 5)]$	Section 1.7 Idea 32 Idea 35
9. Find the value of $M = 6 - a(p + 5d)$ when $a = 4, p = -9, d = 1$.	Section 1.8 Idea 37

1.1 Integers and Absolute Value

Idea **1** The numbers

$$0, 1, 2, 3, 4, 5, 6, 7, 8, 9, 10, \text{ and so on}$$

are called **whole numbers.** Whole numbers can be shown by equally spaced points on a straight line called a **number line.** In Figure 1–1, the arrowhead shows that the number line continues to the right without ending. The three dots under the number line show that the numbers also continue without ending.

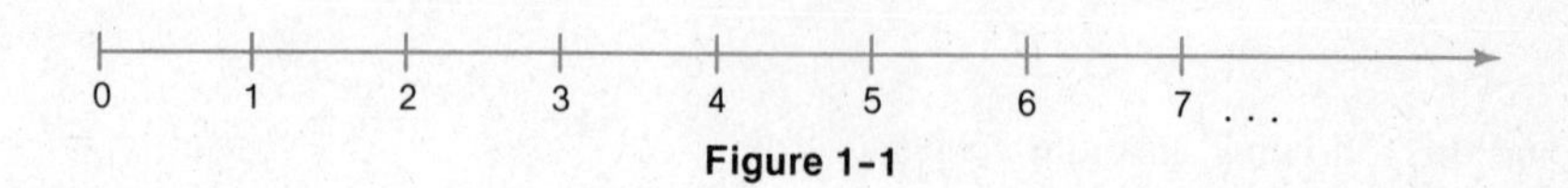

Figure 1-1

Whole numbers cannot be used in some situations. For example, no whole number can represent a temperature of 5° below zero or a loss of 10 yards in a football game. In such situations, integers must be used.

2 The numbers

$$\ldots -5, -4, -3, -2, -1, 0, 1, 2, 3, 4, 5, \ldots$$

are called **integers.** Integers can be shown on a number line as in Figure 1–2. The arrowheads show that the number line continues in both directions without ending.

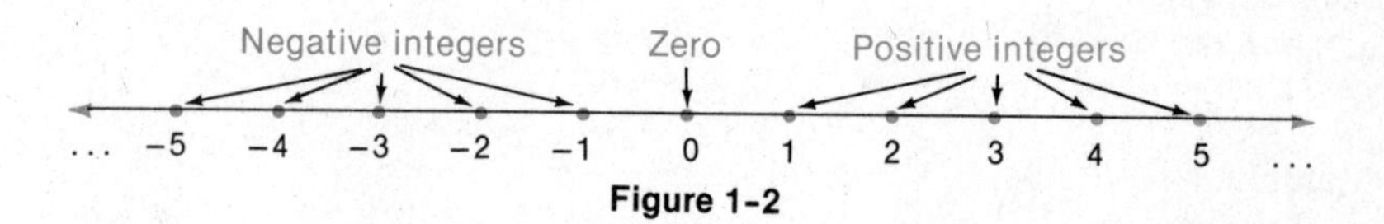

Figure 1-2

Integers to the right of zero are positive (greater than zero). A positive integer may have a plus sign (+) in front of it or it may have no sign at all. Integers to the left of zero are negative (less than zero). A negative integer has a minus sign (−) in front of it. Zero is the only integer that is neither positive nor negative.

3 Positive and negative integers can be used to represent opposite quantities. In football, for example, a gain of five yards can be expressed as 5 or +5 (positive 5), whereas a five-yard loss is expressed as −5 (negative 5). Some other examples are as follows:

1. A profit of \$30 can be represented by 30 or +30; a \$30 loss is expressed as −30.
2. A checking account containing \$55 can be represented by 55 or +55; a checking account overdrawn \$55 is represented by −55.

Practice Problem 1 ***Express each statement as a positive or negative integer.***

a. \$50 profit **b.** \$35 loss

c. 10-yard gain **d.** 15° below zero

NOTE: Answers to Practice Problems are at the end of the section in which they appear.

4 Until now, we have been concerned with the sign of an integer. However, frequently in mathematics we are interested in the numerical value of an integer without regard to its sign. This value is called the **absolute value** of an integer.

Definition of Absolute Value:

The **absolute value** of an integer x, written $|x|$, is the distance or number of units between 0 and x on the number line.

Example 1 Find the following absolute values.

a. $|4|$ **b.** $|-4|$

Solution **1a.** $|4| = 4$ since the number of units between 0 and 4 is 4 (see Figure 1–3).

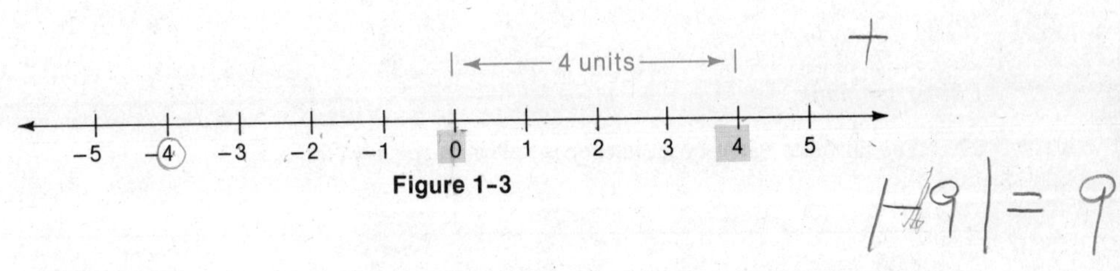

Figure 1-3

1b. $|-4| = 4$ since the number of units between 0 and -4 is 4 (see Figure 1–4).

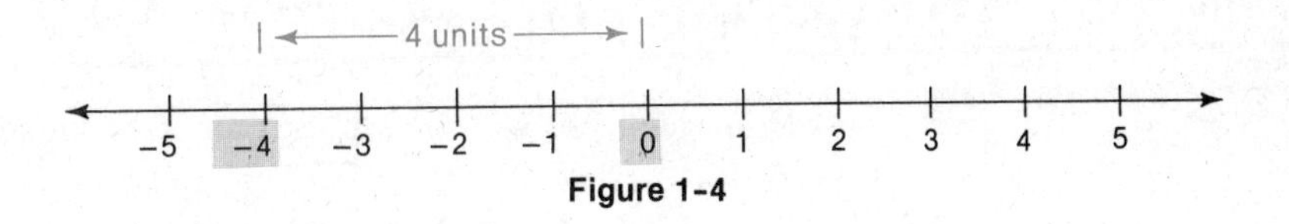

Figure 1-4

NOTE: Since the absolute value of a number simply represents a distance, it will never be negative.

5 When finding the absolute value of an integer, it may be inconvenient to draw a number line. Instead, we can use the fact that

$$|x| = x \quad \text{and} \quad |-x| = x$$

if x is a positive number or zero.

Example 2 Find the following absolute values.

a. $|-40|$ **b.** $|+50|$ **c.** $|727|$

Solution **2a.** $|-40| = 40$ **2b.** $|+50| = 50$ **2c.** $|727| = 727$

Practice Problem 2 ***Find the following absolute values.***

a. $|-10|$ **b.** $|+75|$ **c.** $|3|$ **d.** $|0|$

1.1 Exercises

Express each statement as a positive or negative integer.

1. 300 feet above sea level **2.** \$13 debt **3.** \$100 profit

4. 10–pound gain **5.** \$39 loss **6.** 5–pound loss

7. 300 feet below sea level **8.** 8° above zero **9.** having \$35

10. 13° below zero

Find the following absolute values.

11. $|-9|$ **12.** $|-84|$ **13.** $|-95|$ **14.** $|25|$ **15.** $|35|$ **16.** $|-37|$

17. $|-104|$ **18.** $|+84|$ **19.** $|724|$ **20.** $|74|$ **21.** $|0|$ **22.** $|9|$

23. $|+5|$ **24.** $|+73|$ **25.** $|+324|$ **26.** $|622|$

Fill in the blank.

27. The absolute value of an integer is never ______________.

Answers to Practice Problems **1a.** 50, or +50 **b.** −35 **c.** 10, or +10 **d.** −15 **2a.** 10 **b.** 75 **c.** 3 **d.** 0

1.2 Addition of Integers

6

To add integers, you will need to use your skills in adding and subtracting whole numbers.

7

To Add Two or More Integers Having the Same Sign:

1. Find the sum of their absolute values.
2. In front of this sum, place the common sign.

Example 3 Add.

a. $-3 + (-10)$ **b.** $-100 + (-150)$ **c.** $-8 + (-10) + (-50)$

Solution To add two or more negative integers, find the sum of their absolute values and place a minus sign in front of this sum.

Sign ↓ ↙ Sum of absolute values

3a. $-3 + (-10) = -(3 + 10) = -13$

3b. $-100 + (-150) = -(100 + 150) = -250$

3c. $-8 + (-10) + (-50) = -(8 + 10 + 50) = -68$

NOTE: The "( )" used in Example 3 are parentheses and are used as grouping symbols. They do one of two things.

1. They may indicate that whatever is inside the parentheses is treated as one quantity or number. For example, in the expression $-3 + 10$, -3 is a negative number. However, in the expression $-(3 + 10)$, the sum of 3 and 10 is a negative number.

2. They may separate a signed number from an operation symbol. For example, the parentheses in $-3 + (-10)$ separate the plus sign from the -10. You should *not* write $-3 + -10$.

8 To help you understand why *the sum of two or more negative numbers is a negative number,* consider an example involving money. If you owe Frank \$3 and you owe Dan \$10, you owe a total of \$13. This can be expressed as $-3 + (-10) = -13$.

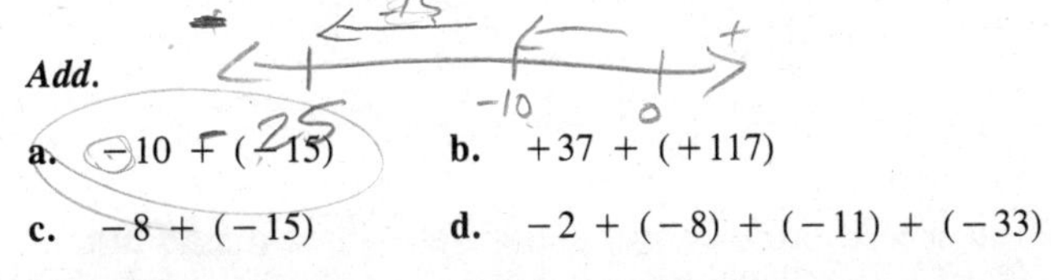

Practice Problem 3 **Add.**

a. $-10 + (-15)$ **b.** $+37 + (+117)$

c. $-8 + (-15)$ **d.** $-2 + (-8) + (-11) + (-33)$

9

To Add Two Integers Having Different Signs:

1. Find the difference between their absolute values by subtracting the smaller absolute value from the larger absolute value.
2. In front of this difference, place the sign of the integer with the larger absolute value.

Example 4 Add.

a. $-60 + 45$ **b.** $-30 + 65$ **c.** $45 + (-60)$ **d.** $815 + (-175)$

Solution To add a positive integer and a negative integer, subtract their absolute values (the larger minus smaller). Place the sign of the number with the larger absolute value in front of this difference.

Sign of the integer with the larger absolute value

Larger absolute value minus smaller

4a. $-60 + 45 = -(60 - 45)$
$= -15$

4b. $-30 + 65 = +(65 - 30) = 35$

4c. $45 + (-60) = -(60 - 45)$
$= -15$

4d. $815 + (-175) = +(815 - 175) = 640$

NOTE: When adding a positive integer and a negative integer, be sure to distinguish between the larger integer and the integer with the larger absolute value. For example, 45 is the larger integer in the problem $-60 + 45$, since a positive integer is greater than a negative integer. However, -60 is the integer with the larger absolute value, since $|-60| = 60$ and $|45| = 45$.

10 In Examples 4a and 4b, notice that the sum of a positive integer and a negative integer having *different absolute values* may be either positive or negative. It depends on which integer has the larger absolute value.

NOTE: The sum of two integers with different signs and the *same absolute value* is always zero. For example, $6 + (-6) = 0$ and $-3 + 3 = 0$.

Practice Problem 4 ***Add.***

a. $-40 + 75$ **b.** $-80 + 10$ **c.** $50 + (-70)$ **d.** $85 + (-15)$

11 There are two basic properties of addition. The first property is the **commutative law for addition.** This law states that the order in which two numbers are added does not affect the sum.

Example 5 Add 45 and -10 in two different ways.

Solution
$$-10 + 45 = +(45 - 10) = 35 \quad \text{or} \quad 45 + (-10) = +(45 - 10) = 35$$

12 The second property is the **associative law for addition.** This law states that the sum of three or more numbers remains the same no matter how they are grouped.

Example 6 Add $-3 + 10 + (-15)$.

Solution
$$\begin{aligned} &-3 + 10 + (-15) \\ &= [-3 + 10] + (-15) \\ &= 7 + (-15) \\ &= -8 \end{aligned} \quad \text{or} \quad \begin{aligned} &-3 + 10 + (-15) \\ &= -3 + [10 + (-15)] \\ &= -3 + (-5) \\ &= -8 \end{aligned}$$

NOTE: The "[ ]" are brackets and are used as grouping symbols. As with parentheses, first simplify inside the brackets and then add the other number or numbers.

13 Suppose you had to add more than two integers having different signs. You could add them two at a time, as in Example 6. However, this can be inconvenient. Instead, use the associative and commutative laws to rearrange the numbers so that they will be easier to add.

To Add Three or More Integers Having Different Signs:

1. Add all negative integers.
2. Add all positive integers.
3. Add the results obtained in steps 1 and 2.

Example 7 Add.

a. $-3 + 8 + (-10) + 40$

b. $-35 + (-10) + 15 + (-30) + 7$

c. $-10 + 18 + (-61) + 30 + (-65) + 42$

Solution To add integers having different signs, first add the negative integers, next add the positive integers, and then add the results.

7a. $-3 + 8 + (-10) + 40$

$= -13 + 48$
$= 35$

7b. $-35 + (-10) + 15 + (-30) + 7$
$= -75 + 22$
$= -53$

7c. $-10 + 18 + (-61) + 30 + (-65) + 42$
$= -136 + 90$
$= -46$

Practice Problem 5 ***Add.***

a. $-2 + 5 + (-16)$ **b.** $8 + (-50) + 95 + (-18)$

c. $-16 + 50 + 8 + (-30) + (-118) + 3$

14 Now that you have reviewed the basic skills required to add integers, read the following suggestions for solving word problems involving the addition of integers. These suggestions will also be useful when solving word problems involving the subtraction, multiplication, and division of integers.

Steps for Solving Word Problems:

1. **Identify knowns and unknowns.** Read the problem until you understand what information is being given and what must be found.
2. **Write a mathematical expression.** Reread the problem and then express the written statement as a mathematical expression. Sometimes it is helpful to first rewrite the original problem as a short statement.
3. **Find the answer.** Simplify the expression written in step 2 and answer the question asked in the original problem.
4. **Check the answer.** Determine (mentally) if the answer is reasonable.

Example 8 Express each statement as an addition problem and then solve.

a. A football player gained 11 yards on an off-tackle play and then lost 13 yards on a broken play. What is his final yardage gain or loss?

b. Ron weighed 260 pounds. Find his weight after the following changes: lost 10 pounds, gained 3 pounds, lost 15 pounds.

Known / Unknown

Solution 8a. **Think:** 11-yard gain + 13-yard loss = ?
Write: $11 + (-13)$
$= -(13 - 11)$
Find: $= -2$
The football player lost two yards.
Check: The football player lost more yardage than he gained. Therefore, a loss of two yards is a reasonable answer.

8b. Think: original weight + 10-lb loss + 3-lb gain + 15-lb loss

$$260 + (-10) + 3 + (-15)$$
$$= 263 + (-25)$$
$$= 238$$

Ron weighs 238 pounds.

Practice Problem 6 ***Express each statement as an addition problem and then solve.***

a. One day the temperature rose 9° from a previous reading of 15° below zero. What is the new temperature?

b. A certain stock had an opening price of 95 (meaning \$95). If the daily changes in the stock were listed as $+2$, -1, $+5$, -6, $+1$, and -3, find the final closing price of the stock.

1.2 Exercises

Add.

1. $-16 + (-10)$

2. $-10 + (-45)$

3. $16 + (-10)$

4. $-25 + 19$

5. $16 + (-30)$

6. $25 + (-45)$

7. $-100 + (-29)$

8. $-100 + 29$

9. $100 + (-30)$

10. $70 + (-85)$

11. $-70 + (-8) + (-13)$

12. $35 + 70$

13. $-35 + (-70)$

14. $-40 + (-10)$

15. $70 + 25$

16. $-3 + (-5) + (-8)$

17. $-6 + (-10) + (-118)$

18. $-30 + 17$

19. $16 + (-39)$

20. $-3 + (-4) + (-8) + (-35)$

21. $-16 + 400$

22. $-131 + (-151)$

23. $-601 + 101$

24. $-8 + (-5) + (-2)$

25. $10 + (-33) + 40$

26. $-6 + (-30) + 40$

27. $-10 + (-80) + 190 + (-70)$

28. $-35 + (-6) + 39 + 51$

29. $-350 + 500 + 600 + (-80)$

30. $-105 + 600 + (-800)$

31. $-350 + 600 + (-100) + 20$

32. $-61 + (-101) + 31 + (-81) + 131$

33. $-183 + 7 + (-190) + (-60) + 5$

34. $-61 + (-8) + 401 + (-71) + (-100) + 160 + (-3)$

35. $-3 + 8 + (-9) + (-6) + 13 + (-10) + 60 + 3$

36. $-8 + 13 + (-14) + (-11) + 18 + (-15) + 65 + 8$

37. $-8 + (-2) + (-10) + (-10) + (-20)$

38. $-5 + (-8) + (-15) + (-2) + (-8)$

39. $-5 + 8 + (-7) + 4$

40. $-10 + 13 + (-8) + 5$

41. $-15 + (-17)$

42. $-11 + (-12)$

43. $-18 + 8$

44. $-19 + 9$

45. $17 + (-20)$

46. $3 + (-9)$

47. $-8 + (-3) + (-4)$

48. $-3 + (-10) + (-5)$

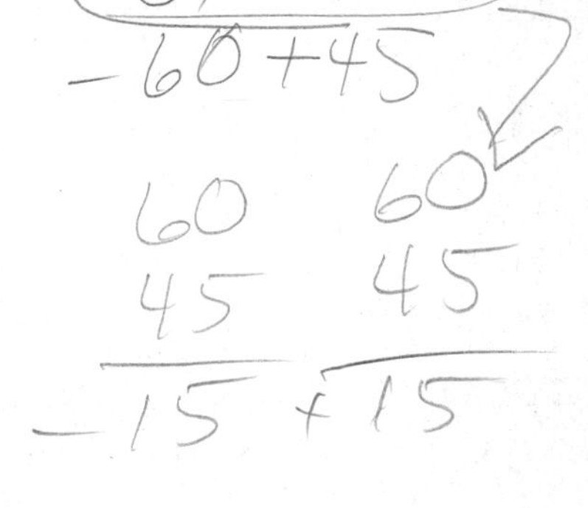

49. $-400 + 100$

50. $-500 + 300$

51. $15 + (-10) + 45$

52. $10 + (-20) + 40$

53. $-8 + 2 + 5 + 13 + (-18)$

54. $-5 + 3 + 7 + (-18) + 2$

55. $-5 + 8 + (-9) + 3 + (-16)$

56. $-9 + 3 + (-15) + 25 + (-1)$

57. $-18 + 3$

58. $-17 + 11$

59. $323 + (-220)$

60. $831 + (-530)$

61. $15 + (-115)$

62. $13 + (-213)$

63. $-18 + 53$

64. $-19 + 81$

Fill in the blanks with one or more words.

65. If you add two or more negative integers, the sign of the sum is ______________.

66. If you add two integers having different signs, the sign of the sum is determined by ______________.

67. The ______________ law allows us to change the order in which two integers are added without affecting the sum.

68. We can regroup a collection of integers without changing its sum by the ______________ law.

Express each statement as an addition problem and then find the answer. Show your work.

69. If the entries in your financial records read \$100 profit, \$30 loss, \$180 loss, and \$50 profit, what is the amount of your final profit or loss expressed as an integer?

70. If Joe records the monthly changes in his weight on a chart that reads "gained 3, lost 4, gained 5, lost 7, and lost 2," what is his final weight gain or loss expressed as an integer?

71. If Joe's original weight was 190 pounds in problem 70, how much does he weigh after all the changes in his weight?

72. The opening price of a certain stock is listed as 55 (meaning \$55). If changes in the stock are listed as $+1$, -2, $+5$, -6, -2, and $+10$ during the day, find the final closing price of the stock.

73. Find the sum of -3 and -5.

74. Find the sum of -4 and -5.

75. Find the sum of -3, 4, 5, -8, 15, -20, and 2.

76. Find the sum of -5, 5, 7, -10, 18, -30 and 3.

Answers to Practice Problems **3a.** -25 **b.** 154 **c.** -23 **d.** -54 **4a.** 35 **b.** -70 **c.** -20 **d.** 70 **5a.** -13 **b.** 35 **c.** -103 **6a.** $-6°$ or 6° below zero **b.** \$93

1.3 Subtraction of Integers

15 To subtract one integer from another, you must be able to add integers. To find $7 - 2$, for example, ask, "What number must be added to 2 to obtain 7?". Thus, $7 - 2 = 5$ because $2 + 5 = 7$.

Example 9 If you buy a book that costs \$7 and give the cashier a \$10 bill, he will probably give you three one-dollar bills in change. In other words, he will have subtracted \$7 from \$10 by determining how much money must be added to \$7 to obtain \$10.

NOTE: In the statement $7 - 2 = 5$, 7 is called the **minuend,** 2 is the **subtrahend,** and 5 is the **difference.** In general, if a, b, and c are any three numbers and $a - b = c$, then a is the minuend, b is the subtrahend, and c is the difference.

16 Before considering how to subtract one integer from another, we must learn how to find the additive inverse of an integer.

Definition of Additive Inverse:

The **additive inverse** or **opposite** of an integer x, denoted by $-x$, is that number which can be added to x so that the sum will be zero.

Based on this definition, the additive inverse of -3 is 3, since $-3 + 3 = 0$. Similarly, the additive inverse of 3 is -3.

Example 10 Find the additive inverses of the following integers.

a. -60 **b.** 160 **c.** 0

Solution To find the additive inverse of any integer except zero, change the sign of that integer.

10 a. The additive inverse of -60 is 60, since $-60 + 60 = 0$.

10 b. The additive inverse of 160 is -160, since $160 + (-160) = 0$.

10 c. The additive inverse of 0 is 0, since $0 + 0 = 0$.

NOTE: The additive inverse of any integer x except zero is written as $-x$. Thus, $-(-3) = 3$ can be read "The additive inverse of -3 is equal to 3." Similarly, $-(-4) = 4$, $-(-5) = 5$, $-(-10) = 10$, and $-(-6) = 6$.

Practice Problem 7 ***Find the additive inverses of the following integers.***

a. 6 **b.** -13 **c.** $+8$ **d.** -101 **e.** 0

17 The rule for subtracting one integer from another is as follows:

Subtraction Rule:

If a and b are two integers, then

$$a - b = a + (-b).$$

In other words, to subtract one integer, b, from another, a, we *add* the additive inverse of b to a.

Example 11 Subtract.

a. $-9 - (-15)$ **b.** $32 - 157$ **c.** $-10 - 15$ **d.** $9 - (-15)$

Solution To subtract one integer from another, add the additive inverse of the subtrahend.

Additive inverse of subtrahend

11a. $-9 - (-15) = -9 + 15 = 6$

Change operation to addition

11b. $32 - 157 = 32 + (-157) = -125$

11c. $-10 - 15 = -10 + (-15) = -25$

11d. $9 - (-15) = 9 + 15 = 24$

NOTE: The answer to a subtraction problem is correct if the subtrahend plus the difference is equal to the minuend. Thus, in Example 11, it is clear that:

$$-9 - (-15) = 6 \quad \text{if} \quad -15 + 6 = -9;$$
$$32 - 157 = -125 \quad \text{if} \quad 157 + (-125) = 32;$$
$$-10 - 15 = -25 \quad \text{if} \quad 15 + (-25) = -10; \text{ and}$$
$$9 - (-15) = 24 \quad \text{if} \quad -15 + 24 = 9.$$

In general,

$$a - b = c \quad \text{if} \quad b + c = a.$$

Practice Problem 8 ***Subtract and check your answers.***

a. $-35 - (-75)$ **b.** $-75 - 10$ **c.** $75 - 100$

18 The "$-$" sign can be used for three different purposes:

1. To indicate subtraction, as in $2 - 3$ (read as "two minus three");
2. To represent a negative number (for example, -3, which is read as "negative three"); and
3. To signify the additive inverse of a number (for example, -3 can be read as "the additive inverse of three").

The meaning of the sign is usually clear from the way in which the problem is written.

19 In Section 1.2, you learned that the order in which two numbers are added does not affect the answer (commutative law for addition). However, the order in which two numbers are subtracted *does* affect the answer. For example, $3 - 7$ is not equal to $7 - 3$, since $3 - 7 = -4$ and $7 - 3 = 4$.

20 Since order is important in subtraction, be careful in identifying the subtrahend (number to be subtracted). This is especially true for word problems that use the word *difference*. **"The difference between a and b"** is written $a - b$, and b is the subtrahend.

Example 12 Solve the following problems. Show all work.

a. A person's net worth is determined by subtracting liabilities from assets. If your liabilities are \$5500 and your assets are \$4800, what is your net worth?

b. The highest temperature recorded in Chicago one winter day was $-12°$. The lowest temperature was $-20°$. Find the difference between the highest and lowest temperatures.

Solution **12a.** Since the liabilities (\$5500) must be subtracted from the assets to determine your net worth, think of the liabilities as the subtrahend.

$$\begin{aligned}\text{Net worth} &= \text{assets} - \text{liabilities}\\ &= 4800 - 5500\\ &= 4800 + (-5500)\\ &= -700\end{aligned}$$

This means that you are \$700 in debt.

12b. Since you want to find the difference between the highest and the lowest temperature, the lowest temperature (or the second number) mentioned is the subtrahend.

$$\begin{aligned}\text{Temperature difference} &= \text{highest} - \text{lowest}\\ &= -12 - (-20)\\ &= -12 + 20\\ &= 8\end{aligned}$$

The difference between the highest and the lowest temperature is 8°.

NOTE: "Subtract a from b" is written $b - a$, and a is the subtrahend. However, "find the difference between a and b" is written $a - b$, and b is the subtrahend.

Practice Problem 9 ***Solve. Show all work.***

a. Subtract -5 from 37.

b. The average temperature in June for a certain city is 85°. On June 7, 1983, the temperature was 80°. Find the difference between the June 7 temperature and the average temperature in June.

1.3 Exercises

Find the additive inverses of the following integers.

1. -10 **2.** 5 **3.** -4 **4.** 10 **5.** $+8$

6. -100 **7.** 0 **8.** -35 **9.** 100 **10.** -16

Subtract and check your answers.

11. $-9 - 12$ **12.** $9 - (-12)$ **13.** $9 - 12$

14. $-9 - (-12)$ **15.** $50 - 75$ **16.** $75 - (-50)$

17. $-13 - (-16)$ **18.** $-16 - (-13)$ **19.** $-135 - 70$

20. $85 - 100$ **21.** $-16 - 32$ **22.** $-14 - (32)$

23. $-32 - (-14)$ **24.** $81 - (-29)$ **25.** $-18 - 50$

26. $35 - 82$ **27.** $-15 - (-35)$ **28.** $-35 - (-15)$

29. $85 - (-100)$ **30.** $16 - 35$ **31.** $0 - 15$

32. $-170 - 335$ **33.** $-911 - (-1100)$ **34.** $-600 - (-190)$

35. $-100 - (-2000)$ **36.** $100 - (-615)$ **37.** $851 - 1135$

38. $150 - 250$ **39.** $-13 - 18$ **40.** $-19 - 35$

41. $-13 - (-18)$ **42.** $-50 - (-60)$ **43.** $-18 - (-10)$

44. $-28 - (-13)$ **45.** $16 - (-16)$ **46.** $10 - (-11)$

47. $-18 - (-18)$ **48.** $-13 - (-13)$ **49.** $-859 - (-359)$

50. $-789 - (-180)$

Fill in the blanks with one or more words.

51. In the problem $35 - 45 = ?$, the subtrahend is ______________.

52. To subtract one integer from another, we add the additive inverse of the ______________ to the ______________.

53. In problem 51, the additive inverse of the subtrahend is ______________.

54. The additive inverse of a positive integer is a ______________ integer. The additive inverse of a negative integer is a ______________ integer.

55. ______________ is its own additive inverse.

Solve. Show your work.

56. The highest temperature in New York one winter day was 15°, and the lowest temperature was $-5°$. Find the difference between the highest and the lowest temperatures.

57. Mount Whitney, California is 14,495 feet above sea level, and El Centro, California is 45 feet below sea level. Find the difference in their elevations by subtracting the lowest from the highest elevation.

58. You are playing a game in which you have a score of -68. How many points must you earn to obtain a score of 35? (Hint: see Idea 17.)

59. Subtract -16 from -50.

60. The latitude of Lima, Peru is 12° south of the equator (expressed as -12). The latitude of Perth, Australia is 32° south of the equator (expressed as -32). Find the difference between Lima's and Perth's latitudes.

61. Subtract -5 from -10.

62. Subtract -10 from -5.

63. Find the difference between -5 and -10.

64. Find the difference between -10 and -5.

65. Subtract -8 from 8.

66. Subtract -10 from 10.

67. Find the difference between 8 and 10.

68. Find the difference between -8 and -10.

Answers to Practice Problems **7a.** -6 **b.** 13 **c.** -8 **d.** 101 **e.** 0 **8a.** 40 **b.** -85 **c.** -25 **9a.** 42 **b.** $-5°$

1.4 Addition and Subtraction of Integers

21 Sometimes in mathematics you must find the value of an expression that contains both the operations of subtraction and addition. To compute this value, you could do the indicated operations in order from left to right. However, you could also proceed as follows:

To Simplify Expressions Containing Subtraction and Addition:

1. Change every operation of subtraction to addition using the fact that $a - b = a + (-b)$, where $-b$ is the additive inverse of b.
2. Solve the resulting addition problem.

Example 13 Find the value of each expression.

a. $-13 + 8 - 17$ **b.** $-35 - 75 + 14 - (-100)$

c. $-18 - 75 + (-10) - 8$

Solution To simplify an expression containing both the operations of subtraction and addition, first change all operations of subtraction to addition by using the subtraction rule (Idea 17). Then, do the addition.

Additive inverse

13a. $-13 + 8 - 17 = -13 + 8 + (-17)$

Change operation to addition

$= -30 + 8$
$= -22$

13b. $-35 - 75 + 14 - (-100) = -35 + (-75) + 14 + 100$
$= -110 + 114$
$= 4$

13c. $-18 - 75 + (-10) - 8 = -18 + (-75) + (-10) + (-8)$
$= -111$

NOTE: If you are having difficulty with a particular topic, it is probably because you have not mastered the skills developed *before* the new topic. For example, to find the value of an expression containing both the operations of subtraction and addition, you must first master how to add integers (Section 1.2) and how to subtract one integer from another (Section 1.3).

Practice Problem 10 ***Find the value of the following expressions.***

a. $-81 - 17 + 100$ **b.** $35 - (-10) + (-18) - 15$

c. $-16 + (-30) - 17 - 8$

22 The concepts of income and expenses in business produce problems involving both addition and subtraction.

Example 14 At the end of June, the ADT Paint Company's cash assets were $800. In July, the company's balance sheet indicated the following cash transactions:

1. Painted John Smith's house and collected $5000.
2. Paid $2500 in employees' salaries.
3. Sold paint supplies and collected $3000.
4. Paid Johnson's Can Company $4000.
5. Paid $500 in rent.

How much cash does the ADT Paint Company have at the end of July?

Solution Incoming cash increases the amount of available cash, and outgoing cash decreases it.

$$\begin{aligned} &800 + 5000 - 2500 + 3000 - 4000 - 500 \\ &\quad = 800 + 5000 + (-2500) + 3000 + (-4000) + (-500) \\ &\quad = 8800 + (-7000) \\ &\quad = 1800 \end{aligned}$$

This implies that the ADT Paint Company has $1800 in cash assets at the end of July.

Practice Problem 11 ***Solve.***

Cazzie's Auto Supply Company has $5000 in cash. In January, his records indicate that he withdrew $600 for personal use, received $6000 for services rendered, paid $2000 to employees, collected $800 from Dr. Watson, and paid $600 for new equipment. If all of the above were cash transactions, what is Cazzie's cash balance?

23 The rules used to find the value of an expression containing both the operations of subtraction and addition can be used to simplify an expression involving only the operation of subtraction.

Example 15 Find the value of each expression.

a. $5 - 7 - 8 - (-13)$ **b.** $-5 - 10 - 8 - 13 - 15$

Solution To find the value of an expression containing only the operation of subtraction, change every operation of subtraction to addition by using the subtraction rule (Idea 17). Next, do the resulting addition problem.

15a. $5 - 7 - 8 - (-13) = 5 + (-7) + (-8) + 13$
$= 18 + (-15)$
$= 3$

15b. $-5 - 10 - 8 - 13 - 15 = -5 + (-10) + (-8) + (-13) + (-15)$
$= -51$

Practice Problem 12 ***Find the value of each expression.***

a. $8 - 15 - 10 - 3$ **b.** $-2 - 7 - 5 - 6 - 5$

1.4 Exercises

Find the value of each expression.

1. $-8 + 16 - 35$

2. $15 - 37 + (-18)$

3. $8 - (-35) + (-69)$

4. $-17 + (-40) - 6$

5. $-17 - (-35) + 15$

6. $-40 + 17 - (-6)$

7. $-16 - 10 + (-14)$

8. $-20 - 5 + 3$

9. $-17 - 8 + (-5) - 4$

10. $50 - 75 + (-8) - (-17)$

11. $60 - 70 - 40 - 110$

12. $-35 - (-15) + 81 - 3$

13. $-40 - (-50) + (-10) - 30$

14. $85 - 2 - 100 - 74$

15. $-14 + (-40) - (-5)$

16. $-117 - 501 - 171 - 34$

17. $-40 + 12 - 5$

18. $-65 - 401 - 101 - 807$

19. $-15 - 20 + 12 - (-8)$

20. $88 - 41 - 381 - 635$

21. $-8 - 8 + 2 - 5 + 1$

22. $-9 - 9 + 3 - 6 + 2$

23. $-3 - 4 - 1 - 5 - 7$

24. $-1 - 8 - 7 - 3 - 5 - 1$

25. $-10 + 15 - 30$

26. $-15 + 20 - 40$

27. $-10 - (-35) + 15 - 18$

28. $-8 - (-15) + 3 - 9$

29. $15 - 35 + (-8) + 5$

30. $8 - 13 + (-7) + 10$

31. $-40 + 15 - 8 + 2 - 10$

32. $-50 + 10 - 8 + 3 - 8$

33. $10 - 30 + (-15) + 18$

34. $8 - 20 + (-9) + 50$

35. $30 + (-10) + 13 + 17 - 18$

36. $20 + (-10) + 11 + 9 - 13$

37. $70 - 95 + (-5) + 31$

38. $60 - 85 + (-5) + 32$

39. $18 - 25 - 13 - 5$

40. $22 - 32 - 10 - 8$

41. $-18 + 10 - 17 + 5 - (-8) - 10 - 30 + (-3) - 8 + 3 - 50$

42. $-17 + 9 - 16 + 4 - (-7) - 20 - 35 + (-3) - 8 + 7 - 40$

Fill in the blanks.

43. To find the value of an expression containing both the operations of addition and subtraction, change every operation of __________ to __________ by using the __________ rule. Next, do the resulting __________ problem.

44. $5 - 18 + 15 - 8 - (-19) = 5 + (\_\_\_\_) + 15 + (\_\_\_\_) + (\_\_\_\_)$

45. The rule for solving a problem involving both addition and subtraction of integers is also used to find the value of an expression containing only the operation of __________.

Solve and show all work.

46. The PDQ Candy Company owes the Davis Chocolate Company \$5000. PDQ's financial records indicate the following cash transactions:

- Collected \$3000 from Kirkwood & Sons.
- Withdrew \$900 for personal use.
- Paid Davis Chocolate Company \$1500.
- Received \$1125 due from Kathy Gross.

What is PDQ's cash balance after the last transaction?

47. Guerin has \$850 in his checking account at the end of April. In May, his records indicate that he wrote checks for \$550 and \$200, paid a \$4 service charge, deposited \$30, and wrote another check for \$130. If the bank pays all of his checks, does he still have any available cash or is his account overdrawn?

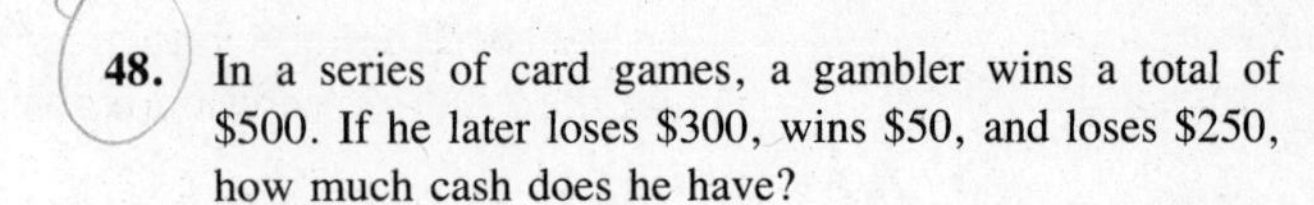

48. In a series of card games, a gambler wins a total of \$500. If he later loses \$300, wins \$50, and loses \$250, how much cash does he have?

49. The ACT Moving Company has \$30,000 in cash assets. ACT's balance sheet indicates that they paid \$8000 for new equipment, paid \$1900 for rent, collected \$575 from a customer for services rendered, withdrew \$15,000 for personal use, and received \$3145 from the sale of used furniture. If ACT's business dealings are all cash transactions, how much available cash does the company have after selling the furniture?

Answers to Practice Problems **10a.** 2 **b.** 12 **c.** −71 **11.** \$8,600 **12a.** −20 **b.** −25

1.5 Multiplication of Integers

24 In arithmetic, multiplication is viewed as repeated addition. That is, $2 \cdot 5$ (read as "2 times 5") means the sum of 2 fives, or $2 \cdot 5 = 5 + 5 = 10$. Similarly, $5 \cdot 2$ means the sum of 5 twos, or

$$5 \cdot 2 = 2 + 2 + 2 + 2 + 2 = 10.$$

This shows that *the product of two positive integers is a positive integer.* Note that $2 \cdot 5 = 5 \cdot 2$, suggesting that there is a commutative law for multiplication. In fact, the order in which you multiply two numbers does not affect the product.

NOTE: Multiplication will be indicated by a dot or parentheses. Thus, 3 times 5 is written as $3 \cdot 5$, $3(5)$, or $(3)(5)$.

25 It is easy to see what happens when a positive integer is multiplied by a negative integer. For example, $(3)(-5)$ means the sum of 3 negative fives, or

$$(3)(-5) = (-5) + (-5) + (-5) = -15.$$

However, when a negative integer is to be multiplied by a positive integer, you must be careful. For example, $(-3)(4)$ is meaningless since you cannot add the number 4 "negative three" times. But, since the order in which two numbers are multiplied does not affect the product, we know that

$$(-3)(4) = 4(-3) = -3 + (-3) + (-3) + (-3) = -12.$$

Therefore, we can think of $(-3)(4)$ as $(4)(-3)$, or the sum of 4 negative threes. As these examples show, *the product of a positive integer and a negative integer is a negative integer.*

26 Now consider the following sequence:

$$\begin{aligned} 4(-2) &= -8 \\ 3(-2) &= -6 \\ 2(-2) &= -4 \\ 1(-2) &= -2 \\ 0(-2) &= 0 \end{aligned}$$

As the first multipliers decrease by 1 (from 4 to 3 to 2, etc.), the products or answers increase by 2 (from -8 to -6 to -4, etc.). If you continued this pattern for multipliers less than zero, you would obtain

$$(-1)(-2) = 2$$
$$(-2)(-2) = 4$$
$$(-3)(-2) = 6$$
$$(-4)(-2) = 8.$$

This sequence suggests that *the product of two negative integers is always positive.* In fact, this is true in our number system.

27 Based on the findings presented in Ideas 24–26, we can state a general rule for multiplying two integers.

To Multiply Any Two Integers:

1. Multiply their absolute values.
2. If the signs of the two integers are the same, the product is positive. If the signs are different, the product is negative.

Example 16 Multiply.

a. $3(-7)$ **b.** $(-30)(7)$ **c.** $(-3)(-7)$ **d.** $(-10)(-9)$

Solution To multiply two integers, multiply their absolute values. The product is a positive number if the signs are the same. The product is a negative number if the signs are different.

Signs different — Product is negative — Multiply absolute values

16a. $3(-7) = -(3 \cdot 7)$
$= -21$

16b. $(-30)(7) = -(30 \cdot 7)$
$= -210$

16c. $(-3)(-7) = +(3 \cdot 7)$
$= 21$

16d. $(-10)(-9) = +(10 \cdot 9)$
$= 90$

NOTE: The product of zero and any integer is zero. For example, $(0)(-7) = 0$ and $(-7)(0) = 0$.

Practice Problem 13 ***Multiply.***

a. $(-4)(8)$ **b.** $(-1)(9)$ **c.** $(5)(-7)$ **d.** $(-6)(-3)$

28 Suppose you wanted to multiply more than two integers. To do this, you could multiply them two at a time as follows:

Example 17

a.
$$\begin{aligned}&(-2)(3)(-5)\\ &= -(2 \cdot 3)(-5)\\ &= (-6)(-5)\\ &= +(6 \cdot 5)\\ &= 30\end{aligned}$$

b.
$$\begin{aligned}&(-2)(-3)(-5)\\ &= +(2 \cdot 3)(-5)\\ &= (6)(-5)\\ &= -(6 \cdot 5)\\ &= -30\end{aligned}$$

However, it is easier simply to count the number of negative signs. Example 17a indicates that a product involving two (or an even number of) negative signs results in a positive answer. Example 17b suggests that a product involving three (or an odd number of) negative signs yields a negative answer. Using this concept, we can rewrite the rule for multiplying integers.

To Multiply Two or More Integers:

1. Multiply their absolute values.
2. If the problem contains an odd number of negative signs, the product is negative. If the problem contains an even number of negative signs, the product is positive.

Example 18 Multiply.

a. $(-2)(-3)(5)(-2)$ b. $(5)(-2)$

c. $(-2)(3)(-5)(2)$ d. $(-2)(-3)(2)(-5)(-3)$

Solution To multiply two or more integers, multiply their absolute values. The product is a positive number if the problem contains an even number of negative signs. The product is a negative number if the problem contains an odd number of negative signs.

3 negative signs   Multiply absolute values

18a. $(-2)(-3)(5)(-2) = -(2 \cdot 3 \cdot 5 \cdot 2)$
$= -60$

18b. $5(-2) = -(5 \cdot 2)$
$= -10$

18c. $(-2)(3)(-5)(2) = +(2 \cdot 3 \cdot 5 \cdot 2)$
$= 60$

18d. $(-2)(-3)(2)(-5)(-3) = +(2 \cdot 3 \cdot 2 \cdot 5 \cdot 3)$
$= 180$

NOTE: The **associative law for multiplication** states that the product of three or more numbers remains the same no matter how they are grouped. For example,

$$(2 \cdot 3) \cdot 5 = 2 \cdot (3 \cdot 5).$$

Practice Problem 14 ***Multiply.***

a. $(-5)(-3)(2)$ c. $(-2)(-2)(-3)(5)(-2)$

b. $(-2)(-2)(3)(-2)$ d. $35(-3)$

1.5 Exercises

Multiply.

1. $(-5)(-2)$ **2.** $(-2)(-3)(5)(-2)$ **3.** $5(-3)$

4. $(-6)(-3)$ **5.** $4(-8)$ **6.** $(-7)(-2)(10)$

7. $(-6)(4)$ **8.** $(2)(-3)(5)(-2)$ **9.** $(-10)(-2)$

10. (3)(−2)(−2)
11. (4)(3)(−2)
12. (−2)(−2)(−2)(−2)
13. (0)(−10)
14. (−10)(5)(−2)
15. (−5)(3)(−2)
16. 17(−8)
17. (−5)(11)
18. (−6)(−9)
19. (−7)(−2)(−2)
20. (−8)(−9)
21. (−2)(−2)(3)(−2)(−5)
22. (15)(−3)(21)
23. (3)(−4)(−2)
24. (−2)(3)(−5)(−2)(3)(5)
25. (−1)(−1)(−4)(−1)(1)(−2)
26. (−1)(−1)(−2)(−1)(1)(−2)
27. 3(−35)
28. 4(−12)
29. (−4)(−7)
30. (−8)(−7)
31. (−2)(−2)(−2)
32. (−3)(−3)(−3)
33. (−2)(−2)(−2)(−2)
34. (−3)(−3)(−3)(−3)
35. (2)(−10)(3)
36. (3)(−8)(2)
37. (−3)(1)(−2)(−2)
38. (−3)(2)(−1)(−1)
39. (2)(−3)(−2)(5)(2)
40. (3)(−2)(−2)(5)(2)

Fill in the blanks.

41. Multiplication is repeated ________________.

42. 5(−6) can be thought of as the sum of 5 ________________.

43. If you multiply a positive integer and a negative integer, the answer is always ________________.

44. The product of two negative integers is always ________________.

45. The product of two or more integers is a ________________ number if the problem contains an odd number of negative signs.

46. The product of two or more integers is a ________________ number if the problem contains an even number of negative signs.

47. To multiply two or more integers, you should multiply their absolute values and then determine the signs.

Express each answer as an integer. Show all work.

48. Mr. Jones has 100 shares of a stock that has decreased \$5 per share in value. If a \$5 decrease in one share is expressed as a −5, what was his total loss?

49. Shirley lost three pounds per week for 10 consecutive weeks. If a three-pound weight loss is represented by −3, how much weight did she lose?

50. The ADT Paint Company laid off 90 employees who each earn \$200 a week. If each person's salary decreases the company's weekly expenses by \$200 (−200), then what is the total decrease in the company's weekly expenses?

51. Find the product of −3, −9, −10, and −10.

52. Find the product of −2, −2, −5, and −5.

53. Multiply 3 and −10.

54. Multiply −3 and 15.

55. Find the product of 2, −3, 5, and 2.

56. Find the product of −2, 3, −5, and −3.

Answers to Practice Problems **13a.** −32 **b.** −9 **c.** −35 **d.** 18 **14a.** 30 **b.** −24 **c.** 120 **d.** −105

1.6 Division of Integers

29 In arithmetic, to divide 12 by 6 $\left(\text{written as } \frac{12}{6}\right)$, we must find an integer c such that $6 \cdot c = 12$. Clearly, c is equal to 2. In other words, $\frac{12}{6} = 2$ because $6 \cdot 2 = 12$. In general, $\frac{a}{b} = c$ if and only if $b \cdot c = a$ for a number c. In this example, a is called the **dividend,** b is the **divisor,** and c is the **quotient.**

30 Let us now use the fact that $\frac{a}{b} = c$ if and only if $b \cdot c = a$ to help us determine how to divide one integer by another.

Example 19 Divide to find the quotient.

a. $\frac{-6}{2}$ **b.** $\frac{-6}{-2}$ **c.** $\frac{6}{-2}$ **d.** $\frac{0}{-4}$ **e.** $\frac{5}{0}$

Solution **19a.** $\frac{-6}{2} = -3$ since $2(-3) = -6$

19b. $\frac{-6}{-2} = 3$ since $-2(3) = -6$

19c. $\frac{6}{-2} = -3$ since $-2(-3) = 6$

19d. $\frac{0}{-4} = 0$ since $-4(0) = 0$

19e. $\frac{5}{0} =$ undefined since division by zero is undefined

NOTE: Zero divided by any nonzero number always equals zero (Example 19d), but division by zero (Example 19e) is meaningless. For example, $\frac{5}{0}$ is said to be undefined since there is no number c such that $0 \cdot c = 5$. If a problem involves division by zero, write "undefined."

31 Based on the results in Example 19, we can state a rule for dividing one integer by another.

To Divide One Integer by Another:

1. Divide their absolute values.
2. If the dividend and divisor have the same sign, the quotient is positive. If the signs are different, the quotient is negative.

Example 20 Divide.

a. $\frac{-30}{2}$ **b.** $\frac{-15}{-3}$ **c.** $\frac{75}{-5}$ **d.** $\frac{-60}{-10}$

Solution To divide one integer by another, divide their absolute values. The quotient is a positive number if the signs are the same. The quotient is a negative number if the signs are different.

Signs different — Divide absolute values

20a. $\dfrac{-30}{2} = -\left(\dfrac{30}{2}\right) = -15$

Signs same — Divide absolute values

20b. $\dfrac{-15}{-3} = +\left(\dfrac{15}{3}\right) = 5$

20c. $\dfrac{75}{-5} = -\left(\dfrac{75}{5}\right) = -15$

20d. $\dfrac{-60}{-10} = +\left(\dfrac{60}{10}\right) = 6$

NOTE: If you have divided correctly, the divisor times the quotient must be equal to the dividend. For example, $\dfrac{-30}{2} = -15$ since $2(-15) = -30$.

Practice Problem 15 **Divide.**

a. $\dfrac{-16}{2}$ **b.** $\dfrac{0}{-16}$ **c.** $\dfrac{-16}{-8}$ **d.** $\dfrac{-15}{0}$

1.6 Exercises

Divide and check your answers.

1. $\dfrac{-16}{-2}$ **2.** $\dfrac{-36}{4}$ **3.** $\dfrac{-15}{5}$ **4.** $\dfrac{0}{-9}$ **5.** $\dfrac{-35}{7}$

6. $\dfrac{16}{-2}$ **7.** $\dfrac{-100}{2}$ **8.** $\dfrac{-9}{0}$ **9.** $\dfrac{-75}{-5}$ **10.** $\dfrac{-28}{4}$

11. $\dfrac{-66}{-6}$ **12.** $\dfrac{-728}{8}$ **13.** $\dfrac{-35}{-35}$ **14.** $\dfrac{-8118}{9}$ **15.** $\dfrac{-35}{35}$

16. $\dfrac{255}{-5}$ **17.** $\dfrac{65}{-5}$ **18.** $\dfrac{0}{7}$ **19.** $\dfrac{-70}{7}$ **20.** $\dfrac{-800}{-20}$

21. $\dfrac{-88}{-4}$ **22.** $\dfrac{19}{0}$ **23.** $\dfrac{-64}{16}$ **24.** $\dfrac{-32}{16}$ **25.** $\dfrac{-16}{-2}$

26. $\dfrac{-18}{-9}$ **27.** $\dfrac{64}{-8}$ **28.** $\dfrac{32}{-2}$ **29.** $\dfrac{500}{-5}$ **30.** $\dfrac{600}{-2}$

31. $\dfrac{-26}{-13}$ **32.** $\dfrac{-39}{-13}$ **33.** $\dfrac{-65}{5}$ **34.** $\dfrac{-75}{5}$ **35.** $\dfrac{-622}{-2}$

36. $\dfrac{-844}{-4}$ **37.** $\dfrac{0}{-1001}$ **38.** $\dfrac{0}{-2001}$ **39.** $\dfrac{-1001}{0}$ **40.** $\dfrac{-2001}{0}$

Fill in the blanks.

41. If you divide a negative integer by a negative integer, the quotient or answer is ________.

42. If you divide a positive integer by a negative integer or a negative integer by a positive integer, the quotient or answer is ________.

43. To divide one integer by another, divide their ________ and then determine the ________ of your final answer.

44. Zero divided by a nonzero integer is ________.

45. Any integer divided by zero is ________.

Express each answer as an integer. Show all work.

46. A gambler lost a total of \$1000 (-1000) in five consecutive hands of poker. Find his average loss per hand by dividing the amount he lost by the number of hands he played.

47. Twenty people lost a total \$20,000 ($-20{,}000$) in a business deal. Find the average amount of each person's loss by dividing the total loss by the number of people involved in the deal.

48. One day in Sheboygan, Wisconsin, the temperature dropped 30° (-30) in 10 hours. Find the average temperature change per hour.

49. Divide 12 by -6.

50. Divide 10 by -2.

51. Divide -6 into -12.

52. Divide -2 into -14.

53. Ten people lost a total of \$1,000 ($-1{,}000$) in a business deal. Find the average amount of each person's loss.

54. One day in Chicago, the temperature dropped 20° (-20) in five hours. Find the average temperature change per hour.

Answers to Practice Problems **15a.** -8 **b.** 0 **c.** 2 **d.** undefined

1.7 Order of Operations

32 Suppose you wanted to simplify an expression containing more than one operation. For example, to simplify $3 + 4 \cdot 2$, you could add first and the answer would be:

$$\begin{aligned} 3 + 4 \cdot 2 &= 7 \cdot 2 \\ &= 14. \end{aligned}$$

However, if you multiplied first, your answer would be:

$$\begin{aligned} 3 + 4 \cdot 2 &= 3 + 8 \\ &= 11. \end{aligned}$$

Obviously, this situation could create many problems. To avoid confusion, mathematicians have agreed upon the following order of operations rule:

Order of Operations Rule (No Grouping Symbols):

To simplify an expression containing more than one operation and no grouping symbols:

1. First, do all multiplication and/or division as you work from left to right.
2. Next, do all addition and/or subtraction as you work from left to right.

Example 21 Simplify.

a. $3 + 4 \cdot 2$ **b.** $-7 - 5 \cdot 2$

c. $35 + (-50) \div 10$ **d.** $10 - 6 \cdot 5 + (-40) - 8 \div (-2)$

Solution To simplify an expression containing more than one operation and no grouping symbols, do all multiplication and/or division before adding and/or subtracting.

21a. $3 + 4 \cdot 2$

$= 3 + 8$ Multiply

$= 11$ Add

21b. $-7 - \boxed{5 \cdot 2}$

$$\begin{aligned} &= -7 - 10 && \text{Multiply} \\ &= -7 + (-10) && \text{Subtraction rule} \\ &= -17 \end{aligned}$$

21c. $35 + \boxed{(-50) \div 10}$

$$\begin{aligned} &= 35 + (-5) && \text{Divide} \\ &= 30 && \text{Add} \end{aligned}$$

21d. $10 - \boxed{6 \cdot 5} + (-40) - \boxed{8 \div (-2)}$

$$\begin{aligned} &= 10 - 30 + (-40) - (-4) && \text{Multiply and divide} \\ &= 10 + (-30) + (-40) + 4 && \text{Subtraction rule} \\ &= 14 + (-70) \\ &= -56 && \text{Add} \end{aligned}$$

NOTE: If an expression contains the operations of multiplication and division only, multiply and divide in order from left to right. For example,

$$\begin{aligned} -6 \cdot 2 \div (-4) &= -12 \div (-4) \\ &= 3. \end{aligned}$$

Practice Problem 16 ***Simplify.***

a. $8 + 5 \cdot 2$ **b.** $-16 - 30 \cdot 2$

c. $-12 - 6(-3) + 40 \div (-10)$ **d.** $-60 \div 15 \cdot 7$

33 In Idea 32, we said to multiply and/or divide before adding and/or subtracting. In some cases, however, a problem may call for adding or subtracting *before* multiplying or dividing. When this happens, parentheses ( ), brackets [ ], or braces { }—which are called **grouping symbols**—are placed around the part of the problem that must be computed out of the order stated in Idea 32.

34 To simplify an expression in which a grouping symbol is used to indicate order of operations, use the following rule:

Order of Operations Rule (One Grouping Symbol): 

To simplify an expression containing more than one operation and only one set of grouping symbols:

1. First, do all operations inside the grouping symbols. Remember to multiply and/or divide before adding and/or subtracting.
2. Next, do all multiplication and/or division as you work from left to right.
3. Finally, do all addition and/or subtraction.

Example 22 Simplify.

a. $40 + 3(8 - 15)$

b. $3 - 6(-5 + 7 \cdot 2)$

c. $-16 + (-80 + 6 \cdot 6) \div 4$

d. $-80 - 20(-8 + 16 \div 8) + 4(-12 - 2 \cdot 5)$

Solution To simplify an expression containing more than one operation and one pair of grouping symbols, first do the operations inside the grouping symbols. Next, follow the order of operation rule (Idea 32).

22a. $40 + 3(8 - 15)$

$= 40 + 3(-7)$	Simplify inside the ( )
$= 40 + (-21)$	Multiply
$= 19$	Add

22b. $3 - 6(-5 + 7 \cdot 2)$

$= 3 - 6(-5 + 14)$	Simplify inside the ( )
$= 3 - 6(9)$	
$= 3 - 54$	Multiply
$= -51$	Subtract

22c. $-16 + (-80 + 6 \cdot 6) \div 4$

$= -16 + (-80 + 36) \div 4$	Simplify inside the ( )
$= -16 + (-44) \div 4$	
$= -16 + (-11)$	Divide
$= -27$	Add

22d. $-80 - 20(-8 + 16 \div 8) + 4(-12 - 2 \cdot 5)$

$= -80 - 20(-8 + 2) + 4(-12 - 10)$	Simplify inside each ( ).
$= -80 - 20(-6) + 4(-22)$	
$= -80 - (-120) + (-88)$	Multiply
$= -80 + 120 + (-88)$	Subtraction Rule
$= -168 + 120$	Add
$= -48$	

NOTE: The expression $40 + 3(8 - 15)$ is read as "forty plus three times the quantity eight minus fifteen."

Practice Problem 17 ***Simplify.***

a. $-20 + 3(-5 - 2)$ **b.** $-4 + (-16 + 3 \cdot 4) \div 2$

c. $-16 + 3(-7 + 2 \cdot 5) - 2(-30 + 8 \cdot 3)$

35 Sometimes an expression may contain a grouping symbol within a grouping symbol (nested symbols). To simplify this type of expression, use the following rule:

Order of Operations Rule (Nested Grouping Symbols):

To simplify an expression containing more than one operation and nested grouping symbols:

1. First, simplify inside the innermost grouping symbol and remember to multiply and/or divide before adding and/or subtracting.
2. Next, repeat step 1 until all symbols of grouping have been eliminated.
3. Finally, do all multiplication and/or division before adding and/or subtracting.

Example 23 Simplify.

a. $3[-10 + 4(2 - 3)] + 2$ **b.** $-6 + 4[(-2 - 3) + (-5 + 3 \cdot 2)]$

c. $-2\{1 + 3[-2 + 5(-6 + 9)]\}$

Solution To simplify an expression containing more than one operation and nested grouping symbols, simplify inside the innermost grouping symbol first.

23a. $3[-10 + 4(2 - 3)] + 2$
$= 3[-10 + 4(-1)] + 2$ Simplify inside the ( ) first
$= 3[-10 + (-4)] + 2$ Simplify inside the [ ]
$= 3[-14] + 2$
$= -42 + 2$ Multiply
$= -40$ Add

23b. $-6 + 4[(-2 - 3) + (-5 + 3 \cdot 2)]$
$= -6 + 4[-5 + (-5 + 6)]$ Simplify inside the ( ) first
$= -6 + 4[-5 + 1]$
$= -6 + 4[-4]$ Simplify inside the [ ]
$= -6 + [-16]$ Multiply
$= -22$ Add

23c. $-2\{1 + 3[-2 + 5(-6 + 9)]\}$
$= -2\{1 + 3[-2 + 5(3)]\}$ Simplify inside the ( ) first
$= -2\{1 + 3[-2 + 15]\}$
$= -2\{1 + 3[13]\}$ Simplify inside the [ ]
$= -2\{1 + 39\}$ Simplify inside the { }
$= -2\{40\}$
$= -80$ Multiply

Practice Problem 18 ***Simplify.***

a. $-3[2 + 4(5 - 7)]$ **b.** $-6 + 3[2 + 3(4 - 6) + (-2 - 3)]$

c. $-4\{2 + 3[-2 + 5(-11 + 5)]\}$

1.7 Exercises

Simplify.

1. $-6 + 2 \cdot 5$
2. $-8 + 30 \div (-5)$
3. $-6 + 6(-3) + 15 \div (-3)$
4. $-6 \cdot 10 \div 2$
5. $-2 - 6 \cdot 5 + (-30) + 6 \div (-2)$
6. $3(-2 - 4)$
7. $(4 + 4)(4 - 4)$
8. $4 + 4(4 - 4)$
9. $-8 + 2(-3 + 5 \cdot 2)$
10. $-5 + (18 + 2 \cdot 3) \div 3$
11. $-18 - 5(-16 + 8 \div 2) + (-6 + 2 \cdot 5)$
12. $3[2 + 6(-2 - 3)]$
13. $-5 + 3[(-5 - 2) + (-7 + 4 \cdot 2)]$
14. $2[-1 + 4\{-2 \cdot (-12 + 3) \cdot 2\} + 5]$
15. $-2 + 3[(-1 + 4 \cdot 2) + (-5)]$
16. $3[-10 + 5(-3 + 1)]$
17. $-11 + 5 \cdot 2 - 55 + 5 - 30 \div 6$
18. $-6 + (-8 + 2 \cdot 2) \div 2 - 4(-8 - 2)$
19. $-60 \div [(-8 + 5) + 4(2 - 4) + 5]$
20. $2\{-3 + 2[-5 + 2(3 + 1)]\}$
21. $-3 + 5 \cdot 2$
22. $-8 + 5 \cdot 3$
23. $-6 - 30 \div 5$
24. $-8 - 60 \div 10$
25. $-10 + 3(2 - 5)$
26. $-15 + 4(8 - 9)$
27. $3(-2 + 5 \cdot 2)$
28. $4(-8 + 2 \cdot 3)$
29. $-8 + (-15 + 2 \cdot 3) \div 3$
30. $-10 + (5 - 4 \cdot 8) \div 3$
31. $8 - 8 \div 2 + 5 \cdot 3 - 6 \div 2$
32. $-8 - 6 \div 2 + 4 \cdot 2 - 8 \div 4$
33. $-3 + 3(-3 - 3 \cdot 2)$
34. $-4 + 4(-4 - 4 \cdot 2)$

35. $5 - 8[2 - 4(3 + 1 \cdot 4)]$

36. $8 - 10[6 - 7(-3 - 1 \cdot 2)]$

37. $3[-8 + 5(-8 + 2 \cdot 3)] - 10$

38. $2[-7 + 3(-6 + 4 \cdot 2)] - 8$

39. $2\{-4 + 2[-6 + 2(3 - 4)]\}$

40. $3\{-2 + 3[-7 + 2(-3 - 2)]\}$

Fill in the blanks.

41. To simplify an expression involving several operations and no grouping symbols, we should _______________ and/or _______________ before _______________ and/or _______________.

42. If a problem only involves multiplication and division and no grouping symbols, we work from _______________ to _______________.

43. To simplify an expression containing several operations and parentheses, we should do all the operation inside the _______________ first. However, you must still remember to _______________ and _______________ before _______________ and _______________.

44. In the problem $(6 + 8) \cdot 2$, the parentheses are used to indicate that you should do the operation of _______________ before you do the operation of _______________.

45. To simplify an expression involving several operations and nested grouping symbols, you should simplify inside the _______________ first.

46. In the problem $2[3 + 5(3 + 2)]$, the parentheses are used as a _______________ and also indicate the operation of _______________.

Express each answer as an integer. Show all work.

47. Dan Johnson buys blue jeans for $15 each and resells them for $23 each. If Dan sells 35 pairs of jeans, find the total profit from the sale. (*Hint:* use the fact that the total profit is equal to the resale price minus the original cost times the number of items sold.

48. In a "going-out-of-business" sale, Cazzie sold three washers at a profit of $30 each, eight dryers at a loss of $10 each, nine suits at a profit of $60 each, and four desks at a loss of $50 each. If profits are represented by a positive integer and losses are represented by a negative integer, find the total profit or loss on these sales.

Answers to Practice Problems **16a.** 18 **b.** −76 **c.** 2 **d.** −28 **17a.** −41 **b.** −6 **c.** 5 **18a.** 18 **b.** −33 **c.** 376

1.8 Formulas

36 Many problems in mathematics, business, physics, and chemistry require the use of formulas. A mathematical **formula** is a rule that uses letters to express a mathematical relationship between two or more quantities. For example, the formula

$$A = L \cdot W$$

indicates that the area of a rectangle is equal to the length of the rectangle times its width.

NOTE: If two or more letters (or a number and a letter) are placed next to each other in a formula, the operation indicated is multiplication. The formula $A = L \cdot W$ is usually written $A = LW$.

37

To Evaluate a Formula:

1. Substitute (replace) the given values for the letters in the formula.
2. Simplify the expression obtained in step 1 by performing the indicated operation or operations (addition, subtraction, multiplication, and/or division).

Example 24 Evaluate.

a. $E = Q - W$ when $Q = -4$ and $W = 16$.

b. $V = S + 4T$ when $S = 6$ and $T = -2$.

c. $P = a + 2(c + 3d)$ when $a = -6$, $c = -10$, $d = 5$.

Solution To evaluate a formula, replace all letters with their given numerical values and then perform the indicated operations.

24a.
$$\begin{aligned} E &= Q - W \\ &= -4 - 16 \\ &= -20 \end{aligned}$$

24b.
$$\begin{aligned} V &= S + 4T \\ &= 6 + 4(-2) \\ &= 6 + (-8) \\ &= -2 \end{aligned}$$

24c.
$$\begin{aligned} P &= a + 2(c + 3d) \\ &= -6 + 2(-10 + 3 \cdot 5) \\ &= -6 + 2(-10 + 15) \\ &= -6 + 2(5) \\ &= -6 + 10 \\ &= 4 \end{aligned}$$

NOTE: In solution 24b, when $T = -2$ is substituted into the formula, the -2 is placed within parentheses to ensure that the operation of multiplication is indicated properly. In general, multiplication of integers a and $-b$ is indicated by the expression $a(-b)$.

Practice Problem 19 ***Evaluate.***

a. $M = a - b + c$ when $a = -6$, $b = 7$, $c = -5$

b. $A = b + mx$ when $m = 4$, $b = -5$, $x = -3$

c. $B = a + 2(c + 3f)$ when $a = -4$, $c = -15$, $f = 4$

38 Some formulas may be written in fractional form, that is, in the form $\frac{a}{b}$, where a is called the **numerator** and b is the **denominator**.

To Evaluate a Formula Written in Fractional Form:

1. First, substitute the given values for the letters in the formula.
2. Next, simplify the numerator and denominator by performing the indicated operations.
3. Finally, divide the denominator into the numerator.

Example 25 Evaluate.

a. $M = \dfrac{a - b}{c - d}$ when $a = 7, b = -5, c = 3, d = 5$

b. $X = \dfrac{md + MD}{m + M}$ when $m = 5, d = -8, M = 2, D = 6$

c. $N = \dfrac{4[x - d]}{d}$ when $x = -6, d = -2$

Solution To evaluate a formula written as a fraction, replace all letters with their given numerical values. Next, simplify the numerator and denominator and then divide.

25a.
$$M = \frac{a - b}{c - d} = \frac{7 - (-5)}{3 - 5} = \frac{12}{-2} = -6$$

25b.
$$X = \frac{md + MD}{m + M} = \frac{5(-8) + 2(6)}{5 + 2} = \frac{-40 + 12}{7} = \frac{-28}{7} = -4$$

25c.
$$N = \frac{4[x - d]}{d} = \frac{4[-6 - (-2)]}{-2} = \frac{4[-4]}{-2} = \frac{-16}{-2} = 8$$

NOTE: The fraction bar (——) is also a grouping symbol. This is illustrated by the fact that

$$\frac{-40 + 12}{7} = (-40 + 12) \div 7.$$

Practice Problem 20 ***Evaluate.***

a. $S = \dfrac{a + b}{a - b}$ when $a = -4, b = -2$

b. $X = \dfrac{mb + mc + md}{m + n}$ when $m = 3, b = -4, c = 2, d = -6, n = -5$

c. $C = \dfrac{5(F - 32)}{9}$ when $F = -4$

1.8 Exercises

Evaluate.

1. $A = \dfrac{K}{L}$ when $K = -60, L = -6$
2. $P = S - C$ when $S = -5, C = -12$
3. $T = L - Bx$ when $L = 10, B = -7, x = -2$

4. $D = KRt$ when $K = -3, R = 5, t = -2$

5. $F = c + p$ when $c = -21, p = 57$

6. $B = a - b + c - d$ when $a = -5, b = 12, c = 13, d = 30$

Fill in the blanks with one or more words.

7. A rule which uses letters to express a mathematical relationship between two or more quantities is called a ________.

8. In the formula $y = mx + b$, mx means m ________ x.

9. The first step in finding the value of a formula is to replace the ________ in the formula with their given numerical values.

10. To simplify an expression written in the form of a fraction $\frac{a}{b}$, you would simplify the ________ and ________ and then ________ the ________ into the ________.

The change in internal energy ("E") is expressed by the formula $E = Q - W$, where Q = change in heat and W = work done. Find a value for "E" given the following information:

11. $Q = -10, W = -30$

12. $Q = 30, W = -10$

13. $Q = 30, W = -100$

14. $Q = 0, W = 30$

15. $Q = 0, W = -30$

16. $Q = -20, W = -40$

Given the formula $m = \frac{x - y}{a - b}$, find the value of m given the following information:

17. $x = -16, y = -2, a = 4, b = 6$

18. $x = -8, y = 2, a = 5, b = 10$

19. $x = 3, y = -12, a = -2, b = -7$

Given the formula $f = \frac{p \cdot q}{p + q}$, find a value for f given the following information:

20. $p = 12, q = -3$

21. $p = 10, q = -5$

22. $p = -6, q = 10$

23. $p = -8, q = 4$

The formula $°F = 32 - \frac{9}{5}°C$ is used to convert a temperature from Celsius (C) to Fahrenheit (F). Find a value for F given:

24. $C = -15°$

25. $C = -20°$

26. $C = 5°$

27. $C = 15°$

28. $C = -10°$

29. $C = -30°$

30. $C = 20°$

31. $C = 10°$

Given the formula $M = \frac{ax + by + cz}{a + b + c}$, find the value of M given the following information:

32. $a = 2, b = 4, c = 6, x = -3, y = -9, z = -7$

33. $a = 10, b = 15, c = 25, x = -5, y = -10, z = 2$

The sum S of the first n terms of an arithmetic progression is given by the formula $S = \frac{n[2a + (n - 1)d]}{2}$, where n = the number of terms, a = the first term, and d = the difference between any two consecutive terms. Find a value for S given the following information:

34. $a = -15, d = 2, n = 30$

35. $a = -5, d = 3, n = 10$

36. $a = -2, d = -3, n = 14$

Answers to Practice Problems **19a.** -18 **b.** -17 **c.** -10 **20a.** 3 **b.** 12 **c.** -20

Summary

Important Terms

1.1

integers
positive integer
negative integer
zero
absolute value, $|x|$

1.2

sum
parentheses, ( )
commutative law for addition
associative law for addition

1.3

minuend
subtrahend
difference
additive inverse

1.5

commutative law for multiplication
product
associative law for multiplication

1.6

dividend
divisor
quotient
undefined

1.7

grouping symbols
braces, { }
brackets, [ ]
nested grouping symbols

1.8

formula

Important Skills

1.1

Representing written statements as integers
Finding the absolute value of an integer

1.2

Adding integers having the same sign
Adding two integers having different signs
Adding more than two integers having different signs
Solving word problems involving addition of integers

1.3

Finding the additive inverse of an integer
Subtracting one integer from another
Solving word problems involving subtraction of integers

1.4

Simplifying expressions involving addition and subtraction
Solving word problems involving addition and subtraction

1.5

Multiplying two or more integers
Solving word problems involving multiplication of integers

1.6

Dividing one integer by another
Solving word problems involving division of integers

1.7

Simplifying expressions involving more than one operation
Simplifying expressions containing one grouping symbol
Simplifying expressions containing nested grouping symbols
Solving word problems involving more than one operation

1.8

Evaluating a formula
Evaluating a formula written as a fraction

Review Exercises

The answers to these problems—and the sections in which similar problems appear—are given in the Answer Section. It will be helpful to review the sections referred to for the problems you did incorrectly.

Find the following absolute values.

1. $|-20|$ **2.** $|60|$ **3.** $|95|$ **4.** $|0|$

Find the additive inverses of the following integers.

5. -20 **6.** 60 **7.** 95 **8.** 0

Perform the indicated operations and simplify.

9. $-6 + (-9) + (-35)$

10. $4 - (-35)$

11. $-3 + 18 \div (-2) + 40 \cdot 2$

12. $(-8)(5)$

13. $\dfrac{35}{-7}$

14. $\dfrac{-60}{-10}$

15. $-45 + 98$

16. $(-5)(2)(-3)(-4)$

17. $8 + (-16) - (-40) + 35$

18. $35 - 85$

19. $-10 - 6 + 32 - 15$

20. $\dfrac{-11116}{0}$

21. $-5[-8 + 3(-4 - 2) + 9(-8 + 2 \cdot 3)]$

Solve. Show all work.

22. The opening price of a certain stock was listed as 115 (meaning \$115). If daily changes in the stock were listed as $+5$, -3, -2, $+10$, -6, $+2$, -5, and $+15$, find the final closing price of the stock.

23. Subtract -15 from 30.

24. In a fire sale, Guerin sold eight coats at a profit of \$16 each, three clocks at a loss of \$4 each, seven bicycles at a loss of \$10 each, and five radios at a profit of \$30 each. Find his total profit or loss on these sales, and express the answer as an integer.

25. Howie lost \$42 in three weeks. Express his average loss per day as an integer.

26. On October 31, Adrienne has a balance of \$500 in her checking account. The chart on the side shows the transactions she made in November. How much was in her account on November 29?

Date	Check Number	Checks	Deposits
Nov. 1	113	\$ 50	
Nov. 3	114	75	
Nov. 7	115	100	
Nov. 10			\$250
Nov. 15	116	300	
Nov. 21	117	73	
Nov. 23			55
Nov. 23	118	53	
Nov. 26			118
Nov. 28	119	172	

27. Find the value of $M = \dfrac{-4[2(d + 3a)]}{a + d}$ when $a = -2$ and $d = 4$.

FRACTIONS AND MIXED NUMBERS

OBJECTIVES

1. Find the prime factorization of a number.
2. Reduce a fraction to its lowest terms.
3. Change a mixed number to an improper fraction.
4. Change an improper fraction to a mixed number.
5. Multiply fractions and/or mixed numbers.
6. Divide fractions and/or mixed numbers.
7. Add two or more fractions.
8. Add mixed numbers.
9. Subtract fractions and/or mixed numbers.
10. Add, subtract, multiply, or divide signed fractions.

PRETEST

EXPLANATION

1. Find the prime factorization.
a. 24 **b.** 77 **c.** 297

Section 2.1
Ideas 5–7

2. Reduce to lowest terms.
a. $\frac{56}{64}$ **b.** $\frac{273}{182}$

Section 2.2
Ideas 10 and 11

3. Write as an improper fraction.
a. $3\frac{1}{3}$ **b.** $9\frac{5}{8}$

Section 2.3
Idea 14

4. Write as a mixed number.
a. $\frac{15}{2}$ **b.** $\frac{97}{8}$

Section 2.3
Idea 15

5. Multiply.
a. $\frac{9}{4} \cdot \frac{14}{15}$ **b.** $45 \cdot 2\frac{1}{30}$

Section 2.4
Ideas 17 and 18

6. Divide.
a. $\frac{27}{16} \div \frac{18}{12}$ **b.** $7\frac{3}{7} \div 2\frac{1}{30}$

Section 2.5
Ideas 20 and 21

7. Add.
a. $\frac{5}{9} + \frac{1}{66}$ **b.** $\frac{3}{4} + \frac{1}{22} + \frac{5}{33}$

Section 2.6
Ideas 27 and 28

8. Add.
a. $5\frac{3}{7} + 6\frac{6}{7}$ **b.** $12\frac{3}{8} + 11\frac{5}{12} + 5\frac{1}{6}$

Section 2.7
Ideas 29 and 30

9. Subtract.
a. $\frac{7}{8} - \frac{3}{10}$ **b.** $4\frac{9}{14} - 1\frac{11}{21}$

Section 2.8
Ideas 32–34

10. Do the indicated operation.
a. $\frac{5}{6} - \frac{3}{10}$ **b.** $\frac{40}{60} \div \left(-\frac{88}{15}\right)$

Section 2.9
Ideas 37 and 39

2.1 Introduction and Prime Factorizations

Idea 1 A number written in the form $\frac{a}{b}$ with b not equal to zero is called a **fraction.** The number above the fraction bar is the **numerator,** and the number below the bar is the **denominator.** For example, in the fraction $\frac{2}{3}$, 2 is the numerator and 3 is the denominator. The number $\frac{2}{3}$ is also called a **proper fraction** since its numerator is smaller than its denominator. However, if the numerator of a fraction is greater than or equal to its denominator, the fraction is called an **improper fraction.** Examples of improper fractions include $\frac{3}{2}$, $\frac{5}{3}$, and $\frac{6}{6}$.

Practice Problem 1 ***Tell whether each number is a proper or an improper fraction and identify the numerator and denominator.***

a. $\frac{3}{7}$ **b.** $\frac{5}{3}$ **c.** $\frac{7}{7}$ **d.** $\frac{11}{13}$

2 In Section 1.6, fractions were used to indicate the operation of division. However, fractions are used more frequently to represent a part of the whole. For example, the rectangle in Figure 2–1 is divided into three equal parts, of which two parts are shaded. That is, $\frac{2}{3}$ (two-thirds) of the rectangle is shaded. Similarly, if you did three out of four problems on a quiz, you completed $\frac{3}{4}$ (three-fourths) of the quiz.

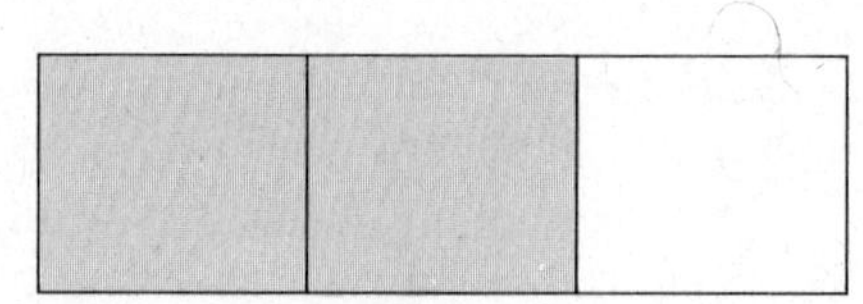

Figure 2-1

NOTE: Instead of saying that you completed $\frac{3}{4}$ of the quiz, you could say that the fraction $\frac{3}{4}$ represents the **fractional part** of the quiz that you completed.

Practice Problem 2 ***What part of each figure is shaded?***

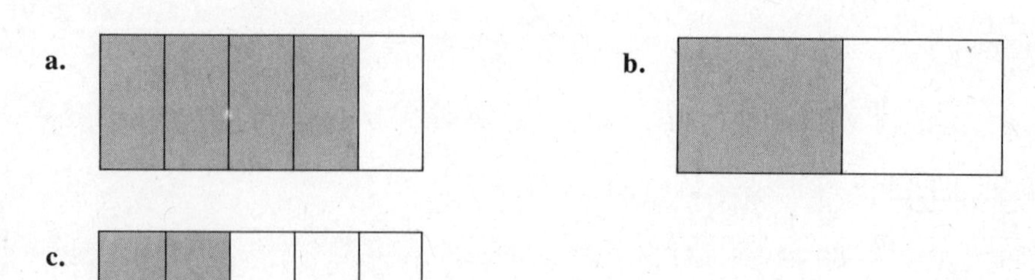

c.

3 Before considering how to reduce, add, subtract, multiply, and divide fractions, we must discuss the concepts of factors, prime numbers, and prime factorizations.

If two or more natural numbers (1, 2, 3, etc.) can be multiplied together to form a product K, each of these numbers is called a **factor** or **divisor** of K. For example,

2 and 3 are the factors of 6 since $2 \cdot 3 = 6$. Other factors of 6 are 1 and 6, since $1 \cdot 6 = 6$.

NOTE: Since 1, 2, 3, and 6 are factors (divisors) of 6, each of these numbers will divide 6 evenly; that is, the division results in a remainder of zero. We can also say that 6 is *divisible* by 1, 2, 3, and 6.

4 A natural number greater than 1 whose only divisors are 1 and itself is called a **prime number.** The primes less than 50 are 2, 3, 5, 7, 11, 13, 17, 19, 23, 29, 31, 37, 41, 43, and 47. Natural numbers greater than 1 that have more than two divisors are called **composite numbers.** For example, 4, 6, 8, 9, and 10 are composite numbers. The number 1 is neither prime nor composite.

Example 1 Tell whether each number is prime or composite.

a. 32 **b.** 61 **c.** 85

Solution A prime number has only two divisors or factors—1 and itself. A composite number has more than two divisors (factors).

1a. 32 is a composite number since 1, 2, 4, 8, 16, and 32 are divisors of 32.

1b. 61 is a prime number since its only divisors are 1 and 61.

1c. 85 is a composite number since 1, 5, 17, and 85 are divisors of 85.

Practice Problem 3 ***Tell whether each number is prime or composite.***

a. 52 **b.** 60 **c.** 71

5 All composite numbers can be factored. That is, they can be written as a product of two natural numbers. If all the factors in this product are prime numbers, this product is the **prime factorization** of the given number. To find the prime factorization of a number, (1) express the number as a product of two natural numbers; and (2) rewrite all composite factors as a product of primes by repeating step 1.

Example 2 Find the prime factorization of each number.

a. 30 **b.** 12 **c.** 11

Solution To find the prime factorization of a number, write the number as a product of two natural numbers and then write each factor as a product of primes.

2a. $30 = 2 \cdot 15 = 2 \cdot 3 \cdot 5$ or $30 = 6 \cdot 5 = 2 \cdot 3 \cdot 5$

2b. $12 = 2 \cdot 6 = 2 \cdot 2 \cdot 3$ or $12 = 4 \cdot 3 = 2 \cdot 2 \cdot 3$

2c. 11 is prime and its only factorization is the trivial factorization $11 = 1 \cdot 11$.

NOTE: Example 2 shows that no matter what factorization of a number you begin with, you will obtain the same prime factors. It is customary to write the prime factorizations of a number in numerical order (smallest factor to largest factor).

Practice Problem 4 ***Find the prime factorization of each number.***

a. 20 **b.** 19 **c.** 50 **d.** 45

6 Sometimes it is difficult to find the prime factorization of a large number by first writing the number as a product of two numbers. When this occurs, it will be helpful to know the following tests of divisibility:

Tests of Divisibility:

1. A natural number whose last digit is an even number (that is, 0, 2, 4, 6, or 8) is divisible by 2.
2. If the sum of the digits of a natural number is divisible by 3, the number is divisible by 3.
3. A natural number whose last digit is 0 or 5 is divisible by 5.

Example 3 Tell whether each number is divisible by 2, 3, and/or 5.

a. 36 **b.** 861 **c.** 520

Solution **3a.** 36 is divisible by 2 since its last digit, 6, is an even number.
36 is divisible by 3 since $3 + 6 = 9$ is divisible by 3.
36 is not divisible by 5 since its last digit is neither 0 nor 5.

3b. 861 is not divisible by 2 since its last digit is not an even number.
861 is divisible by 3 since $8 + 6 + 1 = 15$ is divisible by 3.
861 is not divisible by 5 since its last digit is neither 0 nor 5.

3c. 520 is divisible by 2 since its last digit, 0, is an even number.
520 is not divisible by 3 since $5 + 2 + 0 = 7$ is not divisible by 3.
520 is divisible by 5 since its last digit is 0.

NOTE: A **common factor** of two numbers is a number that is a factor of both numbers. When a number is divisible by two or more numbers having no common factors, the number is divisible by the product of those numbers. For example, 36 is divisible by 2 and 3. Therefore, 36 is divisible by the product $2 \cdot 3 = 6$ since 2 and 3 have no common factors.

Practice Problem 5 ***Tell whether each number is divisible by 2, 3, and/or 5.***

a. 24 **b.** 111 **c.** 120 **d.** 169

7 We can use the tests of divisibility to help us factor. In other words, to find the prime factorization of any natural number, use the following procedure:

Prime Factorization Procedure:

1. Determine if the number is divisible by 2, 3, or 5. If it is, then divide. If it is not, then try the other primes (7, 11, 13, 17, etc.) in numerical order. Any time a divisor multiplied by itself is larger than the number you wish to factor, stop trying to find a divisor since this indicates that the number is prime.
2. Repeat step 1 until your quotient is a prime number.
3. Write the original number as a product of its prime divisors and the final quotient (a prime number).
4. Check to see if the product written in step 3 is equal to the original number.

NOTE: To divide one number into another, write:

2/100 or divisor/dividend
/50 /quotient.

Example 4 Find the prime factorization of each number.

a. 36 **b.** 297 **c.** 222 **d.** 187 **e.** 47

Solution To find the prime factorization of a number, repeatedly divide by prime numbers until you obtain a prime number as a quotient. Next, write the original number as a product of its prime divisors and the final quotient.

4a. 2/36
2/18
3/9
/3

$36 = 2 \cdot 2 \cdot 3 \cdot 3$

4b. 3/297
3/99
3/33
/11

$297 = 3 \cdot 3 \cdot 3 \cdot 11$

4c. 2/222
3/111
/37

$222 = 2 \cdot 3 \cdot 37$

4d. Clearly, 187 is not divisible by 2, 3, or 5. Thus, try 7.
7/187
/26 remainder is 5
Now, try 11.
11/187
/17 $187 = 11 \cdot 17$

4e. Clearly, 47 is not divisible by 2, 3, 5, or 7. Also, you can stop trying to find a divisor larger than 7 since the product of 7 and itself ($7 \cdot 7 = 49$) is larger than the number you are trying to factor (47). Therefore, 47 is prime.

NOTE: With practice you should be able to find the prime factorizations of some numbers mentally.

Practice Problem 6 ***Find the prime factorization of each number.***

a. 60 **b.** 105 **c.** 89 **d.** 143 **e.** 522

2.1 Exercises

Identify the numerator and denominator.

1. $\frac{3}{14}$ **2.** $\frac{6}{8}$ **3.** $\frac{10}{9}$ **4.** $\frac{15}{2}$

What part is shaded?

5.

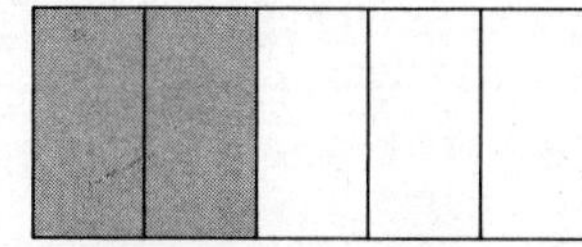

6.

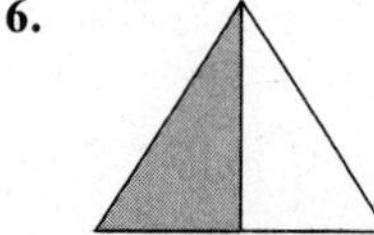

7.

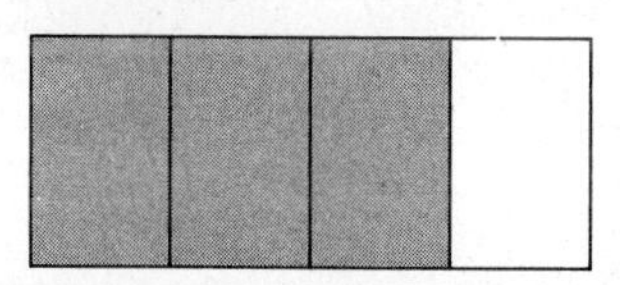

8.

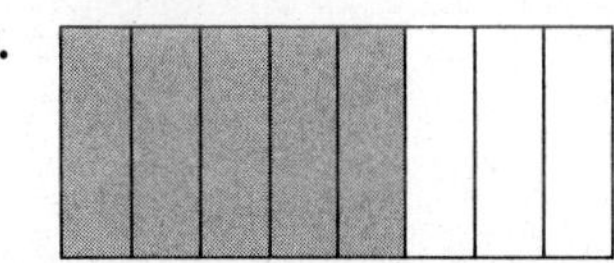

Determine if each number is divisible by 2, 3, and 5.

9. 21 **10.** 120 **11.** 201 **12.** 2754
13. 1755 **14.** 4901 **15.** 685,470 **16.** 16,842

Write the prime factorization of each number.

17. 14 **18.** 61 **19.** 60 **20.** 64 **21.** 39 **22.** 72
23. 42 **24.** 99 **25.** 77 **26.** 90 **27.** 88 **28.** 66
29. 81 **30.** 57 **31.** 25 **32.** 129 **33.** 165 **34.** 198
35. 135 **36.** 121 **37.** 115 **38.** 231 **39.** 462 **40.** 234
41. 1001 **42.** 1599 **43.** 1598 **44.** 323 **45.** 1200 **46.** 1500
47. 848 **48.** 101 **49.** 297 **50.** 169 **51.** 2431 **52.** 1463

Fill in the blanks.

53. The only divisors of 89 are 1 and 89. Therefore, 89 is a ______________ number.

54. Since $3 \cdot 5 = 15$, we know that 3 and 5 are ______________ or ______________ of 15. We also know that ______________ and ______________ will divide ______________ evenly.

55. To find the prime factorization of a number, repeatedly divide by ______________ numbers until the ______________ is a prime number. Next, write the original number as a ______________ of its ______________ divisors and the final ______________.

Solve.

56. Pam completed three questions on a test containing five problems. What fractional part of the test did she complete?

57. In a mathematics class there are 13 men and 17 women. What fractional part of the class is women?

58. Reggie must move five cars. If he moves four of the cars, what fractional part of the job must be completed?

Answers to Practice Problems **1a.** proper fraction (3 = numerator; 7 = denominator) **b.** improper fraction (5 = numerator; 3 = denominator) **c.** improper fraction (7 = numerator; 7 = denominator) **d.** proper fraction (11 = numerator; 13 = denominator) **2a.** $\frac{4}{5}$ **b.** $\frac{1}{2}$ **c.** $\frac{2}{5}$ **3a.** composite **b.** composite **c.** prime **4a.** $2 \cdot 2 \cdot 5$ **b.** 19 (prime) **c.** $2 \cdot 5 \cdot 5$ **d.** $3 \cdot 3 \cdot 5$ **5a.** 2 and 3 **b.** 3 **c.** 2, 3, and 5 **d.** not divisible by 2, 3, or 5 **6a.** $2 \cdot 2 \cdot 3 \cdot 5$ **b.** $3 \cdot 5 \cdot 7$ **c.** 89 (prime) **d.** $11 \cdot 13$ **e.** $2 \cdot 3 \cdot 3 \cdot 29$

2.2 Reducing Fractions

8 More than one fraction can be used to represent the same part of a unit. We say that such fractions are **equal** or **equivalent.** Figure 2–2 shows that $\frac{1}{2}$ and $\frac{2}{4}$ represent the same part of a whole unit. Therefore, $\frac{1}{2}$ and $\frac{2}{4}$ are equal or equivalent.

$\frac{1}{2}$ of the rectangle is shaded

$\frac{2}{4}$ of the rectangle is shaded

Figure 2-2

NOTE: Two fractions are equal if their cross-products are equal. That is, $\frac{a}{b} = \frac{c}{d}$ (when b and d are not equal to zero) if and only if $a \cdot d = b \cdot c$. For example,

$$\frac{1}{2} = \frac{2}{4} \quad \text{since} \quad 1 \cdot 4 = 2 \cdot 2.$$

9 We know that $\frac{1}{2} = \frac{2}{4}$. We can also say that $\frac{1}{2}$ is the simplest form of $\frac{2}{4}$ since 1 is the only number that divides both the numerator and denominator of the fraction $\frac{1}{2}$ evenly. In general, a fraction is said to be reduced to its lowest terms when 1 is the only number that divides the numerator and denominator evenly.

10 To reduce a fraction to its lowest terms, divide both the numerator and denominator by their largest common divisor. In the fraction $\frac{4}{6}$, for example, the largest common divisor of 4 and 6 is 2. Thus,

$$\frac{4}{6} = \frac{4 \div 2}{6 \div 2} = \frac{2}{3}.$$

NOTE: We know that $\frac{4}{6}$ reduced to its lowest terms is $\frac{2}{3}$ since the only common divisor of 2 and 3 is 1, and $\frac{4}{6} = \frac{2}{3}$ since $4 \cdot 3 = 6 \cdot 2$.

11 Sometimes it will be difficult to recognize the largest common divisor of a fraction's numerator and denominator. In such a case, reduce the fraction to its lowest terms by using the following rule:

Rule for Reducing Fractions:

1. Write the prime factorization of the numerator and denominator.
2. Divide out every factor that appears in both the numerator and denominator. If all the factors divide out in the numerator or denominator, the corresponding numerator or denominator is 1.
3. Multiply the remaining factors in the numerator and multiply the remaining factors in the denominator to obtain a new fraction that is equivalent to the original fraction.

In the fraction $\frac{4}{6}$, for example, we write the prime factorization of both the numerator and denominator.

$$\frac{4}{6} = \frac{\cancel{2} \cdot 2}{\cancel{2} \cdot 3} = \frac{2}{3}$$

To divide both the numerator and denominator by 2 (or any number), draw a slanted line through the number in both the numerator and the denominator.

Example 5 Reduce each fraction to its lowest terms.

a. $\frac{42}{36}$ **b.** $\frac{9}{27}$ **c.** $\frac{231}{297}$ **d.** $\frac{25}{49}$ **e.** $\frac{819}{1599}$

Solution To reduce a fraction to its lowest terms, write the prime factorization of the numerator and denominator and then divide out all of their common factors. Next, find the product of the factors remaining in the numerator and place it over the product of the factors remaining in the denominator.

5a. $\frac{42}{36} = \frac{\cancel{2} \cdot \cancel{3} \cdot 7}{\cancel{2} \cdot 2 \cdot \cancel{3} \cdot 3} = \frac{7}{6}$

5b. $\frac{9}{27} = \frac{\cancel{3} \cdot \cancel{3}}{\cancel{3} \cdot \cancel{3} \cdot 3} = \frac{1}{3}$

5c. $\frac{231}{297} = \frac{\cancel{3} \cdot 7 \cdot \cancel{11}}{\cancel{3} \cdot 3 \cdot 3 \cdot \cancel{11}} = \frac{7}{9}$

3/231 7/77 /11 3/297 3/99 3/33 /11

5d. $\frac{25}{49} = \frac{5 \cdot 5}{7 \cdot 7} = \frac{25}{49}$

5e. $\frac{819}{1599} = \frac{\cancel{3} \cdot 3 \cdot 7 \cdot \cancel{13}}{\cancel{3} \cdot \cancel{13} \cdot 41} = \frac{21}{41}$

3/819 3/273 7/91 /13 3/1599 13/533 /41

Example 6 Reduce each fraction to its lowest terms.

a. $\frac{50}{60}$ **b.** $\frac{36}{72}$

Solution If you can recognize the largest common divisor of the numerator and denominator of a fraction, you do not need to find the prime factorization of these numbers.

6a. $\frac{50}{60} = \frac{\cancel{10} \cdot 5}{\cancel{10} \cdot 6} = \frac{5}{6}$

6b. $\frac{36}{72} = \frac{\cancel{36} \cdot 1}{\cancel{36} \cdot 2} = \frac{1}{2}$

Practice Problem 7 ***Reduce each fraction to its lowest terms.***

a. $\frac{16}{24}$ **b.** $\frac{70}{14}$ **c.** $\frac{30}{77}$ **d.** $\frac{65}{169}$ **e.** $\frac{117}{153}$

2.2 Exercises

Determine if the following pairs of fractions are equal.

1. $\frac{1}{2}, \frac{3}{4}$ **2.** $\frac{3}{5}, \frac{30}{50}$ **3.** $\frac{42}{36}, \frac{7}{6}$ **4.** $\frac{231}{297}, \frac{7}{9}$

Reduce each fraction to its lowest terms.

5. $\frac{6}{9}$ **6.** $\frac{10}{15}$ **7.** $\frac{12}{20}$ **8.** $\frac{54}{81}$ **9.** $\frac{36}{48}$ **10.** $\frac{54}{72}$

11. $\frac{16}{24}$ **12.** $\frac{9}{25}$ **13.** $\frac{10}{90}$ **14.** $\frac{35}{91}$ **15.** $\frac{20}{28}$ **16.** $\frac{30}{42}$

17. $\frac{26}{169}$ **18.** $\frac{33}{22}$ **19.** $\frac{77}{121}$ **20.** $\frac{120}{15}$ **21.** $\frac{55}{66}$ **22.** $\frac{52}{78}$

23. $\frac{48}{120}$ **24.** $\frac{26}{39}$ **25.** $\frac{64}{110}$ **26.** $\frac{152}{312}$ **27.** $\frac{66}{210}$ **28.** $\frac{81}{405}$

29. $\frac{215}{258}$ **30.** $\frac{189}{437}$ **31.** $\frac{121}{169}$ **32.** $\frac{53}{159}$ **33.** $\frac{133}{161}$ **34.** $\frac{187}{209}$

35. $\frac{156}{222}$ **36.** $\frac{126}{444}$ **37.** $\frac{115}{135}$ **38.** $\frac{330}{460}$ **39.** $\frac{600}{900}$ **40.** $\frac{400}{800}$

41. $\frac{266}{462}$ **42.** $\frac{1001}{770}$ **43.** $\frac{126}{132}$ **44.** $\frac{192}{352}$ **45.** $\frac{138}{805}$ **46.** $\frac{132}{220}$

47. $\frac{1547}{4641}$ **48.** $\frac{2310}{980}$ **49.** $\frac{1575}{4158}$ **50.** $\frac{252}{3025}$ **51.** $\frac{3100}{4300}$ **52.** $\frac{1700}{1900}$

Fill in the blanks.

53. Two fractions are equivalent or equal if their ________________ are equal.

54. A fraction is reduced to its lowest terms if the only common divisor of the ________________ and ________________ is ________________.

55. To reduce a fraction to its lowest terms, write the ________________ factorization of the ________________ and ________________ and then divide out all their ________________ factors. Next, find the product of the factors remaining in the ________________ and place it over the product of the factors remaining in the ________________.

Solve and express each answer in its simplest form.

56. Rosalind made a $120 profit by selling Tupperware. If she deposits $105 in her savings account and spends the rest, what fractional part of her profits is she saving?

57. In problem 56, what fractional part of her profits has she spent?

58. Reggie, Laura, and Richard went into business together. If Reggie invested $500, Laura invested $1500, and Richard invested $1000, what fractional part of the business does each person own?

59. Joe cut a pie into six equal pieces and served three pieces. What fractional part of the pie was left?

60. Ed bought a dozen eggs and used six of them in a cake. What fractional part of the dozen eggs did he use?

61. Jane got 60 hits out of 140 times at bat. What fractional part of her times at bat did she get a hit?

62. Ron got 70 hits in 200 times at bat. What fractional part of his times at bat did he get a hit?

63. Joe drank two bottles of beer from a pack of six bottles. What fractional part of the six-pack did he drink? What fractional part was left?

64. One day, 10 students were absent from a class of 40 students. What fractional part of the class was absent? What fractional part of the class was present?

Answers to Practice Problems 7a. $\frac{2}{3}$ b. 5 c. $\frac{30}{77}$ d. $\frac{5}{13}$ e. $\frac{13}{17}$

2.3 Improper Fractions and Mixed Numbers

12 A **mixed number** is the sum of a whole number and a proper fraction. Mixed numbers are written without a plus sign. For example, $4\frac{3}{5}$, which is read as "four and three-fifths," is equal to $4 + \frac{3}{5}$. Similarly, the mixed number $7\frac{2}{3}$ (read as "seven and two-thirds") is equal to $7 + \frac{2}{3}$.

13 Mixed numbers can be written as improper fractions, and improper fractions can be expressed as mixed numbers. Examine the rectangles in Figure 2–3. The shaded area can be expressed as the mixed number $1\frac{2}{3}$ since one rectangle plus two-thirds of another rectangle is shaded. This same shaded area can be represented by the improper fraction $\frac{5}{3}$ since each rectangle is divided into three equal parts and five of these parts are shaded. Thus, $1\frac{2}{3} = \frac{5}{3}$ and $\frac{5}{3} = 1\frac{2}{3}$.

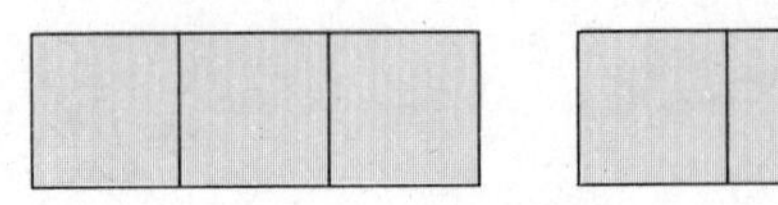

Figure 2-3

14 Another way to show that $1\frac{2}{3} = \frac{5}{3}$ is by writing

$$1\frac{2}{3} = 1 + \frac{2}{3} = \frac{3}{3} + \frac{2}{3} = \frac{3+2}{3} = \frac{5}{3}.$$

There is a shortcut to this procedure.

To Change a Mixed Number to an Improper Fraction:

1. Multiply the denominator of the fraction by the whole number.
2. Add the numerator to the product obtained in step 1.
3. Place this sum over the original denominator.

Example 7 Change each mixed number to an improper fraction.

a. $3\frac{2}{5}$ **b.** $5\frac{3}{7}$ **c.** $15\frac{11}{13}$ **d.** $9\frac{8}{11}$

Solution To change a mixed number to an improper fraction, first multiply the denominator by the whole number. Next, add the numerator. Finally, place this sum over the original denominator.

7a. $3\frac{2}{5} = \frac{5 \cdot 3 + 2}{5} = \frac{15 + 2}{5} = \frac{17}{5}$

7b. $5\frac{3}{7} = \frac{7 \cdot 5 + 3}{7} = \frac{35 + 3}{7} = \frac{38}{7}$

7c. $15\frac{11}{13} = \frac{13 \cdot 15 + 11}{13} = \frac{195 + 11}{13} = \frac{206}{13}$

7d. $9\frac{8}{11} = \frac{11 \cdot 9 + 8}{11} = \frac{99 + 8}{11} = \frac{107}{11}$

Practice Problem 8 ***Change each mixed number to an improper fraction.***

a. $8\frac{3}{5}$ **b.** $6\frac{2}{9}$ **c.** $17\frac{13}{14}$

15 We can show that $\frac{5}{3} = 1\frac{2}{3}$ by using long division. In other words,

$$\frac{5}{3} = 3\overline{)5.} \quad \text{quotient } 1,\ \ 5 - 3 = 2$$

To Change an Improper Fraction to a Mixed Number:

1. Divide the denominator into the numerator.
2. The quotient (Q) is the whole number part of the mixed number. The remainder (R) is the numerator of the fractional part and the original denominator (d) is the denominator.

In the problem above, $\frac{5}{3} = Q\frac{R}{d} = 1\frac{2}{3}$.

Example 8 Change each fraction to a mixed number.

a. $\frac{29}{3}$ **b.** $\frac{9}{2}$ **c.** $\frac{135}{13}$ **d.** $\frac{302}{8}$

Solution To change an improper fraction to a mixed number, divide the denominator into the numerator. The quotient is the whole number part of the mixed number, and the fractional part is the remainder over the original denominator.

8a. $\frac{29}{3} = 9\frac{2}{3}$ — $3\overline{)29}$ gives 9 ← Quotient; $29 - 27 = 2$ ← Remainder; 3 ← Denominator

8b. $\frac{9}{2} = 4\frac{1}{2}$ — $2\overline{)9}$ gives 4; $9 - 8 = 1$

8c. $\dfrac{135}{13} = 10\dfrac{5}{13}$ $\quad 13\overline{)135}$ quotient 10; 13; remainder 5

8d. $\dfrac{302}{8} = 37\dfrac{6}{8} = 37\dfrac{3}{4}$ $\quad 8\overline{)302}$ quotient 37; 24; 62; 56; remainder 6

NOTE: Always reduce the fractional part of a mixed number to its lowest terms.

Practice Problem 9 ***Change each fraction to a mixed number.***

a. $\dfrac{8}{3}$ **b.** $\dfrac{17}{4}$ **c.** $\dfrac{156}{17}$ **d.** $\dfrac{66}{18}$

2.3 Exercises

Change each mixed number to an improper fraction.

1. $2\dfrac{1}{2}$ **2.** $3\dfrac{3}{4}$ **3.** $5\dfrac{7}{8}$ **4.** $6\dfrac{1}{9}$

5. $10\dfrac{2}{9}$ **6.** $7\dfrac{3}{11}$ **7.** $9\dfrac{2}{5}$ **8.** $15\dfrac{2}{3}$

9. $12\dfrac{3}{11}$ **10.** $11\dfrac{11}{13}$ **11.** $14\dfrac{3}{16}$ **12** $21\dfrac{3}{4}$

13. $35\dfrac{1}{9}$ **14.** $72\dfrac{13}{19}$ **15.** $73\dfrac{5}{9}$ **16.** $27\dfrac{3}{7}$

17. $5\dfrac{1}{17}$ **18.** $100\dfrac{18}{29}$ **19.** $13\dfrac{3}{13}$ **20.** $262\dfrac{3}{16}$

Change each fraction to a mixed number.

21. $\dfrac{15}{2}$ **22.** $\dfrac{16}{7}$ **23.** $\dfrac{60}{7}$ **24.** $\dfrac{19}{4}$

25. $\dfrac{35}{3}$ **26.** $\dfrac{33}{4}$ **27.** $\dfrac{62}{9}$ **28.** $\dfrac{75}{8}$

29. $\dfrac{42}{8}$ **30.** $\dfrac{28}{6}$ **31.** $\dfrac{34}{16}$ **32.** $\dfrac{34}{10}$

33. $\dfrac{180}{50}$ **34.** $\dfrac{106}{13}$ **35.** $\dfrac{102}{18}$ **36.** $\dfrac{120}{48}$

37. $\dfrac{152}{11}$ **38.** $\dfrac{135}{115}$ **39.** $\dfrac{169}{65}$ **40.** $\dfrac{148}{111}$

Fill in the blanks.

41. A ______________ is the sum of a whole number and a proper fraction.

42. To change a mixed number to an improper fraction, multiply the denominator times the ______________, add the ______________, and place this sum over the original ______________.

43. To change an improper fraction to a mixed number, divide the ______________ into the ______________. The ______________ is the whole number part of the mixed number and the fractional part is written as the ______________ over the original denominator.

Solve and show all work.

44. The Davis Catering Company must serve 57 pieces of pie at a luncheon. If each pie is divided into five equal pieces, how many pies are needed?

45. Write $6 + \frac{4}{5}$ as an improper fraction.

Answers to Practice Problems 8a. $\frac{43}{5}$ b. $\frac{56}{9}$ c. $\frac{251}{14}$ 9a. $2\frac{2}{3}$ b. $4\frac{1}{4}$ c. $9\frac{3}{17}$ d. $3\frac{2}{3}$

2.4 Multiplication

16 To multiply fractions, multiply numerators, multiply denominators, and reduce (if necessary) the resulting fraction. For example,

$$\frac{1}{2} \cdot \frac{2}{6} = \frac{1 \cdot 2}{2 \cdot 6} = \frac{2}{12} = \frac{1}{6}.$$

We can show that $\frac{1}{2} \cdot \frac{2}{6} = \frac{1}{6}$ by noting that $\frac{1}{2} \cdot \frac{2}{6}$ means $\frac{1}{2}$ *of* $\frac{2}{6}$. In Figure 2–4, two-sixths $\left(\frac{2}{6}\right)$ of the rectangle is shaded. Note that half $\left(\frac{1}{2}\right)$ of the shaded area is equal to one-sixth $\left(\frac{1}{6}\right)$ of the rectangle.

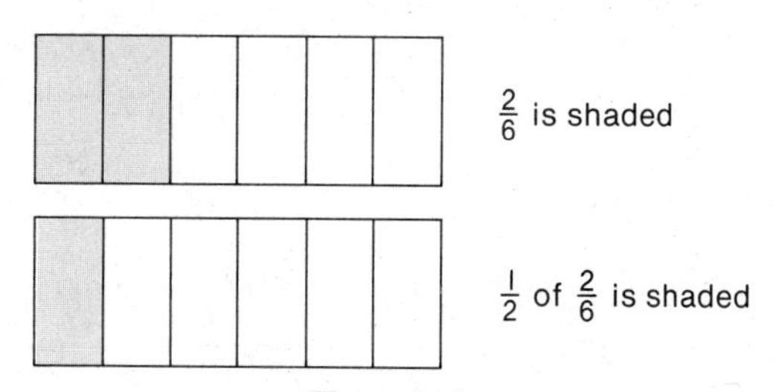

Figure 2-4

17 Sometimes it will be inconvenient to multiply and then reduce. Therefore, to multiply two or more fractions, use the following rule:

Rule for Multiplying Fractions:

1. Write a new fraction which contains the factors of the original numerators over the factors of the original denominators.
2. Reduce the resulting fraction to its lowest terms.

To multiply a whole number N times a fraction, write N as $\frac{N}{1}$ and use the rules for multiplying fractions.

Example 9 Multiply and reduce.

a. $\dfrac{4}{9} \cdot \dfrac{15}{14}$ **b.** $\dfrac{1}{6} \cdot \dfrac{25}{7}$ **c.** $\dfrac{24}{42} \cdot \dfrac{28}{86}$

d. $\dfrac{14}{24} \cdot \dfrac{33}{35} \cdot \dfrac{4}{77}$ **e.** $\dfrac{222}{297} \cdot \dfrac{231}{182}$ **f.** $6 \cdot \dfrac{9}{30}$

Solution To multiply fractions, write the factors of the numerators over the factors of the denominators. Reduce the resulting fraction.

9a.
$$\frac{4}{9} \cdot \frac{15}{14} = \frac{4 \cdot 15}{9 \cdot 14}$$
$$= \frac{\cancel{2} \cdot 2 \cdot \cancel{3} \cdot 5}{\cancel{3} \cdot 3 \cdot \cancel{2} \cdot 7} \quad \text{Factor and divide out like factors}$$
$$= \frac{10}{21} \quad \text{Multiply}$$

9b.
$$\frac{1}{6} \cdot \frac{25}{7} = \frac{1 \cdot 25}{6 \cdot 7}$$
$$= \frac{1 \cdot 5 \cdot 5}{2 \cdot 3 \cdot 7} \quad \text{No common factors}$$
$$= \frac{25}{42}$$

9c.
$$\frac{24}{42} \cdot \frac{28}{86} = \frac{24 \cdot 28}{42 \cdot 86}$$
$$= \frac{\cancel{2} \cdot \cancel{2} \cdot 2 \cdot \cancel{3} \cdot 2 \cdot 2 \cdot \cancel{7}}{\cancel{2} \cdot \cancel{3} \cdot \cancel{7} \cdot \cancel{2} \cdot 43}$$
$$= \frac{8}{43}$$

9d.
$$\frac{14}{24} \cdot \frac{33}{35} \cdot \frac{4}{77} = \frac{14 \cdot 33 \cdot 4}{24 \cdot 35 \cdot 77}$$
$$= \frac{\cancel{2} \cdot \cancel{7} \cdot \cancel{3} \cdot \cancel{11} \cdot \cancel{2} \cdot \cancel{2}}{\cancel{2} \cdot \cancel{2} \cdot \cancel{2} \cdot \cancel{3} \cdot 5 \cdot \cancel{7} \cdot 7 \cdot \cancel{11}}$$
$$= \frac{1}{35}$$

9e.
$$\frac{222}{297} \cdot \frac{231}{182} = \frac{222 \cdot 231}{297 \cdot 182}$$
$$= \frac{\cancel{2} \cdot \cancel{3} \cdot 37 \cdot \cancel{3} \cdot \cancel{7} \cdot \cancel{11}}{\cancel{3} \cdot \cancel{3} \cdot 3 \cdot \cancel{11} \cdot \cancel{2} \cdot \cancel{7} \cdot 13}$$
$$= \frac{37}{39}$$

9f.
$$6 \cdot \frac{9}{30} = \frac{6}{1} \cdot \frac{9}{30} \quad \text{Write 6 as a fraction } \left(\frac{6}{1}\right)$$
$$= \frac{6 \cdot 9}{1 \cdot 30}$$
$$= \frac{\cancel{2} \cdot \cancel{3} \cdot 3 \cdot 3}{1 \cdot \cancel{2} \cdot \cancel{3} \cdot 5}$$
$$= \frac{9}{5}$$

NOTE: If the product of two numbers is 1, the numbers are **reciprocals** of each other. To find the reciprocal of a nonzero number, just interchange the numerator and denominator. For example, the reciprocal of $\frac{3}{4}$ is $\frac{4}{3}$.

Example 10 Multiply and reduce.

a. $\frac{4}{9} \cdot \frac{15}{14}$ **b.** $\frac{4}{5} \cdot \frac{10}{3} \cdot \frac{9}{16}$

Solution If you can recognize the largest common divisor of numbers in the numerator and denominator, you will not need to factor.

10a. $\frac{4}{9} \cdot \frac{15}{14} = \frac{\overset{2}{\cancel{4}} \cdot \overset{5}{\cancel{15}}}{\underset{3}{\cancel{9}} \cdot \underset{7}{\cancel{14}}}$ Divide 4 and 14 by 2; Divide 9 and 15 by 3

$= \frac{10}{21}$

10b. $\frac{4}{5} \cdot \frac{10}{3} \cdot \frac{9}{16} = \frac{\overset{1}{\cancel{4}} \cdot \overset{\overset{1}{\cancel{2}}}{\cancel{10}} \cdot \overset{3}{\cancel{9}}}{\underset{1}{\cancel{5}} \cdot \underset{1}{\cancel{3}} \cdot \underset{\underset{2}{\cancel{4}}}{\cancel{16}}}$ Divide 4 and 16 by 4; Divide 5 and 10 by 5; Divide 3 and 9 by 3; Divide 2 and 4 by 2

$= \frac{3}{2}$

NOTE: When using the techniques shown in Example 10, check to make sure you have reduced your final answer to its lowest terms.

Practice Problem 10 ***Multiply and reduce.***

a. $\frac{8}{12} \cdot \frac{10}{15}$ **b.** $6 \cdot \frac{7}{15}$

c. $\frac{14}{20} \cdot \frac{66}{42} \cdot \frac{2}{55}$ **d.** $\frac{117}{77} \cdot \frac{143}{273}$

18 Frequently in mathematics you must find the product of mixed numbers, fractions, and/or whole numbers.

To Multiply Using Mixed Numbers:

1. Change all mixed numbers to improper fractions.
2. Multiply using the rules given in this section.
3. Write the answer as a mixed number or an improper fraction.

Example 11 Multiply.

a. $\frac{6}{35} \cdot 2\frac{1}{10}$ **b.** $1\frac{22}{43} \cdot 3\frac{12}{39} \cdot 1\frac{1}{10}$ **c.** $60 \cdot \frac{25}{42} \cdot 2\frac{1}{10}$

Solution To multiply using mixed numbers, change all mixed numbers to improper fractions and multiply. Express the final answer as a mixed number or an improper fraction.

11a. $\frac{6}{35} \cdot 2\frac{1}{10} = \frac{6}{35} \cdot \frac{21}{10}$ Change a mixed number to a fraction

$= \frac{\overset{3}{\cancel{6}} \cdot \overset{3}{\cancel{21}}}{\underset{5}{\cancel{35}} \cdot \underset{5}{\cancel{10}}}$ Divide 6 and 10 by 2; Divide 21 and 35 by 7

$= \frac{9}{25}$

11b. $1\frac{22}{43} \cdot 3\frac{12}{39} \cdot 1\frac{1}{10} = \frac{65}{43} \cdot \frac{129}{39} \cdot \frac{11}{10}$

$= \frac{65 \cdot 129 \cdot 11}{43 \cdot 39 \cdot 10}$

$= \frac{\cancel{5} \cdot \cancel{13} \cdot \cancel{3} \cdot \cancel{43} \cdot 11}{\cancel{43} \cdot \cancel{3} \cdot \cancel{13} \cdot 2 \cdot \cancel{5}}$

$= \frac{11}{2}$

$= 5\frac{1}{2}$

11c. $60 \cdot \frac{25}{42} \cdot 2\frac{1}{10} = \frac{60}{1} \cdot \frac{25}{42} \cdot \frac{21}{10}$

$= \frac{60 \cdot 25 \cdot 21}{1 \cdot 42 \cdot 10}$

$= \frac{\cancel{2} \cdot \cancel{2} \cdot \cancel{3} \cdot \cancel{5} \cdot 5 \cdot 5 \cdot 3 \cdot \cancel{7}}{1 \cdot \cancel{2} \cdot \cancel{3} \cdot \cancel{7} \cdot \cancel{2} \cdot \cancel{5}}$

$= \frac{75}{1}$

$= 75$

Practice Problem 11 ***Multiply and express each answer as a mixed number.***

a. $5\frac{2}{5} \cdot 1\frac{9}{21}$ **b.** $\frac{15}{20} \cdot 2\frac{4}{11}$ **c.** $99 \cdot 1\frac{25}{66}$

Example 12 Solve and show all work.

a. Ronnie borrows \$2500 from his mother. If he agrees to repay $\frac{2}{5}$ of the money in three weeks, how much money should his mother receive at that time?

b. If Glenn's car can travel $16\frac{1}{4}$ miles on one gallon of gas, how far will his car travel on $6\frac{4}{5}$ gallons of gas?

Solution **12a.** ***Think:*** Find $\frac{2}{5}$ of \$2500.

$$\frac{2}{5}\cdot 2500 = \frac{2}{5}\cdot\frac{2500}{1}$$

$$= \frac{2\cdot \overset{500}{\cancel{2500}}}{\underset{1}{\cancel{5}}\cdot 1} \quad \text{Divide 5 and 2500 by 5}$$

$$= \frac{1000}{1}$$

$$= 1000$$

Ronnie's mother should receive \$1000.

12b. To find the distance traveled, use the formula:

$$\text{Miles traveled} = (\text{miles per gallon})(\text{gallons of gas used})$$

$$= \left(16\frac{1}{4}\right)\left(6\frac{4}{5}\right)$$

$$= \frac{65}{4}\cdot\frac{34}{5}$$

$$= \frac{\overset{13}{\cancel{65}}\cdot\overset{17}{\cancel{34}}}{\underset{2}{\cancel{4}}\cdot\underset{1}{\cancel{5}}} \quad \begin{array}{l}\text{Divide 5 and 65 by 5}\\ \text{Divide 4 and 34 by 2}\end{array}$$

$$= \frac{221}{2}$$

$$= 110\frac{1}{2}$$

Glenn's car will travel $110\frac{1}{2}$ miles.

Practice Problem 12 ***Solve and show all work.***

a. A real estate broker told Carver that his house is worth $2\frac{1}{2}$ times its original cost. If the original cost was \$24,000, how much is the house worth now?

b. If $\frac{3}{4}$ of the 24 people in a social club paid their dues and $\frac{2}{3}$ of these people were men, how many men paid their dues?

2.4 Exercises

Multiply and reduce to lowest terms.

1. $\frac{6}{25}\cdot\frac{5}{8}$

2. $\frac{6}{34}\cdot\frac{51}{56}$

3. $\frac{4}{10}\cdot\frac{6}{8}$

4. $5\cdot\frac{14}{20}$

5. $\frac{1}{6}\cdot\frac{5}{7}$

6. $10\cdot\frac{6}{4}$

7. $\frac{45}{44}\cdot\frac{11}{24}$

8. $6\cdot\frac{4}{9}$

9. $\frac{36}{99}\cdot\frac{25}{40}\cdot\frac{42}{28}$

10. $\frac{86}{65}\cdot\frac{39}{129}\cdot\frac{12}{69}$

11. $135\cdot\frac{30}{105}$

12. $\frac{14}{60}\cdot\frac{66}{77}$

13. $\frac{65}{56}\cdot\frac{48}{39}\cdot\frac{15}{18}$

14. $\frac{48}{70}\cdot\frac{80}{84}\cdot\frac{154}{192}$

15. $81\cdot\frac{14}{54}$

16. $\frac{234}{222} \cdot \frac{152}{312}$

17. $\frac{75}{1001} \cdot \frac{154}{125}$

18. $\frac{42}{70} \cdot \frac{25}{35} \cdot \frac{14}{6}$

19. $\frac{21}{36} \cdot \frac{27}{42} \cdot \frac{45}{63} \cdot \frac{21}{10}$

20. $\frac{96}{117} \cdot \frac{48}{192} \cdot \frac{104}{114} \cdot 7$

21. $\frac{7}{8} \cdot \frac{8}{7}$

22. $\frac{3}{5} \cdot \frac{5}{3}$

23. $\frac{16}{27} \cdot \frac{18}{24}$

24. $\frac{14}{35} \cdot \frac{21}{10}$

25. $15 \cdot \frac{4}{35}$

26. $12 \cdot \frac{9}{16}$

27. $\frac{56}{52} \cdot \frac{55}{77} \cdot \frac{26}{20}$

28. $\frac{65}{20} \cdot \frac{60}{91} \cdot \frac{75}{2}$

29. $\frac{8}{69} \cdot \frac{23}{4} \cdot 6$

30. $\frac{20}{91} \cdot \frac{77}{15} \cdot 39$

31. $\frac{33}{25} \cdot \frac{9}{10} \cdot \frac{75}{20}$

32. $\frac{72}{39} \cdot \frac{24}{96} \cdot \frac{52}{32}$

33. $\frac{32}{80} \cdot \frac{25}{28}$

34. $\frac{36}{70} \cdot \frac{35}{44}$

35. $\frac{54}{143} \cdot \frac{13}{270}$

36. $\frac{16}{39} \cdot \frac{26}{240}$

37. $\frac{2}{11} \cdot 3300$

38. $\frac{3}{13} \cdot 3900$

39. $\frac{33}{105} \cdot \frac{6}{1155} \cdot \frac{22}{23}$

40. $\frac{8}{9} \cdot \frac{25}{81} \cdot \frac{243}{1000}$

Multiply and express each answer as a mixed number.

41. $\left(1\frac{1}{3}\right)\left(1\frac{1}{2}\right)$

42. $\left(8\frac{1}{2}\right)\left(2\frac{2}{3}\right)$

43. $\left(3\frac{3}{5}\right)(15)$

44. $\left(7\frac{1}{5}\right)(25)$

45. $\left(7\frac{1}{2}\right)\left(4\frac{3}{4}\right)$

46. $\left(2\frac{1}{7}\right)\left(2\frac{4}{5}\right)$

47. $(6)\left(2\frac{1}{12}\right)$

48. $(18)\left(4\frac{1}{12}\right)$

49. $\left(\frac{3}{7}\right)\left(3\frac{2}{9}\right)$

50. $\left(\frac{5}{8}\right)\left(2\frac{1}{2}\right)$

51. $\left(1\frac{2}{7}\right)\left(2\frac{1}{3}\right)\left(2\frac{1}{6}\right)$

52. $\left(2\frac{2}{3}\right)\left(2\frac{1}{4}\right)\left(1\frac{1}{2}\right)$

53. $\left(10\frac{2}{3}\right)\left(\frac{5}{8}\right)$

54. $\left(8\frac{3}{4}\right)\left(\frac{6}{25}\right)$

55. $\left(2\frac{1}{7}\right)\left(\frac{33}{36}\right)\left(4\frac{2}{3}\right)$

56. $\left(2\frac{5}{8}\right)\left(\frac{7}{9}\right)\left(2\frac{2}{5}\right)$

57. $\left(\frac{6}{13}\right)\left(2\frac{1}{6}\right)$

58. $\left(\frac{7}{15}\right)\left(2\frac{1}{7}\right)$

59. $1\frac{1}{5} \cdot 1\frac{3}{8}$

60. $2\frac{1}{17} \cdot 5\frac{1}{10}$

61. $4\frac{1}{6} \cdot 1\frac{2}{5}$

62. $14 \cdot 2\frac{1}{2}$

63. $5\frac{2}{15} \cdot \frac{20}{91}$

64. $\frac{14}{22} \cdot 3\frac{3}{10}$

65. $\frac{77}{164} \cdot 2\frac{25}{49}$

66. $1\frac{2}{3} \cdot \frac{48}{70}$

67. $45 \cdot 2\frac{1}{30}$

68. $2\frac{2}{5} \cdot 2\frac{3}{16}$

69. $\frac{42}{54} \cdot 1\frac{5}{7} \cdot 5$

70. $15\frac{7}{15} \cdot 4\frac{19}{24}$

71. $5\frac{2}{5} \cdot 1\frac{9}{21} \cdot \frac{14}{81} \cdot 6$

72. $\frac{124}{35} \cdot 1\frac{23}{42} \cdot \frac{98}{37} \cdot 2\frac{3}{4} \cdot \frac{3}{155}$

73. $5\frac{5}{9} \cdot 1\frac{11}{25} \cdot 1\frac{27}{64} \cdot \frac{110}{1001} \cdot 3$

Fill in the blanks.

74. To multiply two or more fractions, write the factors of the ________ over the factors of the ________ and ________ the resulting fraction.

75. To multiply a whole number N times a fraction, write N as ________ and then use the rules for multiplying fractions.

76. To multiply using mixed numbers, change the ________ to ________. Next, use the rules for ________ fractions and express the final answer as a ________ number or ________ fraction.

Solve and show all work.

77. If ADK stock costs $\$12\frac{1}{2}$ per share, find the cost of 45 shares.

78. Twenty out of every 45 high school basketball players play college basketball. One out of every 2000 college basketball players play professional basketball. What fractional part of high school basketball players play professional basketball?

79. A warehouse will hold $\frac{2}{9}$ of a ton of sugar. If the warehouse is $\frac{3}{4}$ full, how much sugar is in the warehouse?

80. Cazzie earns \$60 a day. If he works only $\frac{7}{12}$ of the day, how much money will he earn?

81. Joyce's car can travel $15\frac{1}{3}$ miles on a gallon of gas. How far can the car travel on $12\frac{3}{4}$ gallons of gas?

82. Terry and Keith bought a painting. Terry invested \$2000, and Keith invested \$2400. Five years later, they sold the painting for \$13,090. How much money should each man receive if the \$13,090 is divided fairly? (*Hint:* Whatever fractional part of the money each man originally invested, he should receive that same fractional part back from the sale of the painting.)

83. If two-thirds of the 450 students at a school are men, how many students at the school are men?

84. If three-fifths of the graduating class of 175 seniors are women, how many women are graduating?

85. Find the product of $\frac{123}{77}$ and $\frac{82}{49}$.

86. Find the product of $\frac{21}{27}$ and $\frac{198}{15}$.

87. Multiply $\frac{4}{46}$ by $\frac{69}{12}$.

88. Multiply $\frac{14}{9}$ by $\frac{81}{21}$.

89. Multiply $2\frac{2}{5}$ and 240.

90. Multiply $15\frac{1}{2}$ and 420.

91. Pete's car can travel $16\frac{1}{2}$ miles on one gallon of gas. How far can the car travel on $9\frac{1}{3}$ gallons of gas?

92. Bill's car can travel $17\frac{1}{2}$ miles on one gallon of gas. How far can the car travel on $7\frac{3}{5}$ gallons of gas?

93. A company bought 1,000 shares of a stock at $\$25\frac{1}{8}$ per share. Find the total cost of this purchase.

94. Lisa bought 960 shares of a stock at $\$17\frac{3}{8}$ per share. Find the total cost of this purchase.

95. A truck can hold $\frac{2}{3}$ of a ton of salt. If the truck is $\frac{1}{10}$ full, how much salt is in the truck?

96. A storage room can hold $\frac{3}{4}$ of a ton of flour. If the room is $\frac{1}{6}$ full, how much flour is in the room?

Answers to Practice Problems **10a.** $\frac{4}{9}$ **b.** $\frac{14}{5}$ **c.** $\frac{1}{25}$ **d.** $\frac{39}{49}$ **11a.** $7\frac{5}{7}$ **b.** $1\frac{17}{22}$ **c.** $136\frac{1}{2}$
12a. \$60,000 **b.** 12

2.5 Division

19 In Section 1.6, we found that $12 \div 6 = 2$ since $6 \cdot 2 = 12$. Similarly, to find $\frac{3}{5} \div \frac{1}{5}$, we must find a number c such that $\frac{1}{5} \cdot c = \frac{3}{5}$. By inspection, we know that $c = 3$ since $\frac{1}{5} \cdot 3 = \frac{3}{5}$. Thus, $\frac{3}{5} \div \frac{1}{5} = 3$.

20 In most cases, you will not be able to divide fractions by inspection. Therefore, to divide one fraction by another fraction, use the following rule:

Rule for Dividing One Fraction by Another:

1. Keep the same dividend.
2. Multiply the dividend by the reciprocal of the divisor (interchange the numerator and denominator).

NOTE: Writing the reciprocal of a nonzero number is the same as inverting that number. For example, the reciprocal of $\frac{4}{5}$ is $\frac{5}{4}$. Any number multiplied by its reciprocal is 1.

Example 13 Divide and reduce.

a. $\frac{9}{16} \div \frac{15}{8}$

b. $\frac{25}{28} \div 15$

c. $\frac{42}{120} \div \frac{77}{16}$

d. $\frac{10}{1001} \div \frac{35}{1309}$

Solution To divide fractions, multiply the dividend by the reciprocal of the divisor (invert the divisor).

Reciprocal

13a. $\frac{9}{16} \div \frac{15}{8} = \frac{9}{16} \cdot \frac{8}{15}$

$= \frac{\overset{3}{\cancel{9}} \cdot \overset{1}{\cancel{8}}}{\underset{2}{\cancel{16}} \cdot \underset{5}{\cancel{15}}}$ Divide 8 and 16 by 2
Divide 9 and 15 by 3

$= \frac{3}{10}$

13b. $\frac{25}{28} \div 15 = \frac{25}{28} \cdot \frac{1}{15}$

$= \frac{\overset{5}{\cancel{25}} \cdot 1}{28 \cdot \underset{3}{\cancel{15}}}$ Divide 25 and 15 by 5

$= \frac{5}{84}$

Since $15 = \frac{15}{1}$, its reciprocal is $\frac{1}{15}$.

13c. $\dfrac{42}{120} \div \dfrac{77}{16} = \dfrac{42}{120} \cdot \dfrac{16}{77}$

$= \dfrac{42 \cdot 16}{120 \cdot 77}$

$= \dfrac{\cancel{2} \cdot \cancel{3} \cdot \cancel{7} \cdot \cancel{2} \cdot \cancel{2} \cdot 2 \cdot 2}{\cancel{2} \cdot \cancel{2} \cdot \cancel{2} \cdot \cancel{3} \cdot 5 \cdot \cancel{7} \cdot 11}$

$= \dfrac{4}{55}$

13d. $\dfrac{10}{1001} \div \dfrac{35}{1309} = \dfrac{10}{1001} \cdot \dfrac{1309}{35}$

$= \dfrac{10 \cdot 1309}{1001 \cdot 35}$

$= \dfrac{2 \cdot \cancel{5} \cdot \cancel{7} \cdot \cancel{11} \cdot 17}{\cancel{7} \cdot \cancel{11} \cdot 13 \cdot \cancel{5} \cdot 7}$

$= \dfrac{34}{91}$

Practice Problem 13 ***Divide and reduce.***

a. $\dfrac{14}{17} \div \dfrac{28}{51}$ **b.** $50 \div \dfrac{15}{39}$ **c.** $\dfrac{85}{90} \div \dfrac{105}{84}$

21 Sometimes the dividend and/or the divisor in a division problem will be a mixed number.

To Divide Using Mixed Numbers:

1. Change all mixed numbers to improper fractions.
2. Divide using the rules discussed in this section.

Example 14 Divide and reduce.

a. $3\frac{1}{3} \div 5\frac{1}{2}$ **b.** $45 \div 3\frac{3}{5}$ **c.** $5\frac{11}{15} \div 12\frac{9}{10}$

Solution To divide using mixed numbers, first change all mixed numbers to improper fractions and then divide.

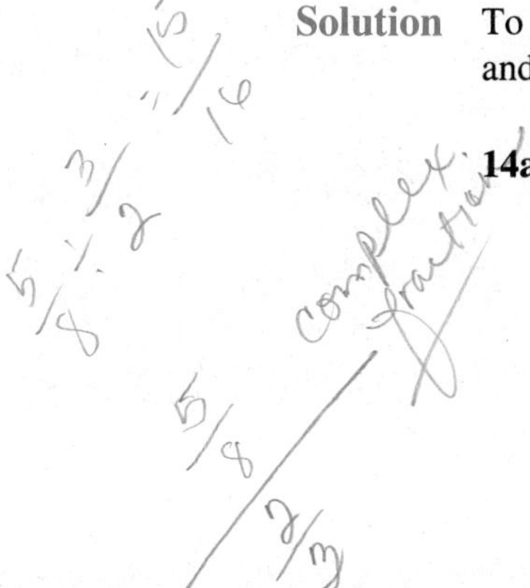

14a. $3\frac{1}{3} \div 5\frac{1}{2} = \dfrac{10}{3} \div \dfrac{11}{2}$ Express mixed numbers as fractions

$= \dfrac{10}{3} \cdot \dfrac{2}{11}$

$= \dfrac{10 \cdot 2}{3 \cdot 11}$ No common factors

$= \dfrac{20}{33}$

14b. $45 \div 3\frac{3}{5} = \frac{45}{1} \div \frac{18}{5}$

$= \frac{45}{1} \cdot \frac{5}{18}$

$= \frac{\overset{5}{\cancel{45}} \cdot 5}{1 \cdot \underset{2}{\cancel{18}}}$ Divide 45 and 18 by 9

$= \frac{25}{2}$

$= 12\frac{1}{2}$

14c. $5\frac{11}{15} \div 12\frac{9}{10} = \frac{86}{15} \div \frac{129}{10}$

$= \frac{86}{15} \cdot \frac{10}{129}$

$= \frac{86 \cdot 10}{15 \cdot 129}$

$= \frac{2 \cdot \cancel{43} \cdot 2 \cdot \cancel{5}}{3 \cdot \cancel{5} \cdot 3 \cdot \cancel{43}}$

$= \frac{4}{9}$

Practice Problem 14 ***Divide and reduce.***

a. $1\frac{2}{9} \div \frac{45}{66}$ **b.** $18 \div 8\frac{2}{11}$ **c.** $13\frac{7}{9} \div 1\frac{17}{27}$

Example 15 Carver's car traveled 325 miles on $12\frac{1}{2}$ gallons of gas. How many miles per gallon (mpg) did his car get?

Solution Car's mpg = miles traveled ÷ gallons of gas used

$= 325 \div 12\frac{1}{2}$

$= \frac{325}{1} \div \frac{25}{2}$

$= \frac{325}{1} \cdot \frac{2}{25}$

$= \frac{325 \cdot 2}{1 \cdot 25}$

$= \frac{\cancel{5} \cdot \cancel{5} \cdot 13 \cdot 2}{1 \cdot \cancel{5} \cdot \cancel{5}}$

$= \frac{26}{1}$

$= 26$

Carver's car gets 26 miles per gallon.

Practice Problem 15 ***If a car traveled $110\frac{1}{2}$ miles on $6\frac{4}{5}$ gallons of gas, how many miles per gallon did the car get?***

2.5 Exercises

Divide and reduce.

1. $\frac{8}{9} \div \frac{16}{54}$

2. $\frac{15}{4} \div \frac{27}{8}$

3. $\frac{43}{10} \div \frac{86}{25}$

4. $\frac{4}{9} \div 6$

5. $\frac{25}{49} \div \frac{4}{9}$

6. $26 \div \frac{39}{7}$

7. $\frac{6}{23} \div \frac{8}{69}$

8. $\frac{27}{16} \div \frac{18}{24}$

9. $\frac{15}{42} \div \frac{1}{60}$

10. $\frac{123}{77} \div \frac{82}{49}$

11. $\frac{66}{12} \div \frac{44}{45}$

12. $\frac{150}{260} \div \frac{78}{30}$

13. $\frac{75}{24} \div \frac{30}{16}$

14. $\frac{72}{55} \div \frac{108}{33}$

15. $\frac{72}{81} \div \frac{96}{83}$

16. $\frac{429}{165} \div \frac{310}{234}$

17. $\frac{2}{51} \div \frac{1}{119}$

18. $\frac{231}{522} \div \frac{297}{174}$

19. $\frac{4}{9} \div 36$

20. $\frac{5}{6} \div 25$

21. $\frac{35}{16} \div \frac{21}{22}$

22. $\frac{14}{18} \div \frac{42}{15}$

23. $34 \div \frac{17}{14}$

24. $42 \div \frac{21}{13}$

25. $\frac{33}{84} \div \frac{22}{60}$

26. $\frac{56}{15} \div \frac{42}{90}$

27. $\frac{13}{28} \div \frac{26}{56}$

28. $\frac{8}{15} \div \frac{16}{30}$

29. $0 \div \frac{3}{4}$

30. $0 \div \frac{5}{6}$

31. $\frac{3}{4} \div 0$

32. $\frac{5}{6} \div 0$

33. $\frac{36}{10} \div \frac{6}{25}$

34. $\frac{18}{8} \div \frac{9}{12}$

35. $\frac{33}{26} \div \frac{22}{91}$

36. $\frac{18}{35} \div \frac{66}{65}$

37. $2100 \div \frac{7}{3}$

38. $3500 \div \frac{7}{3}$

39. $\frac{123}{92} \div \frac{168}{115}$

40. $\frac{170}{175} \div \frac{340}{50}$

41. $\frac{18}{143} \div \frac{27}{209}$

42. $\frac{12}{429} \div \frac{20}{221}$

43. $1\frac{9}{10} \div 1\frac{11}{14}$

44. $3\frac{1}{7} \div 1\frac{4}{11}$

45. $2\frac{1}{2} \div 1\frac{3}{4}$

46. $5\frac{2}{3} \div 2\frac{2}{3}$

47. $3\frac{1}{3} \div 3\frac{1}{3}$

48. $6\frac{1}{5} \div 6\frac{1}{5}$

49. $2\frac{1}{8} \div \frac{34}{8}$

50. $5\frac{1}{6} \div \frac{62}{12}$

51. $250 \div 2\frac{1}{2}$

55. $4\frac{5}{7} \div \frac{1}{14}$

56. $7\frac{1}{2} \div \frac{1}{10}$

57. $6\frac{4}{5} \div 1\frac{7}{10}$

58. $8\frac{1}{4} \div 6\frac{7}{8}$

59. $7\frac{1}{9} \div 0$

60. $8\frac{3}{5} \div 0$

61. $6 \div 2\frac{1}{7}$

62. $3\frac{3}{4} \div 5\frac{5}{6}$

63. $6\frac{2}{9} \div 8$

64. $0 \div 7\frac{2}{3}$

65. $\frac{6}{4} \div 1\frac{1}{2}$

66. $3\frac{1}{8} \div \frac{15}{16}$

67. $5\frac{2}{5} \div \frac{21}{30}$

68. $10\frac{2}{5} \div \frac{13}{15}$

69. $2\frac{1}{5} \div \frac{4}{27}$

70. $\frac{56}{72} \div 1\frac{1}{27}$

71. $\frac{48}{60} \div 1\frac{1}{27}$

72. $7\frac{3}{7} \div 2\frac{8}{35}$

73. $13\frac{4}{9} \div 6\frac{1}{15}$

74. $10\frac{5}{9} \div 1\frac{17}{21}$

75. $12\frac{10}{11} \div 3\frac{13}{33}$

76. $9\frac{5}{8} \div 11\frac{11}{12}$

77. $7\frac{1}{25} \div 13\frac{1}{5}$

78. $29\frac{7}{10} \div 23\frac{1}{10}$

Fill in the blanks.

79. To find the ______________ of a fraction, interchange the numerator and denominator.

80. To divide one fraction by another fraction, multiply the ______________ and the ______________ of the divisor.

81. To divide using mixed numbers, change all ______________ to ______________ and then use the rules for dividing fractions.

Solve and show all work.

82. What is $\frac{99}{120}$ divided by $\frac{66}{132}$?

83. What is $\frac{44}{154}$ divided by $\frac{52}{110}$?

84. Divide $1\frac{1}{65}$ into $1\frac{1}{35}$.

85. Divide $8\frac{2}{5}$ into $5\frac{3}{5}$.

86. Three-fourths of a pound of peanuts must be divided equally among six people. How much will each person receive?

87. Three-fourths of a pound of sugar must be divided equally among nine people. How much will each person receive?

88. Jerry owns 45 acres of land. If he divides this property into smaller lots, each containing $\frac{3}{5}$ of an acre, how many lots will there be?

89. Annette must eat $\frac{3}{4}$ of a pound of meat. If she eats $\frac{1}{8}$ of a pound every day, how many days will the meat last?

90. A car traveled 180 miles on $11\frac{1}{4}$ gallons of gas. How many miles per gallon did the car get?

91. Bill's car can travel $16\frac{1}{4}$ miles on one gallon of gas. If he drives $110\frac{1}{2}$ miles, how much gas will he use? (*Hint:* gallons of gas used = miles driven ÷ miles per gallon.)

92. If you wanted to travel 42 miles on a bike at a constant rate of $15\frac{3}{4}$ miles per hour, how many hours would it take? (*Hint:* hours traveled = miles traveled ÷ miles per hour.)

93. John has one-half of a tank of gas in his car. If he uses one-tenth of a tank of gas to go to and from school, how many trips can he make to and from school?

94. Bill has three-fifths of a pound of coffee. If he uses one-fifth of a pound of coffee a day, how many days will the coffee last?

95. A car traveled 180 miles on $9\frac{2}{7}$ gallons of gas. How many miles per gallon of gas did the car get?

96. A car traveled 200 miles on $10\frac{2}{5}$ gallons of gas. How many miles per gallon of gas did the car get?

Answers to Practice Problems **13a.** $\frac{3}{2}$ **b.** 130 **c.** $\frac{34}{45}$ **14a.** $1\frac{107}{135}$ **b.** $2\frac{1}{5}$ **c.** $8\frac{5}{11}$ **15.** $16\frac{1}{4}$

2.6 Addition

22 The sum of $\frac{1}{5} + \frac{3}{5} = \frac{4}{5}$, as shown by Figure 2–5.

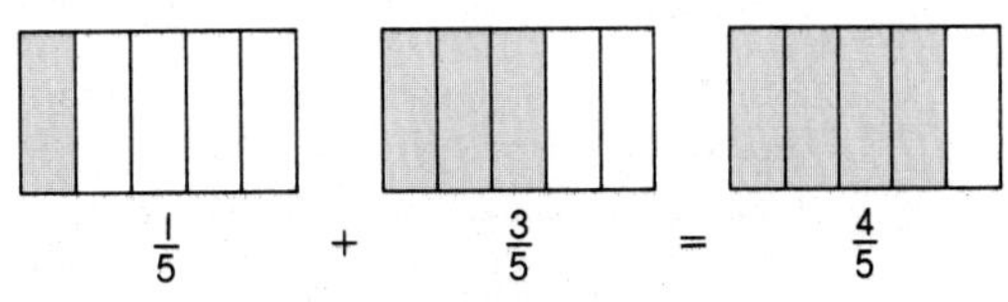

Figure 2-5

Look at this figure to understand the following rule:

> **Rule for Adding Fractions Having a Common Denominator:**
>
> **1.** Add the numerators and place this sum over the common denominator.
>
> **2.** If possible, reduce the fraction obtained in step 1.

Example 16 Add and reduce.

a. $\frac{3}{11} + \frac{4}{11}$ **b.** $\frac{7}{15} + \frac{2}{15} + \frac{3}{15}$ **c.** $\frac{11}{117} + \frac{34}{117}$

Solution To add fractions having a common denominator, write the sum of the numerators over the common denominator and reduce the resulting fraction.

16a.
$$\frac{3}{11} + \frac{4}{11} = \frac{3 + 4}{11} = \frac{7}{11}$$

16b.
$$\frac{7}{15} + \frac{2}{15} + \frac{3}{15} = \frac{7 + 2 + 3}{15} = \frac{12}{15} = \frac{4}{5}$$
Divided 12 and 15 by 3

16c.
$$\frac{11}{117} + \frac{34}{117} = \frac{11 + 34}{117} = \frac{45}{117} = \frac{\cancel{3} \cdot \cancel{3} \cdot 5}{\cancel{3} \cdot \cancel{3} \cdot 13} = \frac{5}{13}$$

Practice Problem 16 ***Add and reduce.***

a. $\frac{3}{10} + \frac{4}{10}$ **b.** $\frac{5}{18} + \frac{1}{18} + \frac{3}{18}$ **c.** $\frac{32}{105} + \frac{38}{105}$

23 To add fractions having different denominators, you must be able to write a fraction as an equivalent fraction with a larger denominator. This can be done by multiplying the numerator and denominator of the fraction by the same nonzero number.

Example 17 Find the missing numerator that makes the fractions equivalent.

a. $\frac{1}{2} = \frac{?}{6}$ **b.** $\frac{1}{3} = \frac{?}{6}$ **c.** $\frac{5}{8} = \frac{?}{88}$

Solution To rewrite a fraction as an equivalent fraction with a larger denominator, divide the original denominator into the new denominator and then multiply the result by the numerator and denominator of the original fraction.

17a. Since $6 \div 2 = 3$, multiply by 3.

$$\frac{1}{2} = \frac{1 \cdot 3}{2 \cdot 3} = \frac{3}{6}$$

17b. Since $6 \div 3 = 2$, multiply by 2.

$$\frac{1}{3} = \frac{1 \cdot 2}{3 \cdot 2} = \frac{2}{6}$$

17c. Since $88 \div 8 = 11$, multiply by 11.

$$\frac{5}{8} = \frac{5 \cdot 11}{8 \cdot 11} = \frac{55}{88}$$

NOTE: Many times you will be able to mentally change a fraction to an equivalent fraction with a larger denominator. Do this by dividing the original denominator into the new denominator and multiplying the result by the numerator and denominator of the original fraction. For example, you can find the missing numerator in the expression $\frac{1}{3} = \frac{?}{6}$ by thinking that 3 divided into 6 is 2, and $2 \cdot 1 = 2$. Thus, $\frac{1}{3} = \frac{2}{6}$. Similarly, you can find the missing numerator in the expression $\frac{1}{2} = \frac{?}{6}$ by thinking that 2 divided into 6 is 3, and $3 \cdot 1 = 3$. Thus, $\frac{1}{2} = \frac{3}{6}$.

Practice Problem 17 ***Find the missing numerator that makes the fractions equivalent. Show all work.***

a. $\frac{3}{7} = \frac{?}{21}$ **b.** $\frac{6}{13} = \frac{?}{39}$ **c.** $\frac{5}{3} = \frac{?}{18}$

Practice Problem 18 ***Find the missing numerator that makes the fractions equivalent (do this mentally).***

a. $\frac{3}{8} = \frac{?}{24}$ **b.** $\frac{5}{9} = \frac{?}{72}$ **c.** $\frac{8}{11} = \frac{?}{77}$

24 We can now find the sum of two fractions that have different denominators. To find the sum of $\frac{1}{2}$ and $\frac{1}{3}$, first rewrite $\frac{1}{2}$ and $\frac{1}{3}$ as equivalent fractions which have a *common* denominator and then add. For example, since $\frac{1}{2} = \frac{3}{6}$ and $\frac{1}{3} = \frac{2}{6}$,

$$\frac{1}{2} + \frac{1}{3} = \frac{3}{6} + \frac{2}{6}$$
$$= \frac{5}{6}.$$

25 The fractions $\frac{1}{2}$ and $\frac{1}{3}$ can be written with a common denominator of 6, 12, 18, or 24 and then added. However, it is more convenient to use the number 6 since it is the smallest or lowest common denominator (LCD) of 2 and 3. The **lowest common denominator** (LCD) of two or more denominators is the smallest number that each denominator divides evenly. The LCD is sometimes referred to as the **least common multiple** (LCM) of the denominators. Remember that a multiple of a natural number is any number that is the product of the given number and some other natural number. For example, 6 is a multiple of 2 since $2 \cdot 3 = 6$.

26 By inspection, we know that the LCD of 2 and 3 is 6. However, you will not always be able to find the LCD by inspection. Therefore, to find the LCD of two or more fractions use the following procedure:

Procedure for Finding the LCD:

1. Write the prime factorization of each denominator.
2. The LCD is the product of the distinct prime factors listed in step 1. Each factor must be used the greatest number of times it occurs in any one factorization.

Example 18 Find the LCD.

a. $\frac{1}{6} + \frac{1}{8}$

b. $\frac{3}{26} + \frac{5}{39}$

c. $\frac{1}{21} + \frac{3}{14} + \frac{5}{18}$

d. $\frac{3}{44} + \frac{5}{36} + \frac{1}{66}$

Solution To find the LCD, write the prime factorization of each denominator. The LCD is the product of the distinct prime factors of the denominators. Each factor is used the greatest number of times it appears in any one denominator.

18a. $6 = 2 \cdot 3$
$8 = 2 \cdot 2 \cdot 2$
$\text{LCD} = 2 \cdot 2 \cdot 2 \cdot 3 = 24$

The factor 2 occurs once in the denominator 6 and three times in the denominator 8, so we put it three times into the LCD. The factor 3 occurs once in the denominator 6, so we put it once into the LCD.

18b. $26 = 2 \cdot 13$
$39 = 3 \cdot 13$
$\text{LCD} = 2 \cdot 3 \cdot 13 = 78$

18c. $21 = 3 \cdot 7$
$14 = 2 \cdot 7$
$18 = 2 \cdot 3 \cdot 3$
$\text{LCD} = 2 \cdot 3 \cdot 3 \cdot 7 = 126$

18d. $44 = 2 \cdot 2 \cdot 11$
$36 = 2 \cdot 2 \cdot 3 \cdot 3$
$66 = 2 \cdot 3 \cdot 11$
$\text{LCD} = 2 \cdot 2 \cdot 3 \cdot 3 \cdot 11 = 396$

Practice Problem 19 ***Find the LCD.***

a. $\frac{1}{9} + \frac{1}{6}$ b. $\frac{1}{10} + \frac{3}{12} + \frac{4}{15}$ c. $\frac{5}{24} + \frac{7}{32}$

27 Based on the information presented in Ideas 25 and 26, we can now state a rule for adding fractions having different denominators.

Rule for Adding Fractions:

1. Find the LCD.
2. Change all fractions to equivalent fractions having the LCD as a common denominator.
3. Add using the rules for adding fractions having a common denominator.

Example 19 Add and reduce.

a. $\frac{1}{6} + \frac{5}{9}$

b. $\frac{1}{10} + \frac{5}{18} + 2$

c. $\frac{5}{36} + \frac{7}{66}$

d. $\frac{5}{42} + \frac{3}{28} + \frac{5}{63}$

Solution To add fractions having different denominators, first find the LCD. Next, express each fraction as an equivalent fraction having the LCD as the common denominator and then add.

19a. $\frac{1}{6} + \frac{5}{9} = \frac{1 \cdot 3}{6 \cdot 3} + \frac{5 \cdot 2}{9 \cdot 2}$ Do this step mentally.

$= \frac{3}{18} + \frac{10}{18}$

$= \frac{13}{18}$

$6 = 2 \cdot 3$
$9 = 3 \cdot 3$
$\text{LCD} = 2 \cdot 3 \cdot 3 = 18$

19b. $\frac{1}{10} + \frac{5}{18} + 2 = \frac{1}{10} + \frac{5}{18} + \frac{2}{1}$

$= \frac{9}{90} + \frac{25}{90} + \frac{180}{90}$

$= \frac{214}{90}$

$= \frac{107}{45}$ Divide 214 and 90 by 2

$10 = 2 \cdot 5$
$18 = 2 \cdot 3 \cdot 3$
$\text{LCD} = 2 \cdot 3 \cdot 3 \cdot 5 = 90$

19c. $\frac{5}{36} + \frac{7}{66} = \frac{55}{396} + \frac{42}{396}$

$= \frac{97}{396}$

$36 = 2 \cdot 2 \cdot 3 \cdot 3$
$66 = 2 \cdot 3 \cdot 11$
$\text{LCD} = 2 \cdot 2 \cdot 3 \cdot 3 \cdot 11 = 396$

19d. $\frac{5}{42} + \frac{3}{28} + \frac{5}{63} = \frac{30}{252} + \frac{27}{252} + \frac{20}{252}$

$= \frac{77}{252}$

$= \frac{\cancel{7} \cdot 11}{\cancel{7} \cdot 36}$

$= \frac{11}{36}$

$42 = 2 \cdot 3 \cdot 7$
$28 = 2 \cdot 2 \cdot 7$
$63 = 3 \cdot 3 \cdot 7$
$\text{LCD} = 2 \cdot 2 \cdot 3 \cdot 3 \cdot 7 = 252$

NOTE: In Example 19c, to write $\frac{5}{36}$ as $\frac{55}{396}$, we divided 36 into 396 and multiplied the result times the numerator and denominator of $\frac{5}{36}$. The division can be done either by long division or by thinking of 396 ÷ 36 as a fraction that must be reduced. That is,

$$\frac{396}{36} = \frac{\cancel{2} \cdot \cancel{2} \cdot \cancel{3} \cdot \cancel{3} \cdot 11}{\cancel{2} \cdot \cancel{2} \cdot \cancel{3} \cdot \cancel{3}} = \frac{11}{1} = 11.$$

The point is that to divide a denominator into the LCD, you can delete all the prime factors that the two numbers have in common, and the product of the factors remaining in the LCD is the quotient. For example, given that

$$\begin{aligned} 36 &= 2 \cdot 2 \cdot 3 \cdot 3 \text{ and} \\ 66 &= 2 \cdot 3 \cdot 11, \\ \text{LCD} &= 2 \cdot 2 \cdot 3 \cdot 3 \cdot 11 = 396. \end{aligned}$$

If you delete all the prime factors that 36 and 396 have in common, a factor of 11 remains in 396 (LCD). Thus, 396 ÷ 36 = 11. Similarly, you can divide 66 into 396 by deleting all their common prime factors (2, 3, and 11) and then noticing that $2 \cdot 3 = 6$ remains in 396. Thus 396 ÷ 66 = 6.

28 Fractions can also be added by using a vertical format, rather than a horizontal one.

Example 20 Add and reduce.

a. $\frac{1}{6} + \frac{1}{8}$ **b.** $\frac{5}{12} + \frac{2}{21} + \frac{1}{14}$

Solution To add fractions having different denominators, first find the LCD. Next, express each fraction as an equivalent fraction having the LCD as the common denominator and then add.

20a.

$$\begin{array}{r} \frac{1}{6} = \frac{4}{24} \\ +\frac{1}{8} = \frac{3}{24} \\ \hline \frac{7}{24} \end{array}$$

$6 = 2 \cdot 3$
$8 = 2 \cdot 2 \cdot 2$
$\text{LCD} = 2 \cdot 2 \cdot 2 \cdot 3 = 24$

20b.

$$\begin{array}{r} \frac{5}{12} = \frac{35}{84} \\ \frac{2}{21} = \frac{8}{84} \\ +\frac{1}{14} = \frac{6}{84} \\ \hline \frac{49}{84} = \frac{7}{12} \end{array}$$

$12 = 2 \cdot 2 \cdot 3$
$21 = 3 \cdot 7$
$14 = 2 \cdot 7$
$\text{LCD} = 2 \cdot 2 \cdot 3 \cdot 7 = 84$

Divide 49 and 84 by 7

Practice Problem 20 ***Add and reduce.***

a. $\frac{1}{9} + \frac{5}{15}$ **b.** $\frac{1}{4} + \frac{5}{6} + \frac{3}{20}$ **c.** $\frac{2}{75} + \frac{7}{30}$

2.6 Exercises

Add and reduce.

1. $\frac{3}{7}+\frac{2}{7}$
2. $\frac{3}{10}+\frac{2}{10}$
3. $\frac{13}{7}+\frac{3}{7}$
4. $\frac{3}{20}+\frac{11}{20}$
5. $\frac{1}{12}+\frac{7}{12}$
6. $\frac{5}{111}+\frac{28}{111}$
7. $\frac{9}{100}+\frac{43}{100}+\frac{13}{100}$
8. $\frac{15}{135}+\frac{20}{135}+\frac{35}{135}+\frac{2}{135}$
9. $\frac{5}{66}+\frac{13}{66}+\frac{15}{66}$
10. $\frac{1}{91}+\frac{3}{91}+\frac{10}{91}$

Find the missing numerator that makes the fractions equivalent.

11. $\frac{3}{8}=\frac{?}{24}$
12. $\frac{5}{7}=\frac{?}{42}$
13. $\frac{1}{30}=\frac{?}{90}$
14. $\frac{3}{35}=\frac{?}{70}$
15. $\frac{5}{17}=\frac{?}{51}$
16. $\frac{7}{3}=\frac{?}{90}$
17. $\frac{1}{15}=\frac{?}{105}$
18. $\frac{13}{18}=\frac{?}{90}$
19. $\frac{7}{36}=\frac{?}{396}$
20. $\frac{3}{37}=\frac{?}{111}$
21. $\frac{1}{18}=\frac{?}{54}$
22. $\frac{1}{16}=\frac{?}{48}$
23. $\frac{5}{53}=\frac{?}{159}$
24. $\frac{7}{49}=\frac{?}{98}$
25. $\frac{5}{6}=\frac{?}{90}$
26. $\frac{4}{15}=\frac{?}{90}$
27. $\frac{7}{30}=\frac{?}{150}$
28. $\frac{13}{50}=\frac{?}{150}$
29. $\frac{3}{14}=\frac{?}{588}$
30. $\frac{5}{12}=\frac{?}{588}$
31. $\frac{2}{49}=\frac{?}{588}$
32. $\frac{7}{42}=\frac{?}{588}$
33. $\frac{5}{12}=\frac{?}{252}$
34. $\frac{8}{21}=\frac{?}{252}$

Find the LCD.

35. $\frac{1}{6}+\frac{2}{9}$
36. $\frac{3}{10}+\frac{7}{15}$
37. $\frac{4}{8}+\frac{3}{10}$
38. $\frac{5}{7}+\frac{3}{14}$
39. $\frac{5}{9}+\frac{7}{10}$
40. $\frac{5}{18}+\frac{7}{12}$
41. $\frac{13}{15}+\frac{5}{36}$
42. $\frac{1}{24}+\frac{5}{36}+\frac{1}{40}$
43. $\frac{3}{25}+\frac{1}{30}+\frac{7}{40}$
44. $\frac{1}{4}+\frac{5}{6}+\frac{2}{15}$
45. $\frac{1}{6}+\frac{5}{8}+\frac{1}{18}$
46. $\frac{3}{10}+\frac{7}{15}+\frac{8}{12}$

Add and reduce.

47. $\frac{1}{9}+\frac{5}{6}$
48. $\frac{3}{4}+\frac{5}{6}$
49. $\frac{1}{3}+\frac{3}{4}$
50. $\frac{5}{12}+\frac{1}{9}$
51. $\frac{7}{8}+\frac{3}{10}$
52. $\frac{8}{14}+\frac{5}{21}$

53. $\frac{5}{16} + \frac{1}{12}$

54. $\frac{9}{8} + \frac{5}{6}$

55. $\frac{4}{15} + \frac{5}{12}$

56. $\frac{5}{26} + \frac{7}{39}$

57. $\frac{3}{10} + \frac{14}{15} + \frac{7}{12}$

58. $\frac{1}{36} + \frac{5}{18} + \frac{7}{24}$

59. $\frac{5}{24} + \frac{7}{36} + \frac{1}{48}$

60. $\frac{17}{22} + \frac{3}{4} + \frac{8}{33}$

61. $\frac{3}{77} + \frac{5}{91} + \frac{8}{49}$

62. $\frac{7}{20} + \frac{3}{10} + \frac{11}{30}$

63. $\frac{3}{30} + \frac{7}{75}$

64. $\frac{3}{48} + \frac{7}{54}$

65. $\frac{1}{15} + \frac{5}{33} + \frac{7}{55}$

66. $\frac{5}{36} + \frac{7}{66}$

67. $\frac{11}{28} + \frac{5}{42}$

68. $\frac{11}{18} + \frac{7}{24}$

69. $\frac{1}{16} + \frac{3}{24}$

70. $\frac{11}{18} + \frac{4}{15}$

71. $\frac{1}{35} + \frac{8}{15}$

72. $\frac{5}{36} + \frac{3}{20}$

73. $\frac{9}{14} + \frac{5}{42}$

74. $\frac{5}{18} + \frac{7}{54}$

75. $\frac{17}{40} + \frac{3}{16}$

76. $\frac{7}{10} + \frac{3}{7}$

77. $\frac{8}{13} + \frac{3}{5}$

78. $\frac{11}{24} + \frac{1}{36}$

79. $\frac{3}{56} + \frac{7}{24}$

80. $\frac{1}{14} + \frac{3}{77}$

81. $\frac{1}{8} + \frac{7}{12} + \frac{5}{18}$

82. $\frac{5}{6} + \frac{9}{10} + \frac{11}{24}$

83. $\frac{3}{4} + \frac{1}{10} + \frac{5}{8}$

84. $\frac{5}{6} + \frac{9}{8} + \frac{1}{12}$

85. $\frac{1}{24} + \frac{5}{86} + \frac{13}{40}$

86. $\frac{3}{25} + \frac{7}{30} + \frac{1}{40}$

87. $\frac{5}{14} + \frac{3}{49} + \frac{7}{12}$

88. $\frac{1}{10} + \frac{11}{25} + \frac{5}{12}$

Fill in the blanks.

89. To add fractions having a common denominator, add the ______________ and place this sum over the ______________.

90. To add fractions having different denominators, find the ______________. Next, express each fraction as an ______________ fraction having the ______________ as the common denominator and then add.

Solve. Show all work.

91. Find the sum of $\frac{3}{10}$, $\frac{7}{8}$, and $\frac{5}{6}$.

92. Find the sum of $\frac{1}{9}$, $\frac{7}{16}$, and $\frac{5}{6}$.

93. Find the sum of $\frac{11}{24}$ and $\frac{5}{42}$.

94. Find the sum of $\frac{5}{36}$ and $\frac{7}{48}$.

95. Jim mixed $\frac{1}{3}$ pound of jellybeans, $\frac{1}{4}$ pound of Spanish peanuts, and $\frac{1}{6}$ pound of candy corn. What is the final weight of the mixture?

96. In one week, Linda jogged $\frac{3}{4}$ mile, $\frac{11}{12}$ mile, and $\frac{9}{10}$ mile. How many miles did she jog altogether?

97. On a hunting trip, Joe spent $\frac{1}{4}$ of his money on gas, $\frac{1}{6}$ of his money on food and $\frac{1}{10}$ of his money on post-cards. What fractional part of his money did he spend altogether?

98. Bob drank $\frac{2}{9}$ of his water on Monday, $\frac{1}{3}$ of his water on Tuesday, and $\frac{1}{6}$ of his water on Wednesday. What fractional part of his water did he drink altogether?

99. A recipe calls for $\frac{1}{4}$ teaspoon of thyme, $\frac{1}{3}$ teaspoon of curry powder, and $\frac{1}{8}$ teaspoon of basil. Find the total amount of herbs and spices used in the recipe.

100. Another recipe calls for $\frac{1}{2}$ cup of flour, $\frac{2}{3}$ cup of milk, and $\frac{1}{6}$ cup of sugar. Find the total amount of ingredients used in this recipe.

101. Find the sum of $\frac{6}{1001} + \frac{7}{1001}$.

102. Find the sum of $\frac{7}{715} + \frac{4}{715}$.

103. Bill owns an eighth of a certain company. If he acquires another sixth of the company, what fractional part of the company will he own?

104. Bob owns a tenth of the stock in a certain company. If he acquires another three-eighths of the stock, what fractional part of the company's stock will he own?

105. If three pieces of wood are glued together that are $\frac{1}{4}$ inch thick, $\frac{1}{10}$ inch thick, and $\frac{1}{8}$ inch thick, what is the total thickness of the wood?

Answers to Practice Problems **16a.** $\frac{7}{10}$ **b.** $\frac{1}{2}$ **c.** $\frac{2}{3}$ **17a.** 9 **b.** 18 **c.** 30 **18a.** 9 **b.** 40 **c.** 56 **19a.** $\frac{5}{18}$ **b.** $\frac{37}{60}$ **c.** $\frac{41}{96}$ **20a.** $\frac{4}{9}$ **b.** $\frac{37}{30}$ **c.** $\frac{13}{50}$

2.7 Addition of Mixed Numbers

29 One way to add using mixed numbers is to use a vertical format.

To Add Mixed Numbers:

1. Add the whole number parts.
2. Add the fractional parts (reduce if necessary).
3. If the sum of the fractional parts is an improper fraction, change it to a mixed number and then add the result to the value obtained in step 1.

Example 21 Add and express each answer as a mixed number.

a. $3\frac{3}{5} + 5\frac{4}{5}$ **b.** $5\frac{2}{9} + \frac{1}{6}$

c. $1\frac{1}{8} + 3\frac{7}{12} + 2\frac{5}{18}$ **d.** $16\frac{5}{24} + 23 + 59\frac{3}{32}$

Solution To add using mixed numbers, add the whole number parts and the fractional parts separately. If possible, the final answer should be written as a mixed number.

21a.

$$\begin{array}{r} 3\frac{3}{5} \\ +5\frac{4}{5} \\ \hline 8\frac{7}{5} \end{array} = 8 + 1\frac{2}{5} = 9\frac{2}{5}$$

Remember: $8\frac{7}{5} = 8 + \frac{7}{5}$

21b.

$$\begin{array}{rcl} 5\frac{2}{9} & = & 5\frac{4}{18} \\ + \ \frac{1}{6} & = & \frac{3}{18} \\ \hline & & 5\frac{7}{18} \end{array}$$

$9 = 3 \cdot 3$
$6 = 2 \cdot 3$
$\text{LCD} = 2 \cdot 3 \cdot 3 = 18$

21c.

$$\begin{array}{rcl} 1\frac{1}{8} & = & 1\frac{9}{72} \\ 3\frac{7}{12} & = & 3\frac{42}{72} \\ +2\frac{5}{18} & = & 2\frac{20}{72} \\ \hline & & 6\frac{71}{72} \end{array}$$

$8 = 2 \cdot 2 \cdot 2$
$12 = 2 \cdot 2 \cdot 3$
$18 = 2 \cdot 3 \cdot 3$
$\text{LCD} = 2 \cdot 2 \cdot 2 \cdot 3 \cdot 3 = 72$

21d.

$$\begin{array}{rcl} 16\frac{5}{24} & = & 16\frac{20}{96} \\ 23 & = & 23 \\ +59\frac{3}{32} & = & 59\frac{9}{96} \\ \hline & & 98\frac{29}{96} \end{array}$$

$24 = 2 \cdot 2 \cdot 2 \cdot 3$
$32 = 2 \cdot 2 \cdot 2 \cdot 2 \cdot 2$
$\text{LCD} = 2 \cdot 2 \cdot 2 \cdot 2 \cdot 2 \cdot 3 = 96$

30 Another way to add using mixed numbers is to change all mixed numbers to improper fractions and then add.

Example 22 Add and express each answer as a mixed number.

a. $3\frac{5}{6} + 4\frac{1}{4}$ **b.** $2\frac{1}{10} + 5\frac{4}{15}$

Solution 22a.

$$\begin{aligned} 3\frac{5}{6} + 4\frac{1}{4} &= \frac{23}{6} + \frac{17}{4} \\ &= \frac{46}{12} + \frac{51}{12} \\ &= \frac{97}{12} \\ &= 8\frac{1}{12} \end{aligned}$$

$6 = 2 \cdot 3$
$4 = 2 \cdot 2$
$\text{LCD} = 2 \cdot 2 \cdot 3 = 12$

22b.

$$\begin{aligned} 2\frac{1}{10} + 5\frac{4}{15} &= \frac{21}{10} + \frac{79}{15} \\ &= \frac{63}{30} + \frac{158}{30} \\ &= \frac{221}{30} \\ &= 7\frac{11}{30} \end{aligned}$$

$10 = 2 \cdot 5$
$15 = 3 \cdot 5$
$\text{LCD} = 2 \cdot 3 \cdot 5 = 30$

Practice Problem 21 ***Add and express each answer as a mixed number.***

a. $2\frac{1}{3} + 4\frac{1}{3}$ **b.** $3\frac{9}{10} + 5\frac{5}{6}$

c. $5\frac{2}{9} + 1\frac{1}{6} + 7\frac{5}{12}$ d. $32\frac{13}{16} + 76\frac{5}{24}$

2.7 Exercises

Add and express each answer as a mixed number.

1. $2\frac{1}{3} + 3\frac{1}{3}$
2. $5\frac{2}{9} + 7\frac{1}{9}$
3. $1\frac{3}{8} + 9$
4. $5\frac{3}{5} + 2\frac{4}{5}$
5. $6\frac{3}{8} + 2\frac{5}{8}$
6. $5\frac{3}{4} + 7\frac{5}{8}$
7. $4\frac{1}{10} + 11\frac{3}{6}$
8. $12\frac{3}{8} + 11\frac{5}{12}$
9. $2\frac{5}{18} + 3\frac{7}{12}$
10. $3\frac{1}{14} + 5\frac{4}{21}$
11. $8\frac{3}{4} + 9\frac{5}{6}$
12. $4\frac{5}{6} + 3\frac{7}{8}$
13. $6\frac{11}{20} + 7\frac{9}{10}$
14. $5\frac{2}{3} + 6\frac{11}{12}$
15. $7\frac{1}{12} + 8\frac{9}{15}$
16. $4\frac{1}{15} + 3\frac{7}{20}$
17. $3\frac{4}{7} + 5$
18. $4\frac{4}{11} + 5$
19. $\frac{3}{10} + 5\frac{5}{6}$
20. $\frac{3}{8} + 4\frac{1}{6}$
21. $\frac{11}{14} + 6\frac{13}{21} + \frac{1}{7}$
22. $\frac{19}{22} + 7\frac{5}{11} + \frac{1}{33}$
23. $8\frac{3}{8} + 9\frac{1}{12} + 6\frac{8}{9}$
24. $5\frac{1}{6} + 7\frac{9}{10} + 8\frac{7}{15}$
25. $8\frac{3}{16} + 5\frac{1}{24} + \frac{5}{12}$
26. $9\frac{3}{18} + 8\frac{1}{12} + \frac{5}{6}$
27. $15 + 9\frac{5}{24} + \frac{7}{36}$
28. $18 + 8\frac{9}{10} + \frac{11}{30}$
29. $100\frac{7}{20} + 89\frac{1}{30}$
30. $99\frac{7}{12} + 103\frac{1}{18}$
31. $18\frac{7}{30} + 19\frac{13}{15} + 33\frac{19}{20}$
32. $6\frac{7}{10} + 1\frac{2}{15}$
33. $7\frac{7}{9} + 3\frac{4}{15}$
34. $10\frac{3}{20} + \frac{9}{25}$
35. $3\frac{5}{12} + 7\frac{8}{15}$
36. $1\frac{3}{14} + 6\frac{8}{21}$
37. $13\frac{1}{18} + 15\frac{7}{24} + 12\frac{1}{12}$
38. $85\frac{2}{15} + 55\frac{7}{20} + 65\frac{3}{4}$
39. $19\frac{2}{21} + 17\frac{6}{49} + 99\frac{3}{28}$
40. $31\frac{11}{20} + 66 + \frac{7}{12}$
41. $85\frac{3}{10} + 91\frac{1}{14} + 87\frac{6}{35}$

Fill in the blanks.

42. To add using mixed numbers, add the ______________ number parts and the ______________ parts separately.

Solve and show all work.

43. Frank mixed $4\frac{1}{3}$ pounds of roasted soybeans, $\frac{1}{4}$ pound of peanuts, and $4\frac{1}{6}$ pounds of raisins. What is the total weight of the mixture?

44. Reggie rode his bike for $\frac{5}{6}$ of an hour, jogged for $4\frac{1}{4}$ hours, and played tennis for $1\frac{7}{15}$ of an hour. How many hours did he exercise?

45. In one week, Amy drove $89\frac{3}{10}$ miles, $45\frac{7}{15}$ miles, $75\frac{7}{12}$ miles, and $135\frac{3}{4}$ miles. How many miles did she drive in all?

46. The perimeter of a triangle is the sum of the length of its three sides. If one side of a triangle is $7\frac{5}{9}$ inches, another side is $9\frac{1}{6}$ inches, and the third side is 11 inches, find the perimeter of the triangle.

47. Find the sum of $8\frac{3}{4}$ inches, $9\frac{1}{6}$ inches, and $11\frac{1}{8}$ inches.

48. Find the sum of $9\frac{3}{10}$ miles, $8\frac{7}{15}$ miles, and $12\frac{1}{6}$ miles.

49. Jack weighs $195\frac{3}{4}$ pounds, Ed weighs $187\frac{2}{3}$ pounds, and Paul weighs $187\frac{5}{8}$ pounds. Find the combined weight of the three men.

50. Betty weighs $113\frac{1}{2}$ pounds, Kathy weighs $105\frac{1}{3}$ pounds, and Jane weighs $122\frac{8}{9}$ pounds. Find the combined weight of the three women.

Answers to Practice Problems **21a.** $6\frac{2}{3}$ **b.** $9\frac{11}{15}$ **c.** $13\frac{29}{36}$ **d.** $109\frac{1}{48}$

2.8 Subtraction

31 Finding the difference between two fractions is very similar to computing the sum of two fractions.

Subtraction Rule for Fractions Having a Common Denominator:

1. Subtract the numerators and place this difference over the common denominator.
2. If possible, reduce the fraction obtained in step 1.

Example 23 Subtract and reduce.

a. $\frac{7}{15} - \frac{3}{15}$ **b.** $\frac{9}{10} - \frac{1}{10}$ **c.** $\frac{25}{91} - \frac{11}{91}$

Solution To subtract one fraction from another fraction having the same denominator, write the difference of the numerators over the common denominator and reduce the resulting fraction.

23a.
$$\frac{7}{15} - \frac{3}{15} = \frac{7-3}{15} = \frac{4}{15}$$

23b.
$$\frac{9}{10} - \frac{1}{10} = \frac{9-1}{10} = \frac{8}{10} = \frac{4}{5}$$ Divide 8 and 10 by 2

23c. $\frac{25}{91} - \frac{11}{91} = \frac{25 - 11}{91}$

$= \frac{14}{91}$

$= \frac{2}{13}$ Divide 14 and 91 by 7

Practice Problem 22 ***Subtract and reduce.***

a. $\frac{3}{5} - \frac{1}{5}$ **b.** $\frac{6}{14} - \frac{4}{14}$ **c.** $\frac{43}{115} - \frac{23}{115}$

32 To subtract one fraction from another fraction having a different denominator, use the following rule:

Rule for Subtracting One Fraction from Another:

1. Find the LCD.
2. Change all fractions to equivalent fractions having the LCD as the common denominator.
3. Subtract using the rule for subtracting fractions having a common denominator.

Example 24 Subtract and reduce.

a. $\frac{3}{4} - \frac{1}{3}$ **b.** $\frac{5}{9} - \frac{2}{15}$ **c.** $\frac{19}{2} - 9$ **d.** $\frac{4}{21} - \frac{2}{15}$

Solution To subtract a fraction from another fraction having a different denominator, first find the LCD. Next, express each fraction as an equivalent fraction having the LCD as a common denominator and then subtract.

24a. $\frac{3}{4} - \frac{1}{3} = \frac{3 \cdot 3}{4 \cdot 3} - \frac{1 \cdot 4}{3 \cdot 4}$ Do this step mentally.

$= \frac{9}{12} - \frac{4}{12}$

$= \frac{5}{12}$

$4 = 2 \cdot 2$
$3 = 3$
$\text{LCD} = 2 \cdot 2 \cdot 3 = 12$

24b. $\frac{5}{9} - \frac{2}{15} = \frac{25}{45} - \frac{6}{45}$

$= \frac{19}{45}$

$9 = 3 \cdot 3$
$15 = 3 \cdot 5$
$\text{LCD} = 3 \cdot 3 \cdot 5 = 45$

24c. $\frac{19}{2} - 9 = \frac{19}{2} - \frac{9}{1}$

$= \frac{19}{2} - \frac{18}{2}$

$= \frac{1}{2}$

24d. $\dfrac{4}{21} - \dfrac{2}{15} = \dfrac{20}{105} - \dfrac{14}{105}$ $\quad$ $21 = 3 \cdot 7$

$15 = 3 \cdot 5$

LCD $= 3 \cdot 5 \cdot 7 = 105$

$= \dfrac{6}{105}$

$= \dfrac{2 \cdot \cancel{3}}{\cancel{3} \cdot 5 \cdot 7}$

$= \dfrac{2}{35}$

NOTE: Fractions can also be subtracted by using a vertical format. For example, subtract $\frac{5}{18}$ from $\frac{9}{10}$.

$$\begin{array}{r} \frac{9}{10} = \frac{81}{90} \\ -\frac{5}{18} = \frac{25}{90} \\ \hline = \frac{56}{90} = \frac{28}{45} \end{array}$$

$10 = 2 \cdot 5$

$18 = 2 \cdot 3 \cdot 3$

LCD $= 2 \cdot 3 \cdot 3 \cdot 5 = 90$

Practice Problem 23 ***Subtract and reduce.***

a. $\frac{3}{4} - \frac{3}{5}$ **b.** $\frac{5}{8} - \frac{3}{6}$ **c.** $\frac{11}{24} - \frac{1}{20}$

33 Mixed numbers can be subtracted in much the same way that they are added.

To Subtract Mixed Numbers:

1. Express the fractional parts as equivalent fractions having the LCD as a common denominator.

2a. When the fraction in the subtrahend (number being subtracted) is *smaller* than the fraction in the minuend, subtract the whole number parts, and then subtract the fractional parts.

2b. When the fraction in the subtrahend is *larger* than the fraction in the minuend, borrow 1 from the whole number part of the minuend and add it to the fractional part. Then subtract the whole number parts and subtract the fractional parts.

3. If possible, express the answer as a mixed number.

Example 25 Subtract and express each answer as a mixed number.

a. $8\frac{3}{7} - 5\frac{1}{7}$ **b.** $9\frac{3}{4} - 7\frac{1}{2}$ **c.** $6\frac{1}{10} - 4\frac{5}{6}$

d. $9\frac{1}{12} - 2\frac{7}{9}$ **e.** $9 - 3\frac{2}{3}$

Solution To subtract using mixed numbers, subtract the whole number parts and then subtract the fractional parts. Whenever the fraction in the subtrahend is larger than the fraction in the minuend, borrow and then subtract.

25a.
$$\begin{array}{r} 8\frac{3}{7} \\ -5\frac{1}{7} \\ \hline 3\frac{2}{7} \end{array}$$

25b.
$$\begin{array}{rcl} 9\frac{3}{4} & = & 9\frac{3}{4} \\ -7\frac{1}{2} & = & 7\frac{2}{4} \\ \hline & & 2\frac{1}{4} \end{array}$$

$4 = 2 \cdot 2$
$2 = 2$
$\text{LCD} = 2 \cdot 2 = 4$

25c.
$$\begin{array}{rcccl} 6\frac{1}{10} & = & 6\frac{3}{30} & = & 5\frac{33}{30} \\ -4\frac{5}{6} & = & 4\frac{25}{30} & = & 4\frac{25}{30} \\ \hline & & & & 1\frac{8}{30} = 1\frac{4}{15} \end{array}$$

$10 = 2 \cdot 5$
$6 = 2 \cdot 3$
$\text{LCD} = 2 \cdot 3 \cdot 5 = 30$

Remember: $6\frac{3}{30} = 5 + 1\frac{3}{30} = 5 + \frac{33}{30} = 5\frac{33}{30}$

25d.
$$\begin{array}{rcccl} 9\frac{1}{12} & = & 9\frac{3}{36} & = & 8\frac{39}{36} \\ -2\frac{7}{9} & = & 2\frac{28}{36} & = & 2\frac{28}{36} \\ \hline & & & & 6\frac{11}{36} \end{array}$$

$12 = 2 \cdot 2 \cdot 3$
$9 = 3 \cdot 3$
$\text{LCD} = 2 \cdot 2 \cdot 3 \cdot 3 = 36$

Remember: $9\frac{3}{36} = 8 + 1\frac{3}{36} = 8 + \frac{39}{36} = 8\frac{39}{36}$

25e.
$$\begin{array}{rcl} 9 & = & 8\frac{3}{3} \\ -3\frac{2}{3} & = & 3\frac{2}{3} \\ \hline & & 5\frac{1}{3} \end{array}$$

Remember: $9 = 8 + 1 = 8 + \frac{3}{3} = 8\frac{3}{3}$

NOTE: Once you understand the concept of borrowing, this technique can be done mentally. For example, you can write $9\frac{3}{36}$ as $8\frac{39}{36}$ by decreasing the original whole number by 1 and by noting that the new fraction is the sum of the old numerator and denominator over the old denominator. In other words,

$$9\frac{3}{36} = (9 - 1)\,\frac{3 + 36}{36} = 8\frac{39}{36}.$$

34 Another way to subtract using mixed numbers is to change all mixed numbers to improper fractions and then subtract. When you use this method, you do not have to borrow.

Example 26 Subtract and express each answer as a mixed number.

a. $6\frac{1}{10} - 4\frac{5}{6}$ **b.** $9 - 3\frac{2}{3}$

Solution

26a. $6\frac{1}{10} - 4\frac{5}{6} = \frac{61}{10} - \frac{29}{6}$ Express mixed numbers as fractions

$= \frac{183}{30} - \frac{145}{30}$ $10 = 2 \cdot 5$, $6 = 2 \cdot 3$, LCD $= 2 \cdot 3 \cdot 5 = 30$

$= \frac{38}{30}$

$= 1\frac{8}{30}$

$= 1\frac{4}{15}$

26b. $9 - 3\frac{2}{3} = \frac{9}{1} - \frac{11}{3}$

$= \frac{27}{3} - \frac{11}{3}$

$= \frac{16}{3}$

$= 5\frac{1}{3}$

Practice Problem 24 ***Subtract and express each answer as a mixed number.***

a. $9\frac{5}{7} - 3\frac{2}{7}$ **b.** $10\frac{7}{8} - 8\frac{1}{6}$

c. $31\frac{7}{15} - 29\frac{11}{20}$ **d.** $299 - 75\frac{3}{4}$

2.8 Exercises

Subtract and reduce.

1. $\frac{3}{4} - \frac{1}{4}$ **2.** $\frac{5}{7} - \frac{4}{7}$ **3.** $\frac{9}{10} - \frac{3}{10}$

4. $\frac{52}{169} - \frac{26}{169}$ **5.** $\frac{2}{3} - \frac{3}{5}$ **6.** $\frac{6}{7} - \frac{2}{3}$

7. $\frac{3}{5} - \frac{3}{10}$ **8.** $\frac{11}{20} - \frac{2}{5}$ **9.** $\frac{5}{6} - \frac{3}{4}$

10. $\frac{7}{9} - \frac{1}{6}$ **11.** $\frac{9}{10} - \frac{4}{15}$ **12.** $\frac{7}{8} - \frac{3}{10}$

13. $\frac{3}{4} - \frac{1}{5}$ **14.** $\frac{2}{3} - \frac{2}{5}$ **15.** $\frac{7}{8} - \frac{1}{3}$

16. $\frac{8}{9} - \frac{1}{2}$ **17.** $\frac{5}{9} - \frac{1}{12}$ **18.** $\frac{5}{6} - \frac{3}{8}$

19. $\frac{9}{10} - \frac{5}{6}$ **20.** $\frac{3}{4} - \frac{1}{10}$ **21.** $\frac{19}{20} - \frac{7}{10}$

22. $\frac{11}{15} - \frac{3}{5}$

23. $\frac{15}{16} - \frac{11}{12}$

24. $\frac{11}{18} - \frac{5}{12}$

25. $\frac{15}{16} - \frac{11}{24}$

26. $\frac{13}{18} - \frac{5}{27}$

27. $\frac{11}{12} - \frac{1}{20}$

28. $\frac{9}{10} - \frac{5}{12}$

29. $\frac{11}{35} - \frac{3}{10}$

30. $\frac{13}{25} - \frac{1}{10}$

31. $\frac{13}{15} - \frac{5}{12}$

32. $\frac{21}{25} - \frac{11}{15}$

33. $\frac{7}{28} - \frac{5}{21}$

34. $\frac{23}{40} - \frac{5}{16}$

35. $\frac{13}{18} - \frac{7}{15}$

36. $\frac{19}{30} - \frac{5}{12}$

37. $\frac{65}{84} - \frac{17}{72}$

38. $\frac{11}{48} - \frac{3}{32}$

39. $\frac{23}{50} - \frac{3}{40}$

40. $\frac{14}{75} - \frac{1}{30}$

41. $\frac{21}{44} - \frac{7}{36}$

42. $\frac{31}{32} - \frac{7}{40}$

Fill in the blanks.

43. $7\frac{3}{5} = 6 + 1\frac{3}{5}$
$= 6 + \underline{?}$
$= \underline{?}$

44. $9\frac{13}{23} = 8 + 1\frac{13}{23}$
$= 8 + \underline{?}$
$= \underline{?}$

45. $5\frac{7}{9} = 4 + 1\frac{7}{9}$
$= 4 + \underline{?}$
$= \underline{?}$

46. $9\frac{5}{11} = 8\frac{?}{11}$

47. $7\frac{3}{10} = 6\frac{?}{10}$

48. $10 = 9\frac{?}{7}$

49. $8\frac{3}{10} = 7\frac{?}{30}$

50. $6\frac{5}{8} = 5\frac{?}{40}$

51. $8 = 7\frac{?}{9}$

52. $6 = 5\frac{?}{6}$

53. $8\frac{9}{13} = 7\frac{?}{65}$

54. $9\frac{11}{15} = 8\frac{?}{60}$

Subtract and express each answer as a mixed number.

55. $8\frac{3}{7} - 5\frac{1}{7}$

56. $9\frac{5}{8} - 6\frac{1}{8}$

57. $5\frac{3}{9} - 2\frac{1}{9}$

58. $6\frac{3}{8} - 4\frac{1}{8}$

59. $8\frac{7}{10} - 6\frac{3}{10}$

60. $5\frac{7}{12} - 2\frac{5}{12}$

61. $8\frac{1}{6} - 2\frac{5}{6}$

62. $9\frac{1}{4} - 6\frac{3}{4}$

63. $8\frac{7}{9} - 5\frac{1}{9}$

64. $5\frac{5}{9} - 3\frac{2}{9}$

65. $9\frac{5}{6} - 7\frac{1}{6}$

66. $7\frac{5}{7} - 4\frac{6}{7}$

67. $9\frac{1}{8} - 6\frac{3}{8}$

68. $11\frac{5}{9} - 5\frac{1}{6}$

69. $4\frac{7}{8} - 1\frac{3}{4}$

70. $9\frac{9}{20} - 7\frac{3}{10}$

71. $2\frac{1}{8} - 1\frac{1}{12}$

72. $6\frac{5}{6} - 2\frac{7}{15}$

73. $6\frac{5}{12} - 3\frac{7}{9}$

74. $13\frac{2}{15} - 9\frac{3}{10}$

75. $5\frac{5}{14} - 2\frac{12}{21}$

76. $9\frac{5}{12} - 2\frac{9}{16}$

77. $41\frac{5}{24} - 13\frac{15}{16}$

78. $6\frac{5}{14} - 3\frac{19}{35}$

79. $11\frac{1}{18} - 9\frac{7}{15}$

80. $55\frac{5}{12} - 30\frac{2}{15}$

81. $73\frac{23}{32} - 15\frac{1}{24}$

82. $15\frac{11}{16} - 9\frac{7}{24}$

83. $8\frac{1}{4} - 6$

84. $9\frac{3}{4} - 7$

85. $8 - 3\frac{2}{3}$

86. $9 - 2\frac{3}{4}$

87. $16 - 7\frac{3}{8}$

88. $5 - 2\frac{1}{5}$

89. $8\frac{3}{14} - 2\frac{13}{18}$

90. $5\frac{1}{12} - 2\frac{13}{27}$

91. $105\frac{1}{36} - 92\frac{19}{20}$

92. $113\frac{1}{32} - 87\frac{11}{20}$

Fill in the blanks.

93. To subtract fractions having a common denominator, subtract the ______________ and place this difference over the common ______________.

94. To subtract fractions having different denominators, find the LCD. Next, express each fraction as an ______________ fraction having the ______________ as the common denominator and then subtract.

95. To subtract using mixed numbers, subtract the ______________ number parts and subtract the ______________ parts. However, if the fraction in the subtrahend is larger than the fraction in the minuend, then before subtracting you should borrow 1 from the ______________ number part of the ______________ and add it to the ______________ part.

Solve and show all work.

96. Subtract $\frac{8}{35}$ from $\frac{11}{14}$.

97. Find the difference between $4\frac{5}{6}$ and $1\frac{1}{4}$.

98. Dan jogged $15\frac{3}{4}$ miles and Frank jogged $12\frac{1}{6}$ miles. How many more miles did Dan jog than Frank?

99. Cazzie has $\frac{7}{8}$ pound of peanuts. If he gives $\frac{1}{6}$ pound of his peanuts to Guerin, how many pounds of peanuts does he have left?

100. On Monday morning, the stock of the ABT Clothing Company opened at \$115 per share and then dropped $\$5\frac{5}{8}$ per share. Find the closing price of ABT stock.

101. Find the difference between 8 and $6\frac{1}{2}$.

102. Find the difference between 9 and $6\frac{1}{4}$.

103. Subtract $\frac{5}{6}$ from $8\frac{5}{8}$.

104. Subtract $\frac{11}{12}$ from $7\frac{3}{10}$.

105. Find the difference between $110\frac{5}{6}$ and $75\frac{1}{12}$.

106. Find the difference between $103\frac{5}{8}$ and $65\frac{1}{16}$.

107. What is $\frac{13}{66}$ minus $\frac{1}{36}$?

108. What is $\frac{5}{44}$ minus $\frac{1}{20}$?

109. John cut off $15\frac{1}{2}$ yards of wire from a piece of wire that was 65 yards. How much wire was left?

110. Betty cut off $\frac{1}{8}$ of a pound of fat from a two-pound steak. How much steak was left?

Answers to Practice Problems **22a.** $\frac{2}{5}$ **b.** $\frac{1}{7}$ **c.** $\frac{4}{23}$ **23a.** $\frac{3}{20}$ **b.** $\frac{1}{8}$ **c.** $\frac{49}{120}$ **24a.** $6\frac{3}{7}$ **b.** $2\frac{17}{24}$ **c.** $1\frac{11}{12}$ **d.** $223\frac{1}{4}$

2.9 Signed Fractions

35 We will use the term **signed fraction** to mean a positive or negative fraction. The rules, definitions, and procedures used to perform operations on integers and fractions also can be extended to signed fractions.

Example 27 Find the additive inverse.

a. $\frac{3}{4}$ **b.** $-\frac{6}{5}$

Solution To find the additive inverse of a number, change its sign.

27a. The additive inverse of $\frac{3}{4}$ is $-\frac{3}{4}$.

27b. The additive inverse of $-\frac{6}{5}$ is $\frac{6}{5}$.

Example 28 Find the absolute value.

a. $\left|-\frac{3}{4}\right|$ **b.** $\left|\frac{3}{4}\right|$

Solution If x is a positive number, then $|x| = x$ and $|-x| = x$.

28a. $\left|-\frac{3}{4}\right| = \frac{3}{4}$ **28b.** $\left|\frac{3}{4}\right| = \frac{3}{4}$

Practice Problem 25 ***Solve.***

a. The additive inverse of $\frac{9}{13}$ is ______.

b. The absolute value of $-\frac{3}{7}$ is ______.

c. The additive inverse of $-\frac{13}{5}$ is ______.

d. The absolute value of $-\frac{6}{5}$ is ______.

36 Signed fractions are added the same way integers are added. It will be helpful to remember that if $\frac{a}{b}$ and $\frac{c}{b}$ are two fractions, then $\frac{a}{b}$ is larger than $\frac{c}{b}$ when a is greater than c. For example, $\frac{4}{5}$ is larger than $\frac{3}{5}$ since 4 is greater than 3.

Example 29 Add and reduce.

a. $-\frac{5}{16} + \left(-\frac{7}{16}\right)$ **b.** $-\frac{3}{4} + \frac{5}{6}$ **c.** $\frac{5}{8} + \left(-\frac{11}{12}\right)$

Solution To add signed fractions, express each fraction as an equivalent fraction having the LCD as a common denominator. Then add using the rule for addition of integers (see Section 1.2).

Common sign ↓ ; Sum of absolute values

29a. $-\frac{5}{16} + \left(-\frac{7}{16}\right) = -\left(\frac{5}{16} + \frac{7}{16}\right)$

$= -\frac{12}{16}$

$= -\frac{3}{4}$ Divided 12 and 16 by 4

29b. $-\frac{3}{4} + \frac{5}{6} = -\frac{9}{12} + \frac{10}{12}$ $4 = 2 \cdot 2;\ 6 = 2 \cdot 3;\ \text{LCD} = 2 \cdot 2 \cdot 3 = 12$

$= +\left(\frac{10}{12} - \frac{9}{12}\right)$ Write a "+" sign. Then write the larger absolute value minus the smaller absolute value.

$= \frac{1}{12}$ Subtracted

The answer is positive since the sign is determined by the number with the largest absolute value.

29c. $\frac{5}{8} + \left(-\frac{11}{12}\right) = \frac{15}{24} + \left(-\frac{22}{24}\right)$ $8 = 2 \cdot 2 \cdot 2$
$12 = 2 \cdot 2 \cdot 3$
$\text{LCD} = 2 \cdot 2 \cdot 2 \cdot 3 = 24$

$= -\left(\frac{22}{24} - \frac{15}{24}\right)$

$= -\frac{7}{24}$

NOTE: If a fraction is negative, the negative sign (1) may be to the left of the fraction, (2) may be in the numerator, or (3) may be in the denominator. That is,

$$-\frac{3}{4} = \frac{-3}{4} = \frac{3}{-4}.$$

Practice Problem 26 ***Add and reduce.***

a. $-\frac{7}{24} + \left(-\frac{11}{24}\right)$ **b.** $-\frac{8}{9} + \frac{1}{12}$ **c.** $\frac{7}{8} + \left(-\frac{3}{10}\right)$

37 Signed fractions are subtracted the same way integers are subtracted.

Example 30 **a.** $\frac{1}{9} - \frac{7}{9}$ **b.** $-\frac{5}{6} - \frac{1}{10}$ **c.** $-\frac{1}{15} - \left(-\frac{5}{12}\right)$

Solution To subtract one signed fraction from another, the minuend remains the same and we add the additive inverse of the subtrahend.

30a. $\frac{1}{9} - \frac{7}{9} = \frac{1}{9} + \left(-\frac{7}{9}\right)$

$= -\frac{6}{9}$

$= -\frac{2}{3}$

30b. $-\frac{5}{6} - \frac{1}{10} = -\frac{5}{6} + \left(-\frac{1}{10}\right)$

$= -\frac{25}{30} + \left(-\frac{3}{30}\right)$

$= -\frac{28}{30}$

$= -\frac{14}{15}$

$6 = 2 \cdot 3$
$10 = 2 \cdot 5$
$\text{LCD} = 2 \cdot 3 \cdot 5 = 30$

30c. $-\frac{1}{15} - \left(-\frac{5}{12}\right) = -\frac{1}{15} + \frac{5}{12}$

$= -\frac{4}{60} + \frac{25}{60}$

$= \frac{21}{60}$

$= \frac{7}{20}$

$15 = 3 \cdot 5$
$12 = 2 \cdot 2 \cdot 3$
$\text{LCD} = 2 \cdot 2 \cdot 3 \cdot 5 = 60$

Practice Problem 27 ***Subtract and reduce.***

a. $\frac{1}{12} - \frac{5}{12}$ **b.** $-\frac{4}{15} - \frac{7}{20}$ **c.** $\frac{5}{36} - \left(-\frac{7}{24}\right)$

38 Multiplying signed fractions is similar to multiplying integers.

Example 31 Multiply and reduce.

a. $\left(-\frac{6}{8}\right)\left(-\frac{12}{15}\right)$ **b.** $\left(-\frac{25}{14}\right)\left(\frac{21}{35}\right)$ **c.** $\left(-\frac{14}{24}\right)\left(\frac{33}{35}\right)\left(-\frac{4}{77}\right)$

Solution To multiply signed fractions, multiply their absolute values and then determine the sign of the product by using the rule for multiplying integers.

2 negative signs

31a. $\left(-\frac{6}{8}\right)\left(-\frac{12}{15}\right) = +\left(\frac{6 \cdot 12}{8 \cdot 15}\right)$ Multiply absolute values of fractions

$= +\left(\frac{\cancel{2} \cdot \cancel{3} \cdot \cancel{2} \cdot \cancel{2} \cdot 3}{\cancel{2} \cdot \cancel{2} \cdot \cancel{2} \cdot \cancel{3} \cdot 5}\right)$

$= \frac{3}{5}$

The product of an even number of negative numbers is positive.

31b. $\left(-\frac{25}{14}\right)\left(\frac{21}{35}\right) = -\left(\frac{\overset{5}{\cancel{25}} \cdot \overset{3}{\cancel{21}}}{\underset{2}{\cancel{14}} \cdot \underset{7}{\cancel{35}}}\right)$ Divide 25 and 35 by 5; Divide 14 and 21 by 7

$= -\frac{15}{14}$

The product of an odd number of negative numbers is a negative number.

31c. $$\left(-\frac{14}{24}\right)\left(\frac{33}{35}\right)\left(-\frac{4}{77}\right) = +\left(\frac{14 \cdot 33 \cdot 4}{24 \cdot 35 \cdot 77}\right)$$
$$= +\left(\frac{\cancel{2} \cdot \cancel{7} \cdot \cancel{3} \cdot \cancel{11} \cdot \cancel{2} \cdot \cancel{2}}{\cancel{2} \cdot \cancel{2} \cdot \cancel{2} \cdot \cancel{3} \cdot 5 \cdot \cancel{7} \cdot 7 \cdot \cancel{11}}\right)$$
$$= \frac{1}{35}$$

Practice Problem 28 ***Multiply and reduce.***

a. $\left(-\frac{4}{9}\right)\left(\frac{15}{4}\right)$ b. $(-6)\left(-\frac{9}{30}\right)$ c. $\left(-\frac{36}{99}\right)\left(-\frac{25}{40}\right)\left(-\frac{42}{28}\right)$

39 To divide, multiply by the reciprocal of the divisor. To find the reciprocal of a signed fraction, interchange the numerator and denominator. The sign of the reciprocal is the same as the sign of the original fraction.

Example 32 Divide and reduce.

a. $-\frac{15}{8} \div \frac{10}{12}$ b. $-\frac{16}{18} \div \left(-\frac{24}{22}\right)$ c. $\frac{25}{28} \div (-15)$

Solution To divide one signed fraction by another, multiply the dividend and the reciprocal of the divisor and then determine the sign of the quotient by using the rules for multiplying integers.

Reciprocal

32a. $$-\frac{15}{8} \div \frac{10}{12} = -\frac{15}{8} \cdot \frac{12}{10}$$
$$= -\left(\frac{15 \cdot 12}{8 \cdot 10}\right)$$ Rule for multiplying signed fractions
$$= -\left(\frac{3 \cdot \cancel{5} \cdot \cancel{2} \cdot \cancel{2} \cdot 3}{\cancel{2} \cdot \cancel{2} \cdot 2 \cdot 2 \cdot \cancel{5}}\right)$$
$$= -\frac{9}{4}$$

The quotient of two numbers having different signs is a negative number.

32b. $$-\frac{16}{18} \div \left(-\frac{24}{22}\right) = -\frac{16}{18}\left(-\frac{22}{24}\right)$$
$$= +\left(\frac{16 \cdot 22}{18 \cdot 24}\right)$$
$$= +\left(\frac{\cancel{2} \cdot \cancel{2} \cdot \cancel{2} \cdot \cancel{2} \cdot 2 \cdot 11}{\cancel{2} \cdot 3 \cdot 3 \cdot \cancel{2} \cdot \cancel{2} \cdot \cancel{2} \cdot 3}\right)$$
$$= \frac{22}{27}$$

The quotient of two numbers having the same sign is a positive number.

32c. $\frac{25}{28} \div (-15) = \frac{25}{28}\left(-\frac{1}{15}\right)$

$= -\left(\frac{\overset{5}{\cancel{25}} \cdot 1}{28 \cdot \underset{3}{\cancel{15}}}\right)$ Divide 25 and 15 by 5

$= -\frac{5}{84}$

Practice Problem 29 ***Divide and reduce.***

a. $-\frac{12}{25} \div \left(-\frac{18}{25}\right)$ b. $-\frac{8}{15} \div \frac{12}{10}$ c. $\frac{20}{30} \div \left(-\frac{88}{15}\right)$

2.9 Exercises

Find the additive inverse.

1. $-\frac{3}{5}$
2. $\frac{4}{9}$
3. $-\frac{13}{8}$

Find the value.

4. $\left|-\frac{3}{7}\right|$
5. $\left|\frac{5}{8}\right|$
6. $\left|-\frac{13}{11}\right|$

Add and reduce.

7. $-\frac{2}{9} + \left(-\frac{3}{9}\right)$
8. $-\frac{3}{10} + \frac{1}{10}$
9. $-\frac{7}{10} + \frac{5}{8}$
10. $-\frac{3}{4} + \left(-\frac{1}{3}\right)$
11. $\frac{5}{8} + \left(-\frac{5}{6}\right)$
12. $-\frac{13}{28} + \frac{1}{42}$

Subtract and reduce.

13. $\frac{5}{14} - \frac{1}{14}$
14. $-\frac{5}{12} - \frac{1}{12}$
15. $-\frac{5}{16} - \left(-\frac{7}{12}\right)$
16. $\frac{7}{10} - \left(-\frac{4}{15}\right)$
17. $-\frac{17}{35} - \left(-\frac{5}{14}\right)$
18. $-\frac{31}{32} - \left(-\frac{1}{24}\right)$

Multiply and reduce.

19. $\left(-\frac{6}{8}\right)\left(-\frac{10}{15}\right)$
20. $\left(-\frac{14}{25}\right)\left(\frac{10}{21}\right)$
21. $\left(\frac{72}{35}\right)\left(-\frac{56}{18}\right)$
22. $\left(-\frac{5}{4}\right)\left(\frac{3}{10}\right)\left(-\frac{16}{9}\right)$
23. $\left(-\frac{14}{20}\right)\left(-\frac{66}{42}\right)\left(-\frac{4}{55}\right)$
24. $-8 \cdot \frac{15}{12}$

Find the reciprocal.

25. $-\frac{3}{5}$
26. $\frac{7}{8}$
27. $-\frac{1}{15}$
28. -9

Divide and reduce.

29. $-\frac{8}{14} \div \left(-\frac{10}{21}\right)$
30. $\frac{18}{25} \div \left(-\frac{27}{10}\right)$
31. $\frac{5}{14} \div (-35)$

32. $\frac{-15}{56} \div \frac{14}{90}$

33. $\frac{12}{25} \div \left(-\frac{1}{10}\right)$

34. $-\frac{80}{44} \div \left(-\frac{60}{33}\right)$

Do the indicated operation.

35. $-\frac{5}{8} + \frac{7}{10}$

36. $\frac{20}{21} - \frac{28}{16}$

37. $-\frac{36}{81} \div \frac{32}{63}$

38. $-\frac{7}{66} - \frac{5}{36}$

39. $8 \div \left(-\frac{20}{5}\right)$

40. $\left(-\frac{12}{16}\right)\left(\frac{18}{15}\right)\left(-\frac{20}{3}\right)$

41. $\frac{3}{14} - \frac{11}{21}$

42. $\frac{9}{20} + \left(-\frac{4}{15}\right)$

43. $-\frac{28}{45} \div \left(-\frac{42}{15}\right)$

44. $\left(-\frac{21}{18}\right)\left(-\frac{20}{14}\right)\left(-\frac{6}{5}\right)$

45. $\left(-\frac{3}{10}\right) + \left(-\frac{5}{12}\right) + \left(-\frac{7}{15}\right)$

Solve. Show all work.

46. Find the sum of $-\frac{5}{8}$ and $\frac{7}{10}$.

47. Find the sum of $-\frac{1}{18}$ and $-\frac{5}{12}$.

48. Find the difference between $\frac{1}{35}$ and $\frac{11}{14}$.

49. Find the difference between $-\frac{11}{20}$ and $\frac{8}{15}$.

50. Find the product of -12 and $\frac{7}{10}$.

51. Find the product of -10 and $\frac{5}{6}$.

52. Divide $\frac{6}{25}$ into $-\frac{8}{15}$.

53. Divide $-\frac{9}{25}$ into $\frac{6}{35}$.

54. Subtract $-\frac{5}{12}$ from $\frac{7}{18}$.

55. Subtract $-\frac{11}{24}$ from $-\frac{5}{18}$.

56. Divide -18 by $-\frac{1}{2}$.

57. Divide -24 by $\frac{1}{3}$.

Answers to Practice Problems **25a.** $-\frac{9}{13}$ **b.** $\frac{3}{7}$ **c.** $\frac{13}{5}$ **d.** $\frac{6}{5}$ **26a.** $-\frac{3}{4}$ **b.** $-\frac{29}{36}$ **c.** $\frac{23}{40}$ **27a.** $-\frac{1}{3}$ **b.** $-\frac{37}{60}$ **c.** $\frac{31}{72}$ **28a.** $-\frac{5}{3}$ **b.** $-\frac{9}{5}$ **c.** $\frac{15}{44}$ **29a.** $\frac{2}{3}$ **b.** $-\frac{4}{9}$ **c.** $-\frac{5}{44}$

Summary

Important Terms

2.1
fraction
numerator
denominator
proper fraction
improper fraction
factor
divisor
prime number
composite number
prime factorization

2.2
equivalent fractions
reduced to lowest terms

2.3
mixed number

2.5
reciprocal

2.6
common denominator
lowest common denominator (LCD)

2.8
borrowing

2.9
signed fractions

Important Skills

2.1

Identifying prime and composite numbers
Writing the prime factorization of a number
Determining if a number is divisible by 2, 3, and/or 5

2.2

Determining if two fractions are equivalent
Reducing a fraction to its lowest terms

2.3

Expressing mixed numbers as improper fractions
Expressing improper fractions as mixed numbers

2.4

Multiplying fractions
Multiplying mixed numbers
Solving word problems involving fractions and mixed numbers

2.5

Dividing one fraction by another
Dividing mixed numbers
Solving word problems involving fractions and mixed numbers

2.6

Adding fractions having a common denominator
Writing a fraction as an equivalent fraction
Finding the LCD of two or more fractions
Adding fractions having different denominators
Solving word problems involving fractions

2.7

Adding mixed numbers
Solving word problems involving mixed numbers

2.8

Subtracting fractions having a common denominator
Subtracting fractions having different denominators
Subtracting mixed numbers
Solving word problems involving fractions and mixed numbers

2.9

Adding, subtracting, multiplying, and dividing signed fractions

Review Exercises

Write the prime factorization.

1. 36 **2.** 121 **3.** 522 **4.** 1001

Reduce.

5. $\frac{12}{20}$ **6.** $\frac{35}{91}$ **7.** $\frac{21}{49}$ **8.** $\frac{231}{297}$

Express as a mixed number.

9. $\frac{17}{5}$ **10.** $\frac{77}{6}$ **11.** $\frac{121}{9}$

Express as an improper fraction.

12. $3\frac{3}{4}$

13. $7\frac{5}{8}$

14. $82\frac{3}{4}$

Do the indicated operation.

15. $\frac{18}{25} \cdot \frac{35}{27}$

16. $-\frac{5}{12} - \frac{4}{15}$

17. $\frac{5}{6} + \frac{7}{9}$

18. $-\frac{25}{56} \div \frac{5}{12}$

19. $2\frac{2}{5} \cdot 2\frac{3}{16}$

20. $\frac{7}{9} - \frac{4}{15}$

21. $8\frac{1}{12} - 4\frac{7}{9}$

22. $\frac{8}{39} \div \frac{36}{26}$

23. $5\frac{9}{10} + 6\frac{5}{6}$

24. $10\frac{5}{9} \div 1\frac{17}{21}$

25. $9 - \frac{2}{3}$

26. $-30 \div \frac{1}{7}$

Solve.

27. If three-fourths of the students in Math 111E passed a test, what fractional part of the class did not pass?

28. Determine what fractional part $3\frac{1}{3}$ is of 15.

29. Kathy owes Betty \$1000. If Kathy gives Betty three-fourths of the money that she owes her, how much money will Betty receive?

30. One day APD stock opened at \$35 and closed at $\$31\frac{3}{8}$. How much did the price of this stock drop?

31. A recipe for fruit punch requires $\frac{1}{4}$ cup of lemon juice, $5\frac{3}{8}$ cups of cranberry juice, $\frac{1}{6}$ cup of lime juice, and $3\frac{5}{12}$ cups of raspberry juice. How much juice is required for the punch?

DECIMALS

OBJECTIVES

1. Write the word name for a decimal.
2. Add two or more decimals.
3. Subtract using decimals.
4. Multiply two or more decimals.
5. Round decimals.
6. Divide using decimals.
7. Change a fraction to a decimal.
8. Change a decimal to a fraction.
9. Add, subtract, multiply, or divide signed decimals.

PRETEST	EXPLANATION
1. Write the word name. **a.** 0.03 **b.** 2.415	Section 3.1 Idea 3
2. Add. **a.** $0.31 + 3.1$ **b.** $5.31 + 8.2 + 39.158$	Section 3.2 Idea 4
3. Subtract. **a.** $8.321 - 6.55$ **b.** $11 - 0.328$	Section 3.2 Idea 5
4. Multiply. **a.** $(0.2)(0.8)$ **b.** $(3.89)(7.3)$	Section 3.3 Idea 7
5. Round to the nearest: **a.** 0.323 (hundredth) **b.** 7.951 (tenth)	Section 3.4 Idea 9
6. Divide. **a.** $76.13 \div 2.3$ **b.** $24.36 \div 12$	Section 3.5 Idea 11
7. Write each fraction as a decimal. **a.** $\frac{3}{5}$ **b.** $\frac{8}{3}$	Section 3.6 Idea 12
8. Write each decimal as a fraction. **a.** 0.35 **b.** 3.125	Section 3.6 Idea 13
9. Perform the indicated operation.	Section 3.7
a. $-0.35 + 0.8$	Idea 16
b. $-3.215 - 6.35$	Idea 17
c. $(-3.2)(0.3)(-11)$	Idea 18
d. $\frac{0.646}{-0.02}$	Idea 19

3.1 Reading and Writing Decimals

Idea **1** A **decimal** or **decimal fraction** is a number that can be written as a fraction whose denominator is 1, 10, 100, or 1000, and so on. Some examples are:

Decimal notation	Fraction notation
$.4 =$	$\frac{4}{10}$
$.071 =$	$\frac{71}{1000}$
$2.47 =$	$\frac{247}{100}$
$5 =$	$\frac{5}{1}$

NOTE: The "." is a decimal point. In a whole number, the decimal point is located just to the right of the digit in the ones place (for example, $5 = 5.$).

2 The digits to the left of the decimal point have values of 1, 10, 100, 1000, and so on. The digits to the right of the decimal point have place values of $\frac{1}{10}, \frac{1}{100}, \frac{1}{1000}$, and so on (see Figure 3–1). Remember, each place is $\frac{1}{10}$ as large as the first place to its left.

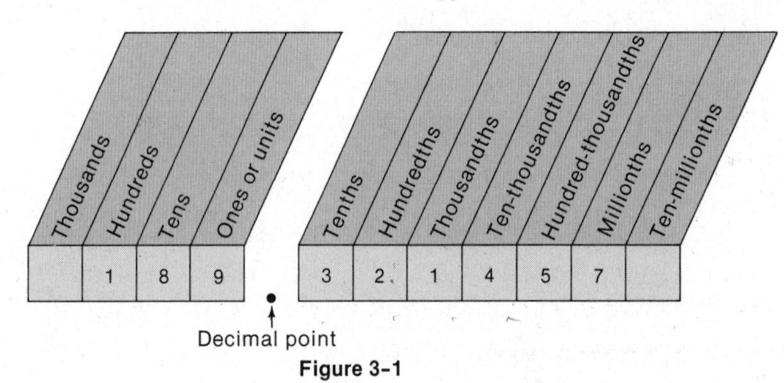

Figure 3-1

NOTE: The **place value** of a digit is determined by where it is in relation to the decimal point. The **value** of each digit, on the other hand, is determined by multiplying the digit by its place value. For example, the place value of the number 2 in Figure 3–1 is hundredths $\left(\frac{1}{100}\right)$, and its value is $2 \cdot \frac{1}{100} = \frac{2}{100}$.

3 The concept of place value is useful in helping us to read or write word names for decimals.

To Write the Word Name for a Decimal:

1. Write the name for the whole number to the left of the decimal point.
2. Write the word "and" for the decimal point.
3. Write the name for the number to the right of the decimal point as if it were a whole number. Then write the name for the place value of the last digit on the right.

NOTE: Omit steps 1 and 2 when a zero or no number appears to the left of the decimal point.

Example 1 Write the word name for each decimal.

a. 0.32 **b.** 0.051 **c.** 2.6 **d.** 17.0031

Solution

(Hundredths place → the 2 in 0.32)

1a. 0.32 = thirty-two hundredths

1b. 0.051 = fifty-one thousandths

1c. 2.6 = two and six tenths

1d. 17.0031 = seventeen and thirty-one ten-thousandths

Practice Problem 1 ***Write the word name for each decimal.***

a. 0.7 **b.** 0.06 **c.** 3.112 **d.** 9.000051

3.1 Exercises

Identify the place value of the 5.

1. 35.01 **2.** 17.056 **3.** 0.513 **4.** 51.731 **5.** 0.0051

6. 3.00051 **7.** 6.00375 **8.** 7.123045 **9.** 175124.8

Write the word name for each decimal.

10. 0.03 **11.** 0.813 **12.** 0.5 **13.** 0.1357 **14.** 0.0404

15. 0.000034 **16.** 7.15 **17.** 6.034 **18.** 17.05 **19.** 0.031

20. 8.158 **21.** 6.703 **22.** 0.051 **23.** 7.7 **24.** 17.0101

Express the following as decimals.

25. Eight and seventeen hundredths

26. Two hundred five thousandths

27. Forty-six ten-thousandths

28. Six hundred and seven hundred-thousandths

29. Ten and five millionths

30. Ninety-four hundredths

Fill in the blanks.

31. A ______________ is a number that can be written as a fraction whose denominator is 1, 10, 100, 1000, and so on.

32. Given the number 3.125, the place value of the 5 is ______________ and its value is ______________.

33. To write a word name for a decimal, write the name for the ______________ number to the left of the decimal point. Write the word "______________" for the decimal point. Finally, write the name for the number to the right of the decimal point as if it were a ______________ number and then write the ______________ for the last digit on the right.

Answers to Practice Problems **1a.** seven tenths **b.** six hundredths **c.** three and one hundred twelve thousandths **d.** nine and fifty-one millionths

3.2 Addition and Subtraction

4 You can only add digits that have the same place value.

To Add Decimals:

1. Write each number so that the decimal points are lined up.
2. Add the numbers as if they were whole numbers.
3. Line up the decimal point in the answer (sum) with the other decimal points.

NOTE: It may be helpful to write zeros to the right of the decimal point so that each number has the same number of digits behind the decimal point. Remember, writing in extra zeros will not change the value of the original number (for example, 3.5 = 3.50).

Example 2 Add.

a. 0.32 + 0.6 + 0.315 **b.** 7.3 + 0.16 + 5.1

c. 17.16 + 5.14 + 9.1345 **d.** 8.37 + 91 + 5.2

Solution To add decimals, line up the decimal points. Add as if you were adding whole numbers and then line up the decimal point in the answer with the other decimal points.

2a.
```
  0 . 320   Line up the "."
  0 . 600
+0 . 315
  1 . 235
```

2b.
```
  7.30
  0.16
+5.10
 12.56
```

2c.
```
  17.1600
   5.1400
+  9.1345
  31.4345
```

2d.
```
   8.37
  91.00
+  5.20
 104.57
```

Practice Problem 2 ***Add.***

a. 5.3 + 6.11 **b.** 0.005 + 0.07 + 0.9

c. 17.8 + 5.81 + 1.857 **d.** 99.5 + 165.35 + 0.316

5 You can subtract a digit only from another digit that has the same place value.

To Subtract Decimals:

1. Write the subtrahend beneath the minuend and line up the decimal points.
2. Subtract the numbers as if they were whole numbers. If necessary, write in zeros.
3. Line up the decimal point in the answer (difference) with the other decimal points.

Example 3 Subtract.

a. 14.97 − 10.54 **b.** 0.6323 − 0.429

c. 16.57 − 8.193 **d.** 91 − 17.34

Solution To subtract decimals, line up the decimal points in the minuend and the subtrahend. Subtract as if you were subtracting whole numbers. Line up the decimal point in the answer with the other decimal points.

3a.
```
  14.97    Line up the "."
- 10.54
   4.43
```

3b.
```
      2
  0.6 3 '23
 -0.4 2  9
  0.2 0  33
```

3c.
```
       4 16
  16. 5 7 '0
 - 8. 1 9  3
   8. 3 7  7
```

3d.
```
  8 10 .9
  9  1 .0 '0
- 1  7 .3  4
  7  3 .6  6
```

Practice Problem 3 ***Subtract.***

a. 5.9 − 3.7 **b.** 0.64315 − 0.5324

c. 35.6 − 29.54 **d.** 18 − 0.008

Example 4 Solve and show all work.

a. Pam bought a shirt for $35.82, a tie for $15.85, and a belt for $18. How much money did she spend?

b. Subtract 15.87 from 39.1.

Solution **4a.** Find the sum of the three items.

```
  35.82
  15.85
+ 18.00
  69.67    Pam spent $69.67
```

4b. The subtrahend is 15.87, and the minuend is 39.1.

```
    8  10
  3 9 . 1 '0
- 1 5 . 8  7
  2 3 . 2  3
```

Practice Problem 4 ***Solve and show all work.***

a. Matt ran 5 miles on Monday, 6.3 miles on Wednesday, 8.35 miles on Friday, and 10.8 miles on Sunday. What was the total distance he ran over the seven-day period?

b. Ernie bought a book for $8.43 and paid for it with a $20 bill. How much change should he receive?

3.2 Exercises

Add.

1. 0.31 + 0.64 **2.** 7.3 + 5.6 **3.** 0.5 + 0.8

4. 0.3 + 0.15 **5.** 9.22 + 8.5 **6.** 7.321 + 0.52

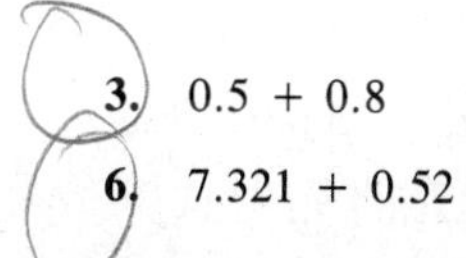

7. 17 + 18.34

8. 9.7 + 8.135

9. 0.71 + 3.578

10. 17.31 + 18.09

11. 5.3 + 2.389

12. 9.51 + 99.531

13. 8.35 + 7.3 + 9.584

14. 8.35 + 0.115 + 16.9 + 13

15. 35.13 + 75.195 + 0.3378

16. 0.8 + 0.95 + 8.192 + 0.7

17. 3.6 + 5.18 + 6 + 0.1891

18. 1.5 + 8.31 + 35.111

19. 8.5 + 3.17 + 18.501 + 98

20. 16.19 + 0.8 + 17.987

Subtract.

21. 15.3 − 8.1

22. 18.5 − 17.3

23. 5.37 − 3.25

24. 0.63 − 0.34

25. 19.1 − 19

26. 0.37 − 0.2

27. 8.7 − 4.623

28. 4.87 − 3.9

29. 84.07 − 83.97

30. 4.006 − 3.94

31. 73.16 − 10.8

32. 96 − 35.1

33. 85 − 11.08

34. 2.07 − 0.007

35. 57.88 − 41

36. 15 − 3.9

37. 97.815 − 17.5

38. 89.58 − 8.731

Perform the indicated operation.

39. 0.51 + 0.8

40. 0.61 + 0.7

41. 3.22 + 8.1 + 9.531

42. 4.54 + 7.1 + 6.351

43. 0.81 − 0.62

44. 0.93 − 0.75

45. 8 − 3.64

46. 6 − 2.52

47. 1.035 + 3.57 + 4

48. 3.125 + 4.63 + 5

49. 6.3 + 8.2 + 9.35 + 0.81

50. 7.2 + 9.3 + 7.45 + 0.63

51. 8357 − 631

52. 9349 − 852

53. 552 + 85 + 1347

54. 675 − 23 + 2356

55. 574.2 − 234.71

56. 837.5 − 236.53

57. 1234.45 − 39.4

58. 2365.83 − 79.6

59. 0.0085 − 0.0063

60. 0.0099 − 0.0078

61. 9.19 + 0.3 + 5.1 + 8.2

62. 6.15 + 0.5 + 6.3 + 8.7

63. 2.395 + 5.74 + 4.3 + 0.963

64. 1.385 + 7.54 + 3.4 + 0.693

65. 0.08 + 1.5834 + 3.463

66. 0.09 + 2.8543 + 4.346

67. 20.5 − 0.5

68. 30.7 − 0.7

69. 33 + 18.7 + 99.35

70. 44 + 19.8 + 73.67

71. 2335.78 − 235.9

72. 3226.87 − 325.9

73. 6.2 + 8.2 + 9.2 + 0.21

74. 2.6 + 8.6 + 5.6 + 0.37

75. 234 − 17.3

76. 335 − 19.7

77. 3374.52 + 831.57

78. 4437.85 − 733.96

79. 87.0468 − 72.3007

80. 78.4068 − 27.7003

81. 175.02 − 139.2

82. 185.03 − 149.3

Fill in the blanks.

83. To add decimals, line up the ______________. Add as if you were adding ______________ numbers and then line up the decimal point in the answer with the other ______________.

84. To subtract using decimals, line up the ______________ in the ______________ and the subtrahend. Subtract as if you were subtracting ______________ numbers and then line up the decimal point in the answer with the other ______________.

85. When adding or subtracting decimals, remember that you can only add or subtract digits having the same ______________.

Solve and show all work.

86. Subtract 15.78 from 33.2.

87. Find the difference between 193.87 and 79.354.

88. During a vacation, Buford's records showed gasoline purchases of 19.75 gallons, 15.4 gallons, 16 gallons, and 13.85 gallons. How many gallons of gas did he buy during his vacation?

89. The perimeter of a triangle is equal to the sum of the length of its sides. Find the perimeter of a triangle if the first side measures 8.75 inches, the second side measures 9.6 inches, and the third side measures 10.375 inches.

90. Rochelle has $331.64 in her savings account. If she withdraws $58.97, what is the new balance in her account?

91. The original price of a shirt is $23.73. If you receive a $4.82 discount, how much does the shirt cost?

92. One week a jogger ran the following distances: 15.3 miles, 18.75 miles, 19 miles, 21.5 miles, 25.375 miles, and 30.25 miles. Find the total distance the jogger ran.

93. Subtract 92.4 from 235.

94. Subtract 16.45 from 335.

95. Find the sum of $3.25, $5.04, and $17.63.

96. Find the sum of $2.35, $4.05, and $71.36.

97. Find the difference between 85.17 and 58.09.

98. Find the difference between 58.71 and 18.04.

99. Find the total of $19.45, $72.53, $18, $55, and $153.45.

100. Find the total of $13.54, $27.35, $81, $75, and $213.17.

101. Find the difference between $131.57 and $19.99.

102. Find the difference between $268.76 and $17.68.

103. Joe has $575.88 in his checking account. If he made deposits of $93.64 and $33, what is his new balance?

104. Susan earns $75 per week from one job and $37.18 per week from another. Find her total weekly salary.

105. Jane earns $85 per week babysitting and $48.50 per week tutoring students in English. Find her total weekly salary.

Answers to Practice Problems **2a.** 11.41 **b.** 0.975 **c.** 25.467 **d.** 265.166 **3a.** 2.2 **b.** 0.11075 **c.** 6.06 **d.** 17.992 **4a.** 30.45 miles **b.** $11.57

3.3 Multiplication

6 A decimal can be expressed as a fraction. This concept will be useful in helping us to develop a rule for multiplying decimals.

Example 5

$$(0.23)(0.5) = \left(\frac{23}{100}\right)\left(\frac{5}{10}\right) = \frac{115}{1000} = 0.115$$

or

0.23	(2 decimal places)
0.5	(1 decimal place)
0.115	(3 decimal places)

NOTE: The number of decimal places in a decimal is the number of digits written to the right of the decimal point.

7 Example 5 suggests that there is a rule for multiplying two or more decimals.

To Multiply Using Decimals:

1. Multiply the numbers as if they were whole numbers.
2. Place the decimal point in the product (answer). The number of decimal places in the product must equal the sum of the number of decimal places in the factors. If necessary, write in zeros.

Example 6 Multiply.

a. (2.3)(5.3) **b.** (131)(1.02)

c. (9.6)(0.0004) **d.** (42.3)(3.5)(0.23)

Solution To multiply decimals, multiply as if you were multiplying whole numbers. Place the decimal point in the product.

6a.

```
   2.3    (1 decimal place)
   5.3    (1 decimal place)
   6 9
 11 5
 12.1 9   (1 + 1 = 2 decimal places)
```

6b.

```
   1 31   (0 decimal places)
   1.02   (2 decimal places)
   2 62
 131 0
 133.62   (0 + 2 = 2 decimal places)
```

6c.

```
      9.6   (1 decimal place)
  0.000 4   (4 decimal places)
 0.0038 4   (1 + 4 = 5 decimal places)
```

Write two zeros to the left of the 3 since the product must have 5 decimal places.

6d.

```
   4 2.3   (1 decimal place)
     3.5   (1 decimal place)
   21 1 5
  126 9
  148.0 5  (1 + 1 = 2 decimal places)

   1 48.05   (2 decimal places)
      0.23   (2 decimal places)
   4 44 15
  29 61 0
  34.05 15   (2 + 2 = 4 decimal places)
```

Practice Problem 5 ***Multiply.***

a. (7.5)(0.33) **b.** (17.5)(35)

c. (0.031)(0.007) **d.** (3.2)(15)(0.075)

Example 7 Solve and show all work.

Barbara was reimbursed 19 cents per mile for mileage for business purposes. If she drove 105 miles, how much money (in dollars) should she be reimbursed?

Solution Recall that 19 cents = \$0.19. Thus,

Reimbursement = (miles driven)(reimbursement per mile)
= (105)(\$0.19)
= \$19.95.

```
  105
 0.19
 9 45
10 5
19.95
```

Barbara should receive \$19.95.

Practice Problem 6 ***Solve and show all work.***

Shirley earns $6.75 an hour. How much will she earn if she works 39 hours?

3.3 Exercises

Place a decimal point correctly in each product. If necessary, write in zeros.

1. (1.74)(0.33) = 5742
2. (17.4)(3.3) = 5742
3. (17.4)(0.33) = 5742
4. (0.174)(0.33) = 5742
5. (1.74)(0.033) = 5742
6. (0.174)(33) = 5742

Multiply.

7. 3.2 × 0.3
8. 0.65 × 0.2
9. 3.4 × 0.32
10. 35 × 0.21
11. 0.32 × 0.005
12. 0.2 × 0.3
13. 15.3 × 5.2
14. 35.4 × 1.05
15. 0.421 × 0.24
16. 77.2 × 0.015
17. 69.5 × 0.34
18. 145.3 × 0.15
19. 0.0529 × 0.0031
20. 89.98 × 7.09
21. 19.7 × 85.1

Multiply.

22. (4.6)(869)
23. (13.75)(9.7)
24. (0.35)(0.271)
25. (22.5)(29.8)
26. (2.51)(0.0031)
27. (0.56)(1.12)
28. (25.32)(35.2)
29. (2.1049)(0.513)
30. (2.3)(0.35)(8.5)
31. (15.1)(17.2)(19.3)
32. (0.15)(1.2)(0.009)
33. (17.5)(0.008)(99)
34. (2.35)(0.4)
35. (6.25)(0.6)
36. (3.15)(3.15)
37. (4.62)(4.62)
38. (41.1067)(0.003)
39. (52.1245)(0.005)
40. (3.51)(0.008)

Fill in the blanks.

41. The number of decimal places in a decimal is the number of ______________ written to the ______________ of the decimal point.
42. To multiply decimals, multiply the numbers as if they were ______________ numbers. Place the ______________ in the product. The number of decimal places in the product must equal the ______________ of the number of decimal places in the ______________.
43. When multiplying decimals, you may need to write ______________ in the product to make sure that you have enough decimal places.

Solve and show all work.

44. John pays a monthly payment of $175.58 for a car. If it takes 36 months to pay for the car, what is the total cost of the car?
45. A coach buys 19 pairs of socks at $2.25 per pair. What is the total cost of the socks?
46. If you earn $7.57 an hour, how much should you earn for a 40-hour week?
47. Rick buys 19 gallons of gas at $1.48 per gallon. What is the total cost of the gas?
48. What is the cost (in dollars) for having 317 pages xeroxed at nine cents per page?
49. A jogger ran 35.8 miles per day for 7 days. How many miles did he run?

50. Bill traveled 500.9 miles per day for 13 days. How many miles did he travel?

51. Find the product of 2.3, 4.5, and 0.06.

52. Find the product of 3.2, 5.4, and 0.08.

53. Multiply 0.894 by 0.32.

54. Multiply 0.498 by 0.23.

55. Joel earns $835.50 per month. If Rich earns twice as much as Joel, how much does Rich earn per month?

56. Dan ran 11.8 miles on Sunday. If Betty ran twice as far as Dan did, how far did Betty run?

57. Find the product of ninety-three thousand and three and six tenths.

58. Find the product of eighty-five thousand and eleven and three tenths.

59. James earns $10.50 per hour. If he works 17 hours, how much will he earn?

60. Gilberto earns $9.75 per hour. If he works 19 hours, how much will he earn?

61. Betty drove 35.7 miles per day for three days. How many miles did she drive?

62. Rich drove 53.7 miles per day for five days. How many miles did he drive?

Answers to Practice Problems **5a.** 2.475 **b.** 612.5 **c.** 0.000217 **d.** 3.6 **6.** $263.25

3.4 Rounding

8 When you want to find an approximate value for a given number, use the concept of **rounding.**

Example 8 Round each number to the nearest tenth.

a. 3.72 **b.** 5.68 **c.** 6.75

Solution **8a.** Since 3.72 is closer to 3.7 than to 3.8 (see Figure 3–2), round *down* to 3.7.

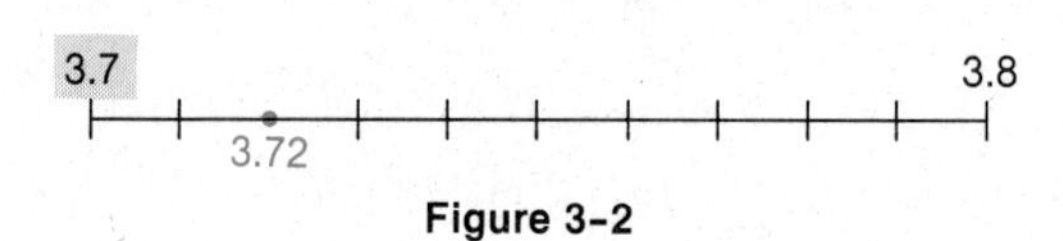

Figure 3-2

8b. Since 5.68 is closer to 5.7 than to 5.6 (see Figure 3–3), round *up* to 5.7.

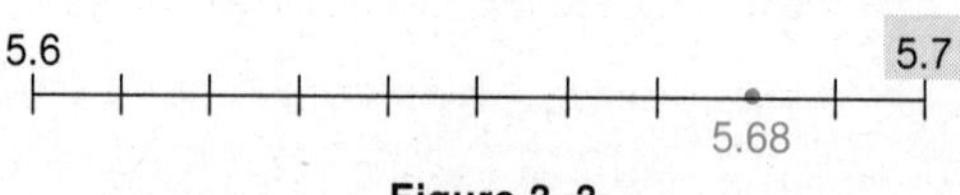

Figure 3-3

8c. Any time a number is halfway between its two closest approximations, round up. For example, 6.75 rounded to the nearest tenth is 6.8 (see Figure 3–4).

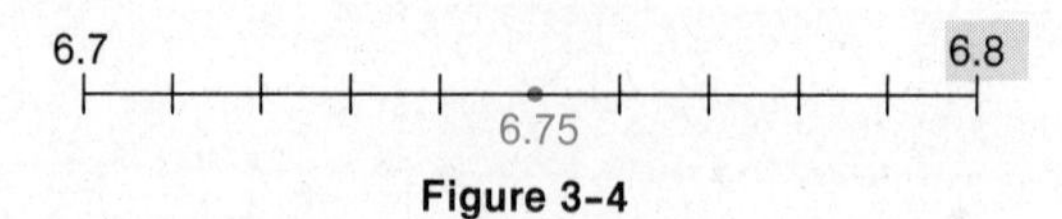

Figure 3-4

NOTE: Instead of writing "6.75 rounded to the nearest tenth is 6.8," write 6.75 $\doteq$ 6.8. The symbol $\doteq$ means *is approximately equal to*.

9 Example 8 suggests a rule for rounding a decimal to a given place value.

Rule for Rounding Decimals:

1. Place your pencil directly below the digit in the given place value.
2. Look at the first digit to the right of a given place value.

3a. If this digit is *less than* 5 (0, 1, 2, 3, or 4), change all digits to the right of the given place value to zeros.

3b. If this digit is *5 or greater* (5, 6, 7, 8, or 9), add 1 to the digit in the given place value and change all digits to the right of it to zeros.

4. Omit any zeros that appear to both the right of the decimal point and to the right of the digit in the given place value.

NOTE: Omitting zeros that are both to the right of the decimal point and to the right of the last digit of a decimal will not change its value (for example, $0.50 = 0.5$).

Example 9

a. Round 7.8691 to the nearest hundredth.

b. Round 10.867542 to the nearest ten-thousandth.

c. Round 1751.79 to the nearest hundred.

d. Round 15.987 to the nearest tenth.

Solution To round a decimal to a given place value, identify the digit to the right of the given place value. If it is 5 or greater, round up. If it is less than 5, round down.

Hundredths place

9a. $7.8\,6\,91 \doteq 7.8700$
$\doteq 7.87$

Ten-thousandths place

9b. $10.867\,5\,42 \doteq 10.867500$
$\doteq 10.8675$

9c. $1751.79 \doteq 1800.00$
$\doteq 1800$

9d. $15.987 \doteq 16.000$
$\doteq 16.0$

Practice Problem 7 ***Solve.***

a. Round 6.761 to the nearest tenth.

b. Round 0.7658 to nearest thousandth.

c. Round 145.77 to the nearest ten.

d. Round 35.7976 to the nearest hundredth.

Example 10

a. Round 0.578 to one decimal place.

b. Round 3.6519 to two decimal places.

Solution To round a decimal to a given number of places, use the same rule for rounding decimals to a given place value.

10a. $0.578 \doteq 0.600$
$\doteq 0.6$

10b. $3.6519 \doteq 3.6500$
$\doteq 3.65$

Practice Problem 8 **Solve.**

a. Round 13.7591 to three decimal places.

b. Round 0.854371 to four decimal places.

3.4 Exercises

Round to the nearest tenth.

1. 0.17 **2.** 0.25 **3.** 0.31

4. 6.178 **5.** 18.117 **6.** 9.35

7. 0.91 **8.** 0.97 **9.** 15.7516

Round to the nearest hundredth.

10. 0.788 **11.** 0.854 **12.** 0.735

13. 0.1666 **14.** 7.658 **15.** 1.379

16. 5.795 **17.** 2.7543 **18.** 7.5582

Round to the nearest thousandth.

19. 0.1456 **20.** 0.7852 **21.** 0.6785

22. 7.8519 **23.** 5.6784 **24.** 1.77777

25. 95.3333 **26.** 8.9595 **27.** 9.99999

Round 1335.3745898 to the nearest:

28. tenth **29.** thousandth **30.** ten

31. hundred **32.** hundredth **33.** millionth

34. thousand **35.** ten-thousandth **36.** one

Fill in the blanks.

37. The symbol $\doteq$ means "is ______________."

38. To round a decimal to a given place value, identify the ______________ to the ______________ of the given place value. If it is less than ______________, round down. If it is ______________ or greater, round up.

39. Rounding a decimal to two decimal places is the same as rounding to the nearest ______________.

Round each decimal to the indicated place.

40. 155.77 (hundreds) **41.** 75.7781 (two decimal places) **42.** 165,781 (thousands)

43. 0.73852 (three decimal places) **44.** 1785.77 (tens) **45.** 4.963 (one decimal place)

Answers to Practice Problems **7a.** 6.8 **b.** 0.766 **c.** 150 **d.** 35.80 **8a.** 13.759 **b.** 0.8544

3.5 Division

10 If you multiply the divisor and the dividend in a division problem by the same nonzero number, the value of the quotient does not change. This concept will be useful in helping us to develop a rule for dividing decimals.

Example 11 Divide: 72 ÷ 2.4.

$$7.2 \div 2.4 = \frac{7.2}{2.4} = \frac{7.2(10)}{2.4(10)} = \frac{72}{24} = 3$$

or

$$2.4\overline{)7.2} = 24\overline{)72.}$$

```
     3.
24)72.
   72
    0
```

11 Example 11 suggests a rule for dividing one decimal by another.

Rule for Division Using Decimals:

1. Rewrite the problem as if you were dividing whole numbers.
2. If necessary, move the decimal point to the right so that the divisor is a whole number.
3. Move the decimal point in the dividend the same number of places to the right.
4. Place the decimal point in the quotient directly above the decimal point in the dividend and then divide.

Example 12 Divide.

a. 9.28 ÷ 3.2 **b.** 38.76 ÷ 12 **c.** 6.936 ÷ 1.20

Solution To divide using decimals, rewrite the problem using the long division format. Move the decimal points to the right the same number of places until the divisor is a whole number. Line up the decimal points in the quotient and the dividend and then divide.

12a. $3.2\overline{)9.28} = 32\overline{)92.8}$ Move the decimal points one place to the right.

```
      2.9
32)92.8
   64
   28 8
   28 8
      0
```

12b. $12\overline{)38.76}$ Since the divisor is a whole number, do not move the decimal point

```
     3.23
12)38.76
   36
    2 7
    2 4
      36
      36
       0
```

12c. $1.20\overline{)6.936} = 12\overline{)69.36}$

```
      5.78
12)69.36
   60
    9 3
    8 4
      96
      96
       0
```

Practice Problem 9 ***Solve.***

a. $8.80 \div 3.2$ **b.** $0.01302 \div 0.21$ **c.** $86.4 \div 72$

Example 13 Solve.

a. $0.48 \div 0.032$ **b.** $1 \div 8$

Solution Sometimes extra zeros must be written in the dividend.

13a. $0.032\overline{)0.48} = 32\overline{)480.}$ Write in a zero as a place holder

$$\begin{array}{r} 15. \\ 32\overline{)480.} \\ \underline{32} \\ 160 \\ \underline{160} \\ 0 \end{array}$$

13b. $8\overline{)1} = 8\overline{)1.000}$ Insert zeros to complete the division

$$\begin{array}{r} 0.125 \\ 8\overline{)1.000} \\ \underline{8} \\ 20 \\ \underline{16} \\ 40 \\ \underline{40} \\ 0 \end{array}$$

Example 14 Divide: $0.314 \div 0.17$

Solution When you do not obtain a remainder of zero, round the quotient to the nearest hundredth.

$$0.17\overline{)0.314} = \begin{array}{r} 1.847 \\ 17\overline{)31.400} \\ \underline{17} \\ 14\ 4 \\ \underline{13\ 6} \\ 80 \\ \underline{68} \\ 120 \\ \underline{119} \\ 1 \end{array} \doteq 1.85$$

NOTE: To round to the nearest hundredth, always carry the division out to the thousandths place in the quotient.

Practice Problem 10 ***Divide. If you do not obtain a remainder of zero, round the quotient to the nearest hundredth.***

a. $4.03 \div 0.032$ **b.** $3 \div 6$ **c.** $0.961 \div 7.5$

Example 15 Solve and show all work.

Marcia borrowed \$846.48. If she pays the loan in 24 equal monthly payments, how much is each payment?

Solution Payments = amount borrowed ÷ number of payments

= $846.48 ÷ 24

= $35.27

```
     35.27
24)846.48
   72
   126
   120
     6 4
     4 8
     1 68
     1 68
        0
```

Practice Problem 11 ***Solve and show all work.***

Larry pays $34.73 a month toward his loan. How many months will it take him to pay a balance of $382.03?

3.5 Exercises

Divide. If you do not obtain a remainder of zero, round the quotient to the nearest hundredth.

1. 431.6 ÷ 13
2. 50.04 ÷ 18
3. 156.619 ÷ 23
4. 19.95 ÷ 105
5. 148.05 ÷ 42.3
6. 0.0384 ÷ 9.6
7. 1.219 ÷ 0.23
8. 57.22 ÷ 3.3
9. 5.722 ÷ 1.74
10. 3.486 ÷ 0.42
11. 0.0735 ÷ 17.5
12. 9.368 ÷ 0.04
13. 0.0035 ÷ 7
14. 65 ÷ 0.013
15. 11 ÷ 55
16. 9.875 ÷ 2.3
17. 7.647 ÷ 1.4
18. 35.3 ÷ 0.06
19. 9 ÷ 5
20. 13.75 ÷ 3.2
21. 0.1571 ÷ 4
22. 0.8215 ÷ 5
23. 282.94 ÷ 4.3
24. 224.74 ÷ 3.9
25. 57.8 ÷ 0.01
26. 65.7 ÷ 0.01
27. 125 ÷ 0.001
28. 375 ÷ 0.001
29. 1.78 ÷ 10
30. 3.87 ÷ 10
31. 11.74 ÷ 0.33
32. 15.66 ÷ 0.44
33. 1.566 ÷ 35.6
34. 1.174 ÷ 35.6
35. 0.054 ÷ 0.11
36. 0.06 ÷ 0.15
37. 15 ÷ 55
38. 20 ÷ 44
39. 287.738 ÷ 6.38
40. 18.8048 ÷ 3.68

Fill in the blanks.

41. If you multiply the divisor and the ______________ in a division problem by the same nonzero number, the value of the ______________ does not change.

42. To divide using decimals, move the decimal point until the ______________ is a whole number. Move the decimal point in the ______________ the same number of places. Place the decimal point in the ______________ directly above the decimal point in the ______________ and then divide.

Solve and show all work.

43. Three men paid a total of $10.26 for lunch. If the cost of the meal is divided equally, how much is each man's share?

44. Joe worked 32 hours and earned $138.56. How much was he paid per hour?

45. Betty earns $8.35 per hour. If her salary one week was $177, how many hours did she work?

46. When Jerry drove 400 miles, he used 12.7 gallons of gas. How many miles per gallon of gas did he get?

47. Adrienne's car gets 23.8 miles per gallon of gas. How many gallons of gas will she need if she drives 202.3 miles?

48. John's odometer read 38,769.1 miles at the beginning of his trip. When the trip ended, it read 39,365.8 miles. If he used 31.5 gallons of gas, how many miles per gallon of gas did he get? Round the answer to the nearest tenth.

49. Divide 0.36 into 1.836.

50. Divide 3.21 into 18.939.

51. Divide 189.55 by 22.3

52. Divide 593.94 by 11.4.

53. Divide four and five tenths into sixteen and fifty-six hundredths.

54. Divide nine and eight tenths by fourteen thousandths.

55. If 24 cartons of orange juice cost $51.60, find the cost of one carton.

56. If 48 cans of cola cost $21.60, find the cost of one can.

57. Linda's car gets 16.5 miles per gallon of gas. How many gallons of gas will she need if she drives 336.6 miles?

58. Bob's car gets 25.4 miles per gallon of gas. How many gallons of gas will he need if he drives 266.7 miles?

59. If Jane pays $52.26 per month on her charge account, how long will it take her to pay a balance of $1,254.24?

60. If John pays $48.17 per month on his charge account, how long will it take him to pay a balance of $1,734.12?

Answers to Practice Problems **9a.** 2.75 **b.** 0.062 **c.** 1.2 **10a.** 125.94 **b.** 0.5 **c.** 0.13 **11.** 11

3.6 Fraction/Decimal Conversions

12 In mathematics, you will sometimes need to change a fraction to a decimal.

To Change a Fraction to a Decimal:

1. Divide the denominator into the numerator.
2. If necessary, round the answer (quotient) to the desired place value.

Example 16 Express each fraction as a decimal. If necessary, round the answer to the nearest hundredth.

a. $\frac{3}{4}$ **b.** $\frac{7}{8}$ **c.** $\frac{7}{6}$ **d.** $\frac{5}{19}$

Solution To change a fraction to a decimal, divide the numerator by the denominator.

16a. $\frac{3}{4} = 0.75$

```
   0.75
4)3.00
  2 8
    20
    20
     0
```

16b. $\frac{7}{8} = 0.875$

```
   0.875
8)7.000
  6 4
    60
    56
     40
     40
      0
```

16c. $\frac{7}{6} \doteq 1.17$

$$\begin{array}{r} 1.166 \\ 6\overline{)7.000} \\ \underline{6} \\ 1\ 0 \\ \underline{6} \\ 40 \\ \underline{36} \\ 40 \\ \underline{36} \\ 4 \end{array}$$

16d. $\frac{5}{19} \doteq 0.26$

$$\begin{array}{r} 0.263 \\ 19\overline{)5.000} \\ \underline{3\ 8} \\ 1\ 20 \\ \underline{1\ 14} \\ 60 \\ \underline{57} \\ 3 \end{array}$$

NOTE: Since we obtained a repeating remainder (4) in Solution 16c, the quotient of 7 and 6 is a repeating decimal. That is, $\frac{7}{6} = 1.16\overline{6}$. The bar indicates the digit that repeats.

Practice Problem 12 ***Express each fraction as a decimal. If necessary, round the answer to the nearest hundredth.***

a. $\frac{2}{5}$ **b.** $\frac{8}{5}$ **c.** $\frac{23}{33}$ **d.** $\frac{2}{3}$

Example 17 Solve and show all work.

Cazzie got 28 hits in 85 times at bat. What is his batting average? Round to the nearest thousandth.

Solution

$$\begin{aligned} \text{Batting average} &= \frac{\text{hits}}{\text{times at bat}} \\ &= \frac{28}{85} \\ &\doteq 0.329 \end{aligned}$$

$$\begin{array}{r} 0.3294 \\ 85\overline{)28.0000} \\ \underline{25\ 5} \\ 2\ 50 \\ \underline{1\ 70} \\ 800 \\ \underline{765} \\ 350 \\ \underline{340} \\ 10 \end{array}$$

Cazzie's batting average is 0.329.

Practice Problem 13 ***Solve and show all work.***

Penny got 16 hits in 49 times at bat. What is her batting average? Round to the nearest thousandth.

13 Sometimes it may be necessary to change a decimal to a reduced fraction.

To Change a Decimal to a Reduced Fraction:

1. Write a fraction. The numerator is the original number without a decimal point. The denominator is the place value of the last digit on the right.
2. Reduce the resulting fraction to its lowest terms.

Example 18 Express each decimal as a fraction.

a. 0.71 **b.** 0.075 **c.** 13.5 **d.** 3.0007 **e.** 249

Solution To express a decimal as a fraction, write the original number (without a decimal point) over the place value of the last digit of the decimal. Reduce this fraction.

18a. $0.71 = \frac{71}{100}$

The denominator is 100 since the place value of the last digit of the decimal is hundredths.

18b. $0.075 = \frac{75}{1000} = \frac{3}{40}$

18c. $13.5 = \frac{135}{10} = \frac{27}{2}$

18d. $3.0007 = \frac{30007}{10{,}000}$

18e. $249 = \frac{249}{1}$

NOTE: Some fractions and decimals are used so often that it may be helpful to memorize the fraction-decimal equivalents shown in Figure 3–5.

TABLE OF EQUIVALENTS

Fraction	$\frac{1}{2}$	$\frac{1}{4}$	$\frac{1}{3}$	$\frac{2}{3}$	$\frac{1}{5}$	$\frac{2}{5}$	$\frac{3}{5}$	$\frac{3}{4}$	$\frac{1}{8}$
Decimal	0.5	0.25	$0.\overline{3}$	$0.\overline{6}$	0.2	0.4	0.6	0.75	0.125

Figure 3-5

Practice Problem 14 ***Express each decimal as a fraction.***

a. 0.15 **b.** 3.62 **c.** 0.0013 **d.** 32

14 In some problems containing both fractions and decimals, to find the solution it may be necessary to change the fractions to decimals or the decimals to fractions.

Example 19 Solve. Express each answer both as a fraction and a decimal.

a. $0.21 + \frac{3}{5}$ **b.** $\frac{9}{32} \div 0.125$

Solution To simplify an expression containing fractions and decimals, express all fractions as decimals or all decimals as fractions (whichever is more convenient) and then do the indicated operations.

19a. *Method 1*

$0.21 + \frac{3}{5}$

$= 0.21 + 0.6$

$= 0.81$ or $\frac{81}{100}$

Method 2

$0.21 + \frac{3}{5}$

$= \frac{21}{100} + \frac{60}{100}$

$= \frac{81}{100}$ or 0.81

19b. $$\begin{aligned} \frac{9}{32} \div 0.125 &= \frac{9}{32} \div \frac{125}{1000} \\ &= \frac{9}{32} \cdot \frac{1000}{125} \\ &= \frac{9 \cdot 1000}{32 \cdot 125} \\ &= \frac{3 \cdot 3 \cdot \cancel{2} \cdot \cancel{5} \cdot \cancel{2} \cdot \cancel{5} \cdot \cancel{2} \cdot \cancel{5}}{\cancel{2} \cdot \cancel{2} \cdot \cancel{2} \cdot 2 \cdot 2 \cdot \cancel{5} \cdot \cancel{5} \cdot \cancel{5}} \\ &= \frac{9}{4} \text{ or } 2.25 \end{aligned}$$

Practice Problem 15 ***Solve. Express each answer both as a fraction and a decimal.***

a. $3\frac{3}{4} + 2.1$ **b.** $\left(\frac{90}{70}\right)(1.05)$

3.6 Exercises

Express each fraction as a decimal. If necessary, round the answer to the nearest hundredth.

1. $\frac{3}{5}$ **2.** $\frac{1}{2}$ **3.** $\frac{1}{4}$ **4.** $\frac{1}{5}$ **5.** $\frac{9}{8}$

6. $\frac{23}{8}$ **7.** $\frac{7}{25}$ **8.** $\frac{19}{20}$ **9.** $\frac{1}{3}$ **10.** $\frac{7}{11}$

11. $\frac{5}{7}$ **12.** $\frac{11}{18}$ **13.** $\frac{17}{40}$ **14.** $\frac{3}{16}$ **15.** $\frac{31}{50}$

16. $\frac{27}{20}$ **17.** $\frac{9}{10}$ **18.** $\frac{5}{16}$ **19.** $\frac{32}{125}$ **20.** $\frac{3}{8}$

21. $\frac{3}{4}$ **22.** $\frac{4}{5}$ **23.** $\frac{7}{2}$ **24.** $\frac{17}{4}$ **25.** $\frac{1}{8}$

26. $\frac{5}{8}$ **27.** $\frac{1}{16}$ **28.** $\frac{1}{13}$ **29.** $\frac{1}{100}$ **30.** $\frac{1}{1000}$

31. $\frac{7}{400}$ **32.** $\frac{9}{400}$ **33.** $\frac{31}{5}$ **34.** $\frac{2}{5}$ **35.** $\frac{3}{16}$

36. $\frac{11}{15}$ **37.** $\frac{7}{6}$ **38.** $\frac{11}{9}$ **39.** $\frac{7}{500}$ **40.** $\frac{19}{200}$

Express each decimal as a reduced fraction.

41. 0.3 **42.** 0.15 **43.** 0.031 **44.** 0.5 **45.** 0.05

46. 1.8 **47.** 2.5 **48.** 3.011 **49.** 8 **50.** 34

51. 5.12 **52.** 1.01 **53.** 6.1103 **54.** 0.80 **55.** 165

56. 0.0005 **57.** 0.125 **58.** 0.0625 **59.** 0.3125 **60.** 0.8125

Express each decimal as a reduced fraction.

61. 0.45 **62.** 0.85 **63.** 0.111 **64.** 0.513 **65.** 13

66. 17 **67.** 0.30 **68.** 0.90 **69.** 3.5 **70.** 6.8

71. 0.225 **72.** 0.155 **73.** 7.01 **74.** 8.001 **75.** 0.36

76. 0.7 **77.** 0.1500 **78.** 0.8000 **79.** 246.8 **80.** 348.5

Fill in the blanks.

81. To change a fraction to a decimal, divide the ______________ into the ______________. If you do not obtain a remainder of zero, ______________ the answer to the desired place value.

82. If you divide one number into another and you obtain a ______________ remainder, this implies that your quotient is a repeating decimal.

83. To express a decimal as a fraction, write the original number (without a ______________) over the ______________ of the last digit of the decimal. ______________ this fraction.

Solve. Express each answer both as a fraction and a decimal.

84. $1.25 - \frac{4}{5}$

85. $\frac{7}{8} + 3.25$

86. $(0.15)\left(\frac{52}{100}\right)$

87. $\frac{3}{16} \div 0.015$

88. $\left(\frac{75}{22}\right)(0.33)$

89. $\frac{11}{5} - \frac{2}{3}(1.8)$

Solve and show all work.

90. Dick got 19 hits in 60 times at bat. What is his batting average? Round to the nearest thousandth.

91. Barbara paid $15 for 35 notebooks. What was the cost of each notebook? Round to the nearest cent (hundredth).

92. John earned $45 for working 11 hours. How much did he earn each hour? Round to the nearest cent.

93. Pam must drill a 0.5625-inch hole through a table. What is the fractional size of the drill she must use?

94. Joe mixed three-fourths of a quart of orange juice and 0.6 of a quart of pineapple juice together to make a pineapple-orange drink. How many quarts was the final mixture?

95. Express $3\frac{1}{4}$ as a decimal.

96. Express $6\frac{1}{5}$ as a decimal.

97. Find the sum of $\frac{3}{4}$ and 6.75.

98. Find the sum of $\frac{4}{5}$ and 3.85.

99. Find the product of 0.17 and $\frac{1}{100}$.

100. Find the product of 0.001 and $\frac{7}{1000}$.

101. Lynn must drill a 0.125-inch hole. What is the fractional size of the drill she must use?

102. John must drill a 0.625-inch hole. What is the fractional size of the drill he must use?

Answers to Practice Problems **12a.** 0.4 **b.** 1.6 **c.** 0.70 **d.** 0.67 **13.** 0.327 **14a.** $\frac{3}{20}$ **b.** $\frac{181}{50}$ **c.** $\frac{13}{10{,}000}$ **d.** $\frac{32}{1}$ **15a.** 5.85 or $\frac{117}{20}$ **b.** 1.35 or $\frac{27}{20}$

3.7 Signed Decimals

15 We will use the term **signed decimal** to mean a positive or a negative decimal. The rules, definitions, and procedures used to perform operations on integers and decimals also can be applied to signed decimals.

Example 20 Find the additive inverse.

a. 3.25 **b.** −0.834

Solution To find the additive inverse of a number, change its sign.

20a. The additive inverse of 3.25 is −3.25.

20b. The additive inverse of −0.834 is 0.834.

Example 21 Find the absolute value.

a. $|-6.8|$ **b.** $|0.326|$

Solution If x is a positive number, then $|x| = x$ and $|-x| = x$.

21a. $|-6.8| = 6.8$ **21b.** $|0.326| = 0.326$

Practice Problem 16 ***Fill in the blanks.***

a. The additive inverse of 0.85 is ______________.

b. $|-0.73| =$ ______________.

c. The additive inverse of -3.17 is ______________.

d. $|3.92| =$ ______________.

16 Signed decimals are added the same way integers are added. It will be helpful to remember that to compare two decimals, first write in extra zeros so that each number has the same number of decimal places. Then, compare the numbers as if they were whole numbers. For example, 0.3 is larger than 0.19 since

$$0.3 = 0.30$$
$$0.19 = 0.19$$

and 30 is larger than 19.

*

Example 22 Add.

a. $-0.35 + (-0.7)$ **b.** $-0.653 + 0.71$ **c.** $-6.85 + 3.98$

Solution To add signed decimals, express each number with the same number of decimal places. Then add using the rule for adding integers.

22a.
$$\begin{aligned} -0.35 + (-0.7) &= -0.35 + (-0.70) \\ &= -(0.35 + 0.70) \\ &= -1.05 \end{aligned} \qquad \begin{array}{r} 0.35 \\ \underline{0.70} \\ 1.05 \end{array}$$

22b.
$$\begin{aligned} -0.653 + 0.71 &= -0.653 + 0.710 \\ &= +(0.710 - 0.653) \\ &= 0.057 \end{aligned} \qquad \begin{array}{r} 0.710 \\ \underline{0.653} \\ 0.057 \end{array}$$

The answer is positive since the sign is determined by the number having the largest absolute value.

22c.
$$\begin{aligned} -6.85 + 3.98 &= -(6.85 - 3.98) \\ &= -2.87 \end{aligned} \qquad \begin{array}{r} 6.85 \\ \underline{3.98} \\ 2.87 \end{array}$$

Practice Problem 17 ***Add.***

a. $-0.39 + 0.8$ **b.** $-3.37 + (-8.32)$ **c.** $0.651 + (-0.72)$

17 Signed decimals are subtracted the same way integers are subtracted.

Example 23 Subtract.

a. $3.84 - 7.95$ **b.** $-0.8 - 0.51$ **c.** $-0.953 - (-8)$

Solution To subtract using signed decimals, keep the same minuend and add the additive inverse of the subtrahend.

23a. $3.84 - 7.95 = 3.84 + (-7.95)$
$= -4.11$

$$\begin{array}{r} 7.95 \\ \underline{3.84} \\ 4.11 \end{array}$$

23b. $-0.8 - 0.51 = -0.8 + (-0.51)$
$= -1.31$

$$\begin{array}{r} 0.80 \\ \underline{0.51} \\ 1.31 \end{array}$$

23c. $-0.953 - (-8) = -0.953 + 8$
$= 7.047$

$$\begin{array}{r} 8.000 \\ \underline{.953} \\ 7.047 \end{array}$$

Practice Problem 18 ***Subtract.***

a. $-0.36 - 0.8$ **b.** $-0.85 - (-0.111)$ **c.** $-3.65 - (-8.3)$

18 Multiplying signed decimals is similar to multiplying integers.

Example 24 Multiply.

a. $(-0.2)(-0.3)$ **b.** $(-3.2)(3.1)$ **c.** $(-0.09)(0.7)(-2)$

Solution To multiply signed decimals, multiply their absolute values and determine the sign of the product.

24a. $(-0.2)(-0.3) = +[(0.2)(0.3)]$
$= 0.06$

The product of an even number of negative signs is a positive number.

24b. $(-3.2)(3.1) = -[(3.2)(3.1)]$
$= -9.92$

$$\begin{array}{r} 3.2 \\ \underline{3.1} \\ 3\,2 \\ \underline{96} \\ 9.92 \end{array}$$

The product of an odd number of negative signs is a negative number.

24c. $(-0.09)(0.7)(-2) = +[(0.09)(0.7)(2)]$
$= 0.126$

Practice Problem 19 ***Multiply.***

a. $(-2.3)(0.2)$ **b.** $(-31.2)(-0.31)$ **c.** $(-0.3)(-0.3)(-0.3)$

19 Dividing signed decimals is similar to dividing integers.

Example 25 Divide.

a. $\dfrac{-0.88}{-0.2}$ **b.** $\dfrac{35}{-0.7}$ **c.** $\dfrac{-3.754}{2.3}$

Solution To divide using signed decimals, rewrite the problem with a divisor that is a whole number. Then use the rule for dividing integers.

25a. $\dfrac{-0.88}{-0.2} = \dfrac{-8.8}{-2}$
$= 4.4$

The quotient of two numbers having the same sign is positive.

25b. $\dfrac{35}{-0.7} = \dfrac{350}{-7} = -50$

The quotient of two numbers having different signs is negative.

25c. $\dfrac{-3.754}{2.3} = \dfrac{-37.54}{23} \doteq 1.63$

$$\begin{array}{r} 1.632 \\ 23\overline{)37.540} \\ \underline{23} \\ 14\,5 \\ \underline{13\,8} \\ 74 \\ \underline{69} \\ 50 \\ \underline{46} \\ 4 \end{array}$$

Practice Problem 20 ***Divide.***

a. $\dfrac{-3.2}{0.8}$ b. $\dfrac{-1}{-.1}$ c. $\dfrac{4.824}{-1.2}$

3.7 Exercises

Find the additive inverse.

1. -3.21 **2.** 0.53 **3.** -8.231

Find the absolute value.

4. $|-0.5|$ **5.** $|0.85|$ **6.** $|-7.28|$

Tell which is the larger number.

7. 0.96, 0.8 **8.** 0.82, 0.9 **9.** 2.06, 2.008

10. 0.95, 0.099 **11.** 8.5, 8.58 **12.** 4.006, 4.04

13. 0.521, 0.52 **14.** 0.99, 1 **15.** 5.08, 5.7

16. 0.85, 0.6

Add.

17. $-0.6 + (-0.31)$ **18.** $-0.009 + 0.05$ **19.** $-10.91 + 18$

20. $-3.21 + (-6.03)$ **21.** $-0.631 + 0.7$ **22.** $-8.31 + (-0.795)$

Subtract.

23. $-0.61 - 0.32$ **24.** $-0.85 - (-0.91)$ **25.** $5.2 - (-8.34)$

26. $-4.51 - (-8)$ **27.** $0.231 - 0.45$ **28.** $-3.25 - 2.051$

Multiply.

29. $(-0.3)(0.5)$ **30.** $(-3.3)(5)$ **31.** $(-0.4)(-0.2)(-0.3)$

32. $(-0.31)(-3.8)$ **33.** $(-6.1)(0.3)$ **34.** $(-0.013)(-13)$

Divide.

35. $\dfrac{8.88}{-0.4}$ **36.** $\dfrac{-0.12}{-0.06}$ **37.** $\dfrac{-90}{.9}$

38. $\dfrac{-0.7}{-0.8}$ **39.** $\dfrac{-1.302}{0.3}$ **40.** $\dfrac{-16.016}{-1.6}$

Perform the indicated operation.

41. $-0.35 + 0.621$
42. $5.32 - 7$
43. $3.2 - (-8.11)$
44. $-0.561 + 0.87$
45. $-0.3333 \div 0.11$
46. $(-0.5)(-0.2)(-10)$
47. $(9.1)(-0.2)$
48. $-5.8 - (-0.65)$
49. $-8.08 \div (-0.4)$
50. $(-3.2)(0.005)(-5)$

Perform the indicated operation.

51. $0.35 + (-0.44)$
52. $0.65 + (-0.88)$
53. $-6.2 + (-3.24)$
54. $-8.4 + (-4.33)$
55. $-0.45 + 6$
56. $-0.35 + 9$
57. $-0.031 + 0.04$
58. $-0.041 + 0.05$
59. $-0.25 - 0.83$
60. $-0.35 - 0.76$
61. $-0.85 - (-0.9)$
62. $-0.35 - (-0.5)$
63. $3.25 - (-0.75)$
64. $4.36 - (-0.64)$
65. $89.5 - 98.35$
66. $78.6 - 97.45$
67. $(-0.3)(-0.7)$
68. $(-0.4)(-0.9)$
69. $(-3.6)(4)$
70. $(-5.4)(6)$
71. $(-0.2)(0.3)(-0.2)$
72. $(-0.3)(0.2)(-0.3)$
73. $(-0.1)(-0.4)(0.3)(-2)$
74. $(-0.2)(-0.5)(0.3)(-4)$
75. $-0.66 \div 0.11$
76. $-0.77 \div 0.7$
77. $-88.8 \div (-2)$
78. $-99.9 \div (-3)$
79. $14 \div (-0.7)$
80. $65 \div (-0.5)$
81. $1.645 \div (-0.35)$
82. $4.355 \div (-0.65)$

Solve. Show all work.

83. Find the sum of -3.25, 6.8, -7.891, and 0.3658.
84. Find the sum of -4.35, 6.7, -3.158, and 0.4769.
85. Subtract 8.51 from -13.6.
86. Subtract 6.39 from -8.7.
87. Find the product of -0.1, -0.2, -0.4, and -0.15.
88. Find the product of -0.2, -0.3, -0.1, and -0.21.
89. Divide -593.94 by 11.4.
90. Divide -390.05 by 14.5.
91. Find the difference between 8.51 and -13.6.
92. Find the difference between 6.39 and -8.7.
93. Divide -1.5 into 0.051.
94. Divide -3.2 into 0.144.

Answers to Practice Problems **16a.** -0.85 **b.** 0.73 **c.** 3.17 **d.** 3.92 **17a.** 0.41 **b.** -11.69 **c.** -0.069 **18a.** -1.16 **b.** -0.739 **c.** 4.65 **19a.** -0.46 **b.** 9.672 **c.** -0.027 **20a.** -4 **b.** 10 **c.** -4.02

Summary

Important Terms

3.1

decimal
decimal point
place value
value of a digit

3.4

round
approximately equal to, $\doteq$

3.7

signed decimal

Important Skills

3.1

Determining the place value of a digit
Writing the word name of a decimal

3.2

Adding decimals
Subtracting decimals
Solving word problems involving decimals

3.3
Multiplying decimals
Solving word problems involving decimals

3.4
Rounding decimals

3.5
Dividing one decimal by another
Solving word problems involving decimals

3.6
Expressing a decimal as a fraction
Expressing a fraction as a decimal
Solving problems containing both fractions and decimals

3.7
Adding, subtracting, multiplying, and dividing signed decimals

Review Exercises

Write the word name.

1. 5.009 **2.** 0.2318 **3.** 0.35

Complete the following tables by filling in the equivalent fraction or decimal.

	Fraction	Decimal
4.	$\frac{61}{20}$	?
5.	?	0.31
6.	$\frac{1}{3}$	?

	Fraction	Decimal
7.	$\frac{11}{19}$	?
8.	?	0.65
9.	?	0.125

Round 1751.85928 to the nearest:

10. thousandth

11. hundred

Perform the indicated operation.

12. $6.54 + 8.3$

13. $9.325 - 8.41$

14. $(7.89)(0.0005)$

15. $86.16 \div 0.19$

16. $-8.79 + (-3.125)$

17. $66 \div (-0.6)$

18. $(-8.3)(1.02)$

19. $(0.25)(35.7)$

20. $0.897 - 0.2345$

21. $0.358 + 0.21$

22. $0.85 - 0.9$

23. $6.58 + 3.2 + 8.101$

Solve.

24. If gas costs $1.45 per gallon, how much will 10.4 gallons of gas cost?

25. Kay saved $369 in one year. Find her average monthly savings.

26. Tom went shopping with $90 in cash. He bought a record for $8.50, a shirt for $28.75, a sweater for $42, and he spent $4.23 for lunch. If Tom paid cash for all items, how much cash does he have left?

27. Vanessa earns $8.80 per hour. If she worked $35\frac{3}{4}$ hours last week, how much did she earn?

PERCENT

OBJECTIVES

1. Express a percent as a decimal.
2. Express a percent as a fraction.
3. Express a decimal as a percent.
4. Express a fraction as a percent.
5. Find a percent of a number.

PRETEST	EXPLANATION
1. Express each percent as a fraction.	Section 4.1
a. 65% **b.** 0.08% **c.** $\frac{3}{8}\%$	Idea 2
2. Express each percent as a decimal.	Section 4.1
a. 65% **b.** 0.08% **c.** $\frac{3}{8}\%$	Ideas 3 and 4
3. Express each decimal as a percent.	Section 4.2
a. 0.45 **b.** 0.0095 **c.** 3.2	Ideas 5 and 6
4. Express each fraction as a percent.	Section 4.2
a. $\frac{3}{4}$ **b.** $\frac{1}{8}$ **c.** $\frac{2}{3}$	Ideas 7 and 8
5. Solve.	Section 4.3
a. Find 85% of $\frac{15}{34}$.	Idea 9
b. Find 0.09% of 35.8.	Idea 9
c. A basketball team won 76% of the 25 games it played. How many games were won?	Idea 10
d. The nursing program at a college has 220 students. If 85% of the students passed the state licensing examination, how many students failed?	Idea 10

4.1 The Meaning of Percent

Idea 1 The word **percent** means *per hundred,* and the percent symbol (%) means *hundredths.* For example, 9% means 9 hundredths $\left(0.09 \text{ or } \frac{9}{100}\right)$.

Example 1

a. $7\% = 7$ hundredths $= \frac{7}{100}$ 　　 $7\% = 7$ hundredths $= 0.07$

b. $13\% = 13$ hundredths $= \frac{13}{100}$ 　　 $13\% = 13$ hundredths $= 0.13$

2 When P is any number,

$$P\% \text{ means } \frac{P}{100} \quad \text{or} \quad P\left(\frac{1}{100}\right) \quad \text{or} \quad P(0.01).$$

We can use this fact to help us express percents as fractions.

To Express a Percent as a Fraction:

1. Replace the percent symbol with $\frac{1}{100}$.
2. Multiply this number by the number preceding the percent symbol and reduce the resulting fraction.

Example 2 Express each percent as a reduced fraction.

a. 27% 　 **b.** $\frac{3}{4}\%$ 　 **c.** 315% 　 **d.** 80% 　 **e.** $66\frac{2}{3}\%$ 　 **f.** 16.4%

Solution To express a percent as a reduced fraction, use the formula $P\% = P\left(\frac{1}{100}\right)$. Reduce the answer to its lowest terms.

2a. $27\% = 27\left(\frac{1}{100}\right) = \frac{27}{100}$

2b. $\frac{3}{4}\% = \frac{3}{4}\left(\frac{1}{100}\right) = \frac{3}{400}$

2c. $315\% = 315\left(\frac{1}{100}\right) = \frac{315}{100} = \frac{63}{20}$

2d. $80\% = 80\left(\frac{1}{100}\right) = \frac{80}{100} = \frac{4}{5}$

2e. $66\frac{2}{3}\% = \left(66\frac{2}{3}\right)\left(\frac{1}{100}\right)$

$= \frac{200}{3} \cdot \frac{1}{100}$

$= \frac{\overset{2}{\cancel{200}} \cdot 1}{3 \cdot \underset{1}{\cancel{100}}}$

$= \frac{2}{3}$

2f. Convert the decimal to a fraction

$16.4\% = 16\frac{2}{5}\%$ $\quad .4 = \frac{4}{10} = \frac{2}{5}$

$= \frac{82}{5} \cdot \frac{1}{100}$

$= \frac{\overset{41}{\cancel{82}} \cdot 1}{5 \cdot \underset{50}{\cancel{100}}}$

$= \frac{41}{250}$

Practice Problem 1 ***Express each percent as a fraction.***

a. 35% **b.** 110% **c.** $\frac{1}{2}\%$ **d.** 3.25%

3 Expressing percents as decimals is done in much the same way that we converted percents to fractions.

To Express a Percent as a Decimal:

1. Replace the percent symbol with 0.01.
2. Multiply this number by the number preceding the percent symbol.

Example 3 Express each percent as a decimal.

a. 27% **b.** 80% **c.** 100% **d.** 0.05% **e.** $2\frac{3}{4}\%$ **f.** $\frac{3}{5}\%$

Solution To express a percent as a decimal, use the formula $P\% = P(0.01)$.

3a. $27\% = 27(0.01)$
$= 0.27$

3b. $80\% = 80(0.01)$
$= 0.80$ or 0.8

3c. $100\% = 100(0.01)$
$= 1$

3d. $0.05\% = 0.05(0.01)$
$= 0.0005$

3e. Convert the fraction to a decimal

$2\frac{3}{4}\% = 2.75\%$ $\quad \frac{3}{4} = .75$

$= 2.75\ (0.01)$

$= 0.0275$

3f. $\frac{3}{5}\% = 0.6\%$

$= 0.6(0.01)$

$= 0.006$

Practice Problem 2 ***Express each percent as a decimal.***

a. 33% **b.** 0.08% **c.** $3\frac{1}{4}\%$ **d.** $\frac{4}{5}\%$

4 Based on the results in Example 3, the rule for changing a percent to a decimal can be restated. That is:

> To change a percent to a decimal, move the decimal point two places to the left and omit the percent symbol.

Example 4 Express each percent as a decimal.

a. 71% **b.** 1.05% **c.** 130% **d.** $66\frac{2}{3}\%$

Solution To express a percent as a decimal, move the decimal point two places to the left and remove the %.

4a. 71% = 71 %
= 0.71

4b. 1.05% = 01.05%
= 0.0105

4c. 130% = 1 30 %
= 1.30

Convert the fraction to a decimal

4d. $66\frac{2}{3}\% = 66.\overline{6}\%$
$= 0.66\overline{6}$
$\doteq 0.67$

Practice Problem 3 ***Express each percent as a decimal (see Example 4).***

a. 15% **b.** 0.05% **c.** 105% **d.** $3\frac{3}{4}\%$

4.1 Exercises

Express each percent as a fraction.

1. 13%	**2.** 15%	**3.** 31%	**4.** 12%	**5.** 55%
6. 331%	**7.** 2%	**8.** 1.06%	**9.** $6\frac{1}{3}\%$	**10.** $\frac{1}{2}\%$
11. 100%	**12.** $\frac{1}{3}\%$	**13.** $37\frac{1}{2}\%$	**14.** 0.02%	**15.** 1.5%
16. $\frac{7}{8}\%$	**17.** 150%	**18.** $8\frac{2}{3}\%$	**19.** 0.004%	**20.** $4\frac{1}{16}\%$
21. 17%	**22.** 19%	**23.** 18%	**24.** 16%	**25.** 115%
26. 102%	**27.** 1.04%	**28.** 1.08%	**29.** $8\frac{1}{3}\%$	**30.** $6\frac{2}{3}\%$
31. $\frac{3}{4}\%$	**32.** $\frac{1}{4}\%$	**33.** 500%	**34.** 200%	**35.** $55\frac{1}{8}\%$
36. $65\frac{5}{8}\%$	**37.** 0.006%	**38.** 0.008%	**39.** 65%	**40.** 75%

Express each percent as a decimal.

41. 13% **42.** 15% **43.** 31% **44.** 12% **45.** 55%

46. 331% **47.** 2% **48.** 1.06% **49.** $6\frac{1}{4}\%$ **50.** $\frac{1}{2}\%$

51. 100% **52.** 0.8% **53.** $\frac{1}{8}\%$ **54.** 851% **55.** 250%

56. 3.5% **57.** $10\frac{3}{4}\%$ **58.** 0.51% **59.** $\frac{17}{5}\%$ **60.** 2.011%

Express each percent as a decimal.

61. 17% **62.** 19% **63.** 16% **64.** 18% **65.** 115%

66. 102% **67.** 1.04% **68.** 1.08% **69.** $8\frac{1}{3}\%$ **70.** $6\frac{2}{3}\%$

71. $\frac{3}{4}\%$ **72.** $\frac{1}{4}\%$ **73.** 500% **74.** 200% **75.** $55\frac{1}{8}\%$

76. $65\frac{5}{8}\%$ **77.** 0.006% **78.** 0.008% **79.** 65% **80.** 75%

Fill in the blanks.

81. The word ______________ means per hundred.

82. To express a percent as a fraction, replace the percent symbol with ______________. Next, multiply this number by the number preceding the ______________ and ______________ the resulting fraction.

83. To express a percent as a decimal, replace the percent symbol with ______________. Next, multiply this number by the number preceding the ______________.

84. To express a percent as a decimal, move the decimal point ______________ places to the ______________ and remove the ______________.

85. If 23% of the students at a certain college jog, this means that ______________ out of every ______________ students jog.

Solve.

86. John saved 25% of his annual income. What fractional part of his income did he save?

87. If 22.5% of the students at a community college attended a basketball game, what fractional part of the student body attended the game?

88. The Second National Bank pays $5\frac{1}{4}\%$ interest on a savings account. Express this interest rate as a decimal.

89. Jim spends $12\frac{1}{2}\%$ of his time each day reading. Express the amount of time he spends reading as a decimal.

90. The price of a coat is discounted $33\frac{1}{3}\%$. What fractional part of the original price must a customer pay?

91. Express 131% as a fraction.

92. Express 231% as a fraction.

93. Change $\frac{3}{8}\%$ to a decimal.

94. Change $\frac{11}{40}\%$ to a decimal.

95. Rich saved $37\frac{1}{8}\%$ of his annual income. What fractional part of his income did he save?

Answers to Practice Problems **1a.** $\frac{7}{20}$ **b.** $\frac{11}{10}$ **c.** $\frac{1}{200}$ **d.** $\frac{13}{400}$ **2a.** 0.33 **b.** 0.0008 **c.** 0.0325 **d.** 0.008 **3a.** 0.15 **b.** 0.0005 **c.** 1.05 **d.** 0.0375

4.2 Writing Decimals and Fractions as Percents

5 In Example 3c, we showed that 100% = 1. We also know that multiplying a number by 1 does not change the value of that number. Using these two facts, we can convert decimals to percents.

Example 5 Express each decimal as a percent.

a. 0.83 **b.** 0.389 **c.** 0.5 **d.** 2

Solution To express a decimal as a percent, write the decimal and 100% as a product and then multiply.

5a. $0.83 = 0.83(1)$
$= 0.83(100\%)$
$= 83\%$

5b. $0.389 = 0.389(1)$
$= 0.389(100\%)$
$= 38.9\%$

5c. $0.5 = 0.5(1)$
$= 0.5(100\%)$
$= 50\%$

5d. $2 = 2(1)$
$= 2(100\%)$
$= 200\%$

6 Example 5 suggests a rule for expressing a decimal as a percent.

Rule for Expressing a Decimal as a Percent:

1. Move the decimal point two places to the right.
2. Write a percent symbol (%).

Example 6 Express each decimal as a percent.

a. 0.91 **b.** 0.7 **c.** 0.145 **d.** 5.3361

Solution To express a decimal as a percent, move the decimal point two places to the right and write a percent symbol.

6a. 0.91 = 0.91
= 91%

6b. 0.7 = 0.70
= 70%

6c. 0.145 = 0.14 5
= 14.5%

6d. 5.3361 = 5.33 61
= 533.61%

Practice Problem 4 ***Express each decimal as a percent.***

a. 0.13 **b.** 0.8 **c.** 0.115 **d.** 2.1374

7 You have learned how to change decimals to percents and fractions to decimals. By combining these two procedures, you can change fractions to percents.

To Express a Fraction as a Percent:

1. Change the fraction to a decimal.
2. Express the decimal as a percent.

Example 7 Express each fraction as a percent.

a. $\frac{3}{4}$ **b.** $\frac{1}{5}$ **c.** $\frac{3}{500}$ **d.** $\frac{5}{2}$

Solution To express a fraction as a percent, first change the fraction to a decimal and then express the decimal as a percent.

7a. $\frac{3}{4} = 0.75 = 75\%$

$$\begin{array}{r} 0.75 \\ 4\overline{)3.00} \\ \underline{2\,8} \\ 20 \\ \underline{20} \end{array}$$

7b. $\frac{1}{5} = 0.2 = 20\%$

$$\begin{array}{r} 0.2 \\ 5\overline{)1.0} \\ \underline{1\,0} \end{array}$$

7c. $\frac{3}{500} = 0.006 = 0.6\%$

$$\begin{array}{r} 0.006 \\ 500\overline{)3.000} \\ \underline{3\,000} \end{array}$$

7d. $\frac{5}{2} = 2.5 = 250\%$

$$\begin{array}{r} 2.5 \\ 2\overline{)5.0} \\ \underline{4} \\ 1\,0 \\ \underline{1\,0} \end{array}$$

NOTE: Some fractions can be expressed as percents without first changing the fraction to a decimal. For example, $\frac{17}{100} = 17\%$ by using the definition of percent. Similarly, $\frac{4}{5} = \frac{4 \cdot 20}{5 \cdot 20} = \frac{80}{100} = 80\%$.

Example 8 Express each fraction to the nearest tenth of a percent.

a. $\frac{1}{3}$ **b.** $\frac{7}{320}$

Solution To express a fraction to the nearest tenth of a percent, first change the fraction to a decimal rounded to the nearest thousandth.

8a. $\frac{1}{3} \doteq 0.333 = 33.3\%$

Thus, $\frac{1}{3} \doteq 33.3\%$

$$\begin{array}{r} 0.3333 \\ 3\overline{)1.0000} \\ \underline{9} \\ 10 \\ \underline{9} \\ 10 \\ \underline{9} \\ 10 \\ \underline{9} \end{array}$$

8b. $\frac{7}{320} \doteq 0.022 = 2.2\%$

Thus, $\frac{7}{320} \doteq 2.2\%$

$$\begin{array}{r} 0.0218 \\ 320\overline{)7.0000} \\ \underline{6\,40} \\ 600 \\ \underline{320} \\ 2800 \\ \underline{2560} \end{array}$$

Practice Problem 5 ***Express each fraction to the nearest tenth of a percent.***

a. $\frac{3}{7}$ **b.** $\frac{2}{3}$ **c.** $\frac{5}{6}$

8 Frequently in business applications you will need to find the *exact* percent equivalent of a fraction rather than an approximate value.

Example 9 Express $\frac{1}{3}$ as a percent.

Solution The fraction $\frac{1}{3}$ is usually represented by a repeating decimal ($0.3\overline{3}$). Therefore, to find its exact percent equivalent, we will:

Step 1. Express the fraction as a decimal that contains a fraction.

$$\frac{1}{3} = 0.33\frac{1}{3} \qquad \begin{array}{r} 0.33 \\ 3\overline{)1.00} \\ \underline{9} \\ 10 \\ \underline{9} \\ 1 \end{array}$$

NOTE: Carry out the division two places and then write the remainder over the divisor.

Step 2. Next, express the decimal $\left(0.33\frac{1}{3}\right)$ as a percent $\left(33\frac{1}{3}\%\right)$. Thus,

$$\frac{1}{3} = 0.33\frac{1}{3} = 33\frac{1}{3}\%.$$

Practice Problem 6 ***Express each fraction as a percent (see Example 9).***

a. $\frac{3}{7}$ **b.** $\frac{2}{3}$ **c.** $\frac{5}{6}$

4.2 Exercises

Express each decimal as a percent.

1. 0.65 **2.** 0.33 **3.** 2.3 **4.** 32 **5.** 0.006
6. 5 **7.** 0.81 **8.** 0.00768 **9.** 0.75 **10.** 0.17
11. 0.6 **12.** 0.8 **13.** 0.4 **14.** 0.1115 **15.** 51.2
16. 0.8842 **17.** 0.007 **18.** 3.05 **19.** $0.33\frac{1}{3}$ **20.** $0.66\frac{2}{3}$
21. 0.13 **22.** 0.38 **23.** 8 **24.** 9 **25.** 0.008
26. 0.009 **27.** 0.5 **28.** 0.9 **29.** 0.111 **30.** 0.222
31. 8.1 **32.** 1.8 **33.** 0.0045 **34.** 0.0065 **35.** 0.6
36. 0.7 **37.** $0.55\frac{1}{2}$ **38.** $0.65\frac{1}{2}$ **39.** $0.8\frac{1}{2}$ **40.** $0.9\frac{3}{4}$

Express each fraction as a percent.

41. $\frac{3}{5}$ **42.** $\frac{1}{2}$ **43.** $\frac{1}{4}$ **44.** $\frac{1}{5}$ **45.** $\frac{9}{8}$
46. $\frac{23}{50}$ **47.** $\frac{7}{25}$ **48.** $\frac{19}{20}$ **49.** $\frac{1}{3}$ **50.** $\frac{7}{11}$
51. $\frac{5}{7}$ **52.** $\frac{11}{18}$ **53.** $\frac{17}{40}$ **54.** $\frac{3}{16}$ **55.** $\frac{31}{50}$
56. $\frac{27}{20}$ **57.** $\frac{9}{10}$ **58.** $\frac{5}{8}$ **59.** $\frac{3}{4}$ **60.** $\frac{1}{16}$
61. $\frac{1}{10}$ **62.** $\frac{11}{20}$ **63.** $\frac{2}{3}$ **64.** $\frac{3}{8}$ **65.** $\frac{11}{40}$
66. $\frac{22}{50}$ **67.** $\frac{7}{10}$ **68.** $\frac{3}{5}$ **69.** $\frac{15}{10}$ **70.** $\frac{8}{5}$

71. $\frac{7}{1000}$ **72.** $\frac{3}{10,000}$ **73.** $\frac{5}{1}$ **74.** $\frac{8}{1}$ **75.** $\frac{1}{40}$

76. $\frac{1}{20}$ **77.** $\frac{5}{16}$ **78.** $\frac{7}{16}$ **79.** $\frac{11}{10}$ **80.** $\frac{23}{20}$

Fill in the blanks.

81. To express a decimal as a percent, move the ______________ two places to the ______________ and write a percent symbol.

82. To express a fraction as a percent, change the ______________ to a ______________ and then express the ______________ as a ______________.

83. The fraction $\frac{2}{3}$ is represented by a repeating decimal. Thus, to find its exact percent equivalent, first express the ______________ as a ______________ which contains a fraction. Next, express the ______________ as a ______________.

Solve.

84. A clothing store advertised that all items would be reduced by $\frac{1}{4}$ of their original cost. Write this discount as a percent.

85. Three-fifths of the employees at a supermarket smoke. What percent of the employees smoke?

86. A student answered 33 out of 40 questions on an exam correctly. What percent of the questions did he answer incorrectly?

87. When Jim purchased a car, he was told that the interest rate was 0.1455. Express this rate as a percent.

88. A certain type of medicine is 0.057 alcohol. What percent of the medicine is alcohol?

89. A survey indicated that 0.155 of all students at a college exercise daily. What percent of the students exercise daily?

90. A survey indicated that 0.205 of all employees at a certain company jog. What percent of the employees jog?

91. Express $\frac{8}{7}$ as a percent.

92. Express $\frac{10}{7}$ as a percent.

93. Express $0.52\frac{1}{8}$ as a percent.

94. Express $0.78\frac{3}{8}$ as a percent.

95. A medicine is 0.187 alcohol. What percent of the medicine is alcohol?

Answers to Practice Problems **4a.** 13% **b.** 80% **c.** 11.5% **d.** 213.74% **5a.** 42.9% **b.** 66.7% **c.** 83.3% **6a.** $42\frac{6}{7}\%$ **b.** $66\frac{2}{3}\%$ **c.** $83\frac{1}{3}\%$

4.3 Finding a Percent of a Number

9 One application of percents involves finding a percent of a number.

To Find a Percent of a Number:

1. Change the percent to a decimal or a fraction.
2. Multiply this value by the number.

Example 10 Solve.

a. Find 35% of 200. **b.** Find 75% of $\frac{10}{9}$. **c.** Find 150% of 20.

d. Find 5.5% of 30.58. **e.** Find $33\frac{1}{3}\%$ of $3\frac{3}{5}$.

Solution To find a percent of a number, change the percent to a decimal or a fraction and then multiply.

10a. Since 200 is a decimal, convert 35% to a decimal.

$$35\% \text{ of } 200 = (0.35)(200) \qquad 35\% = 0.35$$
$$= 70$$

10b. Since $\frac{10}{9}$ is a fraction, convert 75% to a fraction.

$$75\% \text{ of } \frac{10}{9} = \frac{3}{4} \cdot \frac{10}{9} \qquad 75\% = \frac{75}{100} = \frac{3}{4}$$
$$= \frac{\overset{1}{\cancel{3}} \cdot \overset{5}{\cancel{10}}}{\underset{2}{\cancel{4}} \cdot \underset{3}{\cancel{9}}}$$
$$= \frac{5}{6}$$

10c. $150\% \text{ of } 20 = (1.50)(20)$
$= 30$

10d. $5.5\% \text{ of } 30.58 = (0.055)(30.58)$
$= 1.6819$

10e.
$$33\frac{1}{3}\% \text{ of } 3\frac{3}{5} = \frac{1}{3} \cdot \frac{18}{5} \qquad 33\frac{1}{3}\% = \left(33\frac{1}{3}\right)\left(\frac{1}{100}\right)$$
$$= \frac{1 \cdot \overset{6}{\cancel{18}}}{\underset{1}{\cancel{3}} \cdot 5} \qquad = \frac{100}{3} \cdot \frac{1}{100}$$
$$= \frac{6}{5} \text{ or } 1\frac{1}{5} \qquad = \frac{1}{3}$$

NOTE: In a percent problem, the word **of** usually means *multiply*.

Practice Problem 7 ***Solve.***

a. Find 20% of 10. **b.** Find $13\frac{1}{4}\%$ of 702.

c. Find 4.55% of 40.54. **d.** Find 150% of $30\frac{1}{2}$.

10 Being able to find a percent of a number will be helpful in solving certain types of word problems.

To Solve a Word Problem Involving Finding a Percent of a Number:

1. Rewrite the problem in terms of finding a percent of a number.
2. Multiply the percent times the number.
3. Answer the question asked.

Example 11 Solve.

a. Jim bought two quarts of soda pop that cost a total of \$1.50. If the sales tax is 6%, how much sales tax must Jim pay on the pop?

b. If your savings account earns $5\frac{3}{4}\%$ interest per year, how much interest would you earn on \$1000 in one year?

c. Julie wants to purchase a house that costs \$55,000. If the required down payment is 20% of that cost, how much is the required down payment on the house?

d. A \$3000 used car was sold at a 12% discount. Find the selling price.

e. The Davis Real Estate Corporation sold 80 homes in 1981. If the owners of the corporation project that they will sell 20% more homes in 1982 than in 1981, how many homes do they expect to sell in 1982?

Solution To solve a word problem that involves finding a percent of a number, rewrite the problem as finding a percent of a number. Then, solve it and answer the question asked.

11a. Rewrite: Find 6% of \$1.50.

$$\begin{aligned} 6\% \text{ of } \$1.50 &= (0.06)(\$1.50) \\ &= \$0.09 \end{aligned}$$

John must pay \$0.09 or 9¢ sales tax.

11b. Rewrite: Find $5\frac{3}{4}\%$ of \$1000.

$$\begin{aligned} 5\frac{3}{4}\% \text{ of } \$1000 &= (0.0575)(\$1000) \\ &= \$57.50 \end{aligned}$$

You would earn \$57.50.

11c. Rewrite: Find 20% of \$55,000

$$\begin{aligned} 20\% \text{ of } \$55{,}000 &= (0.20)(\$55{,}000) \\ &= \$11{,}000 \end{aligned}$$

Julie needs an \$11,000 down payment.

11d. Since the car was discounted 12%, the selling price is 88% (100% − 12%) of the original price. Rewrite: Find 88% of \$3000.

$$\begin{aligned} 88\% \text{ of } \$3000 &= (0.88)(\$3000) \\ &= \$2640 \end{aligned}$$

The selling price is \$2640.

11e. Rewrite: Find 120% of 80.

$$\begin{aligned} 120\% \text{ of } 80 &= (1.20)(80) \\ &= 96 \end{aligned}$$

The corporation expects to sell 96 homes in 1982.

Alternate method

$$\begin{aligned} \text{Projected sales} &= \text{1981 sales} + 20\% \text{ of the 1981 sales} \\ &= 80 + 20\% \text{ of } 80 \\ &= 80 + (0.20)(80) \\ &= 80 + 16 \\ &= 96 \end{aligned}$$

Practice Problem 8 ***Solve.***

a. In a class of 40 students, 30% of the students earned an A. How many students earned an A?

b. Pete earns \$420 per week. If he plans to save $15\frac{1}{2}\%$ of his weekly salary, how much should he save each week?

c. There are 220 employees at a certain company. If 85% of the employees are men, how many of the employees are women?

4.3 Exercises

Solve.

1. Find 30% of 50.

2. Find 85% of 1000.

3. Find 16% of 90.

4. Find 35% of 20.

5. Find 110% of 40.

6. Find 60% of $\frac{5}{8}$.

7. Find 10% of $8\frac{3}{4}$.

8. Find 15% of $\frac{80}{9}$.

9. Find 112% of $\frac{3}{2}$.

10. Find 20% of $\frac{15}{4}$.

11. Find $8\frac{1}{3}\%$ of 40.

12. Find 3.5% of 10.08.

13. Find 150% of 32.8.

14. Find 70% of $3\frac{1}{14}$.

15. Find $33\frac{1}{3}\%$ of 200.

16. Find $66\frac{2}{3}\%$ of 300.

17. Find 100% of $\frac{8}{9}$.

18. Find 0.05% of 90.

19. Find 33.3% of 50.

20. Find 40% of $35\frac{1}{2}$.

Fill in the blanks.

21. To find a percent of a number, change the ______________ to a fraction or decimal and then ______________ this value by the ______________.

22. To solve a word problem that involves finding a percent of a number, first rewrite the problem as "find a ______________ of a ______________." Then solve the problem and ______________ the question asked.

23. If 35% of a class is men, then ______________ percent of the class is women.

24. If Adrienne receives a 20% increase in pay, her new salary is ______________ percent of her old salary.

Solve.

25. The Eagle City Bank pays an annual interest rate of 6% on all deposits. If Tim deposits \$3600, how much interest will he earn in one year?

26. Betty earns \$1400 per month. If she plans to spend 25% of her monthly salary on rent, how much can she afford to pay for rent?

27. Bill wants to buy a \$450 tape recorder. If he receives a 35% discount, how much will the recorder cost?

28. Joe earns \$20,500 a year. $12\frac{1}{8}\%$ of his annual income is placed in a tax-sheltered annuity plan. How much money does he have invested in this plan? (Round the answer to the nearest cent.)

29. Dave spends 20% of his monthly income on rent and 15% on food. If he earns \$1000 a month, how much does he spend on these items?

30. A tailor increased the price of a tailor-made suit by 12%. If the suit originally cost \$400, what would be the increase in cost to the consumer?

31. In problem 30, what is the new selling price?

32. Dan made 87.5% of the 50 free throws he attempted. How many free throws did he make? (Round the answer to the nearest whole number.)

33. Mary bought a \$170 jacket. If the sales tax was $5\frac{3}{4}\%$, what was the total price of the jacket? (Round to the nearest cent.)

34. Forty-two percent of the 5100 students at a college are women. How many men are enrolled in the school?

35. Visa charges $1\frac{1}{2}\%$ interest per month on a customer's unpaid balance. If you have an unpaid balance of \$500, what is the finance charge?

36. The Ramiro family invested \$4500 at an annual interest rate of 12.25%. How much interest will be earned in one year?

37. A 30-liter mixture of salt and water is 40% salt. How many liters of salt does the mixture contain?

38. The population of a town is 25,000. If 45% of the town's residents attended a football game, how many did *not* attend?

39. John correctly answered 90% of the questions on a 50-item test. How many questions did he answer correctly?

40. Rochelle earns \$850 a month. If she is given a 12% raise, what will her salary be after the raise?

41. A solution is 75% salt. How many pints of salt are in 20 pints of the solution?

42. A solution is 35% alcohol. How many gallons of alcohol are in 50 gallons of the solution?

43. A bank pays $5\frac{3}{4}\%$ yearly interest on a savings plan. How much interest would you earn on \$800 in one year?

44. A bank pays $12\frac{1}{4}\%$ yearly interest on a savings plan. How much interest would you earn on \$1200 in one year?

45. Joe earns \$1500 per month. If he spends 20.5% of his monthly income on food, how much money does he spend on food?

46. Moe earns \$800 per month. If he plans to spend 26.5% of his monthly income on rent, how much can he afford to pay for rent?

47. Find $12\frac{1}{8}\%$ of 450.

48. Find $10\frac{5}{8}\%$ of 650.

49. John wants to buy a \$550 stereo. If he receives a 30% discount, how much will the stereo cost?

50. Ed wants to buy a \$350 couch. If he receives a 15% discount, how much will the couch cost?

51. Sue made 70% of the 80 free throws she attempted. How many free throws did she miss?

52. Ann made 55% of the field goals she attempted. How many field goals did she miss?

53. Find $5\frac{1}{3}\%$ of $7\frac{7}{24}$.

54. Find $8\frac{2}{3}\%$ of $2\frac{22}{39}$.

Answers to Practice Problems **7a.** 2 **b.** 93.015 **c.** 1.84457 **d.** $\frac{183}{4}$ or 45.75 **8a.** 12 **b.** \$65.10 **c.** 33

Summary

Important Terms

4.1

percent
percent symbol, %

Important Skills

4.1

Expressing a percent as a fraction
Expressing a percent as a decimal

4.2

Expressing a decimal as a percent
Expressing a fraction as a percent

4.3

Finding a percent of a number
Solving word problems involving finding a percent of a number

Review Exercises

Complete the following table by filling in the equivalent fraction, decimal, or percent.

	Decimal	Fraction	Percent
1.	0.5		
2.		$\frac{1}{5}$	
3.			25%
4.	0.61		
5.		$\frac{1}{4}$	
6.			80%
7.	0.125		
8.		$\frac{1}{3}$	

	Decimal	Fraction	Percent
9.			$66\frac{2}{3}\%$
10.	0.6		
11.		$\frac{3}{4}$	
12.			37.5%
13.	5		
14.			$\frac{3}{5}\%$
15.		$\frac{7}{500}$	

Solve.

16. Jesse inherited 55% of his father's four-fifths interest in a business. What fractional part of the business does he own?

17. A questionnaire showed that $8\frac{3}{4}\%$ of the 2000 students at a college are over the age of 50. How many students are over 50 years old?

18. A \$435 stereo is marked up 15%. Find the selling price.

19. A real estate broker earned a 7% commission on the sale of a \$150,000 house. How much did the owner collect?

20. The author of a textbook receives a 15% royalty on each copy of her book that is sold. If 4,000 books are sold at \$13.50 per copy, how much money should the author receive?

21. Nancy answered $87\frac{1}{2}\%$ of the questions on a test correctly. What fractional part of the questions did she answer incorrectly?

EXPONENTS

OBJECTIVES

1. Find the value of a number in exponential form.
2. Multiply exponential expressions having the same base.
3. Divide exponential expressions having the same base.
4. Raise a power to a power.
5. Raise a product to a power.
6. Raise a fraction to a power.
7. Express a number in scientific notation.

PRETEST	EXPLANATION
1. Find the value of each exponential expression. **a.** 2^3 **b.** 2^{-3} **c.** $(-2)^3$	Section 5.1 Idea 2
2. Multiply. **a.** $7^{10} \cdot 7^{40}$ **b.** $x^{-18} \cdot x^{30}$	Section 5.2 Idea 4
3. Divide. **a.** $\frac{7^5}{7^{-8}}$ **b.** $\frac{x^{-9}}{x^9}$	Section 5.2 Idea 6
4. Simplify. **a.** $(x^2)^4$ **b.** $(7^{-3})^5$	Section 5.3 Idea 9
5. Simplify **a.** $(3 \cdot 6)^{11}$ **b.** $(-3x)^4$	Section 5.3 Idea 11
6. Simplify. **a.** $\left(\frac{2}{3}\right)^3$ **b.** $\left(\frac{x}{-5}\right)^3$	Section 5.3 Idea 13
7. Express in scientific notation. **a.** 68,000 **b.** 0.000068	Section 5.4 Idea 18

5.1 Computing the Value of a Number in Exponential Form

Idea 1 The number 3^4 (read as "three to the fourth power") is said to be written in **exponential form,** and 3^4 means that 3 is to be used as a factor four times ($3^4 = 3 \cdot 3 \cdot 3 \cdot 3$). The 3 is called the **base** and the 4 is called the **exponent.** In general, an exponent shows how many times the base is to be used as a factor. When a number does not have an exponent, assume that the exponent is 1. For example, $6 = 6^1$.

Example 1 Explain the meaning of each exponential expression. Tell which is the base and which is the exponent.

a. 3^7 **b.** 2^3 **c.** -4^2 d. $(-4)^2$

Solution To explain the meaning of an exponential expression, remember that an exponent tells you how many times the base is to be used as a factor.

1a. 3^7 means $3 \cdot 3 \cdot 3 \cdot 3 \cdot 3 \cdot 3 \cdot 3$. The base is 3 and the exponent is 7.

1b. 2^3 means $2 \cdot 2 \cdot 2$. The base is 2 and the exponent is 3. Read 2^3 as "two to the third power" or "two cubed."

1c. -4^2 means $-(4 \cdot 4)$. The base is 4 and the exponent is 2. Read -4^2 as "the negative of four to the second power" or "the negative of four squared."

1d. $(-4)^2$ means $(-4)(-4)$. The base is -4 and the exponent is 2. Read $(-4)^2$ as "negative four to the second power" or "negative four squared."

NOTE: The base of an exponent is the number immediately to the left and below the exponent. That is,

-4^2 The base is the number 4.
$(-4)^2$ The base is the number -4, which is inside the parentheses.

Practice Problem 1 ***Explain the meaning of each expression. Identify the base and the exponent.***

a. 7^5 **b.** 3^2 **c.** -2^6 **d.** $(-2)^6$

Example 2 Write each product in exponential form.

a. $3 \cdot 3$ **b.** $(-5)(-5)(-5)$ **c.** $7 \cdot 7 \cdot 7 \cdot 7 \cdot 7 \cdot 7$

Solution To write a product in exponential form, remember that an exponent indicates the number of times the base is used as a factor.

2a. $3 \cdot 3 = 3^2$ **2b.** $(-5)(-5)(-5) = (-5)^3$ **2c.** $7 \cdot 7 \cdot 7 \cdot 7 \cdot 7 \cdot 7 = 7^6$

Practice Problem 2 ***Write each product in exponential form.***

a. $8 \cdot 8 \cdot 8 \cdot 8$ **b.** $(-2)(-2)(-2)$ **c.** $10 \cdot 10 \cdot 10 \cdot 10 \cdot 10$

2 Sometimes in mathematics you will need to find the value of a number written in exponential form. If the exponent is an integer, it will be helpful to know that:

1. $a^n = a \cdot a \cdot a \ldots \cdot a$, n factors of a

2. $a^0 = 1$

3. $a^{-n} = \dfrac{1}{a^n}$

where a is any number (not equal to zero) and n is any positive integer.

NOTE: A letter that is used to represent any number or numbers is called a **variable.** For example, a and n are variables.

Example 3 Find the value of each expression.

a. 5^2 b. 4^3 c. $(-3)^4$ d. -3^4 e. $(0.2)^4$

Solution To find the value of an exponential expression, apply the definition of a^n and then multiply.

3a. $5^2 = 5 \cdot 5 = 25$

3b. $4^3 = 4 \cdot 4 \cdot 4 = 64$

3c. $(-3)^4 = (-3)(-3)(-3)(-3) = 81$

3d. $-3^4 = -(3 \cdot 3 \cdot 3 \cdot 3) = -81$

3e. $(0.2)^4 = (0.2)(0.2)(0.2)(0.2) = 0.0016$

Practice Problem 3 ***Find the value of each expression.***

a. 6^2 b. 2^5 c. -5^3 d. $(-5)^3$

Example 4 Find the value of each expression.

a. 3^0 b. $(-5)^0$ c. -5^0 d. $\left(\frac{2}{3}\right)^0$

Solution To raise any number (except zero) to the zero power, remember that $a^0 = 1$.

4a. $3^0 = 1$ **4b.** $(-5)^0 = 1$ **4c.** $-5^0 = -(5^0) = -1$ **4d.** $\left(\frac{2}{3}\right)^0 = 1$

Example 5 Find the value of each expression.

a. 5^{-2} b. 6^{-3} c. $(-3)^{-4}$ d. -3^{-4} e. $(-4)^{-3}$

Solution To raise any number (except zero) to a negative power, remember that $a^{-n} = \frac{1}{a^n}$.

5a. $5^{-2} = \frac{1}{5^2} = \frac{1}{25}$

5b. $6^{-3} = \frac{1}{6^3} = \frac{1}{216}$

5c. $(-3)^{-4} = \frac{1}{(-3)^4} = \frac{1}{81}$

5d. $-3^{-4} = -\frac{1}{3^4} = -\frac{1}{81}$

5e. $(-4)^{-3} = \frac{1}{(-4)^3} = \frac{1}{-64} = -\frac{1}{64}$

Practice Problem 4 ***Find the value of each expression.***

a. 7^0 b. 3^{-3} c. $(-8)^0$ d. $(-2)^{-5}$ e. -2^{-5}

5.1 Exercises

Explain the meaning of each expression.

1. 3^5 **2.** 13^2 **3.** 5^7 **4.** 6^4 **5.** 7^6

Write each product in exponential notation.

6. $3 \cdot 3 \cdot 3 \cdot 3$ **7.** $9 \cdot 9$ **8.** $2 \cdot 2 \cdot 2$

9. $6 \cdot 6 \cdot 6 \cdot 6 \cdot 6$ **10.** $7 \cdot 7 \cdot 7 \cdot 7 \cdot 7$ **11.** $(.5)(.5)$

12. $(-13)(-13)$ **13.** $4 \cdot 4 \cdot 4 \cdot 4 \cdot 4 \cdot 4$

Find the value of each exponential expression.

14. 2^3 **15.** 3^2 **16.** 5^3 **17.** 6^2 **18.** 4^3

19. 7^{-2} **20.** 9^{-3} **21.** 10^{-3} **22.** 4^{-3} **23.** 2^{-4}

24. 5^0 **25.** 9^0 **26.** $(-6)^0$ **27.** -6^0 **28.** 35^0

29. $(0.3)^2$ **30.** $(0.2)^3$ **31.** $(1.1)^3$ **32.** $(-0.5)^2$ **33.** $(-3)^4$

34. $(-5)^3$ **35.** -3^4 **36.** 3^{-4} **37.** $(-6)^{-2}$ **38.** -6^2

39. $(-7)^{-3}$ **40.** -19^0 **41.** -2^{-5} **42.** 2^{-5} **43.** $(0.01)^5$

44. $(-0.4)^3$

Fill in the blanks.

45. An exponent indicates how many times the ______________ is to be used as a factor.

46. Any number (except zero) raised to the zero power is equal to ______________.

47. If you raise a number (except zero) to a negative power, the result is a ______________ in which the numerator is ______________ and the denominator is that number raised to the corresponding ______________ power.

48. A letter that can be used to represent any number is called a ______________.

Answers to Practice Problems **1a.** $7 \cdot 7 \cdot 7 \cdot 7 \cdot 7$, base = 7, exponent = 5 **b.** $3 \cdot 3$, base = 3, exponent = 2 **c.** $-(2 \cdot 2 \cdot 2 \cdot 2 \cdot 2 \cdot 2)$, base = 2, exponent = 6 **d.** $(-2)(-2)(-2)(-2)(-2)(-2)$, base = -2, exponent = 6 **2a.** 8^4 **b.** $(-2)^3$ **c.** 10^5 **3a.** 36 **b.** 32 **c.** -125 **d.** -125 **4a.** 1 **b.** $\frac{1}{27}$ **c.** 1 **d.** $-\frac{1}{32}$ **e.** $-\frac{1}{32}$

5.2 Computing Products and Quotients

3 We can use the definition of a^n, where n is any positive integer, to develop a rule for multiplying exponential expressions.

Example 6

$$\begin{aligned} 5^2 \cdot 5^4 &= (5 \cdot 5)(5 \cdot 5 \cdot 5 \cdot 5) \\ &= 5 \cdot 5 \cdot 5 \cdot 5 \cdot 5 \cdot 5 \\ &= 5^6 \end{aligned}$$

4 A careful analysis of Example 6 suggests a rule for multiplying two or more exponential expressions having the same base.

Rule for Multiplying Exponential Expressions Having the Same Base:

1. The base remains the same.
2. Add the exponents.

This rule is sometimes written as

$$a^m \cdot a^n = a^{m+n},$$

where m and n are integers.

Example 7 Multiply and express each answer with a positive exponent.

a. $5^2 \cdot 5^8$ **b.** $2^{-6} \cdot 2^{16}$ **c.** $x^{-5} \cdot x^{-8}$ **d.** $7 \cdot 7^3 \cdot 7^9$

Solution To multiply exponential expressions having the same base, use the rule $a^m \cdot a^n = a^{m+n}$.

7a. $5^2 \cdot 5^8 = 5^{2+8}$
$= 5^{10}$

7b. $2^{-6} \cdot 2^{16} = 2^{-6+16}$
$= 2^{10}$

7c. $x^{-5} \cdot x^{-8} = x^{-5+(-8)}$
$= x^{-13}$
$= \dfrac{1}{x^{13}}$

7d. $7 \cdot 7^3 \cdot 7^9 = 7^{1+3+9}$
$= 7^{13}$

Note: $7 = 7^1$

Practice Problem 5 ***Multiply and express each answer with a positive exponent.***

a. $3^2 \cdot 3^5$ **b.** $x^{-5} \cdot x^{-66}$ **c.** $7^{-15} \cdot 7^{10}$

d. $b^{-6} \cdot b^{10}$ **e.** $y^5 \cdot y^9 \cdot y^{15} \cdot y$

5 We can develop the rule for dividing exponential expressions in much the same way that we developed the rule for multiplying them.

Example 8

$$\frac{5^6}{5^2} = \frac{\cancel{5} \cdot \cancel{5} \cdot 5 \cdot 5 \cdot 5 \cdot 5}{\cancel{5} \cdot \cancel{5}}$$
$$= 5 \cdot 5 \cdot 5 \cdot 5$$
$$= 5^4$$

6 Careful analysis of Example 8 suggests a rule for dividing exponential expressions having the same base.

Rule for Dividing Exponential Expressions Having the Same Base:

1. The base remains the same.
2. Subtract the exponents.

This rule is also written as

$$\frac{a^m}{a^n} = a^{m-n}$$

where m and n are integers and a is not equal to zero.

Example 9 Divide and express each answer with a positive exponent.

a. $\frac{7^8}{7^2}$ $\quad$ **b.** $\frac{7^2}{7^8}$ $\quad$ **c.** $\frac{x^{13}}{x^{-2}}$ $\quad$ **d.** $\frac{b^{-19}}{b}$

Solution To divide using exponential expressions having the same base, use the rule

$$\frac{a^m}{a^n} = a^{m-n}.$$

9a. $\frac{7^8}{7^2} = 7^{8-2} = 7^6$

9b. $\frac{7^2}{7^8} = 7^{2-8} = 7^{-6} = \frac{1}{7^6}$

9c. $\frac{x^{13}}{x^{-2}} = x^{13-(-2)} = x^{15}$

9d. $\frac{b^{-19}}{b} = b^{-19-1} = b^{-20} = \frac{1}{b^{20}}$

Practice Problem 6 ***Divide and express each answer with a positive exponent.***

a. $\frac{3^{11}}{3^3}$ $\quad$ **b.** $\frac{5}{5}$ $\quad$ **c.** $\frac{x^{-13}}{x^{-21}}$ $\quad$ **d.** $\frac{a^{-7}}{a}$

7 Sometimes you may need to find the product or quotient of exponential expressions having *different* bases.

To find the product or quotient of exponential expressions having different bases, find the value of each expression and then simplify.

Example 10 Find the value of each expression.

a. $\frac{3^3}{4^2}$ $\quad$ **b.** $-3^2 \cdot 2^5$

Solution **10a.** $\frac{3^3}{4^2} = \frac{3 \cdot 3 \cdot 3}{4 \cdot 4} = \frac{27}{16}$

10b. $-3^2 \cdot 2^5 = -9 \cdot 32 = -288$

NOTE: The method used to solve the problems in Example 10 is also useful in finding the sum or difference of exponential expressions. For example, $(-2)^3 + (-2)^4 = -8 + 16 = 8$.

Practice Problem 7 ***Find the value of each expression.***

a. $5^2 - 5^3$ $\quad$ **b.** $-2^3 \cdot 3^2$ $\quad$ **c.** $3^2 + 2^3$

5.2 Exercises

Multiply. Express each answer with a positive exponent.

1. $3^2 \cdot 3^9$
2. $5^6 \cdot 5^8$
3. $8^7 \cdot 8^8$
4. $3^5 \cdot 3^8$
5. $4^3 \cdot 4^{-10}$
6. $7^8 \cdot 7^{-2}$
7. $2^{-5} \cdot 2^{-6}$
8. $9 \cdot 9^{-5}$
9. $n^{10} \cdot n^{-10}$
10. $y^{-1} \cdot y^{-8}$
11. $m \cdot m^6$
12. $p^0 \cdot p^9$
13. $6^{-10} \cdot 6^3$
14. $2^2 \cdot 2^5 \cdot 2^3$
15. $x^4 \cdot x^5 \cdot x^{10}$

Divide. Express each answer with a positive exponent.

16. $\frac{5^{15}}{5^6}$
17. $\frac{5^6}{5^{15}}$
18. $\frac{7^{18}}{7^2}$
19. $\frac{3^5}{3^{11}}$
20. $\frac{x^{11}}{x^2}$
21. $\frac{3^{-3}}{3^5}$
22. $\frac{2^{-8}}{2}$
23. $\frac{a^2}{a^{-19}}$
24. $\frac{n^{-8}}{n^{-2}}$
25. $\frac{m^{-51}}{m^{-10}}$
26. $\frac{y^{10}}{y^{35}}$
27. $\frac{x^{-14}}{x^2}$
28. $\frac{t^5}{t^5}$
29. $\frac{7^9}{7^{15}}$
30. $\frac{x^{-9}}{x^{-10}}$
31. $\frac{n^{-5}}{n^{-7}}$
32. $\frac{n^{-8}}{n}$
33. $\frac{x^{-5}}{x^{-5}}$

Simplify.

34. $\frac{x^{-8}}{x^{-8}}$
35. $2^5 \cdot 2^{-8}$
36. $5^6 \cdot 5^{-9}$
37. $\frac{2^{-3}}{2^{-8}}$
38. $\frac{3^{-5}}{3^{-15}}$
39. $\frac{x^8}{x^{15}}$
40. $\frac{n^{15}}{n^{21}}$
41. $n^{-8} \cdot n^{-12}$
42. $x^{-6} \cdot x^{-10}$
43. $\frac{7^5}{7^{-10}}$
44. $\frac{5^7}{5^{-18}}$
45. $\frac{x^{-8}}{x^{-12}}$
46. $\frac{b^{-3}}{b^{-9}}$
47. $x^{-6} \cdot x^{10}$
48. $x^{-7} \cdot x^{15}$
49. $\frac{x^{-65}}{x^{-10}}$
50. $\frac{n^{-75}}{n^{-3}}$

Fill in the blanks.

51. To multiply exponential expressions having the same base, the ____________ remains the same and we ____________ the exponents.

52. To divide using exponential expressions having the same base, the ____________ remains the same and we ____________ the exponents. Always write the exponent in the ____________ minus the exponent in the ____________.

53. To find the product or quotient of exponential expressions having different bases, compute the value of each ____________ and then do the indicated ____________.

Find the value of each expression.

54. $3^2 \cdot 2^3$ **55.** $7^0 \cdot 7^2$ **56.** $3^2 \cdot 2^4$

57. $2^{-3} + 2^{-3}$ **58.** $\frac{3^4}{5^3}$ **59.** $-8^0 + 8^2$

60. $\frac{-3^4}{(-2)^6}$ **61.** $3^{-2} \cdot 6^2$ **62.** $-3^2 + (-3)^2$

Answers to Practice Problems **5a.** 3^7 **b.** $\frac{1}{x^{71}}$ **c.** $\frac{1}{7^5}$ **d.** b^4 **e.** y^{30} **6a.** 3^8 **b.** 1 **c.** x^8 **d.** $\frac{1}{a^8}$ **7a.** -100 **b.** -72 **c.** 17

5.3 Raising Products, Quotients, and Powers to Higher Powers

8 The next rule we will develop involves raising an exponential expression to a power. This is commonly referred to as **raising a power to a power.**

Example 11

$$\begin{aligned}(2^3)^2 &= 2^3 \cdot 2^3\\ &= 2^{3+3}\\ &= 2^6\end{aligned}$$

9 Example 11 suggests a rule for raising a power to a power.

Rule for Raising a Power to a Power:

1. The base remains the same.
2. Multiply the exponents.

The rule can be written as

$$(a^m)^n = a^{m \cdot n},$$

where m and n are integers.

Example 12 Simplify and express each answer with a positive exponent.

a. $(5^2)^9$ **b.** $(3^2)^{-8}$ **c.** $(x^{-5})^{-3}$ **d.** $(x^{-2})^3$

Solution To raise a power to a power, use the rule $(a^m) = a^{m \cdot n}$.

12a. $$\begin{aligned}(5^2)^9 &= 5^{2 \cdot 9}\\ &= 5^{18}\end{aligned}$$

12b. $$\begin{aligned}(3^2)^{-8} &= 3^{2(-8)}\\ &= 3^{-16}\\ &= \frac{1}{3^{16}}\end{aligned}$$

12c. $$\begin{aligned}(x^{-5})^{-3} &= x^{(-5)(-3)}\\ &= x^{15}\end{aligned}$$

12d. $$\begin{aligned}(x^{-2})^3 &= x^{-2 \cdot 3}\\ &= x^{-6}\\ &= \frac{1}{x^6}\end{aligned}$$

Practice Problem 8 ***Simplify and express each answer with a positive exponent.***

a. $(x^{11})^2$ **b.** $(8^5)^{-2}$ **c.** $(x^{-7})^5$ **d.** $(13^{-2})^{-7}$

10 Let us now develop a rule for raising a product to a power.

Example 13

$$\begin{aligned}(2 \cdot 5)^3 &= (2 \cdot 5)(2 \cdot 5)(2 \cdot 5)\\ &= 2 \cdot 5 \cdot 2 \cdot 5 \cdot 2 \cdot 5\\ &= 2 \cdot 2 \cdot 2 \cdot 5 \cdot 5 \cdot 5\\ &= 2^3 \cdot 5^3\end{aligned}$$

11 Example 13 suggests that to raise a product to a power, raise each factor to that power. This rule can be written as

$(a \cdot b)^n = a^n \cdot b^n$, where n is any integer.

Example 14 Simplify.

a. $(3 \cdot 8)^9$ **b.** $(4x)^2$ **c.** $(-3xyz)^3$

Solution To raise a product to a power, use the formula $(a \cdot b)^n = a^n \cdot b^n$. Also, if variables (or a number and a variable) appear next to each other, the operation indicated is multiplication. For example, $4x = 4 \cdot x$.

14a. $(3 \cdot 8)^9 = 3^9 \cdot 8^9$

14b. $$\begin{aligned}(4x)^2 &= 4^2x^2\\ &= 16x^2\end{aligned}$$

14c. $$\begin{aligned}(-3xyz)^3 &= (-3)^3x^3y^3z^3\\ &= -27x^3y^3z^3\end{aligned}$$

Practice Problem 9 ***Simplify.***

a. $(3 \cdot 7)^{15}$ **b.** $(-4x)^3$ **c.** $(3xy)^4$

12 The last rule for exponential expressions that we will develop involves raising a fraction to a power.

Example 15

$$\begin{aligned}\left(\frac{3}{4}\right)^3 &= \frac{3}{4} \cdot \frac{3}{4} \cdot \frac{3}{4}\\ &= \frac{3 \cdot 3 \cdot 3}{4 \cdot 4 \cdot 4}\\ &= \frac{3^3}{4^3}\end{aligned}$$

13 Example 15 suggests that to raise a fraction to a power, we should raise both the numerator and the denominator to that power. That is,

$\left(\frac{a}{b}\right)^n = \frac{a^n}{b^n}$, where n is any integer and b is not zero.

Example 16 Simplify.

a. $\left(\frac{2}{3}\right)^3$ **b.** $\left(\frac{x}{5}\right)^9$ **c.** $\left(\frac{-3}{x}\right)^3$

Solution To raise a fraction to a power, use the formula $\left(\frac{a}{b}\right)^n = \frac{a^n}{b^n}$.

16a. $\left(\frac{2}{3}\right)^3 = \frac{2^3}{3^3} = \frac{8}{27}$

16b. $\left(\frac{x}{5}\right)^9 = \frac{x^9}{5^9}$

16c. $\left(\frac{-3}{x}\right)^3 = \frac{(-3)^3}{x^3} = \frac{-27}{x^3} = -\frac{27}{x^3}$

Practice Problem 10 ***Simplify.***

a. $\left(\frac{3}{5}\right)^2$ **b.** $\left(\frac{1}{n}\right)^{35}$ **c.** $\left(\frac{m}{-5}\right)^3$

14 To simplify some expressions, we will need to use more than one rule for exponential expressions.

Example 17 Simplify.

a. $(-3x^3)^2$ **b.** $\frac{5^{10} \cdot 5^5 \cdot 5^9}{5^{25}}$

Solution

		Rule
17a.	$(-3x^3)^2 = (-3)^2(x^3)^2$	$(a \cdot b)^n = a^n \cdot b^n$
	$= 9x^6$	$(a^n)^m = a^{n \cdot m}$
17b.	$\frac{5^{10} \cdot 5^5 \cdot 5^9}{5^{25}} = \frac{5^{24}}{5^{25}}$	$a^n \cdot a^m = a^{n+m}$
	$= 5^{-1}$	$\frac{a^m}{a^n} = a^{m-n}$
	$= \frac{1}{5}$	$a^{-n} = \frac{1}{a^n}$

Practice Problem 11 ***Simplify.***

a. $\left(\frac{2}{5}\right)^{-3}$ **b.** $(-4x^2y^5)^3$ **c.** $\left(\frac{x^2}{7}\right)^2$

5.3 Exercises

Simplify. The final answer should not contain negative exponents.

1. $(2^3)^9$ **2.** $(x^5)^{12}$ **3.** $(y^4)^2$ **4.** $(7^{15})^2$

5. $(3^{-2})^9$ **6.** $(x^{-2})^{-4}$ **7.** $(5^2)^{-8}$ **8.** $(n^5)^{-1}$

9. $(x^3)^{-9}$ **10.** $(n^a)^b$ **11.** $(x^{-a})^{-b}$ **12.** $(9^{11})^{10}$

13. $(x^2)^{-6}$ **14.** $(x)^{-9}$ **15.** $(x^{-2})^{-3}$ **16.** $(x^{-6})^{-8}$

17. $(2^5)^{16}$ **18.** $(3^8)^{15}$ **19.** $(2^{-2})^{-2}$ **20.** $(3^{-1})^{-2}$

Simplify.

21. $(3 \cdot 5)^7$ **22.** $(4x)^3$ **23.** $(5n)^{10}$ **24.** $(2 \cdot 7)^8$

25. $(-3p)^3$ **26.** $(-2y)^4$ **27.** $(4xy)^2$ **28.** $(13xyz)^2$

29. $(abcd)^8$
30. $(-5xp)^{10}$
31. $(-7x)^2$
32. $(xyz)^n$
33. $\left(\frac{2}{3}\right)^4$
34. $\left(\frac{x}{5}\right)^2$
35. $\left(\frac{3}{n}\right)^4$
36. $\left(\frac{a}{b}\right)^3$
37. $\left(\frac{-x}{2}\right)^3$
38. $\left(\frac{n}{-7}\right)^2$
39. $\left(\frac{b}{5}\right)^8$
40. $\left(\frac{n}{-3}\right)^4$
41. $\left(\frac{x}{y}\right)^n$
42. $\left(\frac{1}{x}\right)^5$
43. $\left(\frac{n}{-5}\right)^3$
44. $\left(\frac{x}{a}\right)^{15}$
45. $(2x)^5$
46. $(3x)^4$
47. $(-2x)^3$
48. $(-5n)^3$
49. $(xyz)^9$
50. $(xyz)^{11}$
51. $(7x)^2$
52. $(-9x)^2$
53. $\left(\frac{1}{6}\right)^2$
54. $\left(\frac{1}{8}\right)^2$
55. $\left(\frac{-n}{3}\right)^4$
56. $\left(\frac{-x}{2}\right)^4$
57. $\left(\frac{x}{-7}\right)^3$
58. $\left(\frac{n}{-8}\right)^3$
59. $\left(\frac{1}{5}\right)^{-2}$
60. $\left(\frac{1}{10}\right)^{-2}$

Fill in the blanks.

61. To raise a product to a power, raise each ____________ to that ____________.
62. To raise a power to a power, the ____________ remains the same and we ____________ the exponents.
63. To raise a fraction to a power, raise the ____________ and the ____________ to that ____________.

Simplify. The final answer should not contain negative exponents.

64. $(3x^2)^3$
65. $(-2x^2y^3)^4$
66. $(5x^{-2})^3$
67. $(3^a)^{-b}$
68. $\left(\frac{2}{3}\right)^{-3}$
69. $\frac{(m^2)^4}{(m^3)^5}$
70. $\left(\frac{x^2}{-2}\right)^5$
71. $(a^2b^{-4})^n$
72. $\frac{a^5 \cdot a^{19} \cdot a^{15}}{(a^{12})^3}$

Answers to Practice Problems 8a. x^{22} b. $\frac{1}{8^{10}}$ c. $\frac{1}{x^{35}}$ d. 13^{14} 9a. $3^{15} \cdot 7^{15}$ b. $-64x^3$ c. $81x^4y^4$ 10a. $\frac{9}{25}$ b. $\frac{1}{n^{35}}$ c. $-\frac{m^3}{125}$ 11a. $\frac{125}{8}$ b. $-64x^6y^{15}$ c. $\frac{x^4}{49}$

5.4 Scientific Notation

15 Numbers occurring in science are sometimes very large or very small. Because computations with these types of numbers can be difficult, scientists find it very useful to express these numbers in **scientific notation,** which involves writing a number using powers of ten. Before considering how to write numbers in scientific notation, it will be helpful to discuss how to multiply a number by a power of 10.

Example 18

a. $3 \cdot 10^3 = 3 \cdot 1000 = 3000$

b. $3 \cdot 10^2 = 3 \cdot 100 = 300$

c. $3 \cdot 10^1 = 3 \cdot 10 = 30$

d. $3 \cdot 10^0 = 3 \cdot 1 = 3$

e. $3 \cdot 10^{-1} = 3 \cdot \frac{1}{10} = \frac{3}{10} = 0.3$

f. $3 \cdot 10^{-2} = 3 \cdot \frac{1}{100} = \frac{3}{100} = 0.03$

g. $3 \cdot 10^{-3} = 3 \cdot \frac{1}{1000} = \frac{3}{1000} = 0.003$

16 Example 18 implies that if you multiply a decimal by:

1. 10^n, where n is a positive integer, the decimal point moves n places to the right.
2. 10^{-n}, where n is a positive integer, the decimal point moves n places to the left.

Example 19 Find each product mentally.

a. 6.37×10^4 **b.** 0.0035×10^5

c. 5.8×10^{-3} **d.** 6.37×10^{-4}

Solution If n is any positive integer, multiplying a number by 10^n moves the decimal point n places to the right. Multiplying a number by 10^{-n} moves the decimal point n places to the left.

19a. $6.37 \times 10^4 = 63700$ **19b.** $0.0035 \times 10^5 = 350$

19c. $5.8 \times 10^{-3} = 0.0058$ **19d.** $6.37 \times 10^{-4} = 0.000637$

Practice Problem 12 ***Find each product mentally.***

a. 5.01×10^3 **b.** 6.345×10^{-2}

c. 7.385×10^6 **d.** 0.365×10^{-5}

17 The products in Examples 19 a, c, and d are written in scientific notation. In other words, a number is in scientific notation when it is written as a number that is at least 1 but less than 10 (one nonzero digit is to the right of the decimal point) times a power of 10.

NOTE: Until now, we have used "·" to mean multiplication. In scientific notation, however, products are expressed using the "×" symbol.

Example 20

Decimal notation		Scientific notation
365	=	3.65×10^2
0.0365	=	3.65×10^{-2}
485,000,000	=	4.85×10^8
0.00000485	=	4.85×10^{-6}

18 By analyzing Example 20, we can state a rule for expressing a decimal in scientific notation.

Rule for Expressing a Decimal in Scientific Notation:

1a. If the number is 10 or more, it is written as

$$M \times 10^n$$

where M is at least 1 but less than 10 and n is the number of places the decimal point was moved to make M at least 1 but less than 10.

1b. If the number is less than 1, it is written as

$$M \times 10^{-n}$$

where M is at least 1 but less than 10 and n is the number of places the decimal point was moved to make M at least 1 but less than 10.

2. Check to be sure that $M \times 10^n$ or $M \times 10^{-n}$ is equal to the original number.

Example 21 Express each number in scientific notation.

a. 855 **b.** 0.0095

c. 775,000 **d.** 0.00003801

Solution To write a decimal in scientific notation, move the decimal point until there is only one nonzero digit to the left of the decimal point. Next, multiply this new number M times the appropriate power of 10.

21a. $855 = 8.55 \times 10^n = 8.55 \times 10^2$

21b. $0.0095 = 9.5 \times 10^{-n} = 9.5 \times 10^{-3}$

21c. $775{,}000 = 7.75 \times 10^n = 7.75 \times 10^5$

21d. $0.00003801 = 3.801 \times 10^{-n} = 3.801 \times 10^{-5}$

Practice Problem 13 ***Express each number in scientific notation.***

a. 363 **b.** 0.0363 **c.** 17,600,000 **d.** 0.00000329

19 Scientific notation is useful in multiplying or dividing very large or small numbers.

Example 22 Do the indicated operation.

a. (630,000)(0.000003) **b.** $\dfrac{36{,}000{,}000}{0.0009}$

Solution To multiply (or to divide) very large or small numbers, first express the numbers in scientific notation. Then multiply (divide) the decimals and multiply (divide) the exponential expressions. Express the answer without exponents.

22a.
$$\begin{aligned}&(630{,}000)(0.000003)\\ &= (6.3 \times 10^5)(3 \times 10^{-6})\\ &= (6.3 \times 3)(10^5 \times 10^{-6})\\ &= 18.9 \times 10^{-1}\\ &= 1.89\end{aligned}$$

22b.
$$\begin{aligned}&\frac{36{,}000{,}000}{0.0009}\\ &= \frac{3.6 \times 10^7}{9 \times 10^{-4}}\\ &= \frac{3.6}{9} \times \frac{10^7}{10^{-4}}\\ &= 0.4 \times 10^{11}\\ &= 40{,}000{,}000{,}000\end{aligned}$$

Practice Problem 14 **Do the indicated operation.**

a. (75,000,000)(0.0005) **b.** $\dfrac{666{,}600{,}000}{0.011}$

5.4 Exercises

Find each product mentally.

1. 3.72×10^2
2. 3.72×10^{-2}
3. 8.57×10^{-8}
4. 6.7×10^{11}
5. 8.75×10^{-1}
6. 3.25×10^7
7. 0.3×10^6
8. 3.07×10^{-5}
9. 7×10^{-8}
10. -7.35×10^2
11. -6.77×10^{-4}
12. 76.8×10^{10}
13. 8.1×10^4
14. 1.8×10^4
15. 5.35×10^{-4}
16. 7.5×10^{-5}
17. -3.12×10^6
18. -5.61×10^7
19. 8.1×10^{11}
20. 1.8×10^{13}

Express each number in scientific notation.

21. 275
22. 8571
23. 0.000275
24. 0.031
25. 31.8
26. 0.002017
27. 11,000
28. 0.000011
29. 750,000,000
30. 1750
31. 0.0005
32. 7,000,000,000
33. 0.00000619
34. 0.817
35. 378,000,000,000,000,000
36. 875,000,000
37. 625
38. 893
39. 0.0875
40. 0.0832
41. 35,000,000,000
42. 83,000,000
43. 0.000000051
44. 0.0000000032

Fill in the blanks.

45. If you multiply a decimal by 10^n, where n is a positive integer, the decimal point moves ________ places to the ________.

46. If you multiply a decimal by 10^{-n}, where n is a positive integer, the decimal point moves ________ places to the ________.

47. When we write a number greater than 10 in scientific notation, we do so in the form $M \times 10^n$. M is at least ________ but less than ________, and n is the number of places the ________ was moved to make M at least ________ but less than ________.

48. When we write a number less than 1 in scientific notation, we do so in the form $M \times 10^{-n}$. M is at least ________ but less than ________, and n is the number of places the ________ was moved to make M at least ________ but less than ________.

Perform the indicated operation.

49. (0.0044)(7,000)
50. (0.000055)(0.00081)
51. (32,000)(0.005)(200,000)
52. $\dfrac{900{,}000}{0.003}$
53. $\dfrac{0.000016}{40{,}000}$
54. $\dfrac{0.000091}{0.13}$
55. (32,000)(2,000)(30,000)
56. (24,000)(5,000)(1,000)
57. $\dfrac{9{,}000{,}000}{0.0003}$

58. $\dfrac{16{,}000{,}000}{0.004}$

59. $\dfrac{(35{,}000)(0.005)}{25{,}000{,}000}$

60. $\dfrac{(45{,}000{,}000)(0.00009)}{0.0405}$

61. $(-0.0000005)(0.00031)$

62. $(-0.0000061)(0.005)$

Solve.

63. The distance from the earth to the sun is 93,000,000 miles. Express this number in scientific notation.

64. The average diameter of the sun is about 865,000 miles. Express this number in scientific notation.

65. Under certain conditions, hydrogen gas weighs 0.00005190 ounces per cubic inch. Express this number in scientific notation.

66. A fly's wing weighs approximately 0.0000044 pounds. Express this number in scientific notation.

67. A cosmic year is approximately 225,000,000 years long. Express this number in scientific notation.

68. Express 6,590,000,000,000,000,000,000 tons—which is the approximate weight of the earth—in scientific notation.

Answers to Practice Problems **12a.** 5010 **b.** 0.06345 **c.** 7,385,000 **d.** 0.00000365 **13a.** 3.63×10^2 **b.** 3.63×10^{-2} **c.** 1.76×10^7 **d.** 3.29×10^{-6} **14a.** 37,500 **b.** 60,600,000,000

Summary

Important Terms

5.1

exponent
base
exponent
variable

5.4

scientific notation
decimal notation

Important Skills

5.1

Explaining the meaning of an exponential expression
Writing a product in exponential form
Finding the value of an exponential expression

5.2

Multiplying exponential expressions
Dividing one exponential expression by another
Finding sums and differences of exponential expressions

5.3

Raising a power to a power
Raising a product to a power
Raising a fraction to a power
Simplifying expressions involving exponential expressions

5.4

Multiplying by powers of ten
Writing numbers in scientific notation
Finding products and quotients by using scientific notation

Review Exercises

Find the value of each exponential expression.

1. 3^4 **2.** 3^{-4} **3.** $(-3)^4$ **4.** -3^4

Perform the indicated operations and simplify.

5. $(x^{15})(x^{-35})$

6. $\left(\frac{11}{-4}\right)^2$

7. $\frac{x^{-9}}{x^{-12}}$

8. $(x^3)^{10}$

9. $(-2p)^3$

10. $5^{18} \cdot 5^{35}$

11. $2^{-3} + 2^4$

12. $\frac{9}{9^{25}}$

13. $\left(\frac{3}{4}\right)^2$

14. $(2^{-3})^{-5}$

15. $(2x)^{15}$

16. $\frac{2^3}{3^2}$

17. $(3x^2)^4$

18. $\frac{x^5}{x^{-15}}$

19. $\frac{(5^2)^3(5^3)^{-5}}{(5^4)^5}$

Find each product mentally.

20. 3.47×10^8 **21.** 4.06×10^{-7} **22.** -2.47×10^3

Express each number in scientific notation.

23. 15,000,000 **24.** 0.000000312 **25.** 0.000325

Solve.

26. The Gross National Product (GNP) in 1972 for the United States was $1,155,200,000,000. Express the GNP for 1972 in scientific notation.

Perform the indicated operation. Express the answer without exponents.

27. $(6 \times 10^{53})(3.1 \times 10^{-50})$ **28.** $\frac{(82{,}000{,}000)(0.0003)}{0.02}$

6 POLYNOMIALS

OBJECTIVES

1. Identify the like terms of a polynomial.
2. Combine the like terms of a polynomial.
3. Evaluate a polynomial.
4. Add two polynomials.
5. Subtract one polynomial from another.
6. Multiply two or more monomials.
7. Multiply a monomial times a polynomial.
8. Multiply two polynomials.
9. Divide a monomial by a monomial.
10. Divide a polynomial by a monomial.
11. Divide a polynomial by a polynomial.

PRETEST	EXPLANATION
1. Identify the like terms. $-7x - 7x^2 - 8x + 2$	Section 6.1, Idea 5
2. Combine like terms.	Section 6.1
a. $-8x - 9x$	Idea 6
b. $-8x + x + 7$	Idea 6
3. Evaluate.	Section 6.2
a. $6x^2 - 5x$ when $x = -3$	Idea 8
b. $x^3 + 2x^2 + 2$ when $x = 0.2$	Idea 8
4. Add.	Section 6.3
a. $(5x - 6) + (x - 7)$	Ideas 9 and 10
b. $(3y + z - 6x) + (7x - 3y - z)$	Ideas 9 and 10
5. Subtract.	Section 6.4
a. $(4x - 7) - (3x - 8)$	Ideas 11 and 12
b. $(3.4x^2 - 0.7x - 0.3) - (1.5x + 7x^2 + 1)$	Ideas 11 and 12
Multiply.	Section 6.5
6. $(-3x^4)(-7x^5y)$	Idea 14
7. $3xy^2(5x^2y + 3xy - 2)$	Idea 15
8. $(2x - 3)(2x + 3)$	Ideas 16 and 17
Divide.	Section 6.6
9. $\frac{-77x^5y}{11x^2z}$	Idea 19
10. $\frac{10x^6 - 12x^4 + 8x}{2x^3}$	Idea 21
	Section 6.7
11. $(12x^8 + x^4 - 6) \div (3 + 4x^4)$	Idea 23

6.1 Introduction

Idea **1** Expressions like $2x$, $-5xy$, n^3, and 3 are called **monomials.** That is, a monomial is a constant (number), a variable, or the product of a constant and one or more variables. In a monomial, only positive integers or zero may appear as exponents, and a variable must not appear in the denominator of the expression.

2 The sum of two or more monomials is called a **polynomial.** A polynomial can also be just one monomial.

Example 1

a. $3x^2 + 7x - 9$ is a polynomial in x

b. $6x^2 - 10y^2$ is a polynomial in x and y

3 When a polynomial involves the operation of addition, the parts being added are called **terms.** The numerical factor (number part) of a term is called a **numerical coefficient.**

Example 2 List each term of the polynomial and identify its numerical coefficient.

a. $6x^2 - 7x + 9$ **b.** $5x^3 - 6x^2 - x + 8$

Solution To identify the terms and numerical coefficients in a polynomial, rewrite the polynomial using only the operation of addition.

2a. $6x^2 - 7x + 9 = 6x^2 + (-7x) + 9$

The terms are $6x^2$, $-7x$, and 9. The numerical coefficients are 6, -7, and 9, respectively.

2b. $5x^3 - 6x^2 - x + 8 = 5x^3 + (-6x^2) + (-x) + 8$

The terms are $5x^3$, $-6x^2$, $-x$, and 8. The numerical coefficients are 5, -6, -1, and 8, respectively. The numerical coefficient of $-x$ is -1 since $-x = -1 \cdot x$.

NOTE: We can also identify terms and numerical coefficients by remembering that if a "+" or "−" sign appears between two monomials, the "+" or "−" is the sign of the term and the operation is understood to be addition. For example, the terms of the polynomial $5x^2 - 3x$ (*think:* $5x^2$ plus $-3x$) are $5x^2$ and $-3x$.

Practice Problem 1 ***List the terms of each polynomial and identify their numerical coefficients.***

a. $16x^2 + 7x - 6$ **b.** $13p^3 - p^2 - 8p + 7$

c. $15x - 9$ **d.** $7x^2 - 9xy + y^2$

4 Polynomials can be classified by the number of terms they contain. That is, *monomials, binomials,* and *trinomials* are polynomials containing one, two, and three terms, respectively.

Example 3 Tell whether each polynomial is a monomial, binomial, or trinomial.

a. $6x^2 + 9x - 7$ **b.** $3x^2 - 5$ **c.** $3xy^4$

Solution To classify a polynomial, count the terms.

3a. Since $6x^2 + 9x - 7$ contains three terms, it is a trinomial.

3b. Since $3x^2 - 5$ contains two terms, it is a binomial.

3c. Since $3xy^4$ contains one term, it is a monomial.

Practice Problem 2 ***Tell whether each polynomial is a monomial, binomial, or trinomial.***

a. $-7x^3yz$ **b.** $7x^2 - 9x + 2$ **c.** $6x^2 - 6y^2$

5 Two or more terms of a polynomial that contain the same variable(s) with the same exponent(s) are called **like** or **similar terms.**

Example 4 List the like terms.

a. $6x^3 - 8x^2y - 7x^3 + 9x^2y$ **b.** $x^2 + 9 - 16x^2 - 8 + 6x^2$

c. $6x^2 + 6y^2$

Solution To identify the like terms of a polynomial, remember that like terms contain the same variable(s) raised to the same power.

4a. $6x^3 - 8x^2y - 7x^3 + 9x^2y = 6x^3 + (-8x^2y) + (-7x^3) + 9x^2y$

Like terms: $6x^3$, $-7x^3$ and $-8x^2y$, $9x^2y$

4b. $x^2 + 9 - 16x^2 - 8 + 6x^2 = x^2 + 9 + (-16x^2) + (-8) + 6x^2$

Like terms: x^2, $-16x^2$, $6x^2$ and 9, -8

4c. $6x^2 + 6y^2$

Like terms: none

NOTE: All constant terms are like terms. For example, 5 and 3 are like terms.

Practice Problem 3 ***List the like terms.***

a. $6x^3 - 7x^2 + 9x^3$ **b.** $8xy^2 + 8x^2y - 9xy^2$

c. $3y^2 - 6xy + 9x^2$ **d.** $5n^2 - 9m^2 + 8m^2 + 9$

6 Sometimes we can simplify a polynomial by combining or collecting like terms.

To Combine Like Terms:

1. Identify the like terms.
2. Find the sum of their numerical coefficients; then multiply this sum times the common variable factor(s) of the like terms.

NOTE: The rule for collecting like terms is an application of the **distributive law.** This law states that

$$a(b + c) = ab + ac \quad \text{and} \quad (b + c)a = ba + ca,$$

where a, b, and c are any numbers.

Example 5 Combine the like terms.

a. $6x + 3x^2 + 2x$ **b.** $-7x^2 + 9x^2 + 8x^2$

c. $xy - 8xy$ **d.** $-8n^3 - 15n^3$

e. $0.3x^3 + 0.21x^3$ **f.** $-\frac{1}{18}y^5 + \frac{7}{10}y^5$

Solution To combine like terms, find the sum of their numerical coefficients and then place the common variable factor(s) to the right of this sum.

5a. $6x + 3x^2 + 2x = (6 + 2)x + 3x^2$
$= 8x + 3x^2$

5b. $-7x^2 + 9x^2 + 8x^2 = (-7 + 9 + 8)x^2$
$= 10x^2$

5c. $xy - 8xy = 1xy - 8xy$ *Think:* $1xy$ plus $-8xy$
$= (1 - 8)xy$
$= -7xy$

5d. $-8n^3 - 15n^3 = (-8 - 15)n^3$
$= -23n^3$

5e. $0.3x^3 + 0.21x^3 = (0.3 + 0.21)x^3$
$= 0.51x^3$

5f. $-\frac{1}{18}y^5 + \frac{7}{10}y^5 = \left(-\frac{1}{18} + \frac{7}{10}\right)y^5$

$= \left(-\frac{5}{90} + \frac{63}{90}\right)y^5$ $18 = 2 \cdot 3 \cdot 3$
$10 = 2 \cdot 5$
$\text{LCD} = 2 \cdot 3 \cdot 3 \cdot 5 = 90$

$= \frac{58}{90}y^5$

$= \frac{29}{45}y^5$

NOTE: In Solutions 5c and 5d, the sum of the coefficients is written without a plus sign. That is, $1 + (-8)$ is written $1 - 8$ and $-8 + (-15)$ is written as $-8 - 15$. In general, if a and b are positive numbers,

$$a + (-b) = a - b \quad \text{and} \quad -a + (-b) = -a - b.$$

Practice Problem 4 ***Combine the like terms.***

a. $-6x^5 + 4x^5$ **b.** $6xy + 9x^2y - 9xy$

c. $\frac{5}{14}p^5 - \frac{3}{4}p - \frac{1}{10}p^5$ **d.** $-5.3x^3 - 3.32x^3$

7 To complete our introduction to polynomials, we must discuss the concept of degree. The **degree of a term** containing only one variable is indicated by the exponent of the variable. For example, the degree of the term $8x^3$ is 3. If the term contains more than one variable, the degree of the term is the sum of the exponents of the variables. For example, the degree of the term $-6x^2y$ is 3 since the exponents are 2 and 1, and $2 + 1 = 3$.

Example 6 List each term and identify its degree.

a. $6x^2 - 7x + 8$ **b.** $6xy^7 - 3x^4y^3z^3$

Solution **6a.** $6x^2 - 7x + 8$

Term	*Degree*	
$6x^2$	2	
$-7x$	1	$-7x = -7x^1$
8	0	$8 = 8x^0$

6b. $6xy^7 - 3x^4y^3z^3$

Term	*Degree*	
$6xy^7$	8	$1 + 7 = 8$
$-3x^4y^3z^3$	10	$4 + 3 + 3 = 10$

NOTE: The **degree of a polynomial** is defined as the degree of its highest term. Thus the degree of the polynomial $6x^2 - 7x + 8$ is 2. Zero (0) is not assigned a degree.

Practice Problem 5 ***List each term and identify its degree. Also, identify the degree of the polynomial.***

a. $7x^2 - 6xy + 8y^2$ **b.** $8p^4 - 6p^3 + 5p + 9$

6.1 Exercises

Tell whether each expression is a polynomial.

1. $6x^2 - 7x + 9$

2. $5x^3y - 6xy^{-2}$

3. $\dfrac{5}{x - 6}$

4. $-5x^2$

List each term of the polynomial and identify its numerical coefficient.

5. $7x^3 - 7x^2 + 8$

6. $8x^3 + 7xy^2 - x^2y + y^3$

7. $-3x^2 - 7x + 5$

8. $15p^5 - 8p^4 + 3p^3 - p + 2$

Tell whether each polynomial is a monomial, binomial, or trinomial.

9. $7x^3 - 7x^2 + 8$

10. $6x^2 - 9$

11. $3x^3y$

12. $7x^3y - 7xy^3$

13. $-6x^5$

14. $x^2 + xy + y^2$

List the like terms.

15. $7x^3 - 6x^2 + 3x^3$

16. $6y^2 - 6xy + 9y^2$

17. $3x^3 - 7x^2 + 5x^3 - x^2$

18. $6x^3 + 6x^2 + 6x$

19. $3xy^3 - 6xy^3 + xy^3$

20. $-5p^4 - 8p^3 + p^4 + 7p^3$

21. $7x^2 + 9 - 6x^2 + 5x - 2$

22. $3xy^3 - 4x^3y - 2xy^3$

23. $5x^3 + 3x^2 - x^3 + 3x + 5 + x^2 - 9$

Combine the like terms.

24. $5x + 3x$

25. $6y^2 - 3y^2$

26. $-5xy - 7xy$

27. $7x^5 - 9x^5$

28. $-9px + 2px$

29. $-7x + x$

30. $0.5n^2 + 1.71n^2$

31. $6x^3y + 6xy^3$

32. $\dfrac{3}{8}x^3 + \dfrac{5}{12}x^3$

33. $6x^3 + 7x + 3x^3$

34. $7n^2 - 18n^2 + 5n$

35. $-7xy + 8y + 9xy$

36. $-3t^4 - 7t^3 - 6t^4$

37. $6x^2 + 9x^2 + 3x^2$

38. $-7y + 9y + 18y$

39. $6xy - 9xy + xy$

40. $-3y^2 - 8y^2 - 2y^2$

41. $0.3x + 0.51x + x^2$

42. $0.3x^2 - 5.71x^2 + 8$

43. $0.5x^3 + 3x^2 - x^3$

44. $-\dfrac{7}{18}x^2 + \dfrac{1}{12}x^2 + \dfrac{3}{4}x$

45. $\dfrac{5}{24}x^2 - \dfrac{1}{18}x^2 + \dfrac{1}{3}x$

46. $-\dfrac{11}{36}y^2 + \dfrac{13}{36}y - \dfrac{1}{66}y^2$

47. $6xy - 9xy$

48. $9xy - 12xy$

49. $0.3x - 0.51x$

50. $0.6y + 0.61y$

51. $\dfrac{1}{6}x^2y - \dfrac{1}{8}x^2y$

52. $\dfrac{3}{8}xy^2 - \dfrac{3}{10}xy^2$

53. $-9x - 9y + 9x$

54. $-3n - 3n + 3n$

55. $\frac{3}{16}x^2 - \frac{1}{8}y^2 + \frac{5}{24}x^2$

56. $\frac{5}{18}n - \frac{1}{9}m + \frac{7}{36}n$

57. $3.5a - 3.58a$

58. $7.6ay - 7.68ay$

59. $-3.6b + 7.8b$

60. $-6.3b + 9.2b$

61. $\frac{1}{10}x^3 + \frac{3}{15}x^3 + \frac{1}{18}x^3$

62. $\frac{5}{12}x^4 + \frac{3}{18}x^4 + \frac{5}{24}x^4$

63. $0.3x + \frac{1}{5}x$

64. $0.1x + \frac{3}{5}x$

65. $x - \frac{3}{4}x$

66. $n - \frac{5}{6}n$

67. $-5.6y - \frac{1}{2}y$

68. $-3.2x - \frac{3}{10}x$

Determine the degree of each term and the degree of the polynomial.

69. $6x^3 - 7x^2 + x + 9$

70. $3x^3 - 7x^2y + 2x^2y^2$

Fill in the blanks.

71. A ______________ is a constant, a variable, or the product of a constant and one or more variables.

72. The sum of two or more monomials is called a ______________.

73. When a polynomial involves only the operation of addition, the parts being added are called ______________.

74. The numerical part of a term is called its ______________.

75. Two or more terms that contain the same variable(s) with the same exponent(s) are called ______________ or ______________ terms.

76. To combine like terms, find the sum of their ______________ and then place the common variable factor(s) to the right of this sum.

Answers to Practice Problems

1a.

Term	*Coefficient*
$16x^2$	16
$7x$	7
-6	-6

b.

Term	*Coefficient*
$13p^3$	13
$-p^2$	-1
$-8p$	-8
7	7

c.

Term	*Coefficient*
$15x$	15
-9	-9

d.

Term	*Coefficient*
$7x^2$	7
$-9xy$	-9
y^2	1

2a. monomial **b.** trinomial **c.** binomial **3a.** $6x^3$ and $9x^3$ **b.** $8xy^2$ and $-9xy^2$ **c.** none **d.** $-9m^2$ and $8m^2$ **4a.** $-2x^5$ **b.** $-3xy + 9x^2y$ **c.** $\frac{9}{35}p^5 - \frac{3}{4}p$ **d.** $-8.62x^3$

5a.

Term	*Degree*
$7x^2$	2
$-6xy$	2
$8y^2$	2

Polynomial's degree = 2

b.

Term	*Degree*
$8p^4$	4
$-6p^3$	3
$5p$	1
9	0

Polynomial's degree = 4

6.2 Evaluating Polynomials

8 Frequently in mathematics you must find the value of a polynomial for a specific value of a variable. This process is called **evaluating a polynomial** and you can do it by following these steps:

To Evaluate a Polynomial:

1. Replace the variable with its numerical value.
2. Find a value for all exponentials as you work from left to right.
3. Do all multiplication.
4. Do all addition and/or subtraction.

Example 7 Evaluate each polynomial for the given value.

a. $6x^2 - 5x, x = -2$ **b.** $-x^3 - 4x^2 + 3, x = -2$

c. $x^3 + 3x^2 + 3, x = 0.3$ **d.** $6x^3 - 5x^2 + 2, x = \frac{2}{3}$

Solution To evaluate a polynomial, replace the variable with its given value. Next, find a value for all exponentials. Finally, do all multiplication and then do all addition and/or subtraction.

7a.
$$\begin{aligned} 6x^2 - 5x &= 6(-2)^2 - 5(-2) \\ &= 6(4) - 5(-2) \\ &= 24 + 10 \qquad \textit{Think: } 6(4) + (-5)(-2) \\ &= 34 \end{aligned}$$

7b.
$$\begin{aligned} -x^3 - 4x^2 + 3 &= -(-2)^3 - 4(-2)^2 + 3 \\ &= -(-8) - 4(4) + 3 \\ &= 8 - 16 + 3 \\ &= -5 \qquad \textit{Think: } 8 + (-16) + 3 \end{aligned}$$

7c.
$$\begin{aligned} x^3 + 3x^2 + 3 &= (0.3)^3 + 3(0.3)^2 + 3 \\ &= 0.027 + 3(0.09) + 3 \\ &= 0.027 + 0.27 + 3 \\ &= 3.297 \end{aligned}$$

7d.
$$\begin{aligned} 6x^3 - 5x^2 + 2 &= 6\left(\frac{2}{3}\right)^3 - 5\left(\frac{2}{3}\right)^2 + 2 \\ &= 6\left(\frac{8}{27}\right) - 5\left(\frac{4}{9}\right) + 2 \\ &= \frac{\overset{2}{\cancel{6}} \cdot 8}{\underset{9}{\cancel{27}}} - \frac{5 \cdot 4}{9} + 2 \\ &= \frac{16}{9} - \frac{20}{9} + \frac{2}{1} \\ &= \frac{16}{9} - \frac{20}{9} + \frac{18}{9} \\ &= \frac{16 - 20 + 18}{9} \qquad \textit{Recall: } \frac{2}{1} = \frac{2 \cdot 9}{1 \cdot 9} = \frac{18}{9} \\ &= \frac{14}{9} \end{aligned}$$

Practice Problem 6 ***Evaluate each polynomial for the given value.***

a. $-6x^3 - 3x + 3,\ x = -1$ **b.** $3x^3 + 2x^2 - 2,\ x = -0.2$

c. $6x^3 - 3x^2 + 1,\ x = \frac{3}{2}$

Example 8 Evaluate each polynomial for the given value.

a. $2x^2 - 3x - 4.1,\ x = -0.2$

b. $\frac{25}{36}x^3 - \frac{15}{8}x^2 + \frac{1}{18},\ x = \frac{2}{5}$

Solution **8a.**

$$\begin{aligned} 2x^2 - 3x - 4.1 &= 2(-0.2)^2 - 3(0.2) - 4.1 \\ &= 2(0.04) - 3(-0.2) - 4.1 \\ &= 0.08 + 0.6 - 4.1 \\ &= -3.42 \end{aligned}$$

$$\begin{array}{r} 0.08 \\ +0.60 \\ \hline 0.68 \end{array} \qquad \begin{array}{r} -4.10 \\ +0.68 \\ \hline -3.42 \end{array}$$

8b.

$$\begin{aligned} \frac{25}{36}x^3 - \frac{15}{8}x^2 + \frac{1}{18} &= \frac{25}{36}\left(\frac{2}{5}\right)^3 - \frac{15}{8}\left(\frac{2}{5}\right)^2 + \frac{1}{18} \\ &= \frac{25}{36}\cdot\frac{8}{125} - \frac{15}{8}\cdot\frac{4}{25} + \frac{1}{18} \\ &= \frac{\overset{1}{\cancel{25}}\cdot\overset{2}{\cancel{8}}}{\underset{9}{\cancel{36}}\cdot\underset{5}{\cancel{125}}} - \frac{\overset{3}{\cancel{15}}\cdot\overset{1}{\cancel{4}}}{\underset{2}{\cancel{8}}\cdot\underset{5}{\cancel{25}}} + \frac{1}{18} \\ &= \frac{2}{45} - \frac{3}{10} + \frac{1}{18} \\ &= \frac{4}{90} - \frac{27}{90} + \frac{5}{90} \\ &= \frac{4 - 27 + 5}{90} \\ &= -\frac{18}{90} \\ &= -\frac{1}{5} \end{aligned}$$

$45 = 3 \cdot 3 \cdot 5$
$10 = 2 \cdot 5$
$18 = 2 \cdot 3 \cdot 3$
$\text{LCD} = 2 \cdot 3 \cdot 3 \cdot 5 = 90$

NOTE: Capital letters can be used to represent a polynomial. For example, the polynomial $2x^2 + 3x - 4.1$ could be written as $P(x) = 2x^2 + 3x - 4.1$, where $P(x)$ is read as "P of x." If we wanted the value of the polynomial when $x = 2$, we might state the request in this manner: "Given $P(x) = 2x^2 + 3x$, find $P(2)$."

Practice Problem 7 ***Evaluate each polynomial for the given value.***

a. $3x^2 - 5x,\ x = -0.1$ **b.** $\frac{27}{2}x^2 - \frac{1}{2}x + \frac{5}{36},\ x = \frac{1}{6}$

6.2 Exercises

Evaluate each polynomial for $x = 2$.

1. $3x^2 - 5x$ **2.** $x^3 - 9x^2$ **3.** $-4x - 7$

Evaluate each polynomial for $x = -3$.

4. $x^3 - 4x^2$ **5.** $6x^2 - 7x + 1$ **6.** $-3x^4 + 5$

Evaluate each polynomial for $x = -2$.

7. $6x^3 - 7x^2 + 5$
8. $3x^2 - 5x + 2$
9. $5x^3 - 7x - 1$
10. $-3x^3 - 2x^2 - x$

Evaluate each polynomial for $x = 0.3$.

11. $x^2 + 5x$
12. $x^3 - 2x^2$
13. $3x^2 - 4$

Evaluate each polynomial for $x = -0.1$.

14. $7x^3 + 6x^2 + 3x$
15. $3x^2 - 2x + 3.7$

Evaluate each polynomial for $x = \frac{2}{3}$.

16. $6x^2 + 2x$
17. $x^3 - x^2$
18. $12x^3 + 3$

Evaluate each polynomial for $x = -\frac{3}{4}$.

19. $6x^2 - x + 2$
20. $\frac{16}{15}x^3 - \frac{1}{2}x + \frac{1}{10}$

Evaluate each polynomial for the given value.

21. $5x^4 - 6x^3 + 7x^2 - 5x,\ x = -1$
22. $-3x^3 + 4x^2 - 5,\ x = -3$
23. $0.5x^2 - 0.4x + 8.2,\ x = 0.2$
24. $3x^3 - 6x^2 + 4,\ x = \frac{2}{3}$
25. $\frac{14}{15}x^2 - \frac{1}{9}x + \frac{1}{15},\ x = -\frac{3}{7}$
26. $5x^3 - 0.5x^2 + 3.2x,\ x = -0.2$

Fill in the blanks.

27. When we find the value of a polynomial for a specific value of the variable, this process is called ______________ the polynomial.
28. To evaluate a polynomial, replace the ______________ with its given value. Next, find a value for all ______________. Finally, do all ______________ and then do the ______________ and/or subtraction.

Solve. Show all work.

29. Suppose that the cost (in dollars) of manufacturing p kitchen tables can be approximated by the polynomial $5p^2 - 400p + 15{,}500$. Find the cost of manufacturing 20 tables by evaluating the polynomial when $p = 20$.
30. The amount of profit (in dollars) for selling 2000 toys at x dollars per toy can be approximated by the polynomial $-400x^2 + 6800x - 12{,}000$. Find the earned profits if all 2000 toys are sold at $10 per toy.
31. The cost in dollars of manufacturing x units of a perfume is $x^3 - 18x^2 + 500x - 50$. Find the cost of manufacturing 20 units of the perfume by evaluating the polynomial when $x = 20$.
32. The consumer demand for a certain brand of blue jeans is $-300p + 16{,}000$ pairs per month, with the price being p dollars per pair. Find the consumer demand for the jeans when the price is $45 per pair.
33. Evaluate $0.005x^2 - 0.35x + 15$ when $x = 50$.
34. Evaluate $0.005x^2 - 0.35x + 15$ when $x = 70$.
35. The cost in dollars of manufacturing x units of cologne is $x^3 - 18x^2 + 500x - 50$. Find the cost of manufacturing 50 units by evaluating the polynomial when $x = 50$.
36. Evaluate the polynomial in problem 35 for $x = 80$ to find the cost of manufacturing 80 units of cologne.
37. Evaluate the polynomial $0.4x^2 - 4.5x + 2$ for $x = -\frac{1}{2}$.
38. Evaluate the polynomial $0.6x^2 - 3.8 + 1$ for $x = -\frac{1}{4}$.
39. Evaluate $\frac{6}{25}x^3 - \frac{3}{10}x^2 + \frac{7}{15}$ for $x = -\frac{5}{3}$.
40. Evaluate $\frac{8}{21}x^2 - \frac{4}{35}x - \frac{7}{10}$ for $x = -\frac{7}{2}$.

Answers to Practice Problems **6a.** 12 **b.** -1.944 **c.** $\frac{29}{2}$ **7a.** 0.53 **b.** $\frac{31}{72}$

6.3 Addition

9 Adding polynomials is similar to collecting like terms.

To Add Polynomials:

1. Combine the like terms.
2. If necessary, rearrange the terms in decreasing powers of the variable (this is called writing the polynomial in *descending order*).

NOTE: The commutative and associative laws for addition allow us to rearrange the terms of a polynomial. (For an explanation of these laws, see Section 1.2.)

Example 9 Add.

a. $(5x^2 - 7x + 2) + (x^2 + 9x - 7)$

b. $(7x^2 - 3x - 1) + (3x^3 - 9x^2 + 13x + 7)$

c. $(-5y^2 + 3xy + 7x^2) + (7y^2 - 4xy + 3x^2)$

Solution To add polynomials, combine the like terms and then arrange the resulting terms in descending order.

9a. $(5x^2 - 7x + 2) + (x^2 + 9x - 7)$
$= (5 + 1)x^2 + (-7 + 9)x + (2 - 7)$ Combine like terms
$= 6x^2 + 2x + (-5)$
$= 6x^2 + 2x - 5$ *Think:* $2x + (-5) = 2x - 5$

9b. $(7x^2 - 3x - 1) + (3x^3 - 9x^2 + 13x + 7)$
$= (7 - 9)x^2 + (-3 + 13)x + (-1 + 7) + 3x^3$
$= -2x^2 + 10x + 6 + 3x^3$
$= 3x^3 - 2x^2 + 10x + 6$ Rearrange the terms

9c. $(-5y^2 + 3xy + 7x^2) + (7y^2 - 4xy + 3x^2)$
$= (-5 + 7)y^2 + (3 - 4)xy + (7 + 3)x^2$
$= 2y^2 + (-1xy) + 10x^2$
$= 10x^2 - xy + 2y^2$

NOTE: A polynomial containing two or more variables is said to be written in *descending order* when (1) the variables in each term are in alphabetical order and (2) the terms are arranged in decreasing powers of the first variable (see Solution 9c).

Practice Problem 8 ***Add.***

a. $(5x^2 + 7x - 3) + (-6x^2 - 9x - 5)$

b. $(7x^2 + 4x^3 + 9) + (-x^2 - 7x - 11)$

c. $(-5y + 7z - 6x) + (8x - 3y + 6z)$

Example 10 Add.

a. $(0.6x^2 - 6.3x + 0.08) + (0.3x^2 + 2.35x - 3)$

b. $\left(\frac{3}{5}x^2 + \frac{1}{8}x - \frac{7}{10}\right) + \left(\frac{1}{5}x^2 + \frac{3}{10}x + 2\right)$

Solution **10a.** $(0.6x^2 - 6.3x + 0.08) + (0.3x^2 + 2.35x - 3)$

$$= (0.6 + 0.3)x^2 + (-6.3 + 2.35)x + (0.08 - 3)$$
$$= 0.9x^2 + (-3.95x) + (-2.92)$$
$$= 0.9x^2 - 3.95x - 2.92$$

10b. $\left(\frac{3}{5}x^2 + \frac{1}{8}x - \frac{7}{10}\right) + \left(\frac{1}{5}x^2 + \frac{3}{10}x + 2\right)$

$$= \left(\frac{3}{5} + \frac{1}{5}\right)x^2 + \left(\frac{1}{8} + \frac{3}{10}\right)x + \left(-\frac{7}{10} + 2\right)$$
$$= \frac{4}{5}x^2 + \left(\frac{5}{40} + \frac{12}{40}\right)x + \left(-\frac{7}{10} + \frac{20}{10}\right)$$
$$= \frac{4}{5}x^2 + \frac{17}{40}x + \frac{13}{10}$$

NOTE: If possible, add the fractions mentally.

Practice Problem 9 ***Add.***

a. $(0.31x^3 - 0.5x^2 + 5) + (3.7x^3 - 0.3x^2 - 4.08)$

b. $\left(\frac{5}{14}x^2 + \frac{3}{8}x - 5\right) + \left(\frac{1}{14}x^3 + \frac{1}{10}x^2 - \frac{6}{8}x\right)$

10 Polynomials can also be added by using a vertical or column format. This format will be useful when we discuss multiplication of polynomials.

Example 11 Add $(6x^3 - 7x^2 + 5) + (5x^2 - 9x - 7)$.

Solution To add polynomials by using a vertical format, arrange each polynomial in descending order. Next, write the like terms in the same column and then add.

$$\begin{array}{rrrr} 6x^3 & -\ 7x^2 & & +\ 5 \\ & 5x^2 & -\ 9x & -\ 7 \\ \hline 6x^3 & -\ 2x^2 & -\ 9x & -\ 2 \end{array}$$

Practice Problem 10 ***Add.***

a. $(3x + 5) + (6x - 7)$ **b.** $(x^5 + 5x - 8) + (x^3 - 7x + 5)$

6.3 Exercises

Arrange each polynomial in descending order.

1. $6x + 5x^2 - 7$

2. $8y - 7z + 8x$

3. $n - 5 + 6n^3 - 7n^2$

4. $x^2y^2 - 9x^3 - 6y^2 + 7xy^2$

5. $6x + 7x^3 - 9x^5 + 7x^4 - 8x^6 + 5$

Add.

6. $(3x + 2) + (5x + 7)$

7. $(3n + 5) + (2n - 7)$

8. $(x - 5) + (2x - 5)$

9. $(-5x + 3) + (2x - 8)$

10. $(7x^2 - 9) + (-6x^2 + 3)$

11. $(3y + 7x) + (-7y - 8x)$

12. $(x^2 - 7x + 1) + (x^2 + 9x + 3)$

13. $(n^3 + 5n^2 - 7) + (-4n^3 - 15n^2 - 9)$

14. $(6x^2 + 7xy + 2y^2) + (x^2 - 4xy - 5y^2)$

15. $(3x - 7y + 8z) + (-5x - 8y + 2z)$

16. $(3x^2 - 7x + 5) + (3x^3 - 7x^2 + 5x - 2)$

17. $(4s^3 - 3s^2 - 2) + (5s^2 + 7s + 2)$

18. $(7xy - 8x^2 + 3y^2) + (xy - 13y^2 + 5x^2)$

19. $(-8ab + 5a^2b + 9b^2) + (2a^2b - 10b^2 + 10ab)$

20. $(5x^3 - 7x^2y + 9xy^2 - y^3) + (3x^2 + 9x^2y - 7xy^2 + y^3)$

21. $(8x^4 + 7x^2 - 8x - 5) + (-3x^3 - 18x^2 - 8x + 2)$

22. $(-8x^2 - 7x - 1) + (-7x^2 - 7x + 5)$

23. $(-3x^3 - 7x^2 + 8x - 9) + (5x^2 + 6x^2 - 5x + 9)$

24. $(3x + 5) + (-7x + 9) + (8x - 5)$

25. $(3c^2 - 4ac + 5a^2) + (c^2 - 8ac) + (5c^2 - 8a^2)$

26. $(0.3x + 2.1) + (0.31x + 3.2)$

27. $(3.5x + 0.5) + (1.8x - 2)$

28. $(0.8x^2 + 0.3x - 0.5) + (0.3x^2 - 0.51x + 3.5)$

29. $(3x^2 - 0.5xy - 3.4y^2) + (0.3x^2 + 4.8xy - 0.41y^2)$

30. $(x^2 + 0.8x - 0.35) + (x^2 - 0.17x + 4)$

31. $(0.19x^2 - 3x + 0.8) + (8x - x^2 - 3.7)$

32. $\left(\frac{3}{5}x + \frac{1}{3}\right) + \left(\frac{1}{5}x + \frac{1}{3}\right)$

33. $\left(\frac{4}{7}x - 9\right) + \left(\frac{15}{21}x + \frac{3}{8}\right)$

34. $\left(x^2 + \frac{5}{12}x - \frac{3}{10}\right) + \left(6x^2 + \frac{1}{18}x + \frac{1}{10}\right)$

35. $\left(\frac{5}{36}x^2 - \frac{1}{18}x - \frac{5}{11}\right) + \left(\frac{7}{66}x^2 + \frac{4}{15}x + 3\right)$

36. $\left(\frac{7}{9}x^3 - \frac{3}{4}x^2 + \frac{4}{9}\right) + \left(\frac{1}{4}x^2 + \frac{3}{8}x + \frac{5}{6}\right)$

37. $\left(\frac{3}{10}x^2 - \frac{7}{10}x + \frac{3}{22}\right) + \left(\frac{1}{10}x^2 + \frac{7}{15}x + \frac{1}{33}\right)$

Fill in the blanks.

38. When the terms of a polynomial are arranged in decreasing powers of a variable, the terms of the polynomial are said to be in ________ order.

39. To write a polynomial containing two or more variables in descending order, we first write the variables of each term in ________ order. Next, arrange the terms in decreasing powers of the ________ variable.

40. To add polynomials, combine the ________ terms and then arrange the terms in ________ order.

Solve.

41. Find the sum of $(2x^2 + 5x - 7)$ and $(3x^2 - 8x + 8)$.

42. Find the sum of $(3x^2 - 6x + 8)$ and $(4x^2 + 8x - 10)$.

43. Add $\left(\frac{3}{4}x^2 - \frac{3}{10}\right)$ and $\left(\frac{5}{6}x^2 - \frac{1}{8}\right)$.

44. Add $\left(\frac{1}{6}x^2 - \frac{7}{15}\right)$ and $\left(\frac{5}{8}x^2 - \frac{1}{6}\right)$.

45. Find the sum of $(8.5x - 1.4y)$ and $(-7.05x + 4.58y)$.

46. Find the sum of $(5.8x - 4.1y)$ and $(-3.06x + 7.85y)$.

47. Add $(3x - 5y + 6z)$ and $(-5.1x + 5y - z)$.

48. Add $(4x - 6y + 7z)$ and $(-6.2x + 6y - z)$.

Answers to Practice Problems **8a.** $-x^2 - 2x - 8$ **b.** $4x^3 + 6x^2 - 7x - 2$ **c.** $2x - 8y + 13z$ **9a.** $4.01x^3 - 0.8x^2 + 0.92$ **b.** $\frac{1}{14}x^3 + \frac{16}{35}x^2 - \frac{3}{8}x - 5$ **10a.** $9x - 2$ **b.** $x^5 + x^3 - 2x - 3$

6.4 Subtraction

11 Before stating a rule for subtracting one polynomial from another, we must define the additive inverse of a polynomial. The **additive inverse of a polynomial** is the polynomial that results when you replace the coefficient in each term with its additive inverse. For example, the additive inverse of $2x - 5$ is $-2x + 5$. Similarly, the additive inverse of $x^2 - 7x + 4$ is $-x^2 + 7x - 4$.

Rule for Subtracting One Polynomial from Another:

1. The minuend (first polynomial) remains the same.
2. Add the additive inverse of the subtrahend (second polynomial) to the minuend.

Example 12 Subtract.

a. $(6x^2 - 7x + 5) - (4x^2 - 9x + 8)$

b. $(5x^3 - 7x^2 + 5) - (8 + x^2 - 6x^3 - 8x)$

Solution To subtract using polynomials, add the additive inverse of the subtrahend.

12a. $(6x^2 - 7x + 5) - (4x^2 - 9x + 8)$

$= (6x^2 - 7x + 5) + (-4x^2 + 9x - 8)$ Additive inverse of subtrahend

$= 2x^2 + 2x - 3$ Combine like terms

12b. $(5x^3 - 7x^2 + 5) - (8 + x^2 - 6x^3 - 8x)$

$= (5x^3 - 7x^2 + 5)(-8 - x^2 + 6x^3 + 8x)$

$= 11x^3 - 8x^2 - 3 + 8x$

$= 11x^3 - 8x^2 + 8x - 3$ Rearrange the terms

Practice Problem 11 ***Subtract.***

a. $(6x^2 - 8x + 5) - (x^2 + 7x + 6)$ b. $(-7y^2 - 5xy - 9x^2) - (5y^2 - 5xy - 10x^2)$

Example 13 Subtract.

a. $(3.5x^2 - 0.71x - 0.3) - (8x^2 + 0.3x - 0.56)$

b. $\left(\frac{3}{5}x^2 - \frac{5}{6}x - \frac{1}{3}\right) - \left(\frac{1}{5}x^2 - \frac{7}{10}x - \frac{3}{9}\right)$

Solution **13a.** $(3.5x^2 - 0.71x - 0.3) - (8x^2 + 0.3x - 0.56)$

$= (3.5x^2 - 0.71x - 0.3) + (-8x^2 - 0.3x + 0.56)$

$= (3.5 - 8)x^2 + (-0.71 - 0.3)x + (-0.3 + 0.56)$

$= -4.5x^2 - 1.01x + 0.26$

13b. $\left(\frac{3}{5}x^2 - \frac{5}{6}x - \frac{1}{3}\right) - \left(\frac{1}{5}x^2 - \frac{7}{10}x - \frac{3}{9}\right)$

$= \left(\frac{3}{5}x^2 - \frac{5}{6}x - \frac{1}{3}\right) + \left(-\frac{1}{5}x^2 + \frac{7}{10}x + \frac{3}{9}\right)$

$= \left(\frac{3}{5} - \frac{1}{5}\right)x^2 + \left(-\frac{5}{6} + \frac{7}{10}\right)x + \left(-\frac{1}{3} + \frac{3}{9}\right)$

$= \frac{2}{5}x^2 + \left(-\frac{25}{30} + \frac{21}{30}\right)x + \left(-\frac{3}{9} + \frac{3}{9}\right)$

$= \frac{2}{5}x^2 - \frac{4}{30}x + 0$

$= \frac{2}{5}x^2 - \frac{2}{15}x$

Practice Problem 12 ***Subtract.***

a. $(3.2x^2 - 0.75) - (1.8x^2 - 0.35)$ **b.** $\left(\frac{3}{4}x^2 - \frac{1}{18}x + 3\right) - \left(\frac{5}{4}x^2 + \frac{4}{15}x - \frac{1}{2}\right)$

12 Subtraction can also be done by using a vertical or column format. This format will be useful when we discuss division of polynomials.

Example 14 Subtract.

a. $(6x - 5) - (8x + 3)$ **b.** Subtract $(4 - 7x + x^2)$ from $(3x^2 - 10x - 9)$

Solution To subtract polynomials by using a vertical format, first arrange the terms in descending order. Next, write the subtrahend directly below the minuend and align the like terms. Finally, change the sign of each term in the subtraction and then add.

14a.

$$\begin{array}{rcr} 6x & - & 5 \\ & \ominus & \\ \ominus 8x & + & 3 \\ \hline -2x & - & 8 \end{array}$$

14b.

$$\begin{array}{rcrcr} 3x^2 & - & 10x & - & 9 \\ & \oplus & & \ominus & \\ \ominus x^2 & - & 7x & + & 4 \\ \hline 2x^2 & - & 3x & - & 13 \end{array}$$

NOTE: It may be helpful to draw a circle around the new signs.

Practice Problem 13 ***Subtract.***

a. $(6a - b) - (3a + 4b)$ **b.** Subtract $(3n^2 - 7n + 4)$ from $(-n^2 - 8n + 7)$.

6.4 Exercises

Find the additive inverse.

1. $2x - 6$

2. $-3x^3 + 5x$

3. $-4x - 5$

4. $6x^3 - 7x^2 + 5x$

5. $x - 3y + 4z$

Subtract.

6. $(3x - 7) - (2x + 7)$

7. $(9x^2 - 6) - (3x^2 - 4)$

8. $(7x - 3y) - (4x - 5y)$

9. $(6a + 9) - (2a - 3)$

10. $(x + 5) - (2x + 8)$

11. $(6x - 7) - (9x - 9)$

12. $(7x^3 - 9x + 3) - (5x^3 + 9x - 9)$

13. $(3x^2 - 8x + 1) - (5x^2 - 9x + 5)$

14. $(6x - 3y + 5) - (2x - 3y - 8)$

15. $(3y^2 - 7xy + 2x^2) - (2y^2 - 7xy - y^2)$

16. $(a^2 - 5a + 2) - (a^2 - 9a - 7)$

17. $(7n^2 - n + 2) - (5 + n - 3n^2)$

18. $(x^3 - 7x^2 + 5) - (3x^2 - 5x + 7)$

19. $(8y^2 + 7y - 8) - (10y^2 + 9y - 10)$

20. $(3 + 5b - 6b^2) - (7 + 3b - 9b^2)$

21. $(6x^3 - 7x^2 + 3) - (3x^3 + 5x^2 - 7)$

22. $(5x^2 - 9x + 5) - (7x^3 + 5x^2 - 9x + 11)$

23. $(xy^2 + 7xy + 3) - (-xy^2 + 9xy - 7)$

24. Subtract $(11a - 9b + 7c)$ from $(9a - 7b + 5c)$

25. Subtract $(7a - 3b + 8c)$ from $(7c - 13b + 5a)$

26. $(0.3x^2 - 7) - (0.2x^2 - 9)$

27. $(4.1x + 0.7) - (3.2x + 3)$

28. $(0.31x^2 - 0.71x - 2) - (0.37x^2 + 3.1x - 5)$

29. Subtract $(7.3x^2 + 0.3x - 0.35)$ from $(3.7x^2 - 0.5x + 0.5)$

30. $(0.3x - 4.1y + 3.1z) - (0.5x + 3.15y + 5z)$

31. $\left(\frac{3}{5}x - \frac{1}{3}\right) - \left(\frac{1}{5}x + \frac{1}{3}\right)$

32. $\left(\frac{5}{9}x^2 - \frac{2}{7}\right) - \left(\frac{2}{9}x^2 + 2\right)$

33. $\left(\frac{3}{8}a^2 - \frac{2}{5}a + \frac{1}{10}\right) - \left(\frac{1}{6}a^2 - \frac{4}{5}a + 2\right)$

34. $\left(\frac{3}{7}x^2 - \frac{7}{9}x + \frac{3}{8}\right) - \left(-\frac{4}{7}x^2 - \frac{1}{15}x + \frac{1}{8}\right)$

35. $\left(\frac{5}{14}x^2 - \frac{1}{12}x + \frac{1}{6}\right) - \left(\frac{1}{10}x^2 - \frac{3}{10}x + \frac{1}{9}\right)$

Fill in the blanks.

36. To find the additive inverse of a polynomial, change the ____________ of each ____________ of the polynomial.

37. To subtract one polynomial from another, add the ____________ of the ____________ to the ____________.

38. To subtract one polynomial from another by using a vertical format, arrange the ____________ in descending order. Next, write the subtrahend directly below the ____________ and align the ____________ terms. Finally, change the ____________ of each term in the ____________ and then add.

Subtract.

39. $\begin{array}{r} 5x + 2 \\ \underline{3x + 1} \end{array}$

40. $\begin{array}{r} 7x + 8 \\ \underline{5x + 2} \end{array}$

41. $\begin{array}{r} -4x - 3 \\ \underline{-3x + 3} \end{array}$

42. $\begin{array}{r} -6x - 5 \\ \underline{-2x + 5} \end{array}$

43. $\begin{array}{r} 4.4x^2 - 2.8y + 4.7 \\ \underline{8.2x^2 + 4.2y - 0.09} \end{array}$

44. $\begin{array}{r} 5.5x^2 - 3.9y + 5.8 \\ \underline{9.1x^2 + 4.5y + 0.08} \end{array}$

Solve.

45. Subtract $(7x^2 + 2x - 4)$ from $(7x^3 - 3x^2 + 7)$.

46. Subtract $(8x^2 + 3x - 4)$ from $(8x^3 - 6x^2 + 9)$.

47. Find the difference between $\left(\frac{1}{2}x^2 - \frac{3}{8}x + 4\right)$ and $\left(\frac{3}{2}x^2 - \frac{1}{10}x + \frac{1}{5}\right)$.

48. Find the difference between $\left(\frac{1}{4}x^2 - \frac{1}{12}x + 5\right)$ and $\left(\frac{5}{4}x^2 - \frac{5}{8}x + \frac{1}{4}\right)$.

Answers to Practice Problems **11a.** $5x^2 - 15x - 1$ **b.** $x^2 - 12y^2$ **12a.** $1.4x^2 - 0.4$ **b.** $-\frac{1}{2}x^2 - \frac{29}{90}x + \frac{7}{2}$ **13a.** $3a - 5b$ **b.** $-4n^2 - n + 3$

6.5 Multiplication

13 We can find the product of two monomials by using the commutative law, the associative law, and the multiplication rule for exponential expressions.

Example 15

$$\begin{aligned}(-6x^2)(5x^4) &= -6 \cdot 5 \cdot x^2 \cdot x^4 \\ &= -30 \cdot x^6 \\ &= -30x^6\end{aligned}$$

14 Example 15 suggests a rule for multiplying two or more monomials.

Rule for Multiplying Monomials:

1. Multiply the numerical coefficients.
2. Find the product of the exponential expressions and place this value to the right of the numerical coefficient.

Example 16 Multiply.

a. $(-9x^5)(4x^3)$ **b.** $(-7xy)(8x^3)(-2y)$

c. $\left(\frac{35}{9}n^5\right)\left(\frac{15}{14}n^4\right)$ **d.** $(0.3x^3y^2)(0.2x^2y^4)$

Solution To multiply two or more monomials, multiply the numerical coefficients and then use the multiplication rule for exponential expressions.

16a. $(-9x^5)(4x^3) = -36x^{5+3}$
$= -36x^8$

16b. $(-7xy)(8x^3)(-2y) = 112x^{1+3}y^{1+1}$
$= 112x^4y^2$

16c. $\left(\frac{35}{9}n^5\right)\left(\frac{15}{14}n^4\right) = \frac{\overset{5}{\cancel{35}} \cdot \overset{5}{\cancel{15}}}{\underset{3}{\cancel{9}} \cdot \underset{2}{\cancel{14}}}n^{5+4}$
$= \frac{25}{6}n^9$

16d. $(0.3x^3y^2)(0.2x^2y^4) = 0.06x^{3+2}y^{2+4}$
$= 0.06x^5y^6$

Practice Problem 14 ***Multiply.***

a. $(5x^7)(-8x^{10})$ **b.** $(6xy^2)(-7x)$

c. $(3.1m^4)(0.2m^5)$ **d.** $\left(\frac{8}{12}x^5\right)\left(\frac{10}{15}x^3y\right)$

15 To find the product of a monomial and a polynomial, we will use the distributive law.

To Multiply a Polynomial by a Monomial:

1. Write the products you obtain when you multiply each term of the polynomial by the monomial.
2. Express the products in step 1 as a sum and then simplify.

Example 17 Multiply.

a. $5x^2(3x + 5)$ **b.** $3x^4(x^2 - 7x + 5)$

c. $-5n^2(3n^2 - 7n + 2)$ **d.** $-1.5x^7y(0.3xy - 4)$

Solution To multiply a monomial times a polynomial, use the distributive law.

17a. $5x^2(3x + 5) = 5x^2(3x) + 5x^2(5)$ Distributive law
$= 15x^3 + 25x^2$ Multiply monomials

17b. $3x^4(x^2 - 7x + 5) = 3x^4(x^2) + 3x^4(-7x) + 3x^4(5)$
$= 3x^6 - 21x^5 + 15x^4$

17c. $-5n^2(3n^2 - 7n + 2) = (-5n^2)(3n^2) + (-5n^2)(-7n) + (-5n^2)(2)$
$= -15n^4 + 35n^3 - 10n^2$

17d. $-1.5x^7y(0.3xy - 4) = (-1.5x^7y)(0.3xy) + (-1.5x^7y)(-4)$
$= -0.45x^8y^2 + 6x^7y$

Practice Problem 15 ***Multiply.***

a. $5x^3(7x^5 - 8x^2 + 2)$ **b.** $-3x^2(x^2 - 7x + 2)$

c. $\frac{4}{9}x\left(\frac{5}{7}x^3 - \frac{6}{14}x^2 + \frac{1}{6}\right)$

16 The distributive law can also be used to find the product of two polynomials.

To Multiply Polynomials:

1. Write the products obtained when you multiply each term of the first polynomial by the second polynomial.
2. Express the products in step 1 as a sum.
3. Multiply and then combine like terms.

Example 18 Multiply

a. $(x + 3)(x + 4)$ **b.** $(2x - 5y)(3x + 4y)$

c. $(5x - 3)(x^2 - 7x + 3)$

Solution To multiply one polynomial times another, use the distributive law. Remember that *each term* of the first polynomial must be multiplied by *each term* of the second polynomial.

18a. $(x + 3)(x + 4) = x(x + 4) + 3(x + 4)$ Distributive law
$= x^2 + 4x + 3x + 12$ Multiply
$= x^2 + 7x + 12$ Combine like terms

18b. $(2x - 5y)(3x + 4y) = 2x(3x + 4y) + (-5y)(3x + 4y)$
$= 6x^2 + 8xy - 15xy - 20y^2$
$= 6x^2 - 7xy - 20y^2$

18c. $(5x - 3)(x^2 - 7x + 3) = 5x(x^2 - 7x + 3) + (-3)(x^2 - 7x + 3)$
$= 5x^3 - 35x^2 + 15x - 3x^2 + 21x - 9$
$= 5x^3 - 38x^2 + 36x - 9$

Practice Problem 16 ***Multiply.***

a. $(2x - 7)(3x - 5)$ **b.** $(4x - 7)(3x^2 + 2x - 5)$

c. $(2x^2 + y^2)(x^4 - 3x^2y^2 - y^4)$

17 We can multiply two polynomials by using a vertical or column format.

Example 19 Multiply.

a. $(3x + 5)(6x^2 - x + 1)$ **b.** $(3x^2 - 5x + 1)(2x^2 + 3x - 5)$

Solution To multiply two polynomials by using a vertical format, multiply every term in the polynomial at the top by every term in the polynomial at the bottom. Then, combine like terms.

19a.

$$\begin{array}{rrrr} & 6x^2 & - x & + 1 \\ & & 3x & + 5 \\ \hline 18x^3 & - 3x^2 & + 3x & \\ & + 30x^2 & - 5x & + 5 \\ \hline 18x^3 & + 27x^2 & - 2x & + 5 \end{array}$$

Think: $3x\,(6x^2 - x + 1)$ → $18x^3 - 3x^2 + 3x$; $5\,(6x^2 - x + 1)$ → $30x^2 - 5x + 5$

19b.

$$\begin{array}{rrrrr} 3x^2 & - 5x & + 1 & & \\ 2x^2 & + 3x & - 5 & & \\ \hline 6x^4 & - 10x^3 & + 2x^2 & & \\ & + 9x^3 & - 15x^2 & + 3x & \\ & & - 15x^2 & + 25x & - 5 \\ \hline 6x^4 & - x^3 & - 28x^2 & + 28x & - 5 \end{array}$$

Practice Problem 17 ***Multiply.***

a. $(3x - 2)(x^2 - 7x + 5)$ **b.** $(2x + y - 2)(3x - y + 2)$

6.5 Exercises

Multiply.

1. $(-3x^5)(5x^4)$
2. $(-6x)(7x)$
3. $(5n^5)(-3n^4)$
4. $(5m^2)(-8m^3)$
5. $(-9xy)(5x^3y)$
6. $(-3xy)(-5x)$
7. $(-8x)(-7x^4)$
8. $(5ab)(-7a^3)$
9. $(-3x)(-3y)$
10. $(-9ab^2)(7a^3)$
11. $(2x)(-3x^2)(-5x^3)$
12. $(-5x^2)(-2x^3)(-3y)$
13. $(3.1x^3)(0.3x^5)$
14. $(-0.2x)(-0.4x^5)$
15. $(-5n^5)(2.4n^7)$
16. $(-0.2x^4)(0.2x^3)(-0.2x)$
17. $(-3.7n^3)(0.11n^4)$
18. $\left(\frac{2}{3}x\right)\left(\frac{4}{7}x^3\right)$
19. $\left(\frac{6}{25}x^4\right)\left(-\frac{5}{8}x^3\right)$
20. $\left(-\frac{6}{34}n\right)\left(-\frac{51}{56}n\right)$
21. $(81x^5)\left(-\frac{14}{54}x^2\right)$
22. $\left(-\frac{14}{60}a^2b\right)\left(\frac{66}{77}a^3\right)$
23. $2x\,(x + 5)$
24. $7x^2\,(5x + 4)$
25. $7x^2\,(x^3 - 7x^2 + 2)$
26. $3xy\,(7x^2y + 5xy^2 - 3)$
27. $-3x^2\,(5x^2 - 4x + 2)$
28. $-7np^2\,(3np - 5np^2 + 3p^4)$
29. $4n^5\,(3n^2 - 7n - 8)$
30. $-5x\,(-7x^2 - 7xy + 3y^2)$
31. $0.2x^4\,(0.3x^2 - 3.1x)$
32. $2.2n^4\,(0.3n^4 + 5)$
33. $-0.5x\,(3.1x^2 - 0.2x + 2.1)$
34. $2.1p^4\,(3.1p^3 - 5.2p^2)$
35. $\frac{4}{9}x^2\left(\frac{5}{9}x - \frac{15}{14}\right)$
36. $-\frac{9}{10}n\left(\frac{1}{2}n^2 - \frac{15}{21}\right)$
37. $(x + 3)(x + 5)$
38. $(3x^2 + 5)(3x^2 + 5)$
39. $(7x - 2y)(3x - 5y)$
40. $(2x - 3)(3x - 2)$
41. $(x - 2)(x^2 + 5x + 2)$
42. $(a - b)(a^2 + 2ab + b^2)$
43. $(4n - 3)(6n^2 - 7n + 4)$
44. $(3a + 2)(5a^2 - a + 2)$
45. $(7n^2 - 5n + 1)(n^2 + 4n - 7)$
46. $\left(2x - \frac{1}{3}\right)\left(5x - \frac{3}{4}\right)$
47. $(x^2 - 5x + 2)(x^2 + 3x + 1)$
48. $(x + 0.5)(x^2 - 0.2)$

Fill in the blanks.

49. To multiply two or more monomials, multiply the ______________. Next, find the product of the ______________ and place this value to the right of the ______________.

50. To multiply a monomial times a polynomial we use the ______________ law.

51. To multiply one polynomial times another, write the products obtained when each term of the first polynomial is ______________ by the second polynomial. Next, express these products as a ______________, multiply, and then combine the ______________ terms.

Perform the indicated operation.

52. $(3xy)(-6x^2y)$

53. $(4xy)(-8x^3y)$

54. $(-6x)(-7y)$

55. $(-8x)(-9y)$

56. $(0.2x^2)(-0.4x^2)$

57. $(0.3x^3)(-0.2x^5)$

58. $\left(-\frac{8}{21}x^4\right)\left(\frac{35}{44}xy\right)$

59. $\left(-\frac{6}{25}x^5\right)\left(\frac{30}{33}xy^2\right)$

60. $3x^2(5x^2 - 7x + 3)$

61. $4y^3(6y^2 - 8y + 5)$

62. $-4x(6x^2 - 7xy + y^2)$

63. $-8x(2x^2 - 6xy + y^2)$

64. $5xy(6x^2y + 3xy - 6)$

65. $6xy(5x^2y + 4xy - 6)$

66. $-6xy^2(-4xy - 5xy^2 - 6y^2)$

67. $-2xy^2(-5xy - 6xy^2 - 3y^2)$

68. $0.3x^2(0.2x^2 - 4.1)$

69. $0.2x^3(0.3x^2 - 3.1)$

70. $2.1b^2(5b^2 - 0.4b + 0.05)$

71. $4.1c^2(6c^2 - 0.3c + 0.06)$

72. $\frac{6}{15}x^3\left(\frac{10}{12}x^2 - 30\right)$

73. $\frac{4}{10}x^4\left(\frac{6}{8}x^2 - 20\right)$

74. $\frac{1}{5}x^2y^2\left(\frac{3}{8}xy - \frac{5}{7}xy^2\right)$

75. $\frac{3}{7}a^2b^2\left(\frac{1}{4}ab - \frac{7}{8}ab^2\right)$

76. $(x - y)(x + y)$

77. $(x + 4)(x - 4)$

78. $(x + y)(x + y)$

79. $(x + 4)(x + 4)$

80. $(x + 3)(x + 3)$

81. $(x + 5)(x + 5)$

82. $(4x - 1)(3x^2 - 5x - 2)$

83. $(3x - 1)(4x^2 - 6x - 3)$

84. $(3x + 0.5)(4x - 0.2)$

85. $(4x + 0.3)(2x - 0.5)$

86. $(x + 2.1)(0.5x^2 - 3.1x + 4)$

87. $(x + 3.1)(0.6x^2 - 4.2x + 5)$

88. $(x^2 + x + 1)(x^2 - x + 1)$

89. $(x^4 + x^2 + 2)(x^4 - x^2 + 1)$

90. $(3x + 2y - 4z)(x - 2y + 3z)$

91. $(2x + 3y - z)(x - 3y + 5z)$

92. $(x - y)(x^2 + xy + y^2)$

93. $(x + y)(x^2 - xy + y^2)$

Solve.

94. Find the product of $(x - 7)$ and $(4x^2 + 3x - 6)$.

95. Find the product of $(x - 2)$ and $(3x^2 + 2x - 7)$.

96. Multiply $3x^2 - 6x + 2$ by $-6x$.

97. Multiply $4x^2 - 5x + 1$ by $-3x$.

98. Find the product of $-0.7x^3$ and $-\frac{1}{2}x$.

99. Find the product of $-0.6x^3$ and $-\frac{3}{5}x$.

Answers to Practice Problems **14a.** $-40x^{17}$ **b.** $-42x^2y^2$ **c.** $0.62m^9$ **d.** $\frac{4}{9}x^8y$ **15a.** $35x^8 - 40x^5 + 10x^3$ **b.** $-3x^4 + 21x^3 - 6x^2$ **c.** $\frac{20}{63}x^4 - \frac{4}{21}x^3 + \frac{2}{27}x$ **16a.** $6x^2 - 31x + 35$ **b.** $12x^3 - 13x^2 - 34x + 35$ **16c.** $2x^6 - 5x^4y^2 - 5x^2y^4 - y^6$ **17a.** $3x^3 - 23x^2 + 29x - 10$ **b.** $6x^2 + xy - 2x - y^2 + 4y - 4$

6.6 Division

18 We can find the quotient of two monomials by using the quotient rule for exponentials and the fact that since

$$\text{if} \quad \frac{a}{b} \cdot \frac{c}{b} = \frac{a \cdot c}{b \cdot d} \quad \text{then} \quad \frac{a \cdot c}{b \cdot d} = \frac{a}{b} \cdot \frac{c}{d}.$$

Example 20

$$\begin{aligned} \frac{16x^5}{4x^2} &= \frac{16 \cdot x^5}{4 \cdot x^2} \\ &= \frac{16}{4} \cdot \frac{x^5}{x^2} \\ &= 4 \cdot x^3 \\ &= 4x^3 \end{aligned}$$

19 Example 20 suggests a rule for dividing one monomial by another.

Rule for Dividing a Monomial by a Monomial:

1. Write the numerical coefficients as a quotient and write the exponential expressions having the same base as a quotient. Also, write any exponential expressions that appear only in the numerator or denominator as a quotient.
2. Express the quotients in step 1 as a product and then simplify.

Example 21 Divide.

a. $\frac{-35x^6}{5x^2}$ b. $\frac{42x^5y^2}{77x^3y^5}$ c. $\frac{n^5}{2n^4m^2}$ d. $\frac{-35x^7y}{0.5xz}$

Solution To divide one monomial by another, write the numerical coefficients as a quotient and write the exponential expressions as a quotient. Next, express these quotients as a product and then simplify.

21a. $\frac{-35x^6}{5x^2} = \frac{-35}{5} \cdot \frac{x^6}{x^2}$

$= -7x^4$ *Think:* $\frac{x^6}{x^2} = x^{6-2} = x^4$

21b. $\frac{42x^5y^2}{77x^3y^5} = \frac{\overset{6}{\cancel{42}}}{\underset{11}{\cancel{77}}} \cdot \frac{x^5}{x^3} \cdot \frac{y^2}{y^5}$

$= \frac{6}{11} \cdot x^2 \cdot \frac{1}{y^3}$ *Recall:* $\frac{y^2}{y^5} = y^{-3} = \frac{1}{y^3}$

$= \frac{6x^2}{11y^3}$ *Think:* $\frac{6}{11} \cdot \frac{x^2}{1} \cdot \frac{1}{y^3}$

21c.

$$\begin{aligned} \frac{n^5}{2n^4m^2} &= \frac{1}{2} \cdot \frac{n^5}{n^4} \cdot \frac{1}{m^2} \\ &= \frac{1}{2} \cdot n \cdot \frac{1}{m^2} \\ &= \frac{n}{2m^2} \end{aligned}$$

21d. $$\frac{-35x^7y}{0.5xz} = \frac{-35}{0.5} \cdot \frac{x^7}{x} \cdot \frac{y}{z}$$
$$= -70 \cdot x^6 \cdot \frac{y}{z}$$
$$= \frac{-70x^6y}{z}$$
$$= -\frac{70x^6y}{z}$$

NOTE: Your answer is correct if the quotient times the divisor equals the dividend. That is, $\frac{-35x^6}{5x^2} = -7x^4$ since $-7x^4 \cdot 5x^2 = -35x^6$.

Practice Problem 18 ***Divide.***

a. $\frac{-48x^9}{-12x^2}$ **b.** $\frac{35x^5y^2}{91x^8y}$ **c.** $\frac{35x^3}{0.5xy}$

20 To find the quotient of a polynomial and a monomial, we will use the definition for addition of fractions. That is, since

$$\frac{a}{b} + \frac{c}{b} = \frac{a + c}{b} \qquad \text{then} \qquad \frac{a + c}{b} = \frac{a}{b} + \frac{c}{b}.$$

Example 22 $$\frac{16x^3 + 8x^2 + 4x}{2x} = \frac{16x^3}{2x} + \frac{8x^2}{2x} + \frac{4x}{2x}$$
$$= 8x^2 + 4x + 2$$

21 Example 22 suggests a rule for dividing a polynomial by a monomial.

Rule for Dividing a Polynomial by a Monomial:

1. Write the quotients obtained when each term of the polynomial is divided by the monomial.
2. Express the quotients in step 1 as a sum and then simplify.

Example 23 Divide.

a. $\frac{8x^4 + 12x^3}{2x^2}$ **b.** $\frac{15x^7 - 30x^6 + 40x^2}{10x^4}$

c. $\frac{15x^6 - 10x^4 + 25x^2}{-5x^2}$ **d.** $\frac{35x^2y^5 + 45x}{-5xy}$

Solution To divide a polynomial by a monomial, write each term of the polynomial over the monomial. Express these quotients as a sum and then divide.

23a. $$\frac{8x^4 + 12x^3}{2x^2} = \frac{8x^4}{2x^2} + \frac{12x^3}{2x^2}$$
$$= 4x^2 + 6x$$

Check: $\frac{8x^4 + 12x^3}{2x^2} = 4x^2 + 6x$ since $2x^2\,(4x^2 + 6x) = 8x^4 + 12x^3$

23b. $$\frac{15x^7 - 30x^6 + 40x^2}{10x^4} = \frac{15x^7}{10x^4} + \frac{-30x^6}{10x^4} + \frac{40x^2}{10x^4}$$
$$= \frac{3}{2}x^3 - 3x^2 + \frac{4}{x^2}$$

23c. $$\frac{15x^6 - 10x^4 + 25x^2}{-5x^2} = \frac{15x^6}{-5x^2} + \frac{-10x^4}{-5x^2} + \frac{25x^2}{-5x^2}$$
$$= -3x^4 + 2x^2 - 5$$

23d. $$\frac{35x^2y^5 + 45x}{-5xy} = \frac{35x^2y^5}{-5xy} + \frac{45x}{-5xy}$$
$$= -7xy^4 - \frac{9}{y}$$

Practice Problem 19 **Divide.**

a. $\frac{6x^3 - 4x^2 + 8x}{2x}$ **b.** $\frac{8a^6b^6 - 12a^4b + 18b^6}{-6a^2b^2}$

6.6 Exercises

Divide.

1. $\frac{-8x^5}{-2x^2}$
2. $\frac{-16n^8}{4n}$
3. $\frac{-6x^5y^5}{9x^2y^2}$
4. $\frac{6.5x}{1.5x^3}$
5. $\frac{-21x^2y}{-3xy^2}$
6. $\frac{35ab^3}{91ab}$
7. $\frac{n^6m}{5n^2}$
8. $\frac{66x^8}{0.3x^5}$
9. $\frac{-15xy^5}{5x^3y^3}$
10. $\frac{35a^8}{-7a^9}$
11. $\frac{35x^5}{0.5x^2}$
12. $\frac{14a^3b^2}{4ab^3}$
13. $\frac{x^7y^8}{5x^6y^2}$
14. $\frac{0.75x^6}{-0.5x}$
15. $\frac{55x^4y}{65x^2}$
16. $\frac{100a^3b^3c^3}{-10abc}$
17. $\frac{-51x^3}{-68x}$
18. $\frac{85x^3y}{-17xy}$
19. $\frac{6x^5 + 8x^3}{2x^2}$
20. $\frac{9a^3 - 6a^2}{3a}$
21. $\frac{8x^6 + 6x^4 - 4x^2}{2x}$
22. $\frac{10x^6 - 70x^4 - 15x^2}{5x^2}$
23. $\frac{25x^8 - 15x^7}{-5x^3}$
24. $\frac{14x^6 - 21x^5 + 35x^3}{-7x^2}$
25. $\frac{9a^3 - 6a^2b^4}{3a^2b^2}$
26. $\frac{8x^3y^5 - 4xy}{-4xy^3}$
27. $\frac{24x^3 - 18x^2 + 10x}{6x^2}$
28. $\frac{16n^5 - 48n^3 + 72n^2}{12n^3}$
29. $\frac{9n^5m^2 - 18n^4m}{15n^4m^2}$
30. $\frac{80x^7y^4 - 70x^3y^6}{-10b^5}$
31. $\frac{45a^4 + 15a^3b^2 - 96b^4}{-3ab}$
32. $\frac{20b^4 - 30b^2 + 10b}{-10b^5}$
33. $\frac{6x^8y^6 - 8x^6y^4 + 4x^4y^2 + 12x^2y + 18x}{4x^4y^4}$

Fill in the blanks.

34. To divide one monomial by another, write the ______________ coefficients as a quotient and write the ______________ expressions as a quotient. Express these quotients as a ______________ and then simplify.
35. To divide a polynomial by a monomial, write the quotients obtained when each term of the ______________ is divided by the ______________. Express these quotients as a ______________ and then simplify.

Perform the indicated operation.

36. $\dfrac{-111x^6}{-37x^2}$ **37.** $\dfrac{-155x^8}{-31x^2}$ **38.** $\dfrac{222x}{-3x^4}$

39. $\dfrac{333x}{-3x^5}$ **40.** $\dfrac{8a^3b}{-2a^2b}$ **41.** $\dfrac{6x^3y}{-2x^2y}$

42. $\dfrac{16x^5y^2}{24x^5y}$ **43.** $\dfrac{12x^8y^3}{18x^8y}$ **44.** $\dfrac{-15x^9}{0.3x^6}$

45. $\dfrac{-65x^8}{0.5x^2}$ **46.** $\dfrac{-7.5x^6}{-1.5x^8}$ **47.** $\dfrac{-9.9x^{11}}{-3.3x^{15}}$

48. $\dfrac{77x^2y}{121xy^2}$ **49.** $\dfrac{85a^2b}{119ab^2}$ **50.** $\dfrac{45x^3y^3}{-35xy}$

51. $\dfrac{55a^3b^3}{-65ab}$ **52.** $\dfrac{57x^2y^2}{-76x^3y}$ **53.** $\dfrac{90a^2b^2}{72a^3b}$

54. $\dfrac{-8.5x^4}{-50x}$ **55.** $\dfrac{-6.5x^8}{-50x}$ **56.** $\dfrac{36x^5 - 18x^4 + 12x^3}{6x^2}$

57. $\dfrac{15x^6 - 25x^4 + 35x^3}{5x^2}$ **58.** $\dfrac{9x^3 - 6x^2}{3x}$ **59.** $\dfrac{10x^5 - 8x^3}{2x}$

60. $\dfrac{14x^5 - 28x^2 + 35x}{-7x}$ **61.** $\dfrac{15x^6 - 18x^4 + 21x}{-3x}$ **62.** $\dfrac{16x^4 - 8x^2 + 4x}{4x^3}$

63. $\dfrac{20a^4 - 8a^2 + 4a}{4a^3}$ **64.** $\dfrac{9x^3 - 6x^2y^4}{3x^2y^2}$ **65.** $\dfrac{6x^4 - 9x^3y^5}{3x^3y^3}$

66. $\dfrac{30x^3 - 18x^2 + 15x}{6x^2}$ **67.** $\dfrac{40x^4 - 32x^3 + 12x}{8x^3}$ **68.** $\dfrac{9x^5y^2 - 18x^4y}{15x^4y^2}$

69. $\dfrac{18x^6y^3 - 24x^5y^2}{15x^5y^3}$ **70.** $\dfrac{45x^4 + 15x^3y^2 - 96x^3}{-3xy}$ **71.** $\dfrac{50x^5 + 20x^4y^3 - 55x^4}{-5xy}$

72. $\dfrac{x^{15} + x^{13} + x^{11} - x^9}{-x^3}$ **73.** $\dfrac{x^{16} + x^{14} + x^{12} - x^{10}}{-x^4}$ **74.** $\dfrac{-9a - 6b + 15c}{-3}$

75. $\dfrac{-6x - 9y + 15z}{-3}$

76. Divide $20x^4 - 30x^2 + 10x$ by $-10x$.

77. Divide $30x^5 - 40x^3 + 20x^2$ by $-5x^6$.

78. Divide $-77x^8$ into $66x^5$.

79. Divide $-55x^9$ into $44x^2$.

80. Divide $13abc^2$ into $39a^3bc^2 - 65ab^2c^3 - 78a^2b^2c^2$.

81. Divide $15xyz^2$ into $45x^3yz^2 - 60xy^2z^3 - 75x^2y^2z^2$.

Answers to Practice Problems **18a.** $4x^7$ **b.** $\dfrac{5y}{13x^3}$ **c.** $\dfrac{70x^2}{y}$ **19a.** $3x^2 - 2x + 4$ **b.** $-\dfrac{4}{3}a^4b^4 + \dfrac{2a^2}{b} - \dfrac{3b^4}{a^2}$

6.7 Dividing a Polynomial by a Polynomial

22 When we must find the quotient of two polynomials in which the divisor is not a monomial, we will use a long division format similar to the one used to divide whole numbers.

Example 24 $(x^2 + 6x + 8) \div (x + 2)$

Solution

$$\begin{array}{r} x \\ x+2\overline{)x^2+6x+8} \end{array}$$ Divide x into x^2

$$\begin{array}{r} x \\ x+2\overline{)x^2+6x+8} \\ \underline{x^2+2x} \end{array}$$ Multiply x times $x + 2$

$$\begin{array}{r} x \\ x+2\overline{)x^2+6x+8} \\ \underline{x^2+2x} \\ 4x+8 \end{array}$$ Subtract and bring down the next term

$$\begin{array}{r} x+4 \\ x+2\overline{)x^2+6x+8} \\ \underline{x^2+2x} \\ 4x+8 \\ \underline{4x+8} \end{array}$$ Divide x into $4x$ and then multiply the result times $x + 2$

Check: $(x^2 + 6x + 8) \div (x + 2) = x + 4$ since $(x + 4)(x + 2) = x^2 + 6x + 8$.

23 Example 24 suggests a rule for dividing a polynomial by a polynomial.

Rule for Dividing a Polynomial by a Polynomial:

1. Arrange the polynomials in descending powers of a variable.
2. Set up the problem in the long division format. Also, write in a zero as the coefficient of any missing term.
3. To obtain the first term of the quotient, divide the first term of the divisor into the first term of the dividend.
4. Multiply the value obtained in step 3 times each term in the divisor.
5. Place the product obtained in step 4 beneath the dividend (align the like terms). Then, subtract and bring down the next term. This polynomial is your new dividend.
6. Repeat steps 3–5 until you obtain a remainder of zero or until the degree of the remainder is less than the degree of the divisor.

NOTE: To check your answer, multiply the quotient and the divisor and add the remainder. This result must be equal to the dividend.

Example 25 Divide.

a. $(6x^2 + x - 15) \div (3x + 5)$ **b.** $(a^3 + 1) \div (a + 1)$

c. $(4 - 11x + 5x^2) \div (x - 2)$ **d.** $\dfrac{5xy + 6x^2 - 8y^2}{-2y + 3x}$

Solution To divide a polynomial by a polynomial, arrange each polynomial in descending power of a variable. Next, write the polynomials in a long division format and then divide as if you were dividing using whole numbers.

25a.

$$\begin{array}{r|l} & 2x - 3 \\ \hline 3x + 5 & 6x^2 + x - 15 \\ & \underline{6x^2 + 10x} \\ & \quad -9x - 15 \\ & \quad \underline{-9x - 15} \end{array}$$

25b.

$$\begin{array}{r|l} & a^2 - a + 1 \\ \hline a + 1 & a^3 + 0a^2 + 0a + 1 \\ & \underline{a^3 + a^2} \\ & \quad -a^2 + 0a \\ & \quad \underline{-a^2 - a} \\ & \qquad a + 1 \\ & \qquad \underline{a + 1} \end{array}$$

⟵ Insert zeros as the coefficients of the missing terms

25c.

$$\begin{array}{r|l} & 5x - 1 \\ \hline x - 2 & 5x^2 - 11x + 4 \\ & \underline{5x^2 - 10x} \\ & \quad -x + 4 \\ & \quad \underline{-x + 2} \\ & \qquad 2 \end{array}$$

⟵ Arrange the dividend in descending powers of x

The quotient is $5x - 1$ with a remainder of 2. The answer can be written as $5x - 1$, R2 or $5x - 1 + \dfrac{2}{x - 2}$ $\left(\text{quotient} + \dfrac{\text{remainder}}{\text{divisor}}\right)$.

25d.

$$\begin{array}{r|l} & 2x + 3y \\ \hline 3x - 2y & 6x^2 + 5xy - 8y^2 \\ & \underline{6x^2 - 4xy} \\ & \quad 9xy - 8y^2 \\ & \quad \underline{9xy - 6y^2} \\ & \qquad -2y^2 \end{array}$$

⟵ Arrange the dividend and divisor in descending powers of x

Thus, $\dfrac{5xy + 6x^2 - 8y^2}{-2y + 3x} = 2x + 3y + \dfrac{-2y^2}{-2y + 3x}$

Practice Problem 20 **Divide.**

a. $(x^2 + 6x + 8) \div (x + 4)$

b. $(x^3 - 27) \div (x - 3)$

c. $\dfrac{12x^3 + 10y^3 - 11x^2y + 17xy^2}{3x - 2y}$

6.7 Exercises

Divide.

1. $(x^2 + 5x + 6) \div (x + 2)$

2. $(x^2 + 3x - 10) \div (x - 2)$

3. $(9x^2 - 9x - 11) \div (3x - 5)$

4. $(8x^2 - 22x + 15) \div (2x - 3)$

5. $(3x^2 - 5x - 2) \div (x - 2)$

6. $(2x^2 - 9x - 18) \div (x - 6)$

7. $(10x^2 - xy - 3y^2) \div (y + 2x)$

8. $(24x^2 - xy - 3y^2) \div (y + 3x)$

9. $\dfrac{15x^3 - x^2y - 8xy^2 - 2y^3}{3x + y}$

10. $\dfrac{12x^3 - 10x^2y + 8xy^2 - 3y^3}{2x - y}$

11. $\dfrac{4x^3 + 27x + 31x^2 + 42}{x + 7}$

12. $\dfrac{4x^3 + 12x + 11x^2 + 12}{x + 2}$

13. $(4x^2 - 9y^2) \div (2x + 3y)$

14. $(16x^2 - 25y^2) \div (4x - 5y)$

15. $(x^2 + 3x - 5) \div (x - 3)$

16. $(x^2 + 5x - 12) \div (x - 6)$

17. $(6x^2 - 13x - 6) \div (2x - 3)$

18. $(15x^2 - 11x - 2) \div (3x - 2)$

19. $\dfrac{6a^3 - 28a + 3a^2 + 15}{2a - 3}$

20. $\dfrac{15x^3 - 8x - 31x^2 - 4}{3x - 2}$

21. $(x^2 - y^2) \div (x - y)$

22. $(a^2 - b^2) \div (a + b)$

23. $(x^3 - y^3) \div (x - y)$

24. $(x^3 + y^3) \div (x + y)$

25. $\dfrac{2a^5 - 3a^4 + 5a^2 - 5a - 5}{2a^2 - a - 1}$

26. $\dfrac{2x^5 + x^3 - 3x^2 - 7x + 5}{x^2 + 2x - 1}$

27. $\dfrac{3x^4 + 5x^2 - 2}{2 + x^2}$

28. $\dfrac{2m^4 - 5m^2 + 3}{-1 + m^2}$

29. $(x^2 - 4) \div (x - 2)$

30. $(10a^2 - ab - 3b^2) \div (b + 2a)$

31. $(a^3 - 8) \div (a - 2)$

32. $(3x^2 + 13x - 10) \div (-x - 5)$

33. $\dfrac{6x^3 - 10x^2 - 2x + 2}{2x + 2}$

34. $\dfrac{25x^2 - 4}{5x + 2}$

35. $\dfrac{7 - 9x + 10x^2}{2x + 1}$

36. $\dfrac{a^3 - 1}{a - 1}$

37. $\dfrac{12x^8 + x^4 - 6}{3 + 4x^4}$

38. $\dfrac{x^3 - x^2 + x - 1}{x + 1}$

39. $\dfrac{9t^2 + 3 + 9t}{3t + 1}$

40. $\dfrac{10a^3 + 8a - 16a^2 + 2}{5a^2 - 3a + 1}$

Fill in the blanks.

41. To divide a polynomial by a polynomial, arrange each polynomial in ______________ powers of a variable. Next, write the polynomials in a ______________ format and then divide as if you were dividing using ______________ numbers.

Solve.

42. Divide $x - 2$ into $3x^2 - 5x - 2$.

43. Divide $2x + 3$ into $2x^2 - x - 6$.

44. Divide $a^2 - b^2$ by $a - b$.

45. Divide $a^2 - b^2$ by a + b.

Answers to Practice Problems **20a.** $x + 2$ **b.** $x^2 + 3x + 9$ **c.** $4x^2 - xy + 5y^2 + \dfrac{20y^3}{3x - 2y}$

Summary

Important Terms

6.1

monomial
polynomial
terms
numerical coefficient
binomial
trinomial
like terms
combine like terms
distributive law
degree of a term
degree of a polynomial

6.2

evaluate a polynomial

6.3

descending order

Important Skills

6.1

Identifying the terms of a polynomial
Identifying the numerical coefficient of a term
Classifying polynomials as monomials, binomials, or trinomials
Identifying like terms
Combining like terms
Identifying the degree of a term and a polynomial

6.2

Evaluating polynomials
Solving word problems by evaluating a polynomial

6.3

Adding polynomials
Arranging polynomials in descending order

6.4

Finding the additive inverse of a polynomial
Subtracting one polynomial from another

6.5

Multiplying two or more monomials
Multiplying a monomial and a polynomial
Multiplying one polynomial times another

6.6

Dividing one monomial by another
Dividing a polynomial by a monomial

6.7

Dividing one polynomial by another

Review Exercises

Identify the like terms.

1. $6x^2 - 7x - 16x^2 + 3x$

2. $x^3 - 5x^2y - 3x^2 + x^2y$

Combine the like terms.

3. $5x^2 + 3x - 8x^2$

4. $0.4x + 5x^2 + 0.51x$

5. $\frac{5}{8}x^3y + \frac{7}{12}x^3y$

6. $-\frac{5}{36}x^2 + \frac{1}{36}y^2 - \frac{1}{66}x^2$

Evaluate.

7. $-3x^3 + 4x^2 - 5$ when $x = -2$

8. $3x^3 + 6x^2 + 4$ when $x = \frac{2}{3}$

Perform the indicated operations.

9. $\left(\frac{4}{15}x + \frac{1}{10}\right) + \left(\frac{3}{10}x - \frac{11}{12}\right)$

10. $\frac{-35x^4y^2}{7x^2y^4}$

11. $(-7x^5y^2)(5.2xy^3)$

12. $(7x^2 - 8y^2) - (2x^3 - 9y^2)$

13. $(6x^2 - 4x + 2) + (2x^3 - 8x^2 - 4)$

14. $(x^3 - 8) \div (x - 2)$

15. $-0.5x^2(3x^2 - 0.1x + 4)$

16. $\frac{18a^4 - 24a^3 + 15a}{-6a^2}$

17. $(3x - 2)(x^2 - 8x + 3)$

18. $\frac{(-3x^2y)^3}{36xy^3z}$

19. $(-6y^2 - 5xy + 9x^2) - (5y^2 + 5xy + x^2)$

Solve.

20. An efficiency study indicates that a worker at the Pace Clock Company who arrives at work at 7:00 A.M. should have assembled $-x^3 + 5x^2 + 7x$ clocks x hours later. If Howie arrived at work at 7:00 A.M., how many clocks should he have assembled by 10:00 A.M.?

LINEAR EQUATIONS AND INEQUALITIES

OBJECTIVES

1. Determine if a value is the solution of an equation.
2. Solve equations by using the addition principle.
3. Solve equations by using the multiplication or division principle.
4. Solve equations using the addition and the multiplication or division principle.
5. Solve equations containing a grouping symbol.
6. Solve equations containing fractions.
7. Solve equations for a particular letter.
8. Solve inequalities by using the addition principle.
9. Solve inequalities by using the multiplication or division principle.
10. Solve inequalities by using the addition and the multiplication or division principle.

PRETEST

EXPLANATION

1. Is the given value the solution?
   a. $x - 0.37 = 0.2, x = 0.57$
   b. $20 = 8 - 2(9 - 3x), x = -3$

Section 7.1
Idea 3

2. Solve.
   a. $x - 3 = 1$ **b.** $x - 1.1 = -3.35$

Section 7.2
Idea 6

3. Solve.
   a. $\frac{x}{-6} = 3$ **b.** $15x = 25$

Section 7.2
Ideas 7–9

4. Solve.
   a. $\frac{5}{6}x + 2 = 17$ **b.** $6x - 2 = 8 - 14x$

Section 7.3
Ideas 10 and 11

5. Solve.
   a. $0.2(x - 0.1) = 3$ **b.** $4x - 6 = 2(x + 5)$

Section 7.4
Idea 12

6. Solve.
   a. $5x - 3 = \frac{1}{10} + \frac{3}{4}$ **b.** $\frac{2x}{3} + x = 5$

Section 7.4
Idea 13

7. Solve for x.
   a. $ax + bx = c$ **b.** $P = a(x + d)$

Section 7.5
Ideas 15 and 16

8. Solve.
   a. $x + 11 < -8$ **b.** $x + 1 \leq -8$

Section 7.6
Idea 19

9. Solve.
   a. $-0.5x > 15$ **b.** $\frac{2}{3}x \geq 10$

Section 7.6
Ideas 21–23

10. Solve.
   a. $x + 10 \leq 7x + 20$ **b.** $15 \geq 2(x + 3)$
   c. $\frac{4 - 2x}{3} \leq \frac{21}{12} + \frac{5x - 3}{4}$

Section 7.7
Idea 24

7.1 Introduction

Idea 1 An **equation** is a mathematical statement that two quantities are equal. A **linear** or **first-degree** equation containing one variable is an equation in which the variable has an exponent of one.

Example 1 **a.** $2x + 5 = 7$ Linear equation in one variable

b. $6x + 2y = 8$ Linear equation in two variables

NOTE: The expression to the left of the equal sign is called the left side or left member of the equation. The expression to the right is called the right side or right member of the equation.

2 An equation may be either a true or false statement. For example, the equation $x + 6 = 9$ is a true statement when $x = 3$. We call 3 the **solution** or **root** of the equation since $x = 3$ makes the equation a true statement.

3 **To Determine if a Value is a Solution of an Equation:**

1. Replace the variable with its given (or obtained) value.
2. Do the indicated operations on both sides of the equation.
3. If the same number appears on both sides of the equation, the solution is correct. Otherwise, it is incorrect.

Example 2 Determine if the given value is a solution.

a. $2x - 9 = -15, x = -3$

b. $3x + 2.11 = 5x - 3.2, x = 0.5$

c. $3(x - 2) - 1 = 2 - 5(x + 5), x = -2$

d. $5x - 3 = \frac{1}{10} + \frac{3}{4}x, x = \frac{1}{2}$

Solution To determine if a given value is a solution of an equation, replace the variable with its given value and then determine if the equation is a true or false statement.

2a.
$$\begin{aligned} 2x - 9 &= -15, x = -3 \\ 2(-3) - 9 &= -15 \\ -6 - 9 &= -15 \\ -15 &= -15 \end{aligned}$$
True

Thus, -3 is the correct solution.

2b.
$$\begin{aligned} 3x + 2.11 &= 5x - 3.2, x = 0.5 \\ 3(0.5) + 2.11 &= 5(0.5) - 3.2 \\ 1.5 + 2.11 &= 2.5 - 3.2 \\ 3.61 &= -0.7 \end{aligned}$$
False

Thus, 0.5 is not the correct solution.

2c.

$$\begin{aligned} 3(x-2)-1 &= 2-5(x+5),\ x=-2 \\ 3(-2-2)-1 &= 2-5(-2+5) \\ 3(-4)-1 &= 2-5(3) \\ -12-1 &= 2-15 \\ -13 &= -13 \end{aligned}$$

True

Thus, -2 is the correct solution.

2d.

$$\begin{aligned} 5x-3 &= \frac{1}{10}+\frac{3}{4}x,\ x=\frac{1}{2} \\ 5\left(\frac{1}{2}\right)-3 &= \frac{1}{10}+\frac{3}{4}\cdot\frac{1}{2} \\ \frac{5}{2}-3 &= \frac{1}{10}+\frac{3}{8} \\ \frac{5}{2}-\frac{6}{2} &= \frac{4}{40}+\frac{15}{40} \\ -\frac{1}{2} &= \frac{19}{40} \end{aligned}$$

False

Thus, $\frac{1}{2}$ is not the solution.

Practice Problem 1 ***Determine if the given value is the solution.***

a. $8x-3=-2x+4,\ x=-2$

b. $3.7x-5=-6.87,\ x=3.1$

c. $\frac{5n}{6}+\frac{49}{27}=\frac{12n+7}{27},\ n=-4$

4 There are different types of linear equations.

Example 3 **a.** $x+5=7$

Since 2 is the only solution, $x+5=7$ is a conditional equation. A **conditional equation** is an equation that is true for only certain values of the variable.

b. $2x=x+x$

Since every value of x is a solution to this equation, it is an identity. An **identity** is an equation that is always a true statement for every value of the variable.

c. $x=x+2$

Since no value of x is a solution to this equation, it is said to have no solution. An equation has **no solution** when it is always a false statement for every value of the variable.

7.1 Exercises

Identify the linear equations in one variable.

1. $3x+2y=6$

2. $x+6=8$

3. $x^2+9=16$

4. $3n+2=5n+2$

5. $x^2+y^2=16$

Determine if the given value is a solution.

6. $x + 2 = 6, x = 4$

7. $2x + 5 = 9, x = 4$

8. $10x = 25, x = \frac{5}{2}$

9. $x - 8 = 10, x = -2$

10. $4x - 2(3 - x) = 12, x = 3$

11. $y - 8.4 = -6.5, y = 1.9$

12. $\frac{1}{3} + n = \frac{5}{6}, n = \frac{4}{3}$

13. $b + \frac{5}{12} = \frac{1}{10}, b = -\frac{3}{8}$

14. $4 = \frac{2x}{5} - 6, x = 25$

15. $\frac{3a}{5} + \frac{4}{7} = \frac{a}{5} + \frac{3}{8}, x = 4$

16. $0.023x = -46, x = -20$

17. $-36 + x = 8.3, x = 11.9$

18. $9x + 7 = 5x - 13, x = -5$

19. $5x + 4 = 3x - 2, x = 3$

20. $0.35n - 1.7 = 1.85n + 4, n = -3.8$

21. $6x + 2 = -5x, x = -3$

22. $4x = 2(12 - 2x), x = 3$

23. $5(3x - 2) = 35, x = -2$

24. $10 - 7x = 4(11 - 6x), x = 4$

25. $4x - 5(3 + 2x) = 3, x = 5$

26. $\frac{x}{3} - \frac{x}{2} = 2, x = 12$

27. $\frac{x}{6} - \frac{2x}{9} = -\frac{x}{18}, x = 5$

Fill in the blanks.

28. A linear or ______________ equation in one variable is an equation in which the highest power of the variable is one.

29. If a value for a variable makes an equation a true statement, this value is called a root or ______________ of the equation.

30. To determine if a given value is the solution of an equation, replace the ______________ with its given value. If this results in a ______________ statement, the solution is ______________. Otherwise, it is ______________.

Determine if the given value is a solution.

31. $x + 5.6 = 2.8, x = -2.8$

32. $x + 3.04 = 2.96, x = -0.08$

33. $x + \frac{1}{2} = \frac{5}{2}, x = 4$

34. $x + \frac{3}{4} = \frac{23}{4}, x = 20$

35. $2.5x - 3.8 = -7.9, x = -1.64$

36. $3.75x + 0.125 = -0.125, x = -0.0067$

37. $-73 = 24x + 31, x = -4\frac{1}{3}$

38. $-48 = 36x + 42, x = -2\frac{1}{2}$

39. $7 = \frac{2x}{5} + 3, x = -10$

40. $9 = \frac{3x}{4} + 6, x = -4$

41. $7 - 9x - 12 = 3x + 5 - 8x, x = -\frac{5}{2}$

42. $13 - 11x - 17 = 5x + 4 - 10x, x = -\frac{4}{3}$

43. $10x - 2(3 + 4x) = 7 - (x - 2), x = -5$

44. $7x - 3(2x - 5) = 6(2 + 3x) - 31, x = -2$

45. $\frac{2(x - 3)}{5} - \frac{3(x + 2)}{2} = \frac{7}{10}, x = -3$

46. $\frac{5(x - 4)}{6} - \frac{2(x + 4)}{9} = \frac{5}{18}, x = -5$

Answers to Practice Problems **1a.** -2 is *not* the solution **b.** 3.1 is *not* the solution **c.** -4 is the solution

7.2 Solving Equations Containing One Operation

5 To solve (or find the solution to) a linear equation containing one variable, your goal is to isolate the variable on one side of the equation and a constant on the opposite side. This can be done by using the addition, multiplication, and/or division principles of equality.

6 The **addition principle of equality** states that if the same number is added to both sides of an equation, the resulting sums are equal. Using this principle, we can solve equations of the form

$$x + a = b$$

by adding the additive inverse of a to both sides of the equation.

Example 4 Solve: $x - 7 = -11$

Solution To isolate x on the left side of the equation, add 7 to both sides.

$$x - 7 = -11$$
$$x - 7 + 7 = -11 + 7 \qquad \text{Add 7 to both sides}$$
$$x + 0 = -4 \qquad \text{Do this mentally}$$
$$x = -4$$

To check, let $x = -4$ (See Idea 3, Section 7.1.)

$$x - 7 = -11$$
$$-4 - 7 = -11$$
$$-11 = -11 \qquad \text{True}$$

Thus, -4 is the solution.

NOTE: Since $x - 7 = -11$ and $x = -4$ have the same solution, they are called **equivalent equations.** Also, a conditional linear equation in one variable has only one solution.

Example 5 Solve.

a. $x + 9 = -11$ **b.** $x - 0.11 = -0.5$ **c.** $x - \frac{2}{9} = \frac{1}{6}$

Solution To solve an equation of the form $x + a = b$, add the additive inverse of a to both sides of the equation.

5a.

$$x + 9 = -11$$
$$x + 9 + (-9) = -11 + (-9)$$
$$x = -20$$

Check:

$$x + 9 = -11$$
$$-20 + 9 = -11$$
$$-11 = -11$$

5b.

$$x - 0.11 = -0.5$$
$$x - 0.11 + 0.11 = -0.5 + 0.11$$
$$x = -0.39$$

Check:

$$x - 0.11 = -0.5$$
$$-0.39 - 0.11 = -0.5$$
$$-0.5 = -0.5$$

5c.

$$x - \frac{2}{9} = \frac{1}{6}$$
$$x - \frac{2}{9} + \frac{2}{9} = \frac{1}{6} + \frac{2}{9}$$
$$x = \frac{3}{18} + \frac{4}{18}$$
$$x = \frac{7}{18}$$

Check:

$$x - \frac{2}{9} = \frac{1}{6}$$
$$\frac{7}{18} - \frac{2}{9} = \frac{1}{6}$$
$$\frac{7}{18} - \frac{4}{18} = \frac{1}{6}$$
$$\frac{3}{18} = \frac{1}{6}$$
$$\frac{1}{6} = \frac{1}{6}$$

Practice Problem 2 ***Solve.***

a. $x + 5 = -9$ **b.** $x + 0.35 = 0.1$ **c.** $x - \frac{3}{10} = -\frac{5}{6}$

7 The **division principle of equality** states that if both sides of an equation are divided by the same nonzero number, the resulting quotients are equal. Using this principle, we can solve equations of the form

$$ax = b$$

by dividing both sides of the equation by a (the coefficient of x).

Example 6 Solve: $2x = 6$

Solution To isolate x on the left side, divide both sides by 2.

$$2x = 6$$
$$\frac{2x}{2} = \frac{6}{2} \quad \text{Divide both sides by 2}$$
$$1x = 3 \quad \text{Do this mentally}$$
$$x = 3$$

Check:

$$2x = 6$$
$$2(3) = 6$$
$$6 = 6$$

Example 7 Solve.

a. $-9 = 3x$ **b.** $-0.6x = 3.6$ **c.** $-27x = -36$

Solution To solve an equation of the form $ax = b$, divide both sides by a.

7a.

$$-9 = 3x$$
$$\frac{-9}{3} = \frac{3x}{3}$$
$$-3 = x$$

Check:

$$-9 = 3x$$
$$-9 = 3(-3)$$
$$-9 = -9$$

We could rewrite $-9 = 3x$ as $3x = -9$ and then solve for x. Also, remember that the equation $x = -3$ and the equation $-3 = x$ both mean that x is equal to -3.

7b.

$$-0.6x = 3.6$$
$$\frac{-0.6x}{-0.6} = \frac{3.6}{-0.6}$$
$$x = -6$$

Check:

$$-0.6x = 3.6$$
$$-0.6(-6) = 3.6$$
$$3.6 = 3.6$$

7c.

$$-27x = -36$$
$$\frac{-27x}{-27} = \frac{-36}{-27}$$
$$x = \frac{4}{3}$$

Check:

$$-27x = -36$$
$$-27\left(\frac{4}{3}\right) = -36$$
$$-36 = -36$$

Practice Problem 3 ***Solve.***

a. $-6x = 12$ **b.** $0.2x = 0.84$ **c.** $9x = 15$

8

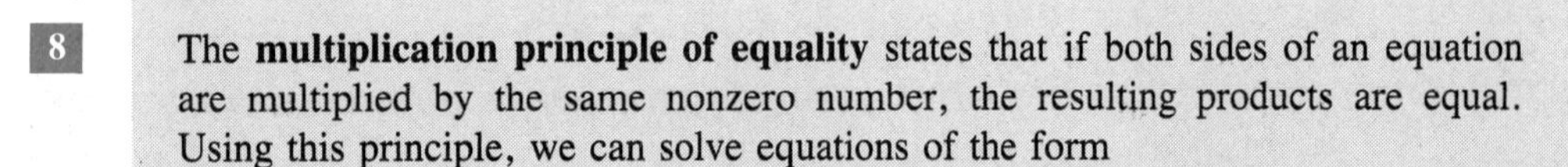

The **multiplication principle of equality** states that if both sides of an equation are multiplied by the same nonzero number, the resulting products are equal. Using this principle, we can solve equations of the form

$$\frac{x}{a} = b$$

by multiplying both sides of the equation by a (the denominator).

Example 8 Solve: $\frac{x}{4} = 6$

Solution To isolate x on the left side of the equation, multiply both sides by 4.

$$\frac{x}{4} = 6$$
$$4 \cdot \frac{x}{4} = 4 \cdot 6 \quad \text{Multiply both sides by 4}$$
$$\frac{\not{4} \cdot x}{\not{4}} = 24 \quad \text{Do this mentally}$$
$$x = 24$$

Check:

$$\frac{x}{4} = 6$$
$$\frac{24}{4} = 6$$
$$6 = 6$$

Example 9 Solve.

a. $\frac{x}{5} = -6$ **b.** $-8 = \frac{x}{-2}$

c. $\frac{n}{0.2} = 0.3$ **d.** $\frac{x}{14} = \frac{6}{7}$

Solution To solve an equation of the form $\frac{x}{a} = b$, multiply both sides by a.

9a.

$$\frac{x}{5} = -6$$
$$5\left(\frac{x}{5}\right) = 5(-6)$$
$$x = -30$$

Check:

$$\frac{x}{5} = -6$$
$$\frac{-30}{5} = -6$$
$$-6 = -6$$

9b.

$$-8 = \frac{x}{-2}$$
$$-2(-8) = -2\left(\frac{x}{-2}\right)$$
$$16 = x$$

Check:

$$-8 = \frac{x}{-2}$$
$$-8 = \frac{16}{-2}$$
$$-8 = -8$$

9c.

$$\frac{n}{0.2} = 0.3$$
$$0.2\left(\frac{n}{0.2}\right) = 0.2\,(0.3)$$
$$n = 0.06$$

Check:
$$\frac{n}{0.2} = 0.3$$
$$\frac{0.06}{0.2} = 0.3$$
$$0.3 = 0.3$$

9d.

$$\frac{x}{14} = \frac{6}{7}$$
$$14 \cdot \frac{x}{14} = 14 \cdot \frac{6}{7}$$
$$x = 12$$

Check:
$$\frac{x}{14} = \frac{6}{7}$$
$$\frac{12}{14} = \frac{6}{7}$$
$$\frac{6}{7} = \frac{6}{7}$$

Practice Problem 4 **Solve.**

a. $\frac{x}{-6} = -2$ **b.** $\frac{x}{4} = -0.5$ **c.** $\frac{n}{12} = \frac{7}{2}$

9

The multiplication principle can also be used to solve equations of the form

$$ax = b.$$

To solve such equations, multiply both sides of the equation by the reciprocal of a.

Example 10 Solve.

a. $\frac{3}{4}x = -18$ **b.** $-\frac{6}{25}x = \frac{21}{10}$ **c.** $3x = -15$

Solution To solve an equation of the form $ax = b$, multiply both sides by the reciprocal of a. Remember, to find the reciprocal of a number, just invert that number.

10a.

$$\frac{3}{4}x = -18$$
$$\frac{4}{3}\left(\frac{3}{4}x\right) = \frac{4}{3}(-18)$$
$$1x = -24 \quad \text{Do this mentally}$$
$$x = -24$$

Check:
$$\frac{3}{4}x = -18$$
$$\frac{3}{4}(-24) = -18$$
$$-18 = -18$$

10b.

$$-\frac{6}{25}x = \frac{21}{10}$$
$$-\frac{25}{6}\left(-\frac{6}{25}x\right) = -\frac{25}{6}\left(\frac{21}{10}\right)$$
$$x = -\frac{\overset{5}{\cancel{25}} \cdot \overset{7}{\cancel{21}}}{\underset{2}{\cancel{10}} \cdot \underset{2}{\cancel{6}}}$$
$$x = -\frac{35}{4}$$

Check:
$$-\frac{6}{25}x = \frac{21}{10}$$
$$-\frac{6}{25}\left(-\frac{35}{4}\right) = \frac{21}{10}$$
$$+\frac{\overset{3}{\cancel{6}} \cdot \overset{7}{\cancel{35}}}{\underset{5}{\cancel{25}} \cdot \underset{2}{\cancel{4}}} = \frac{21}{10}$$
$$\frac{21}{10} = \frac{21}{10}$$

10c.

$$3x = -15$$
$$\frac{1}{3}(3x) = \frac{1}{3}(-15)$$
$$x = -5$$

Check:

$$3x = -15$$
$$3(-5) = -15$$
$$-15 = -15$$

Example 11 Solve: $-x = 2$

Solution To solve for x, multiply by -1.

$$-x = 2$$
$$(-1)(-x) = (-1)(2)$$
$$x = -2$$

Check:

$$-x = 2$$
$$-(-2) = 2$$
$$2 = 2$$

Practice Problem 5 ***Solve.***

a. $\frac{1}{2}x = -4$ **b.** $-\frac{4}{9}x = \frac{10}{12}$ **c.** $-n = -2$

7.2 Exercises

Solve by using the addition principle and then check your answer.

1. $x + 5 = 10$
2. $x - 6 = -10$
3. $x - 7 = -13$
4. $x - 9 = -2$
5. $x + 8 = 3$
6. $8 = x + 13$
7. $x - 8 = -3$
8. $5 + x = -9$
9. $3 + x = -15$
10. $x + 5.2 = 4.8$
11. $x - 3 = -4.1$
12. $x + 0.7 = -0.9$
13. $x - 2.4 = -5.11$
14. $x + 9 = -11$
15. $x + \frac{1}{2} = -\frac{1}{2}$
16. $x - \frac{5}{6} = \frac{7}{10}$
17. $x + \frac{3}{10} = -\frac{5}{8}$
18. $\frac{5}{18} = x + \frac{7}{12}$
19. $x + 9 = -35$
20. $x + 5.7 = 9$
21. $x - \frac{1}{14} = -\frac{1}{21}$

Solve by using the multiplication or division principle and then check your answer.

22. $5x = 10$
23. $-6x = -12$
24. $4x = -16$
25. $-15x = 105$
26. $10x = 30$
27. $8x = -12$
28. $\frac{x}{3} = 5$
29. $\frac{x}{-4} = -7$
30. $\frac{x}{5} = -8$
31. $\frac{x}{0.3} = -0.3$
32. $\frac{x}{-1.2} = -5$
33. $\frac{3}{4}x = -18$
34. $\frac{4}{9}x = 20$
35. $-\frac{3}{5}x = \frac{6}{25}$
36. $\frac{1}{2}x = -10$
37. $1.2x = -3.6$
38. $0.1x = 5$
39. $-0.5x = -1.25$
40. $25x = -35$
41. $-2.4x = 96$
42. $91x = 35$

Fill in the blanks.

43. To solve an equation of the form $x + a = b$, add the ______________ of a to both sides.

44. To solve an equation of the form $ax = b$, ______________ both sides by a or multiply both sides by the ______________ of a.

Solve.

45. $x + 3.5 = 9$ **46.** $0.3x = -3$ **47.** $\frac{x}{-4} = 6$ **48.** $\frac{9}{16}x = -\frac{6}{10}$

49. $-x = -6$ **50.** $45x = 135$ **51.** $\frac{x}{-0.2} = 10$ **52.** $\frac{9}{25} = x + \frac{1}{35}$

53. $3 = -\frac{1}{5}x$ **54.** $8 = -\frac{1}{4}x$ **55.** $\frac{2}{3}x = 10$ **56.** $\frac{3}{4}x = 9$

57. $-\frac{2}{7}x = -6$ **58.** $-\frac{3}{5}x = -9$ **59.** $\frac{3}{8}x = \frac{3}{8}$ **60.** $\frac{2}{9}x = \frac{2}{7}$

61. $\frac{4}{9}x = \frac{10}{27}$ **62.** $\frac{6}{49}x = \frac{15}{14}$ **63.** $-\frac{1}{5}x = \frac{3}{10}$ **64.** $-\frac{1}{4}x = \frac{7}{10}$

65. $\frac{x}{-7} = -5$ **66.** $\frac{x}{-3} = -8$ **67.** $\frac{x}{1.2} = 5$ **68.** $\frac{x}{1.4} = 5$

69. $\frac{x}{9} = -\frac{7}{6}$ **70.** $\frac{x}{6} = -\frac{11}{9}$ **71.** $\frac{x}{35} = \frac{15}{25}$ **72.** $\frac{x}{49} = \frac{14}{21}$

73. $\frac{x}{-0.3} = 0.05$ **74.** $\frac{x}{-0.2} = -0.08$ **75.** $\frac{x}{5} = -8.2$ **76.** $\frac{x}{3} = -7.4$

77. $0.12x = -13.2$ **78.** $0.11x = -12.1$ **79.** $-5x = 18.5$ **80.** $-8x = 33.6$

81. $x + 0.4 = -\frac{3}{10}$ **82.** $x + 0.3 = -\frac{1}{10}$ **83.** $0.4x = -\frac{4}{25}$ **84.** $0.8x = -\frac{8}{25}$

85. $\frac{8}{35}x = \frac{12}{45}$ **86.** $\frac{12}{49}x = \frac{18}{35}$ **87.** $\frac{x}{-20} = -\frac{3}{8}$ **88.** $\frac{x}{-18} = -\frac{5}{27}$

Answers to Practice Problems **2a.** -14 **b.** -0.25 **c.** $-\frac{8}{15}$ **3a.** -2 **b.** 4.2 **c.** $\frac{5}{3}$ **4a.** 12 **b.** -2 **c.** 42 **5a.** -8 **b.** $-\frac{15}{8}$ **c.** 2

7.3 Solving Equations Containing More Than One Operation

10 Sometimes you will need to use more than one principle to solve an equation. When this occurs, you should usually use the addition principle first.

Example 12 Solve.

a. $-3x + 5 = -10$ **b.** $\frac{x}{2} - 10 = -30$

c. $\frac{3}{5}x + 8 = 2$ **d.** $-2x + 6x - 8 = -18$

Solution To solve an equation of the form $ax + b = c$ or $\frac{x}{a} + b = c$, first add the additive inverse of b to both sides. Next, solve the resulting equation.

12a.

$$\begin{aligned} -3x + 5 &= -10 \\ -3x + 5 + (-5) &= -10 + (-5) \\ -3x &= -15 \\ \frac{-3x}{-3} &= \frac{-15}{-3} \\ x &= 5 \end{aligned}$$

Check:

$$\begin{aligned} -3x + 5 &= -10 \\ -3(5) + 5 &= -10 \\ -15 + 5 &= -10 \\ -10 &= -10 \end{aligned}$$

12b.

$$\frac{x}{2} - 10 = -30$$
$$\frac{x}{2} - 10 + 10 = -30 + 10$$
$$\frac{x}{2} = -20$$
$$2\left(\frac{x}{2}\right) = 2(-20)$$
$$x = -40$$

Check:
$$\frac{x}{2} - 10 = -30$$
$$\frac{-40}{2} - 10 = -30$$
$$-20 - 10 = -30$$
$$-30 = -30$$

12c.

$$\frac{3}{5}x + 8 = 2$$
$$\frac{3}{5}x + 8 + (-8) = 2 + (-8)$$
$$\frac{3}{5}x = -6$$
$$\frac{5}{3}\left(\frac{3}{5}x\right) = \frac{5}{3}(-6)$$
$$x = -10$$

Check:
$$\frac{3}{5}x + 8 = 2$$
$$\frac{3}{5}(-10) + 8 = 2$$
$$-6 + 8 = 2$$
$$2 = 2$$

12d. When like terms appear on the same side of an equation, combine them before applying any of the principles.

$$-2x + 6x - 8 = -18$$
$$4x - 8 = -18$$
$$4x - 8 + 8 = -18 + 8$$
$$4x = -10$$
$$\frac{4x}{4} = -\frac{10}{4}$$
$$x = -\frac{5}{2}$$

Check:
$$-2x + 6x - 8 = -18$$
$$-2\left(-\frac{5}{2}\right) + 6\left(-\frac{5}{2}\right) - 8 = -18$$
$$5 + (-15) - 8 = -18$$
$$-18 = -18$$

Practice Problem 6 **Solve.**

a. $-4x + 9 = -3$ **b.** $\frac{1}{2}x - 8 = -13$

c. $x - 0.3x + 5 = -2$

11 Quite often, variables will appear on both sides of an equation.

To Solve Linear Equations Having Variables on Both Sides:

1. Combine like terms on each side of the equation.
2. Isolate all terms containing the variable on one side of the equation and all constants on the opposite side by using the addition principle. Remember to combine like terms.
3. Solve the resulting equation by using the multiplication or division principle.

Example 13 Solve.

a. $6x = 2x + 8$ **b.** $x + 20 = 7x + 10$

c. $1.9x - 7.8 + 5.3x = 3 + 1.8x$

Solution To solve a linear equation having variables on both sides, combine like terms on each side. Next, use the addition principle to get all variables on the same side and then solve the resulting equation.

13a.

$$\begin{aligned} 6x &= 2x + 8 \\ 6x + (-2x) &= 2x + 8 + (-2x) \\ 4x &= 8 \\ \frac{4x}{4} &= \frac{8}{4} \\ x &= 2 \end{aligned}$$

Check:

$$\begin{aligned} 6x &= 2x + 8 \\ 6(2) &= 2(2) + 8 \\ 12 &= 4 + 8 \\ 12 &= 12 \end{aligned}$$

13b.

$$\begin{aligned} x + 20 &= 7x + 10 \\ x + 20 + (-x) &= 7x + 10 + (-x) \\ 20 &= 6x + 10 \\ 20 + (-10) &= 6x + 10 + (-10) \\ 10 &= 6x \\ \frac{10}{6} &= \frac{6x}{6} \\ \frac{5}{3} &= x \end{aligned}$$

Check:

$$\begin{aligned} x + 20 &= 7x + 10 \\ \frac{5}{3} + 20 &= 7\left(\frac{5}{3}\right) + 10 \\ 1\frac{2}{3} + 20 &= \frac{35}{3} + 10 \\ 21\frac{2}{3} &= 11\frac{2}{3} + 10 \\ 21\frac{2}{3} &= 21\frac{2}{3} \end{aligned}$$

13c.

$$\begin{aligned} 1.9x - 7.8 + 5.3x &= 3 + 1.8x \\ 7.2x - 7.8 &= 3 + 1.8x \quad \text{Combine like terms} \\ 7.2x - 7.8 + (-1.8x) &= 3 + 1.8x + (-1.8x) \\ 5.4x - 7.8 &= 3 \\ 5.4x - 7.8 + 7.8 &= 3 + 7.8 \\ 5.4x &= 10.8 \\ \frac{5.4x}{5.4} &= \frac{10.8}{5.4} \\ x &= 2 \end{aligned}$$

Check:

$$\begin{aligned} 1.9x - 7.8 + 5.3x &= 3 + 1.8x \\ 1.9(2) - 7.8 + 5.3(2) &= 3 + 1.8(2) \\ 3.8 - 7.8 + 10.6 &= 3 + 3.6 \\ 6.6 &= 6.6 \end{aligned}$$

Practice Problem 7 ***Solve.***

a. $13x + 5 = 8x + 40$ **b.** $0.4x = x + 8$

c. $x + 2x - 7 = 3x + x - 5$

7.3 Exercises

Solve. Check your answers.

1. $2x + 9 = 189$ **2.** $10x + 8 = -17$ **3.** $0.2x - 0.9 = 0.27$ **4.** $\frac{x}{2} - 10 = -20$

5. $\frac{1}{2}x + 4 = 1$ **6.** $\frac{3}{5}x + 7 = -14$ **7.** $-\frac{6}{7}x + 5 = -5$ **8.** $\frac{x}{0.3} - 0.55 = 0.2$

9. $\frac{x}{-2} - 8 = -10$
10. $2x + 3x = -15$
11. $7x - 9x = -4$
12. $x + 0.3x = 39$

13. $4x = 2x - 12$
14. $8x + 40 = 2x$
15. $x = 13.2 + 0.2x$

Solve. Check your answers.

16. $13x + 5 = 8x + 40$
17. $6x - 2 = 8 - 14x$
18. $x + 2x - 7 = 4x - 5$

19. $x + 1 = 16 - x$
20. $10x + 4 = 7x + 12$
21. $10x - 12 = 6x + 12$

22. $0.8x + 0.6 = 0.2 - x$
23. $4.2x + 0.4 = 6.4 - 16.8x$
24. $8x + 9 = -2x - 16$

25. $6x + 5 = -3x - 10$
26. $2.2x - 5 + 4.5x = 1.7x - 20$
27. $0.17x - 1.7 + 0.18x = 1.85x + 4$

28. $10x + 1 = -15$
29. $8x + 2 = -12$
30. $x + 0.1x = 0.33$

31. $x + 0.2x = 0.48$
32. $-0.2x + 0.42 = 0.8$
33. $-0.3x + 0.24 = 0.6$

34. $5x - 8x = -15$
35. $4x - 6x = -12$
36. $15x - 5 = -20$

37. $18x - 6 = -24$
38. $\frac{x}{3} - 8 = -3.2$
39. $\frac{x}{2} - 6 = -4.4$

40. $\frac{x}{0.2} + 3.1 = -6.9$
41. $\frac{x}{0.3} + 9.3 = -20.7$
42. $\frac{2}{5}x + 3 = 7$

43. $\frac{3}{4}x + 6 = 9$
44. $13 = -\frac{2}{9}x + 3$
45. $41 = -\frac{4}{5}x + 25$

46. $6x + 4 = 4x - 2$
47. $5x + 3 = 2x - 9$
48. $4x - 7 - x = 8 - 2x$

49. $5x - 10 - x = 4 - 3x$
50. $y + 15 = 18y - 19$
51. $y + 18 = 15y - 24$

52. $10x - 8 = 4x$
53. $8x - 10 = 2x$
54. $6.4 - 0.4x = 0.4x$

55. $1.2 - 0.3x = 0.3x$
56. $25x - 91 = -10x$
57. $39x - 77 = -10x$

58. $10x - 6 - 8x = 7 - x + 2$
59. $11x - 9 - 8x = 8 - x + 3$
60. $-9x - 5 = -5x + 5$

61. $-6x - 8 = -2x + 10$
62. $0.7x - 0.8 = 0.8 - 0.9x$
63. $0.5x - 0.7 = 0.7 - 0.9x$

64. $3.4x = -18 + 3x$
65. $4.3x = -15 + 4x$
66. $0.3x - 0.6 = -0.485$

67. $0.2x - 0.4 = -0.382$
68. $\frac{1}{2}x + 0.3 = -0.2$
69. $\frac{1}{4}x + 0.2 = -0.6$

70. $0.37x + 0.013 = -0.098$
71. $0.22x + 0.011 = -0.143$
72. $-3x + 8x + 17 = -8$

73. $-2x + 6x + 11 = -5$
74. $-36 = 18x - 9$
75. $-24 = 12x - 8$

76. $-1.2x - 8 = 16$
77. $1.5x - 7 = 23$
78. $3.5x - 17 = 18.5x + 40$

79. $2.3x - 15 = 15.3x + 39.6$
80. $x + 0.6 - 0.2x = 2x$
81. $x + 0.24 - 0.4x = 3x$

Fill in the blanks.

82. To solve an equation of the form $ax + b = c$ or $\frac{x}{a} + b = c$, first add the ______________ of b to both sides of the equation. Next, solve the resulting equation.

83. To solve an equation having variables on both sides, ______________ like terms on each side of the equation. Next, use the ______________ principle to get all variables on the same side and then solve the resulting equation.

Answers to Practice Problems **6a.** 3 **b.** -10 **c.** -10 **7a.** 7 **b.** $-\frac{40}{3}$ **c.** -2

7.4 Solving Equations Involving Grouping Symbols or Fractions

12 Sometimes an equation may contain grouping symbols.

To Solve Linear Equations Containing Grouping Symbols:

1. Remove the grouping symbol by performing the indicated operation.
2. Solve the resulting equation.

Example 14 Solve.

a. $8x = 4(12 - 3x)$ **b.** $0.5(x + 2) = 0.25$

c. $7(5x - 1) - 18x = 12x - (3 - x)$

Solution To solve an equation containing a grouping symbol, perform the indicated operation to remove the grouping symbol and then solve the resulting equation.

14a.

$$\begin{aligned} 8x &= 4(12 - 3x) \\ 8x &= 48 - 12x \quad \text{Multiply} \\ 8x + 12x &= 48 - 12x + 12x \\ 20x &= 48 \\ \frac{20x}{20} &= \frac{48}{20} \\ x &= \frac{12}{5} \end{aligned}$$

Check:

$$\begin{aligned} 8x &= 4(12 - 3x) \\ 8\left(\frac{12}{5}\right) &= 4\left(12 - 3 \cdot \frac{12}{5}\right) \\ \frac{96}{5} &= 4\left(12 - \frac{36}{5}\right) \\ \frac{96}{5} &= 4\left(\frac{24}{5}\right) \\ \frac{96}{5} &= \frac{96}{5} \end{aligned}$$

14b.

$$\begin{aligned} 0.5(x + 2) &= 0.25 \\ 0.5x + 1 &= 0.25 \\ 0.5x + 1 + (-1) &= 0.25 + (-1) \\ 0.5x &= -0.75 \\ \frac{0.5x}{0.5} &= \frac{-0.75}{0.5} \\ x &= -1.5 \end{aligned}$$

Check:

$$\begin{aligned} 0.5(x + 2) &= 0.25 \\ 0.5(-1.5 + 2) &= 0.25 \\ 0.5(0.5) &= 0.25 \\ 0.25 &= 0.25 \end{aligned}$$

14c.

$$\begin{aligned} 7(5x - 1) - 18x &= 12x - (3 - x) \\ 35x - 7 - 18x &= 12x - 3 + x \\ 17x - 7 &= 13x - 3 \quad \text{Combine like terms} \\ 17x - 7 + (-13x) &= 13x - 3 + (-13x) \\ 4x - 7 &= -3 \\ 4x - 7 + 7 &= -3 + 7 \\ 4x &= 4 \\ \frac{4x}{4} &= \frac{4}{4} \\ x &= 1 \end{aligned}$$

Check:

$$\begin{aligned} 7(5x - 1) - 18x &= 12x - (3 - x) \\ 7(5 \cdot 1 - 1) - 18 \cdot 1 &= 12 \cdot 1 - (3 - 1) \\ 7(5 - 1) - 18 &= 12 - 2 \\ 7(4) - 18 &= 10 \\ 28 - 18 &= 10 \\ 10 &= 10 \end{aligned}$$

Practice Problem 8 ***Solve.***

a. $4x - 6 = 2(x + 5)$ **b.** $20 = 8 - 2(9 - 3x)$ **c.** $0.4(x - .1) = 6$

13 Frequently you will need to solve an equation containing one or more fractions.

To Solve Linear Equations Containing Fractions:

1. Multiply both sides of the equation by the LCD of all terms of the equation.
2. Solve the resulting equation.

Example 15 Solve.

a. $\frac{x}{4} + \frac{x}{3} + \frac{x}{2} = 26$ **b.** $\frac{2x}{3} = 6 + 2x$ **c.** $\frac{3}{4}x + \frac{1}{3}x = x - \frac{5}{3} - \frac{2}{3}x$

Solution To solve an equation containing fractions, remove all fractions by multiplying both sides by the LCD of all denominators. Solve the resulting equation.

15a.

$$\begin{aligned} \frac{x}{4} + \frac{x}{3} + \frac{x}{2} &= 26 \\ {}^{*}\,12\left(\frac{x}{4} + \frac{x}{3} + \frac{x}{2}\right) &= 12\,(26) \\ 12\left(\frac{x}{4}\right) + 12\left(\frac{x}{3}\right) + 12\left(\frac{x}{2}\right) &= 12(26) \\ 3x + 4x + 6x &= 312 \\ 13x &= 312 \\ \frac{13x}{13} &= \frac{312}{13} \\ x &= 24 \end{aligned}$$

Check:

$$\begin{aligned} \frac{x}{4} + \frac{x}{3} + \frac{x}{2} &= 26 \\ \frac{24}{4} + \frac{24}{3} + \frac{24}{2} &= 26 \\ 6 + 8 + 12 &= 26 \\ 26 &= 26 \end{aligned}$$

***NOTE:** Multiplying both sides of an equation by the LCD means multiplying each *term* of the equation by the LCD.

15b.

$$\begin{aligned} \frac{2x}{3} &= 6 + 2x \\ 3\left(\frac{2x}{3}\right) &= 3\,(6 + 2x) \\ 2x &= 18 + 6x \\ 2x + (-6x) &= 18 + 6x + (-6x) \\ -4x &= 18 \\ \frac{-4x}{-4} &= \frac{18}{-4} \\ x &= -\frac{9}{2} \end{aligned}$$

Check:

$$\begin{aligned} \frac{2x}{3} &= 6 + 2x \\ \frac{2\left(-\frac{9}{2}\right)}{3} &= 6 + 2\left(-\frac{9}{2}\right) \\ \frac{-9}{3} &= 6 + (-9) \\ -3 &= -3 \end{aligned}$$

15c.

$$\frac{3}{4}x + \frac{1}{3}x = x - \frac{5}{3} - \frac{2}{3}x$$
$$12\left(\frac{3}{4}x + \frac{1}{3}x\right) = 12\left(x - \frac{5}{3} - \frac{2}{3}x\right)$$
$$9x + 4x = 12x - 20 - 8x$$
$$13x = 4x - 20$$
$$13x + (-4x) = 4x - 20 + (-4x)$$
$$9x = -20$$
$$\frac{9x}{9} = -\frac{20}{9}$$
$$x = -\frac{20}{9}$$

Check:

$$\frac{3}{4}x + \frac{1}{3}x = x - \frac{5}{3} - \frac{2}{3}x$$
$$\frac{3}{4}\left(-\frac{20}{9}\right) + \frac{1}{3}\left(-\frac{20}{9}\right) = -\frac{20}{9} - \frac{5}{3} - \frac{2}{3}\left(-\frac{20}{9}\right)$$
$$-\frac{\overset{1}{\cancel{3}} \cdot \overset{5}{\cancel{20}}}{\underset{1}{\cancel{4}} \cdot \underset{3}{\cancel{9}}} + \left(-\frac{20}{27}\right) = -\frac{20}{9} - \frac{5}{3} + \frac{40}{27}$$
$$-\frac{5}{3} + \left(-\frac{20}{27}\right) = -\frac{20}{9} - \frac{5}{3} + \frac{40}{27}$$
$$-\frac{45}{27} + \left(-\frac{20}{27}\right) = -\frac{60}{27} - \frac{45}{27} + \frac{40}{27}$$
$$-\frac{65}{27} = -\frac{65}{27}$$

Practice Problem 9 **Solve for x.**

a. $\frac{3x}{4} + \frac{3}{4} = \frac{2x}{3}$ **b.** $3x + \frac{13}{25} = 13$

c. $\frac{7}{8}x - \frac{1}{4} + \frac{3}{4}x = \frac{1}{16} + x$

14 You may not always be able to find only one solution for a linear equation in one variable.

Example 16 Solve.

a. $x = x + 2$ **b.** $2x - 2(x - 6) = 12$

Solution **16a.**

$$x = x + 2$$
$$x + (-x) = x + 2 + (-x)$$
$$0 = 2 \qquad \text{False}$$

Since the variable was eliminated and we obtained a false statement, the original equation is a false statement and has no solution.

16b.

$$2x - 2(x - 6) = 12$$
$$2x - 2x + 12 = 12$$
$$12 = 12$$
$$12 + (-12) = 12 + (-12)$$
$$0 = 0 \qquad \text{True}$$

Since the variable was eliminated and we obtained a true statement, the original equation is an identity and every number is a solution.

7.4 Exercises

Solve. Check each answer.

1. $3(x - 6) = -15$

2. $5(x - 2) = -10$

3. $2x - 3(2 + 2x) = -6$

4. $9x - 2(5 + 5x) = 8$

5. $10x + 5 = 3(4x + 5)$

6. $6x + 6 = 2(6x + 9)$

7. $9 + x = 5(9 - 7x)$

8. $10 - 3x = 4(11 - 5x)$

9. $3(2x + 3) = 27$

10. $4(2x - 3) = 28$

11. $45 = 5(3x + 2)$

12. $12 = 3(5x - 2)$

13. $2(3 + 4x) - 9 = 53$

14. $3(5 + 3x) - 8 = 38$

15. $5x - (2x + 8) = 24$

16. $7x - (3x + 8) = 16$

17. $6x - 6 = 6(7 - x)$

18. $6x - 20 = 10(x - 4)$

19. $12 - 4(3x - 1) = 4$

20. $20 - 6(2x - 1) = 2$

21. $7(x - 2) = 5(x + 4)$

22. $9(x + 2) = 3(x - 2)$

23. $4(2x + 1) = 2(7x + 7)$

24. $5(2x + 3) = 3(6x + 6)$

25. $2x + 2 = -0.5(4x - 8)$

26. $2x + 12 = -0.25(12x - 72)$

27. $\frac{1}{2}(4x - 6) = -9$

28. $\frac{1}{4}(4x - 8) = -8$

29. $\frac{3}{4}(8x - 4) = \frac{1}{3}(6x - 6)$

30. $\frac{1}{5}(5x - 10) = \frac{3}{2}(4x - 6)$

31. $\frac{4}{5}(10x - 10) = -4x$

32. $\frac{3}{7}(14x - 21) = -9x$

33. $\frac{x}{2} + \frac{x}{3} = -15$

34. $\frac{x}{3} + \frac{x}{4} = -21$

35. $\frac{x}{3} = 7 - \frac{x}{4}$

36. $\frac{x}{5} = 6 - \frac{x}{3}$

37. $\frac{x}{2} - \frac{x}{5} - 6 = 0$

38. $\frac{x}{3} - \frac{x}{7} = 12$

39. $\frac{x}{4} = \frac{x}{6} + 3$

40. $\frac{x}{6} = \frac{x}{8} + 9$

41. $\frac{x}{4} = 3 - \frac{x}{2}$

42. $\frac{x}{5} = 14 - \frac{x}{2}$

43. $\frac{x}{6} - \frac{x}{4} = \frac{x}{3}$

44. $\frac{x}{6} - \frac{x}{9} = \frac{x}{3}$

45. $\frac{x}{10} + \frac{x}{15} = \frac{x}{2} - 10$

46. $\frac{x}{8} + \frac{x}{6} = \frac{x}{4} - 1$

47. $\frac{x + 3}{10} = \frac{x + 3}{4}$

48. $\frac{x + 2}{8} = \frac{x + 2}{12}$

49. $\frac{x}{3} - \frac{1}{3} = -6$

50. $\frac{x}{4} - \frac{1}{4} = -3$

51. $\frac{2x - 2}{5} = 10 + 3x$

52. $\frac{3x - 3}{4} = 8 + 2x$

53. $\frac{3}{8}x - 5 = \frac{1}{4}$

54. $\frac{1}{6}x - 4 = \frac{1}{2}$

55. $\frac{5x}{2} + \frac{49}{9} = \frac{12x + 7}{9}$

56. $\frac{3x}{2} + \frac{25}{6} = \frac{10x + 8}{6}$

57. $\frac{1}{3}x + \frac{3}{4}x + 5 = \frac{1}{6}x$

58. $\frac{1}{2}x + \frac{3}{5}x + 1 = \frac{1}{10}x$

59. $\frac{1}{4}x + \frac{1}{3}x = \frac{1}{6}x - 5$

60. $\frac{1}{5}x + \frac{1}{3}x = \frac{1}{30}x - 1$

61. $\frac{1}{3}(2x + 5) = \frac{3}{5}$

62. $\frac{1}{4}(2x + 6) = \frac{2}{3}$

Fill in the blanks.

63. To solve an equation containing grouping symbols, ______________ to remove the grouping symbols and then solve the resulting equation.

64. To solve an equation containing fractions, multiply both sides by the ______________ of all terms of the equation. Next, solve the resulting equation.

Solve.

65. $5(x + 4) - 4(x + 3) = 2$

66. $4x - 6 = 2(x + 5)$

67. $5(x - 3) = 15$

68. $3x = 5(12 + 3x)$

69. $7(3x + 6) = 11 - (x + 2)$

70. $\frac{1}{3}(6y - 9) = \frac{1}{2}(8y - 4)$

71. $2 = 5 - 8(3 + 2x)$

72. $0.4(x - 0.1) = 6$

73. $34x - 6(x - 5) = 4 + 8(3x + 3)$

74. $20 - 6(2x - 1) = -10$

75. $\frac{x}{2} + \frac{x}{4} = -6$

76. $\frac{3x}{4} - \frac{3}{4} = \frac{2x}{3}$

77. $\frac{x - 2}{5} + \frac{x}{3} = \frac{1}{5}$

78. $6x - \frac{2}{5} = 8$

79. $\frac{2x - 6}{8} = 10$

80. $\frac{11}{4} = 2 - \frac{3}{10}x$

81. $\frac{3}{8} = \frac{x - 1}{3}$

82. $\frac{x}{-3} + 15 = 2$

83. $\frac{6x}{7} + \frac{8}{5} = \frac{6x}{5} + \frac{4}{7}$

84. $\frac{4 - 2}{3} = \frac{21}{12} - \frac{5x - 3}{4}$

85. $\frac{2}{3}x + 40 = \frac{x}{2} - \frac{x}{6} - \frac{2x}{9}$

86. $\frac{1}{2}\left(\frac{1}{3}x + 4\right) + \frac{x}{4} = 17$

87. $1 - 3(x + 1) = 2(2 - 3x)$

88. $20 - 6(2x - 1) = 2$

Answers to Practice Problems **8a.** 8 **b.** 5 **c.** 15.1 **9a.** -9 **b.** $\frac{104}{25}$ **c.** $\frac{1}{2}$

7.5 Literal Equations

15 A **literal equation** is an equation in which a letter may represent a variable or a constant. For example,

$$2x + 3y = 7 \qquad \text{and} \qquad ax + b = c$$

are literal equations. We will solve literal equations by using the same methods discussed in Sections 7.2, 7.3, and 7.4.

To Solve a Literal Equation for a Particular Letter:

1. Simplify both sides by combining like terms, removing grouping symbols, and/or removing fractions.
2. Isolate all terms containing the letter you are solving for on one side of the equation and all other terms on the opposite side by using the addition principle. Remember to combine like terms.
3. Solve the resulting equation by using the multiplication or division principle.

Example 17 Solve.

a. $ax - b = c$, for x

b. $A = LW$, for W

c. $P = 2(L + W)$, for L

d. $A = \frac{h}{2}(a + b)$, for b

e. $ax + bx = c$, for x

f. $a(x - d) = cx + b$, for x

Solution To solve a literal equation, apply the same principles and procedures used to solve a linear equation in one variable.

17a.

$$ax - b = c$$
$$ax - b + b = c + b \quad \text{Add } b$$
$$ax = c + b$$
$$\frac{ax}{a} = \frac{c + b}{a} \quad \text{Divide by } a$$
$$x = \frac{c + b}{a}$$

17b.

$$A = LW$$
$$\frac{A}{L} = \frac{LW}{L} \quad \text{Divide by } L$$
$$\frac{A}{L} = W$$

17c.

$$P = 2(L + W)$$
$$P = 2L + 2W \quad \text{Multiply}$$
$$P + (-2W) = 2L + 2W + (-2W) \quad \text{Add } -2W$$
$$P - 2W = 2L$$
$$\frac{P - 2W}{2} = \frac{2L}{2} \quad \text{Divide by 2}$$
$$\frac{P - 2W}{2} = L$$

17d.

$$A = \frac{h}{2}(a + b)$$
$$2 \cdot A = 2 \cdot \frac{h}{2}(a + b) \quad \text{Multiply by 2}$$
$$2A = ha + hb$$
$$2A + (-ha) = ha + hb + (-ha) \quad \text{Add } -ha$$
$$2A - ha = hb$$
$$\frac{2A - ha}{h} = \frac{hb}{h} \quad \text{Divide by } h$$
$$\frac{2A - ha}{h} = b$$

17e.

$$ax + bx = c$$
$$(a + b)x = c \quad \text{Combine like terms}$$
$$\frac{(a + b)x}{(a + b)} = \frac{c}{(a + b)} \quad \text{Divide by } (a + b)$$
$$x = \frac{c}{a + b}$$

17f.

$$a(x - d) = cx + b$$
$$ax - ad = cx + b \quad \text{Multiply}$$
$$ax - ad + (-cx) = cx + b + (-cx) \quad \text{Add } -cx$$
$$(a - c)x - ad = b \quad \text{Combine like terms}$$
$$(a - c)x - ad + ad = b + ad \quad \text{Add } ad$$
$$\frac{(a - c)x}{(a - c)} = \frac{b + ad}{(a - c)} \quad \text{Divide by } a - c$$
$$x = \frac{b + ad}{a - c}$$

Practice Problem 10 ***Solve for*** x.

a. $3x + 2y = 6$ **b.** $abx = 1$

c. $\frac{x}{a} + \frac{1}{2} = b$ **d.** $ax + b = cx + d$

16 Formulas are often expressed as literal equations. For example, $\mathbf{d = rt}$ is a distance formula (d = distance, r = rate, and t = time). But sometimes it may be more convenient to express a formula in a different form. For example, suppose we know the distance d and the time t and we wish to find the rate r. In this case, we could solve the equation $d = rt$ for r to obtain a more useful formula.

Example 18 Solve for r: $d = rt$.

Solution

$$d = rt$$

$$\frac{d}{t} = \frac{rt}{t} \quad \text{Divide by } t$$

$$\frac{d}{t} = r$$

Practice Problem 11 ***Solve.***

a. $A = \frac{1}{2}bh$, for h (formula for area of a triangle)

b. $P = 2(L + W)$, for W (formula for perimeter of a rectangle)

7.5 Exercises

Solve.

1. $d = rt$, for t

2. $I = Prt$, for t

3. $A = \frac{1}{2}bh$, for b

4. $P = 2b + f$, for b

5. $h = \frac{11}{2}(p + 40)$, for p

6. $S = (a + l)\frac{n}{2}$, for n

7. $S = (a + l)\frac{n}{2}$, for l

8. $F = \frac{9}{5}C + 32$, for C

9. $ax - bx = 2$, for x

10. $ax + x = 9$, for x

11. $I = \frac{E}{R + r}$, for r

12. $\frac{1}{F} = \frac{1}{U} + \frac{1}{V}$, for U

13. $A = P(1 + rt)$, for r

14. $C = \frac{5}{9}(F - 32)$, for F

15. $4(x - a) = d$, for x

16. $A = 1 + abx$, for x

17. $V = lwh$, for l

18. $F = \frac{W}{2} + \frac{2Wx}{L}$, for x

19. $ax - b = c(x + d)$, for b

20. $L = \frac{Mt - g}{t}$, for t

21. $acx - ax - a = 1$, for x

22. $V = \frac{1}{3}bh$, for b

Solve for x.

23. $x + y = 5$

24. $7x - y = 7$

25. $\frac{x}{3} + \frac{y}{4} = 1$

26. $3x + 5y = -7x - 9$
27. $3x - 5a = c$
28. $3bx + ax = 9$
29. $5(x - a) + 2 = 3(x + 4)$
30. $p(x + b) = c$

Solve.

31. $\frac{1}{a} + \frac{1}{b} = \frac{1}{c}$, for b
32. $\frac{1}{x} + \frac{1}{y} = \frac{1}{z}$, for y
33. $I = Prt$, for r
34. $I = Prt$, for P
35. $ax + bx = 6$, for x
36. $ay + by = 8$, for y
37. $a(b + x) = 3$, for x
38. $b(a + x) = 4$, for x
39. $A = \frac{1}{2}h(a + b)$, for a
40. $A = \frac{1}{2}h(a + b)$, for b

Fill in the blanks.

41. A ______________ equation is an equation in which a letter may represent a variable or a ______________.

42. To solve a literal equation, remember that the letter you are solving for is treated as a ______________. All other letters are treated as ______________.

Answers to Practice Problems 10a. $\frac{6 - 2y}{3}$ b. $\frac{1}{ab}$ c. $ab - \frac{a}{2}$ d. $\frac{b - d}{a - c}$ 11a. $\frac{2A}{b}$ b. $\frac{P - 2L}{2}$

7.6 Solving Inequalities Involving One Operation

17 An **inequality** is a mathematical statement that one quantity is less than or greater than another.

Example 19 The following are linear inequalities.

a. $x + 2 < 5$ This is read as "x plus two is less than five." The $<$ means *less than.*

b. $2x + 3 \leq 9$ $\leq$ means *less than or equal to.*

c. $x - 7 > 6$ $>$ means *greater than.*

d. $2x \geq 10$ $\geq$ means *greater than or equal to.*

NOTE: On a number line:

1. $a < b$ means a is to the left of b

2. $b > a$ means b is to the right of a

Clearly, $a < b$ and $b > a$ are equivalent statements.

18 The methods used to solve linear inequalities are very similar to the procedures we used to solve linear equations in Sections 7.2 and 7.3. That is, to solve a linear inequality, our goal is to isolate the variable on one side of the inequality symbol and a constant on the opposite side. This can be done by using the addition, multiplication, and/or division principles of inequality.

19 The **addition principle of inequality** states that if we add the same number to both sides of an inequality, we will obtain an equivalent inequality. This principle is used to eliminate a term from one side of an inequality.

Example 20 Solve: $x + 2 < 9$.

Solution To solve the inequality, add -2 to both sides.

$$x + 2 < 9$$
$$x + 2 + (-2) < 9 + (-2)$$
$$x < 7$$

The solution $x < 7$ means that all numbers less than 7 (6, 4.5, ½, -2, and so on) are solutions of $x + 2 < 9$. That is, if we replace x with any number less than 7, we will have a true statement. For example, since $1 < 7$,

$$x + 2 < 9$$
$$1 + 2 < 9$$
$$3 < 9. \quad \text{True}$$

Thus, 1 is a solution.

NOTE: We can represent all the solutions to an inequality by using a number line. This is called **graphing the solution.** For example, the solution $x < 7$ can be represented by Figure 7–1. The hollow circle means that 7 is not a solution. The arrow to the left of the circle indicates that all numbers to the left of 7 are solutions.

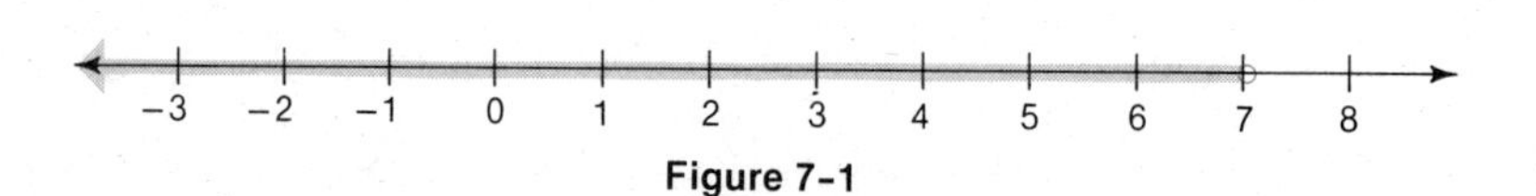

Figure 7-1

Example 21 Solve.

a. $x - 9 \geq -11$ **b.** $8 < x + 4$

Solution To eliminate a term from one side of an inequality, add the additive inverse of that term to both sides of the inequality.

21a.

$$x - 9 \geq -11$$
$$x - 9 + 9 \geq -11 + 9 \quad \text{Add 9}$$
$$x \geq -2$$

The solid dot in Figure 7–2 means that -2 is a solution. The arrow to the right of the dot implies that numbers to the right of -2 are solutions.

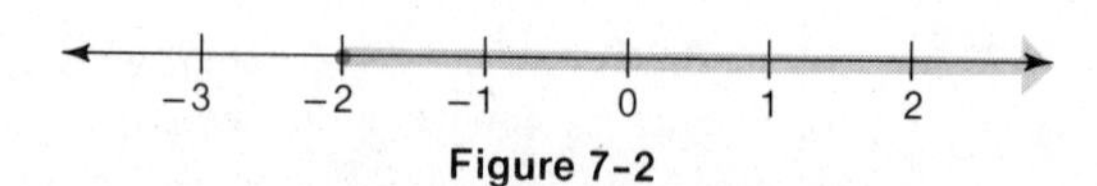

Figure 7-2

21b.

$$8 < x + 4$$
$$8 + (-4) < x + 4 + (-4)$$
$$4 < x$$
$$x > 4 \qquad \text{Rewrite the inequality}$$

We will usually write the variable on the left side. Thus, remember that $a < x$ and $x > a$ mean the same thing. The solution to this problem is shown in Figure 7–3.

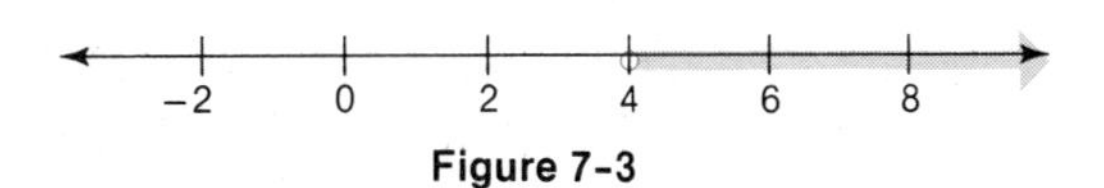

Figure 7-3

Practice Problem 12 ***Solve. Graph each solution.***

a. $x + 5 \leq -9$ **b.** $x - 2 > -8$ **c.** $18 \geq x + 19$

20 Before stating the division principle of inequality, consider the true inequality

$$-6 < 8.$$

If we divide both sides by 2, we get another true inequality:

$$-6 < 8$$
$$\frac{-6}{2} < \frac{8}{2}$$
$$-3 < 4.$$

However, to obtain a true inequality when we divide both sides by -2, we must also reverse the direction of the inequality symbol:

$$-6 < 8$$
$$\frac{-6}{-2} > \frac{8}{-2} \qquad \text{Change } < \text{ to } >$$
$$3 > -4.$$

21 Based on the results in Idea 20, we can state the **division principle of inequality.**

Division Principle of Inequality:

1. If we divide both sides of an inequality by a positive number, we will obtain an equivalent inequality.
2. If we divide both sides of an inequality by a negative number *and the direction of the inequality symbol is reversed,* we will obtain an equivalent inequality.

This principle is used to eliminate the numerical coefficient of a term containing a variable.

Example 22 Solve.

a. $3x < 18$ **b.** $-10x \leq 25$

c. $3.2x \leq -13.76$ **d.** $18 > -6x$

Solution To eliminate the numerical coefficients of a variable, divide both sides of the inequality by that numerical coefficient.

22a. $3x < 18$

$\frac{3x}{3} < \frac{18}{3}$ Divide by 3

$x < 6$ See Figure 7–4

Figure 7-4

22b. $-10x \leq 25$

$\frac{-10x}{-10} \geq \frac{25}{-10}$ Divide by -10 and reverse the $\leq$

$x \geq -\frac{5}{2}$ or $-2\frac{1}{2}$ See Figure 7–5

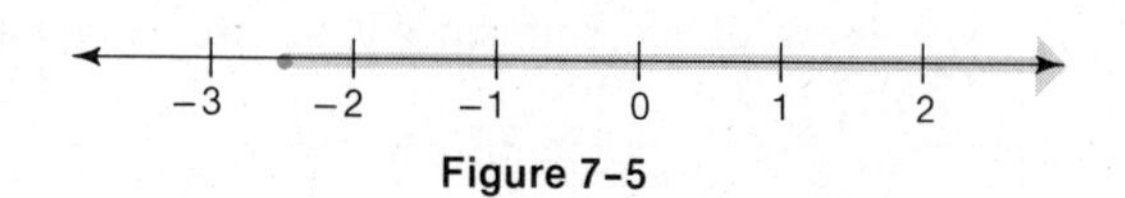

Figure 7-5

22c. $3.2x \leq -13.76$

$\frac{3.2x}{3.2} \leq \frac{-13.76}{3.2}$ Divide by 3.2

$x \leq -4.3$ See Figure 7–6

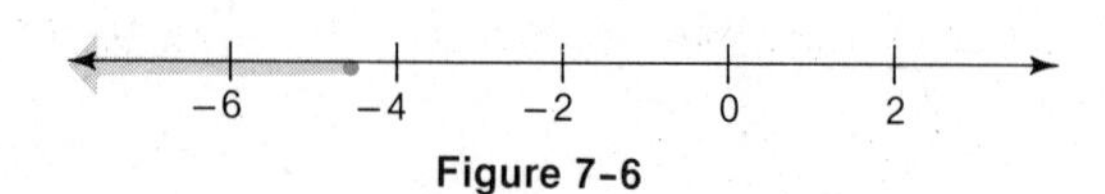

Figure 7-6

22d. $18 > -6x$

$\frac{18}{-6} < \frac{-6x}{-6}$ Divide by -6 and reverse the $>$

$-3 < x$

$x > -3$ Rewrite the inequality. See Figure 7–7.

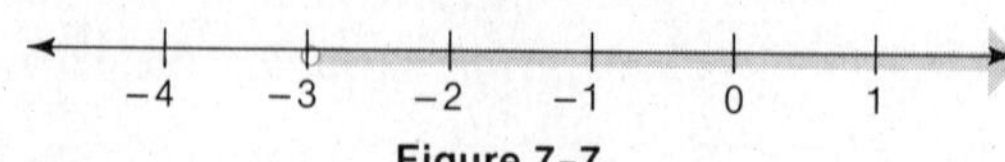

Figure 7-7

Practice Problem 13 ***Solve. Graph each solution.***

a. $6x < -12$ **b.** $-0.3x < 9$ **c.** $-8x \leq 12$

22 The multiplication principle of inequality is similar to the division principle of inequality. To illustrate this, consider the true inequality

$$-3 < 5.$$

If both sides are multiplied by 2, we will obtain another true inequality. However, to obtain a true inequality when both sides are multiplied by -2, we must also reverse the direction of the inequality symbol:

$$-3 < 5$$
$$(-2)(-3) > (-2)(5)$$
$$6 > -10.$$

23 Considering the results in Idea 22, we can state the **multiplication principle of inequality.**

Multiplication Principle of Inequality:

1. If we multiply both sides of an inequality by a positive number, we will obtain an equivalent inequality.
2. If we multiply both sides of an inequality by a negative number *and the direction of the inequality symbol is reversed,* we will obtain an equivalent inequality.

This principle is used to clear an inequality of fractions and to eliminate the numerical coefficient of a term containing a variable.

Example 23 Solve.

a. $\frac{x}{3} < 10$ **b.** $\frac{x}{-0.2} \geq 5$

c. $\frac{4}{9}x \leq \frac{10}{21}$ **d.** $-3x > 12$

Solution To eliminate the denominator of a fraction or a numerical coefficient, use the multiplication principle.

23a.

$$\frac{x}{3} < 10$$

$$3 \cdot \frac{x}{3} < 3 \cdot 10 \qquad \text{Multiply by 3}$$

$$x < 30 \qquad \text{See Figure 7–8}$$

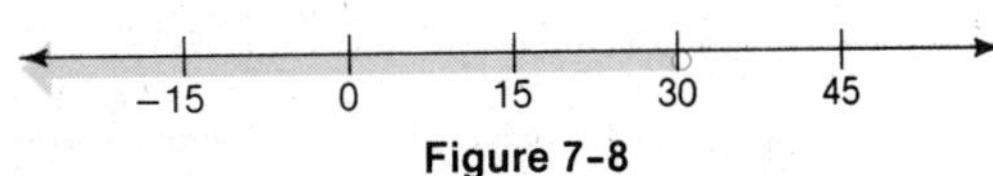

Figure 7-8

23b.

$$\frac{x}{-0.2} \geq 5$$

$$-0.2\left(\frac{x}{-0.2}\right) \leq -0.2\,(5) \qquad \text{Multiply by } -0.2 \text{ and reverse the } \geq$$

$$x \leq -1 \qquad \text{See Figure 7–9}$$

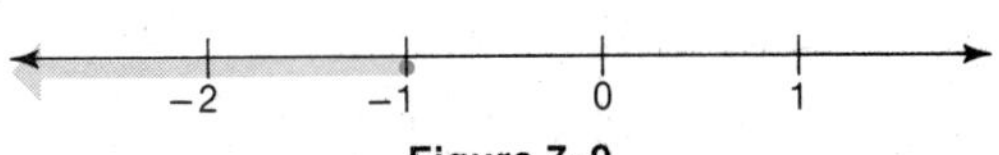

Figure 7-9

23c.

$$\frac{4}{9}x \le \frac{10}{21}$$

$$\frac{9}{4} \cdot \frac{4}{9}x \le \frac{9}{4} \cdot \frac{10}{21} \qquad \text{Multiply by } \frac{9}{4}$$

$$x \le \frac{\overset{3}{\cancel{9}} \cdot \overset{5}{\cancel{10}}}{\underset{2}{\cancel{4}} \cdot \underset{7}{\cancel{21}}}$$

$$x \le \frac{15}{14} \text{ or } 1\frac{1}{14} \qquad \text{See Figure 7–10}$$

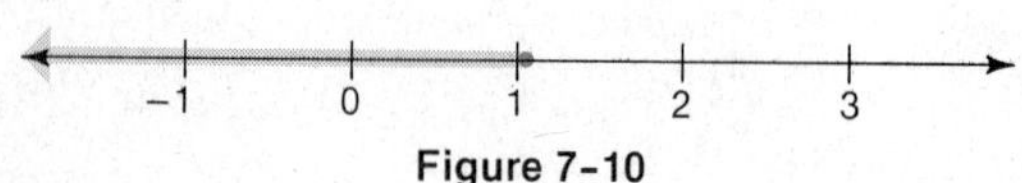

Figure 7-10

23d.

$$-3x > 12$$

$$-\frac{1}{3}(-3x) < -\frac{1}{3}(12) \qquad \text{Multiply by } -\frac{1}{3} \text{ and reverse the } >$$

$$x < -4 \qquad \text{See Figure 7–11}$$

Figure 7-11

Practice Problem 14 **Solve. Graph each solution.**

a. $\frac{x}{-5} > -4$ **b.** $6x \le 15$ **c.** $\frac{35}{18} < \frac{25}{42}x$

7.6 Exercises

Solve by using the addition principle. Graph each solution.

1. $x + 7 < -2$ **2.** $x - 8 > 2$ **3.** $x + 8 \le 2$ **4.** $8 < x - 2$

5. $x + 0.7 \le 0.11$ **6.** $x - 0.2 \ge -0.5$ **7.** $x - 1.5 < 3.2$ **8.** $x - 5 \le -9$

9. $7 + x \ge -3$ **10.** $x + \frac{2}{3} < \frac{5}{3}$ **11.** $x - \frac{5}{6} \ge \frac{1}{10}$ **12.** $x + \frac{1}{18} \le -\frac{3}{10}$

Solve by using the multiplication or division principle. Graph each solution.

13. $4x < 8$ **14.** $5x < -10$ **15.** $-9x \ge -18$ **16.** $15x \ge -25$

17. $-6x \le 33$ **18.** $18x \le -9$ **19.** $0.3x \le 3$ **20.** $-2.3x > 0.69$

21. $1.2x < 1.44$ **22.** $\frac{x}{5} \le -3$ **23.** $\frac{x}{-6} \le 5$ **24.** $\frac{x}{-2} \ge -8$

25. $\frac{2}{3}x < -10$ **26.** $\frac{1}{2}x < -7$ **27.** $-\frac{3}{4}x > 2$ **28.** $\frac{4}{9}x \le \frac{6}{7}$

29. $-\frac{1}{5}x > \frac{3}{5}$ **30.** $\frac{x}{-0.1} \le 1$ **31.** $\frac{x}{0.2} > 0.1$ **32.** $0.5 \le \frac{x}{-5}$

33. $-0.5x \le 0.85$

Solve. Graph each solution.

34. $x + 0.2 < 0.7$

35. $-0.3x \le 6$

36. $-0.4x \le 8$

37. $\frac{3}{4}x > 9$

38. $\frac{3}{5}x > 6$

39. $-\frac{1}{4}x < 2.5$

40. $-\frac{1}{5}x < 3.2$

41. $\frac{x}{-3} \le -5$

42. $\frac{x}{-4} \le -2$

43. $x - \frac{3}{8} < \frac{1}{6}$

44. $x - \frac{3}{4} < \frac{5}{6}$

45. $x + 0.3 \le 2$

46. $x + 0.4 \le 3$

47. $x + 1 < -2$

48. $x + 4 < -6$

49. $x + \frac{3}{4} > -0.25$

50. $x + \frac{7}{4} > -0.25$

51. $6x \le 1.2$

52. $4x \le 1.6$

53. $-2x \ge 10$

54. $-5x \ge 20$

55. $14x < 20$

56. $12x < 30$

57. $-0.3x \le 0.36$

58. $-0.2x \le 0.24$

59. $\frac{x}{2} > -5.4$

60. $\frac{x}{4} > -3.5$

61. $\frac{x}{-5} \le 2$

62. $\frac{x}{-4} \le 6$

63. $\frac{x}{1.2} > 0.53$

64. $\frac{x}{3.4} > 0.48$

65. $6 < \frac{x}{2}$

66. $8 < \frac{x}{2}$

67. $\frac{x}{-2} \ge 4.3$

68. $\frac{x}{-3} \ge 2.4$

69. $\frac{x}{0.3} < -0.4$

70. $\frac{x}{0.6} < -0.3$

71. $\frac{3}{7}x < 6$

72. $\frac{5}{11}x < 10$

73. $\frac{1}{2}x > 4$

74. $\frac{1}{3}x > 2$

75. $-\frac{3}{4}x \ge 6$

76. $-\frac{4}{5}x \ge 8$

77. $\frac{6}{25}x \le \frac{9}{35}$

78. $\frac{4}{25}x \le \frac{6}{45}$

79. $-2x \le \frac{4}{7}$

80. $-3x < \frac{6}{7}$

81. $0.2x \le \frac{1}{5}$

82. $0.3x \le \frac{20}{3}$

83. $-0.1x \le \frac{1}{10}$

84. $-0.3x \le \frac{3}{10}$

Fill in the blanks.

85. An ______________ is a mathematical statement that one quantity is less than or ______________ than another.

86. To solve an inequality, proceed as if you were solving a ______________ in one variable. The only exception is that if you multiply or ______________ both sides of the inequality by a ______________ number the ______________ symbol must be reversed.

87. The solution $x > 1$ is represented by the following figure.

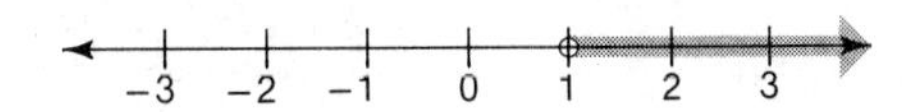

The hollow circle means that 1 is ______________ a solution. The arrow to the right of the circle implies that all numbers to the right of ______________ are ______________.

Answers to Practice Problems

12a. $x \le -14$

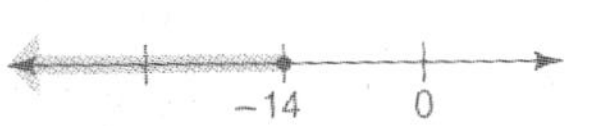

b. $x > -6$

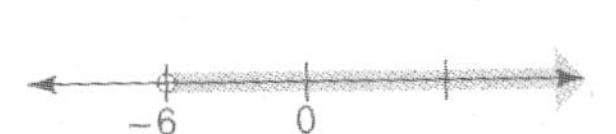

c. $x \le -1$

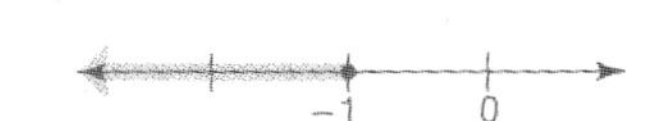

13a. $x < -2$

b. $x > -30$

c. $x \geq -\frac{3}{2}$

14a. $x < 20$

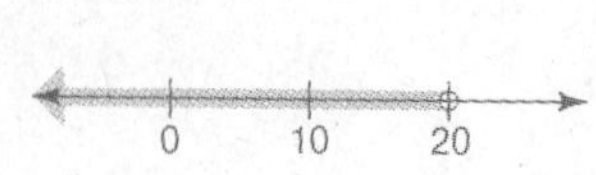

b. $x \leq \frac{5}{2}$

c. $x > \frac{49}{15}$

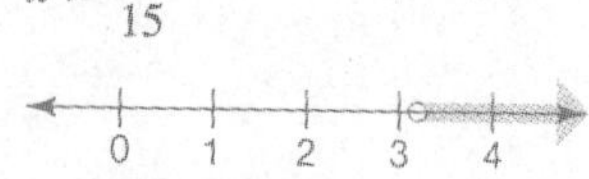

7.7 Solving Inequalities Involving More Than One Operation

24 Quite often you may need to use more than one principle to solve a linear inequality.

Example 24 Solve.

a. $-12x - 1 < -17$ **b.** $7x + 4 > 2x - 6$

c. $7(3x + 6) \leq 11 - (x + 2)$ **d.** $\frac{x+3}{4} - \frac{1}{4} > \frac{x-2}{3}$

Solution To solve a linear inequality, proceed as if you were solving a linear equation. The only exception is that if you multiply or divide both sides of the inequality by a negative number, the inequality symbol must be reversed.

24a.

$$-12x - 1 < -17$$

$$-12x - 1 + 1 < -17 + 1 \qquad \text{Add 1}$$

$$-12x < -16$$

$$\frac{-12x}{-12} > \frac{-16}{-12} \qquad \text{Divide by } -12$$

$$x > \frac{4}{3} \text{ or } 1\frac{1}{3} \qquad \text{See Figure 7–12}$$

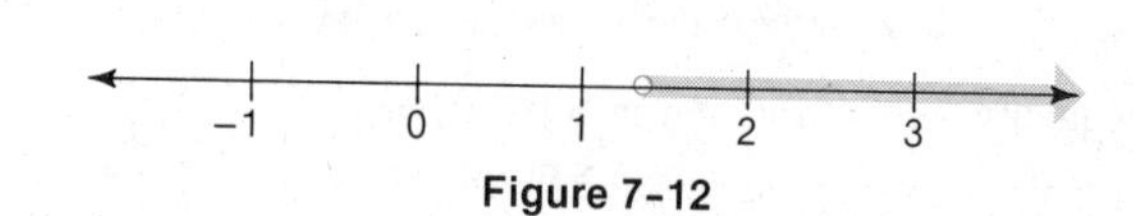

Figure 7-12

24b.

$$7x + 4 > 2x - 6$$

$$7x + 4 + (-2x) > 2x - 6 + (-2x) \qquad \text{Add } -2x$$

$$5x + 4 > -6$$

$$5x + 4 + (-4) > -6 + (-4) \qquad \text{Add } -4$$

$$5x > -10$$

$$\frac{5x}{5} > \frac{-10}{5} \qquad \text{Divide by 5}$$

$$x > -2 \qquad \text{See Figure 7–13}$$

Figure 7-13

24c.

$$7(3x + 6) \le 11 - (x + 2)$$
$$21x + 42 \le 11 - x - 2 \qquad \text{Multiply}$$
$$21x + 42 \le 9 - x \qquad \text{Combine like terms}$$
$$21x + 42 + x \le 9 - x + x \qquad \text{Add } x$$
$$22x + 42 \le 9$$
$$22x + 42 + (-42) \le 9 + (-42) \qquad \text{Add } -42$$
$$22x \le -33$$
$$\frac{22x}{22} \le \frac{-33}{22} \qquad \text{Divide by 22}$$
$$x \le -\frac{3}{2} \text{ or } -1\frac{1}{2} \qquad \text{See Figure 7–14}$$

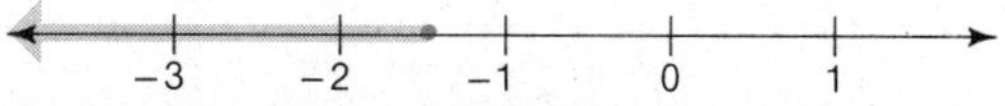

Figure 7-14

24d.

$$\frac{x + 3}{4} - \frac{1}{4} > \frac{x - 2}{3}$$
$$12\left(\frac{x + 3}{4} - \frac{1}{4}\right) > 12\left(\frac{x - 2}{3}\right) \qquad \text{Multiply by LCD} = 12$$
$$3x + 9 - 3 > 4x - 8$$
$$3x + 6 > 4x - 8 \qquad \text{Combine like terms}$$
$$3x + 6 + (-3x) > 4x - 8 + (-3x) \qquad \text{Add } -3x$$
$$6 > x - 8$$
$$6 + 8 > x - 8 + 8 \qquad \text{Add 8}$$
$$14 > x$$
$$x < 14 \qquad \text{Rewrite the inequality. See Figure 7–15.}$$

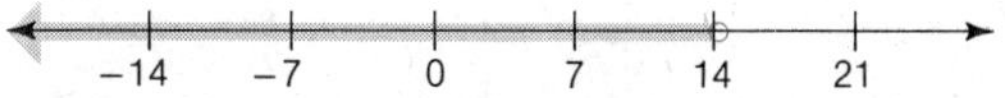

Figure 7-15

Practice Problem 15 ***Solve. Graph each solution.***

a. $6x + 5 \ge -13$ **b.** $x \le 13.2 + 0.2x$

c. $20 > 8 - 2(9 + 4x)$ **d.** $\frac{2x}{4} + \frac{x}{3} \ge 2$

At this point, we should consider two special cases for inequalities.

Example 25 Solve.

a. $x < x + 2$ **b.** $x > x + 2$

Solution **25a.**

$$x < x + 2$$
$$x + (-x) < x + 2 + (-x)$$
$$0 < 2 \qquad \text{True}$$

Since the variable was eliminated and we obtained a true statement, every number is a solution of the inequality.

25b.

$$x > x + 2$$
$$x + (-x) > x + 2 + (-x)$$
$$0 > 2 \quad \text{False}$$

Since the variable was eliminated and the resulting statement was false, there is no solution.

7.7 Exercises

Solve. Graph each solution.

1. $4x + 1 \leq 9$

2. $5x + 2 \leq 12$

3. $6x + 1 > 13$

4. $7x + 1 > 8$

5. $\frac{x}{2} + 1 < 4$

6. $\frac{x}{3} + 2 < 5$

7. $\frac{x}{-4} - 2 \geq -4$

8. $\frac{x}{-3} - 3 \geq -2$

9. $\frac{2}{5}x + 3 < 7$

10. $\frac{3}{4}x + 6 < 9$

11. $\frac{7}{5}x - 4 \leq -11$

12. $\frac{8}{7}x - 14 \leq -54$

13. $7x \leq 5x - 4$

14. $9x \leq 6x - 9$

15. $7x + 7 > 3 + 9x$

16. $5x + 28 > 7 + 2x$

17. $0.3x - 0.7 - 0.1x \geq 0.9 - 0.2x$

18. $0.5x - 0.2 - 0.1x \geq -2.3 - 0.3x$

19. $-0.9x - 0.5 < -0.5x + 0.5$

20. $-1.1x - 0.4 < -0.5x + 0.4$

21. $12 - 5x \leq x$

22. $16 - 3x \leq x$

23. $4(x + 2) < 16$

24. $3(x - 5) < 9$

25. $6(x + 2) \leq -12$

26. $2(x - 3) \leq 6$

27. $3(x + 3) > -3(2x + 3)$

28. $2(2x + 2) > -2(x + 2)$

29. $-4 \geq 3(2x + 1) + 11$

30. $8 \geq 2(3x - 4) + 2x$

31. $9 - 4x < 5(0 - 8x)$

32. $10 - 7x < 4(11 - 6x)$

33. $\frac{x}{6} + \frac{x}{4} \leq \frac{7}{2}$

34. $\frac{x}{15} + \frac{x}{9} \leq \frac{8}{5}$

35. $\frac{x}{2} > \frac{x}{2} + 6$

36. $\frac{x}{3} > \frac{x}{7} + 12$

37. $\frac{x - 2}{10} + \frac{x}{6} \geq \frac{1}{10}$

38. $\frac{x + 2}{8} + \frac{x}{10} \geq \frac{1}{8}$

39. $\frac{2x - 1}{3} + \frac{3x}{4} < \frac{5}{6}$

40. $\frac{3x - 2}{4} + \frac{3x}{8} < \frac{3}{4}$

41. $\frac{x}{3} + 20 < \frac{x}{4} - \frac{x}{12} - \frac{x}{9}$

42. $\frac{x}{5} + 10 < \frac{x}{3} - \frac{x}{10} - \frac{x}{5}$

43. $6(5 - 4x) \leq 3(4x - 2) - 7(6 + 8x)$

44. $5(3 - 2x) \geq 8(3x - 4) - 4(1 + 7x)$

45. $5x + 1 < -11$

46. $-7x + 14 \leq -7$

47. $10x + 4 > -2$

48. $-12x + 40 \geq 10$

49. $5x + 5x \leq -20$

50. $-7x + x \leq -12$

51. $0.1x + 0.11x \leq 21$

52. $8x + 56 < 14 + 2x$

53. $42 - 30x \geq -16x - 14$

54. $0.35x - 1.7 \leq 1.85x + 4$

55. $16x + 1 > 4x + 25$

56. $x \leq 15 + 4x$

57. $4x < 2(12 - 2x)$

58. $2(2x + 3) > 14$

59. $40 > 5(-3x + 2)$

60. $7(5x - 2) < 6(6x - 1)$

61. $\frac{x}{5} + 3 \geq 5$

62. $7 \leq \frac{x}{3} + \frac{x}{4}$

63. $\frac{2x + 4}{2} - \frac{4}{3} < \frac{2x - 4}{3}$

64. $\frac{x}{7} - 10 \leq -\frac{x}{3}$

65. $\frac{x}{3} + 2 \leq 2 + \frac{2x}{6}$

66. $\frac{x + 2}{7} > \frac{x - 3}{2}$

Answers to Practice Problems

15a. $x \geq -3$

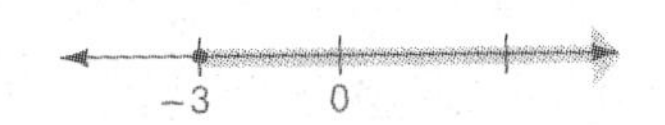

b. $x \leq 16.5$

0 4 8 12 16

c. $x > -\frac{15}{4}$

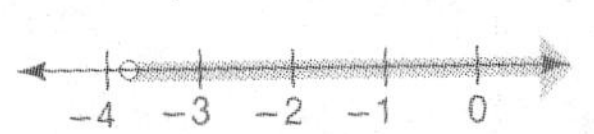

d. $x \geq \frac{12}{5}$

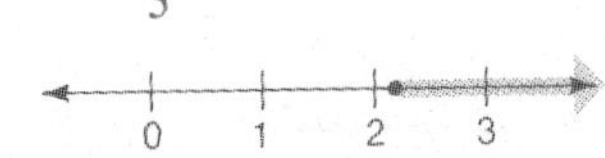

Summary

Important Terms

7.1

equation
linear equation
solution of an equation
conditional equation
identity

7.2

addition principle of equality
division principle of equality
multiplication principle of equality

7.5

literal equation

7.6

inequality
linear inequality
less than, $<$
less than or equal to, $\leq$
greater than, $>$
greater than or equal to, $\geq$
addition principle of inequality
division principle of inequality
multiplication principle of inequality

Important Skills

7.1

Determining if a given value is a solution of an equation

7.2

Solving linear equations by using the addition principle
Solving linear equations by using the division principle
Solving linear equations by using the multiplication principle

7.3

Solving linear equations by using more than one principle
Solving linear equations having variables on both sides

7.4

Solving linear equations containing grouping symbols
Solving linear equations containing fractions

7.5

Solving literal equations

7.6

Solving linear inequalities by using the addition principle
Solving linear inequalities by using the division principle
Solving linear inequalities by using the multiplication principle

7.7

Solving linear inequalities by using more than one principle

Review Exercises

Is the given value the solution?

1. $3x - 9 = -39, x = -10$

2. $3x + 2.11 = 5x - 3.2, x = 0.2$

3. $7(5x + 1) - 18x = 12x, x = -\frac{7}{5}$

Solve for x. For all inequalities, graph the solution.

4. $x + 5 = -2$

5. $-5x < 10$

6. $\frac{2}{3}x = -18$

7. $-6x = 18$

8. $\frac{x}{3} - 9 = 2$

9. $\frac{x}{6} - \frac{x}{10} = 1$

10. $x - 0.7 = -0.31$

11. $\frac{x}{-3} + 4 < 1$

12. $ax + b = c$

13. $x = 0.2x - 16$

14. $8(x + 3) = 4$

15. $12x - \frac{3}{4} = \frac{4x}{3}$

16. $\frac{1}{x} = \frac{1}{y} + \frac{1}{z}$

17. $x + 7 < 1$

18. $\frac{1}{5}x \geq -5$

19. $x - 0.1 \geq 3.52$

20. $4(3x - 5) - (x - 1) = 5x - 5$

21. $2(x - 7) \leq -2x + 8$

22. $\frac{x}{4} + \frac{x}{3} < 26 - \frac{x}{2}$

23. $ax + b = cx + 2b$

24. $\frac{3}{4}x + \frac{1}{3}x = 6\left(x + \frac{5}{36} - \frac{1}{72}x\right)$

WORD PROBLEMS

OBJECTIVES

1. Express written phrases as algebraic expressions.
2. Solve word problems by using equations.
3. Write a written phrase as a ratio.
4. Solve ratio and proportion problems.
5. Solve percent problems by using the percent formula.
6. Solve simple interest problems by using equations.
7. Solve mixture problems by using equations.
8. Solve uniform motion problems by using equations.

PRETEST	EXPLANATION
1. If the sum of two numbers is 12, how would you represent each number?	Section 8.1 Ideas 1 and 2
2. If ten less than twice a number is 50, find the number.	Section 8.1 Ideas 3 and 4
3. Find the ratio of eight inches to eight feet.	Section 8.2 Idea 5
4. If eight pens cost \$10, how much would 20 pens cost?	Section 8.2 Ideas 11 and 12
5. Jane got 90% of the questions correct on a 50-item test. How many items did she get wrong?	Section 8.3 Ideas 13–16
6. Dan invested \$7000 at 10%. How much money must he invest at 16% so that his total investment will yield an annual income of 12%?	Section 8.4 Ideas 17 and 18
7. How many liters of pure acid must be added to 200 liters of a 10% acid solution to obtain a solution that is 20% acid?	Section 8.5 Idea 19
8. A truck leaves a depot at 35 mph. Two hours later, a car leaves the same depot at 55 mph. How many hours will it take the car to pass the truck?	Section 8.6 Ideas 20–22

8.1 Introduction

Idea 1 Many applied or word problems can be solved by using equations. However, before we learn how to solve word problems, it is important to review how to change written statements to algebraic expressions.

Example 1 Change each phrase to an algebraic expression. Let x be the number.

a. six more than a number.

b. a number decreased by 3

c. five less than a number

d. three-fourths of a number

e. the quotient of a number and six

f. the product of a number and two

g. the sum of a number and four

h. a number increased by seven

i. four times the sum of a number and eight

Solution

1a. $x + 6$ — More than indicates addition

1b. $x - 3$ — Decreased by indicates subtraction

1c. $x - 5$ — Less than indicates subtraction

1d. $\frac{3}{4}x$ — A fractional part of a number indicates multiplication

1e. $\frac{x}{6}$ — Quotient indicates division

1f. $2x$ — Product indicates multiplication

1g. $x + 4$ — Sum indicates addition

1h. $x + 7$ — Increased by indicates addition

1i. $4(x + 8)$ — Times indicates multiplication

NOTE: It may be helpful to think of two or more numbers when you want to write the correct expression. For example, you can translate the phrase "six more than a number" into an algebraic expression by thinking:

Number	Six more than a number	Answer
2	$2 + 6$	8
3	$3 + 6$	9
x	$x + 6$	$x + 6$

In other words, to find six more than a number, just add 6 to that number.

Practice Problem 1 ***Change each phrase to an algebraic expression. Let x be the number.***

a. one-half of a number

b. a number squared

c. the sum of a number and nine

d. six times a number

e. a number decreased by two

f. four more than twice a number

g. the reciprocal of a number

2 To solve word problems, you must also be able to write two or more unknown quantities in terms of the same variable.

Example 2 Express each unknown in terms of the same variable.

a. If x = the length of a piece of board, how would you represent the length of a second piece of board that is two feet longer than the first?

b. If d = the cost of one tire, how would you represent the cost of four tires?

c. Let x = an integer. How would you represent the next consecutive integer?

d. Let n = an even integer. How would you represent the next consecutive even integer?

e. Let n = an integer. How would you represent the next two consecutive integers?

f. Let x = the number of men in the class. How would you represent the number of women in the class if the number of women is five less than twice the number of men?

Solution **2a.** $x + 2$ = length of second piece

2b. $4d$ = cost of four tires

2c. The numbers 2 and 3 are examples of two consecutive integers. This implies that given any integer, the next consecutive integer is found by adding 1 to that number. Thus, $x + 1$ = the next consecutive integer.

2d. The numbers 6 and 8 are examples of two consecutive even integers. This implies that given any even integer, the next consecutive even integer is found by adding 2 to that number. Thus, $n + 2$ = the next consecutive even integer.

2e.
n = an integer
$n + 1$ = the next consecutive integer
$n + 2$ = the third consecutive integer

2f. $2x - 5$ = the number of women

NOTE: Always check to see if you have represented the unknowns correctly by replacing the variable with a number. For example, x and $x + 1$ represent two consecutive integers since if $x = 12$, then $x + 1 = 13$.

Practice Problem 2 ***Express each unknown in terms of the given variable.***

a. If x = an odd integer, how would you represent the next two consecutive odd integers?

b. If b = Pam's age, how would you represent her age 10 years from now?

c. Let m = the measure of the first angle of a triangle. If the second angle is three times as large as the first angle and the third angle is 30° less than the second angle, how would you represent the measure of the second and third angles?

Example 3 Represent each unknown in terms of the same variable.

a. If a house costs $18,000 more than a lot, how could you represent the cost of each?

b. The sum of two numbers is 35. How would you represent the two numbers?

c. The ABT construction company has 23 employees. If some of the workers are skilled and the rest are unskilled, how would you represent the number of skilled workers and the number of unskilled workers?

d. Train A leaves Detroit at 9:00 A.M. for Chicago. Train B leaves Chicago at 10:00 A.M. for Detroit. When the trains pass each other, how many hours has each train traveled?

Solution To represent two or more unknowns, let any letter represent one of the unknowns. Next, express all other unknowns in terms of that letter.

3a. x = cost of the lot
$x + 18{,}000$ = cost of the house

3b. n = first number
$35 - n$ = second number

It may be helpful to think of two or more numbers to represent these unknowns.

First number	Second number	Check
2	$35 - 2 = 33$	$2 + 33 = 35$
5	$35 - 5 = 30$	$5 + 30 = 35$
n	$35 - n$	$n + (35 - n) = 35$

Clearly, when we are given the sum of two numbers (or quantities) and we represent one number with a variable, the other number is represented by the given sum minus that variable.

3c. x = number of skilled workers
$23 - x$ = number of unskilled workers

3d. t = hours train A travels before passing train B. Since train B leaves one hour later, it travels one hour less than train A. Thus, $t - 1$ = hours train B travels before passing train A.

NOTE: The concept presented in Example 3b will be helpful in solving many word problems. Thus, we suggest that you review the solution to Example 3b carefully.

Practice Problem 3 ***Represent each unknown in terms of the same variable.***

a. Richard is two years older than Reggie. How would you represent each man's age?

b. Dan invested $9000 in two real estate developments. How would you represent the amount of money he invested in each development?

c. Ernie leaves his house at 11:00 A.M. on a bike. Two hours later, Jerry leaves Ernie's house in a car. How would you represent the number of hours each person has traveled when Jerry catches up to Ernie?

3 The third skill that must be reviewed before we show the complete solution to a word problem is how to change written statements into equations.

Example 4 Write an equation for each problem.

a. If 10 plus three times a number is 40, what is the number?

b. If two-thirds of a number is 20, what is the number?

c. If the sum of two consecutive integers is 19, what are the numbers?

d. Four times a number is equal to the product of three and that number decreased by four. What is the number?

e. Jim earned twice as much money as Joe. If their earnings totaled \$960, how much money did each man earn?

f. A class contains 35 students. The number of girls is five less than three times the number of boys. How many boys and how many girls are in the class?

Solution To change a word problem to an equation, first represent all unknowns in terms of the same letter. Next, rewrite the original problem as a short statement and then translate the words to algebraic expressions or symbols.

4a. x = the number

ten	plus	three times a number	is	40
↓	↓	↓	↓	↓
10	+	$3x$	=	40

4b. x = the number

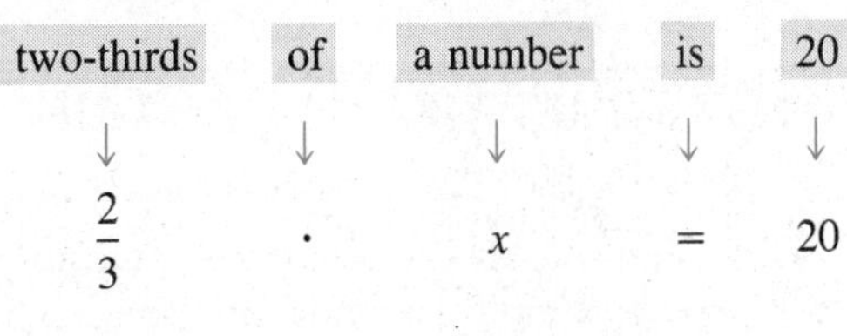

That is, $\frac{2}{3}x = 20$.

4c. x = an integer; $x + 1$ = the next consecutive integer

integer	plus	next consecutive integer	is	19
↓	↓	↓	↓	↓
x	+	$(x + 1)$	=	19

4d. x = the number

four times a number	equals	the product of three and a number decreased by four
↓	↓	↓
$4x$	=	$3(x - 4)$

4e. x = Joe's earnings; $2x$ = Jim's earnings

Jim's earnings	plus	Joe's earnings	is	\$960
↓	↓	↓	↓	↓
$2x$	+	x	=	960

4f. x = number of boys; $3x - 5$ = number of girls

number of boys	plus	number of girls	is	35
↓	↓	↓	↓	↓
x	+	$(3x - 5)$	=	35

Practice Problem 4 ***Write an equation for each of the following.***

a. If 12 more than twice a number is 50, what is the number?

b. The sum of two consecutive even integers is 70. What are the numbers?

c. Jane types 55 words per minute. If Jane only types two-fifths as fast as Jim, how many words per minute can Jim type?

4 Now that we have reviewed the basic skills required to solve word problems, we can present several steps that can be used to solve word problems.

To Solve Word Problems:

1. **Represent unknowns.** Read the problem until you understand what is given and what must be found. Represent each numerical value that you are asked to find in terms of the same letter.
2. **Write an equation.** Reread the problem and then change the written statements into an equation.
3. **Solve the equation and determine all solutions.** Solve the equation written in step 2 and then use this answer to find the quantity or quantities asked for in the original problem.
4. **Check your solution.** Determine if the values obtained in step 3 meet the conditions given in the original problem.

Example 5 Solve by using equations.

a. Three times a number decreased by four is equal to twice the sum of that number and six. What is the number?

b. The sum of two consecutive even integers is 70. Find the integers.

c. A basketball coach wants to buy practice jerseys for her team. If each jersey costs $9.50, how many jerseys can she buy with $136?

d. Cazzie makes hats and sells them to earn extra money. The cost for making one hat is $2 and he has fixed costs of $120 a month. How many hats can he make next month for $270?

Solution To solve a word problem, read the problem carefully and then (1) represent the unknowns, (2) write an equation, (3) solve it and find all solutions, and (4) check your answers.

5a. **Step 1:** x = the number

Step 2:

3 times a number	decreased by	4	is	2 times the sum of that number and 6
$3x$	$-$	4	$=$	$2(x + 6)$

Step 3:

$$3x - 4 = 2(x + 6)$$
$$3x - 4 = 2x + 12$$
$$x - 4 = 12$$
$$x = 16$$

The number is 16.

Check: The answer checks since 3 times 16 decreased by 4 is 44 and 2 times the sum of 16 and 6 is 44.

5b. **Step 1:**

$$x = \text{an even integer}$$
$$x + 2 = \text{next consecutive even integer}$$

Step 2:

first integer	plus	second integer	is	70
x	$+$	$(x + 2)$	$=$	70

Step 3:

$$x + (x + 2) = 70$$
$$2x + 2 = 70$$
$$2x = 68$$
$$x = 34$$
$$x + 2 = 36$$

The numbers are 34 and 36.

Check: The answers check since 34 and 36 are consecutive even integers and $34 + 36 = 70$.

5c. **Step 1:** n = the number of jerseys the coach can buy

Step 2:

total cost of jerseys	equals	amount she can spend
$9.50n$	$=$	136

Step 3: $9.50n = 136$

$n \doteq 14.3$ Divided both sides by 9.50

The coach can buy 14 jerseys.

Check: The answer checks since 14 jerseys will cost 14 (\$9.50) = \$133 and the remaining \$3 is not enough to buy another jersey.

5d. **Step 1:** n = number of hats that can be made

Step 2:

fixed cost	plus	cost for making n hats	is	\$270
120	$+$	$2n$	$=$	270

Step 3:

$$120 + 2n = 270$$
$$2n = 150$$
$$n = 75$$

Cazzie can make 75 hats.

Check: The answer checks since the cost of making 75 hats is \$120 + 2 (\$75) = \$120 + \$150 = \$270.

Practice Problem 5 ***Solve by using equations.***

a. Joan has a piece of material that is 48 inches long. If it must be cut into two pieces so that one piece is 12 inches longer than the other, how long should each piece be?

b. Two-fifths of a number plus one-half of that same number is equal to -9. What is the number?

c. Davis Sporting Goods makes basketball uniforms. The manufacturing cost of each uniform is \$10. If the company has fixed costs of \$100,000 a year, how many uniforms can be made next year for \$250,000?

8.1 Exercises

Change each phrase to an algebraic expression. Let x be the number.

1. 10 more than a number

2. a number decreased by eight

3. two-thirds of a number

4. the quotient of a number and six

5. the product of a number and two

6. the sum of a number and twice that number

7. six less than a number
8. eight more than a number
9. the product of a number and six
10. the product of a number and 10
11. eight less than a number
12. nine less than a number
13. 50 less than a number
14. 60 less than a number
15. the difference between a number and seven
16. the difference between a number and nine
17. the difference between a number and 80
18. the difference between a number and 60
19. the quotient of a number and 12
20. the quotient of a number and 20
21. a number divided into 10
22. a number divided into eight
23. subtract two from a number
24. subtract nine from a number
25. a number increased by 90
26. a number increased by 80
27. nine times a number
28. six times the sum of a number and two
29. the sum of three and the quotient of a number and two
30. the difference of a number and 30
31. the sum of four times a number and seven
32. five times the difference of a number and six
33. twice a number subtracted from two divided by four
34. twice a number subtracted from one divided by five
35. eight more than twice a number
36. six more than three times a number
37. 10 less than half a number
38. eight less than one-third of a number
39. nine times the sum of a number and seven
40. eight times the sum of a number and six
41. five more than one-fourth a number
42. six more than one-fifth a number
43. the sum of a number and the quotient of two and six
44. the difference of a number and the quotient of two and three

Express each unknown in terms of the same variable.

45. If x = John's monthly salary and Jim earns $130 more a month than John, how would you represent Jim's earnings?
46. If b = the cost of one chair, how would you represent the cost of six chairs?
47. If l = the length of a piece of board, how would you represent the length of a second piece of board that is 13 feet shorter than the first piece?
48. If t = the hours train A traveled, and train B traveled 11 hours longer than train A, how would you represent the hours train B traveled?
49. If n = an odd integer, how would you represent the next consecutive odd integer?
50. If n = an even integer, how would you represent the next consecutive even integer?
51. If d = the balance remaining in Steve's checking account, how would you represent his new balance if he withdrew $35?
52. If c = the cost of a coat, how would you represent the cost of a second coat that costs $100 more than the first coat?
53. If s = Joe's salary, how would you represent Sue's salary if she earns $75 more than Joe?
54. If b = Jill's bank balance, how would you represent her new balance after a deposit of $525?
55. If the sum of two numbers is 10 and x = the first number, how would you represent the second number?
56. If the sum of two numbers is 30 and x = the first number, how would you represent the second number?
57. If the sum of two numbers is 40 and x = the first number, how would you represent the second number?
58. If the sum of two numbers is 11 and x = the first number, how would you represent the second number?
59. If x = the number of players on Team A, how would you represent the number of players on Team B if Team B has half as many players as Team A?
60. If n = the number of people on Team A, how would you represent the number of people on Team C if Team C has one-third as many players as Team A.
61. If one number is twice as large as another, how would you represent the two numbers?
62. If one number is four times as large as another, how would you represent the two numbers?
63. If Joe is eight inches taller than John, how would you represent each man's height?
64. If Sue is two inches taller than Jane, how would you represent each woman's height?

65. Tom's savings account contains \$1025 less than Dan's savings account. How would you represent the amount of money each man has in his account?

66. Frank's car is worth \$587 less than Ralph's car. How would you represent how much each man's car is worth?

67. The width of a rectangle is three inches longer than its length. How would you represent the length and the width of the rectangle?

68. The length of a rectangle is eight inches longer than its width. How would you represent the length and the width of the rectangle?

69. The sum of two numbers is 60. How would you represent each number?

70. If the sum of two numbers is eight, how would you represent the two numbers?

71. Joe has \$1000. If he gives part of the money to his brother and the rest to his sister, how much money did each person receive?

72. Rich has \$80,000. If Guerin inherits part of the money and Cazzie inherits the rest, how much money should each man receive?

73. If b = the length of the base of a triangle, how would you represent its altitude if the altitude is five less than twice the length of the base?

74. If the sum of two numbers is 75, how would you represent the two numbers?

75. If one number is four times as large as another, how would you represent each number?

76. If the sum of two numbers is 10, how would you represent each number?

77. The ABC nursery school employs 50 people. If some of the employees are teachers and the rest are teacher-aides, how would you represent the number of teachers and the number of teacher-aides?

78. John is ten years older than Sue. How would you represent each person's age?

79. Vanessa has a balance of \$300 in her savings account. If she deposits x dollars into her account, how would you represent her new balance?

80. One number exceeds another by 12. How would you represent the two numbers?

81. Ernie invested \$5,000. If he invested some of his money in stocks and the rest in real estate, how would you represent the amount he invested in stocks and the amount he invested in real estate?

82. Two cars travel in opposite directions from the state university. If in one hour they are 50 miles apart, how would you represent the distance each car traveled?

83. If x gallons of water is drained from a tank containing 10 gallons of water, how would you represent the amount of water remaining in the tank?

84. The second angle of a triangle is five times as large as the measure of the first angle. The third angle is two degrees smaller than the first angle. How would you represent the measure of the three angles of the triangle?

85. How would you represent three consecutive integers if x = the first integer?

Solve by using equations.

86. If seven is added to three times a number and the result is 22, what is the number?

87. The sum of two numbers is 60. If one number exceeds the other by 10, what are the numbers?

88. A number plus twice that number is 30. What is the number?

89. The sum of two consecutive integers is 35. Find the numbers.

90. The sum of two consecutive even numbers is 70. What are the numbers?

91. Four times the sum of nine and a number is 16. Find the number.

92. Four times a number is equal to the product of three and that number decreased by four. What is the number?

93. When eight is subtracted from six times a number, the result is 76. Find the number.

94. When 37 is subtracted from three times a number, the result is -22. Find the number.

95. If seven is added to three-fifths of a number, the result is 13. Find the number.

96. If eight is added to two-fifths of a number, the result is two. Find the number.

97. The current price of a chair is \$75 less than three times its cost in 1970. If the current price is \$1200, what was the cost in 1970?

98. The current price of a table is \$18 less than three times its cost in 1965. If the current price is \$207, what was the cost in 1965?

99. The sum of two consecutive integers is 79. Find the numbers.

100. The sum of two consecutive integers is 91. Find the numbers.

101. The sum of two consecutive odd numbers is 188. Find the numbers.

102. The sum of two consecutive odd numbers is 168. Find the numbers.

103. Find three consecutive even numbers whose sum is 36.

104. Find three consecutive even numbers whose sum is 60.

105. The first of two numbers is 10 more than the other. If their sum is 56, find the numbers.

106. The first of two numbers is eight more than the other. If their sum is 46, find the numbers.

107. In a school election, Howard received 140 more votes than Gene. If the total number of votes cast was 384, how many votes did each candidate get?

108. Laura earns twice as much money as Dick. If their combined salaries total $615, how much does each person earn?

109. A class contains 40 students. If three-fourths of the students are women, how many women are in the class?

110. A house and lot cost $55,000. If the lot cost $15,000 less than the house, how much does each cost?

111. The second angle of a triangle is five times as large as the measure of the first angle. The third angle is 40° greater than the first angle. Find the measure of the three angles. (Hint: the sum of the measures of the three angles must be 180°.)

112. Guerin makes model airplanes and sells them to make extra money. The cost for making one plane is $5, and he has fixed costs of $50 a month. How many planes can he make for $155?

113. The ABC car rental company charges $20.95 per day plus $0.17 (17 cents) per mile to rent a Ford. If Jim rented a car for one day and his total bill was $50.02, how many miles did he travel?

114. The Carlson car rental company rents compact cars for $19.95 per day and $0.18 per mile. If a car was rented for two days, how many miles was it driven if the total bill was $75.90?

115. Rich has 125 shares of the PDQ paint company. If Rich has only one-fifth as many shares as Ron, how many shares of PDQ stock does Ron have?

116. A class contains 39 students. If the number of girls is five less than three times the number of boys, how many boys and how many girls are in the class?

117. The sum of three consecutive odd numbers is 117. What are the numbers?

118. A record-breaking total of 1100 people attended a local basketball game. The admission was $2 for students and $3 for nonstudents. If total gate receipts were $2800, how many student tickets and how many nonstudent tickets were sold?

119. The daily payroll of the Pace Education Center is $775. The teachers earn $50 per day and the tutors earn $25 per day. If the center employs 21 people, find the number of teachers and the number of tutors employed.

Answers to Practice Problems **1a.** $\frac{1}{2}x$ or $\frac{x}{2}$ **b.** x^2 **c.** $x + 9$ **d.** $6x$ **e.** $x - 2$ **f.** $2x + 4$ **g.** $\frac{1}{x}$ **2a.** $x + 2, x + 4$ **b.** $b + 10$ **c.** $3m, 3m - 30$ **3a.** Reggie's age $= x$, Richard's age $= x + 2$ **b.** $x, 9000 - x$ **c.** Number of hours Jerry traveled $= x$, Number of hours Ernie traveled $= x + 2$ **4a.** $2x + 12 = 50$ **b.** $x + (x + 2) = 50$ **c.** $55 = \frac{2}{5}x$ **5a.** 18 inches, 30 inches **b.** -10 **c.** 15,000

8.2 Ratio and Proportion

5 We can compare numbers in different ways. One way is by using a ratio. A **ratio** is the quotient of two numbers or quantities. That is,

$$\text{the ratio of } a \text{ to } b \text{ is } a \div b \quad \text{or} \quad \frac{a}{b}.$$

This ratio may also be written as $a:b$.

Example 6 Class A contains 25 boys and class B contains 35 boys.

a. What is the ratio of 25 boys to 35 boys?

b. What is the ratio of 35 boys to 25 boys?

Solution The ratio of a to b is written as $\frac{a}{b}$ or $a:b$.

6a. The ratio of 25 boys to 35 boys is $\frac{25 \text{ boys}}{35 \text{ boys}} = \frac{5}{7}$. The ratio can also be written as 5:7. This means that for every five boys in class A there are seven boys in class B. It also means that class A has $\frac{5}{7}$ as many boys as class B.

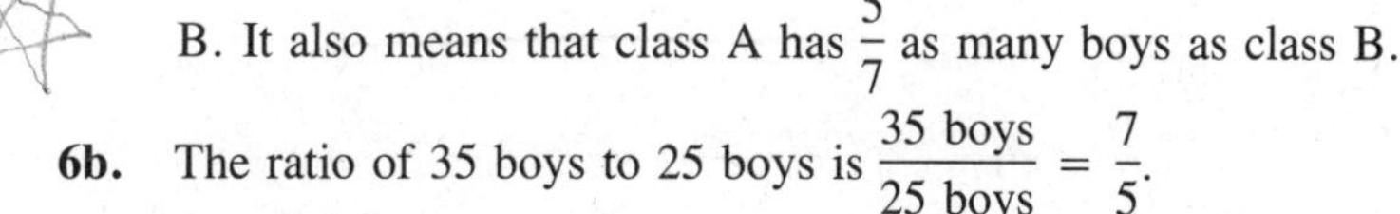

6b. The ratio of 35 boys to 25 boys is $\frac{35 \text{ boys}}{25 \text{ boys}} = \frac{7}{5}$.

NOTE: When a ratio of two quantities contains the same unit of measurement, write the ratio without the unit of measure. Also, always reduce ratios to their lowest terms.

Example 7 **a.** What is the ratio of six hours to three days?

b. What is the ratio of 30 miles to two hours?

Solution When comparing quantities expressed in different units of measure, first express the quantities in the same unit. If this is not possible, write the units with the ratio.

question #51

→ If you can.

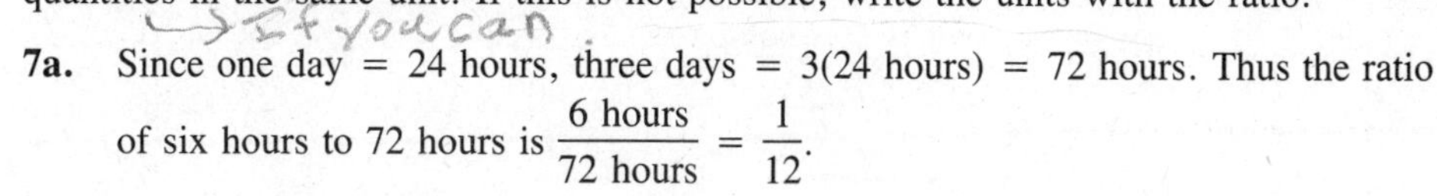

7a. Since one day = 24 hours, three days = 3(24 hours) = 72 hours. Thus the ratio of six hours to 72 hours is $\frac{6 \text{ hours}}{72 \text{ hours}} = \frac{1}{12}$.

7b. The ratio of 30 miles to two hours is $\frac{30 \text{ miles}}{2 \text{ hours}} = 15 \frac{\text{miles}}{\text{hours}}$. This ratio could be read as 15 miles per hour.

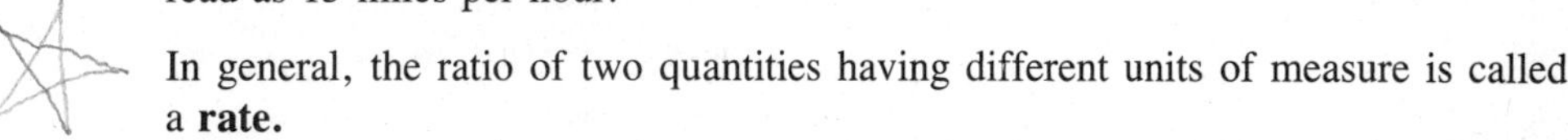

In general, the ratio of two quantities having different units of measure is called a **rate**.

Practice Problem 6 ***Write each statement as a ratio.***

a. The ratio of 11 to 13.

b. The ratio of 16 boys to 24 girls.

c. The ratio of \$30 to three hours.

6 Ratios can be helpful in solving certain types of word problems.

Example 8 Solve.

a. A 3-oz (ounce) jar of lemon pepper costs 98¢. Find the cost per ounce.

b. A 32-oz box of rice costs \$1.80, but a 48-oz box of the same brand of rice costs \$2.40. Which quantity is the better buy?

Solution To find the cost per unit of an item, express the ratio of cost to weight as a fraction, divide, and write the units with the ratio.

8a. The ratio of cost to weight is 98¢ to 3 oz.

$$\frac{98¢}{3 \text{ oz}} \doteq 32.7 \frac{¢}{\text{oz}} \quad \text{Round to the nearest tenth}$$

This implies that the ratio is approximately 32.7¢ to 1 oz. That is, the cost of lemon pepper is 32.7¢ per oz.

8b. *Ratio of $1.80 to 32 oz*

$$\frac{\$1.80}{32 \text{ oz}} \doteq 0.06 \frac{\$}{\text{oz}}$$

The cost is approximately $0.06 (6¢) per oz.

Ratio of $2.40 to 48 oz

$$\frac{\$2.40}{48 \text{ oz}} = 0.05 \frac{\$}{\text{oz}}$$

The cost is $0.05 per oz.

The 48-oz box is a better buy since it costs less per unit.

NOTE: Since the units being compared are not expressed in the same unit of measure, you must write the units with the ratio.

Example 9 Solve.

a. Two numbers have a ratio of 3 to 5. If this sum is 64, what are the numbers?

b. If Paula and Joan divide $5400 in a ratio of 5:4, how much money should each woman receive?

Solution When a ratio problem involves an equation, use the same methods for solving word problems presented in Section 8.1. However, remember that two numbers in the ratio a to b can be expressed as ax and bx, where x is any nonzero number. This can be done since

$$a:b = \frac{a}{b} = \frac{ax}{bx} = ax:bx,$$

where x is not zero.

9a. Step 1: $3x$ = the first number $5x$ = the second number

Step 2:

first number	+	second number	=	64
$3x$	+	$5x$	=	64

Step 3:

$$\begin{aligned} 3x + 5x &= 64 \\ 8x &= 64 \\ x &= 8 \\ 3x &= 24 \\ 5x &= 40 \end{aligned}$$

The numbers are 24 and 40.

Check: The ratio of 24 to 40 is 3:5 and 24 + 40 = 64.

9b. Step 1: $5x$ = dollars Paula should receive
$4x$ = dollars Joan should receive

Step 2:

Paula's share	+	Joan's share	=	$5400
$5x$	+	$4x$	=	5400

Step 3:

$$\begin{aligned} 5x + 4x &= 5400 \\ 9x &= 5400 \\ x &= 600 \\ 5x &= 3000 \\ 4x &= 2400 \end{aligned}$$

Paula and Joan should receive $3000 and $2400, respectively.

Check: The ratio of $3000 to $2400 is 5 to 4, and $3000 + $2400 = $5400.

Practice Problem 7 ***Solve.***

a. An 8-oz jar of mayonnaise costs 99¢. Find the cost per ounce.

b. A 16-oz jar of honey costs $1.50 and a 24-oz jar of the same brand of honey costs $2.40. Which is the better buy?

c. Reggie and Matt divided the profits from the sale of their business in a ratio of 5:8. If the total profits were $29,250, how much money did each man receive?

7 Now that we understand the concept of ratio, we can discuss proportion. A **proportion** is a statement that two ratios are equal.

Example 10 The following equations are proportions.

a. $\frac{3}{4} = \frac{6}{8}$ **b.** $\frac{100}{500} = \frac{1}{5}$

NOTE: Proportions are usually written in the form $\frac{a}{b} = \frac{c}{d}$. However, they can also be written as a:b = c:d (read as "*a* is to *b* as *c* is to *d*").

8 In the proportion $\frac{a}{b} = \frac{c}{d}$, *a* and *d* are the first and fourth terms of the proportion. These terms are called the **extremes.** The second and third terms of the proportion are *b* and *c*. These are called the **means.**

Example 11 Identify the means and the extremes.

a. $\frac{1}{3} = \frac{4}{12}$ **b.** 5:6 = 10:12

Solution The first and fourth terms are the extremes. The second and third terms are the means.

	Extremes	Means
11a.	1,12	3,4
11b.	5,12	6,10

Practice Problem 8 ***Identify the means and the extremes.***

a. $\frac{7}{15} = \frac{21}{45}$ **b.** 5:10 = 1:2

9 Every proportion obeys the property that the product of the extremes is equal to the product of the means.

Property of Proportions:

In the proportion

$$\frac{a}{b} = \frac{c}{d},$$

$a \cdot d = b \cdot c$. This procedure is sometimes called the **cross-multiplication rule.**

NOTE: The property of proportions can also be stated as "the product of the means is equal to the product of the extremes."

10 The property of proportions is useful in determining if two ratios are equal.

Example 12 Determine if the following ratios are equal.

a. $\frac{3}{4} = \frac{6}{9}$ **b.** $4:11 = 8:22$

Solution Two ratios are equal if the product of the extremes is equal to the product of the means.

12a. The product of the extremes ($3 \cdot 9 = 27$) is not equal to the product of the means ($4 \cdot 6 = 24$). Thus, $\frac{3}{4}$ is not equal to $\frac{6}{9}$.

12b. The product of the extremes ($4 \cdot 22 = 88$) is equal to the product of the means ($11 \cdot 8 = 88$). Thus, $4:11 = 8:22$.

Practice Problem 9 ***Determine if the ratios are equal.***

a. $\frac{35}{91} = \frac{5}{13}$ **b.** $2:8 = 3:6$

11 We can also use the property of proportions to solve a proportion. To **solve a proportion** means to find the value of the letter that makes the two ratios equal.

Example 13 Solve each proportion.

a. $\frac{6}{13} = \frac{x}{26}$ **b.** $\frac{3}{5}:\frac{6}{25} = 7:x$ **c.** $\frac{3x}{x + 8} = \frac{3}{5}$

Solution To solve a proportion, use the property of the proportions and then solve the resulting equations.

13a.

$$\frac{6}{13} = \frac{x}{26}$$
$$6 \cdot 26 = 13 \cdot x \quad \text{Cross-multiply}$$
$$156 = 13x$$
$$12 = x$$

Check:

$$\frac{6}{13} = \frac{x}{26}$$
$$\frac{6}{13} = \frac{12}{26}$$
$$6 \cdot 26 = 13 \cdot 12$$
$$156 = 156$$

13b.

$$\frac{3}{5}:\frac{6}{25} = 7:x$$
$$\frac{\frac{3}{5}}{\frac{6}{25}} = \frac{7}{x}$$
$$\frac{6}{25} \cdot \frac{7}{1} = \frac{3}{5} \cdot x$$
$$\frac{42}{25} = \frac{3}{5}x$$
$$\frac{5}{3} \cdot \frac{42}{25} = \frac{5}{3} \cdot \frac{3}{5}x$$
$$\frac{\overset{1}{\cancel{5}} \cdot \overset{14}{\cancel{42}}}{\underset{1}{\cancel{3}} \cdot \underset{5}{\cancel{25}}} = x$$
$$\frac{14}{5} = x$$

Check:

$$\frac{3}{5}:\frac{6}{25} = 7:x$$
$$\frac{\frac{3}{5}}{\frac{6}{25}} = \frac{7}{\frac{14}{5}}$$
$$\frac{6}{25} \cdot \frac{7}{1} = \frac{3}{5} \cdot \frac{14}{5}$$
$$\frac{42}{25} = \frac{42}{25}$$

13c.

$$\frac{3x}{x + 8} = \frac{3}{5}$$
$$3(x + 8) = 5(3x)$$
$$3x + 24 = 15x$$
$$24 = 12x$$
$$2 = x$$

Check:

$$\frac{3x}{x + 8} = \frac{3}{5}$$
$$\frac{3(2)}{2 + 8} = \frac{3}{5}$$
$$\frac{6}{10} = \frac{3}{5}$$
$$10 \cdot 3 = 6 \cdot 5$$
$$30 = 30$$

Practice Problem 10 ***Solve each proportion.***

a. $\frac{6}{10} = \frac{x}{20}$ **b.** $0.3:3 = x:30$

12 Word problems involving proportions can be solved in much the same way that we solved word problems in Section 8.1.

To Solve Word Problems Involving Proportions:

1. Represent the unknown by a letter.
2. Write a proportion. Make sure that the units occupy corresponding positions in the two ratios. For example,
$$\frac{\text{hours}}{\text{dollars}} = \frac{\text{hours}}{\text{dollars}}.$$
3. Solve the proportion (at this point drop the units) and answer the question asked.

Example 14 Solve.

a. Rich drove 120 miles on eight gallons of gas. At this rate, how many gallons of gas will he need to travel 250 miles?

b. One mile is equal to 1.6 kilometers. If Jim traveled 88 kilometers, how many miles did he travel?

c. Curtis won \$513 in a lottery that pays 171 to 2 odds (for every \$2, you bet you can win \$171). How much money did he bet?

d. Cazzie and Guerin formed a real estate corporation. Cazzie invested \$11,000 and Guerin invested \$9000. If the corporation is now worth \$90,000, what is Guerin's share in dollars?

Solution To solve a word problem involving proportions, (1) represent the unknown, (2) write a proportion, and (3) solve the proportion and determine the solution.

14a. Step 1: x = gallons of gas needed to travel 250 miles

Step 2: $$\frac{8 \text{ gallons}}{120 \text{ miles}} = \frac{x \text{ gallons}}{250 \text{ miles}}$$

Note: Ratio of units must be the same on both sides of the equation.

Step 3: $\frac{8}{120} = \frac{x}{250}$

$8 \cdot 250 = 120 \cdot x$

$2000 = 120x$

$16.6 \doteq x$ Round to the nearest tenth

Rich needs approximately 16.6 gallons of gas to travel 250 miles.

14b. Step 1: x = distance traveled in miles

Step 2: $\frac{1 \text{ mile}}{1.6 \text{ kilometers}} = \frac{x \text{ miles}}{88 \text{ kilometers}}$

Step 3: $\frac{1}{1.6} = \frac{x}{88}$

$(1)(88) = (1.6)(x)$

$88 = 1.6x$

$55 = x$

Jim traveled 55 miles.

14c. Step 1: x = the number of dollars Curtis bet

Step 2: $\frac{\$171 \text{ pay-off}}{\$2 \text{ bet}} = \frac{\$513 \text{ pay-off}}{x \text{ bet}}$

Step 3: $\frac{171}{2} = \frac{513}{x}$

$171 \cdot x = 2 \cdot 513$

$171x = 1026$

$x = 6$

Curtis bet $6.

14d. Step 1: x = Guerin's share of the business in dollars

Step 2: $\frac{\$9000 \text{ (Guerin's share)}}{\$20{,}000 \text{ (total)}} = \frac{x \text{ (Guerin's share)}}{\$90{,}000 \text{ (total)}}$

Step 3: $\frac{9000}{20{,}000} = \frac{x}{90{,}000}$

$\frac{9}{20} = \frac{x}{90{,}000}$ Reduce

$9 \cdot 90{,}000 = 20 \cdot x$

$810{,}000 = 20x$

$40{,}500 = x$

Guerin's share of the business is $40,500.

Practice Problem 11 ***Solve.***

a. A basketball team won eight of its first 11 games. At this rate, how many games would you expect them to win out of their first 22 games?

b. A recipe for 50 ice cream sodas requires four quarts of ice cream. How many quarts of ice cream would be needed to make 275 ice cream sodas?

c. If one inch represents 78 miles on a map, how many miles would 4.5 inches represent?

8.2 Exercises

Write each phrase as a ratio in fractional form.

1. 20 miles to 30 miles

2. 60 feet to 90 feet

3. 73 cents to 110 cents

4. 80 men to 120 men

5. 30 inches to 14 inches

6. five days to five hours

7. 12 minutes to three hours

8. five dollars to eight quarters

9. 14 feet to two minutes

10. 70 miles to five hours

11. 30 dollars to two hours

12. 90 dogs to 45 boys

Write each ratio using colon notation (a:b).

13. ratio of 5 to 9

14. ratio of 6 to 15

15. $\frac{3}{7}$

16. $\frac{35}{91}$

Find the cost per unit.

17. A 16-oz can of beans costs 40¢. What is the cost per ounce?

18. A 10-oz can of soup costs 30¢. What is the cost per ounce?

19. A three-pound box of rice costs $2.40. What is the cost per pound?

20. A five-pound bag of flour costs $1.32. What is the cost per pound?

Identify the extremes and the means.

21. $\frac{5}{10} = \frac{15}{30}$

22. $\frac{7}{9} = \frac{21}{27}$

23. $12:13 = 60:65$

24. $3:5 = 9:15$

Determine if the following ratios are equal.

25. $\frac{13}{9} = \frac{26}{18}$

26. $\frac{18}{32} = \frac{36}{96}$

27. $2:5 = 4:25$

28. $7:13 = 49:91$

Find x in the following proportions.

29. $\frac{2}{3} = \frac{6}{x}$

30. $\frac{5}{7} = \frac{x}{21}$

31. $\frac{9}{18} = \frac{x}{54}$

32. $\frac{8}{x} = \frac{12}{8}$

33. $\frac{15}{25} = \frac{x}{100}$

34. $\frac{x}{9} = \frac{30}{24}$

35. $2:3 = x:6$

36. $7:14 = 21:x$

37. $\frac{4}{9}:\frac{6}{35} = 14:x$

38. $\frac{5}{6}:14 = x:\frac{4}{15}$

39. $\frac{0.3}{3} = \frac{x}{5}$

40. $\frac{4}{x} = \frac{0.48}{5.4}$

41. $\frac{7.7}{11} = \frac{x}{5}$

42. $\frac{x}{1} = \frac{19.05}{7.5}$

43. $6:9 = 3\frac{1}{3}:x$

44. $\frac{24}{18}:\frac{30}{28} = x:42$

45. $\frac{x+1}{x+2} = \frac{14}{8}$

46. $\frac{9}{2x+7} = \frac{10}{4x+6}$

47. $\frac{x-2}{2} = \frac{3x-5}{7}$

48. $\frac{3x+4}{8x-2} = \frac{4}{6}$

49. $13:26 = 39:x$

Fill in the blanks.

50. A ________ is the quotient of two numbers or quantities.

51. When comparing quantities expressed in different units of measure, first express the quantities in the ________ unit. If this is not possible, write the ________ with the ratio.

52. A ________ is a statement that two ratios are equal.

53. The property of ________ states that the ________ of the extremes is equal to the ________ of the means.

Solve.

54. A 64-oz bottle of a cola beverage costs $1.50, while a 48-oz bottle of the same brand of cola costs $1.20. Which is the better buy?

55. Two numbers have a ratio of 4:5. If their sum is 45, what are the numbers?

56. A farmer wants to plant 80 acres of corn and soybeans in a ratio of 7 to 9. How many acres of each must he plant?

57. Jim earned $25.50 in six hours. At this rate, how much could he earn in 15 hours?

58. John used 12 gallons of gas on a 420-mile trip. At this rate, how many gallons of gas can he expect to use on a 1050-mile trip?

59. In keeping with equal opportunity employment guidelines, a company tries to hire people on a nondiscriminatory basis. If the ratio of men to women must be 7:4 and the company must hire 165 people, how many men and how many women must be hired?

60. Dr. Cowsen believes that the ideal student-teacher ratio for a day care center is 17:2. If she opens a day care center with an enrollment of 187 students, how many teachers should she employ if she maintains her ideal student-teacher ratio?

61. Joyce wants to invest $15,000 in stocks, bonds, and real estate in a ratio of 3:4:5. How much money should she invest in each?

62. In a certain city of Illinois, the ratio of Republicans to Democrats is 2:5. If there are 1662 Republicans, how many Democrats are there?

63. A profit of $22,000 on the sale of a painting was split between Phyllis and Pam in a ratio of 4 to 7. How much did each woman receive?

64. The real estate tax rate in a certain city is $98.19 per $1000 of assessed valuation. Find the real estate tax on a piece of property assessed at $80,000.

65. A father divided $120,000 among his three children in the ratio 3:5:7. How much did each child receive?

66. A recipe for 60 people uses four cups of milk. How many cups of milk are needed for a recipe for 150 people?

67. A recipe for 40 people uses eight tablespoons of honey. How many tablespoons of honey are needed for a recipe for 100 people?

68. John drove 246 miles on 12 gallons of gas. At this rate, how many gallons of gas will he need to travel 861 miles?

69. Bill drove 170 miles on 10 gallons of gas. At this rate, how far could he drive on 35 gallons of gas?

70. One mile is equal to 1.6 kilometers. If Linda traveled 24 kilometers, how many miles did she travel?

71. One mile is equal to 1.6 kilometers. If Dave traveled 144 miles, how many kilometers did he travel?

72. In four days, Betty ran 50 miles. At this rate, how long would it take her to run 125 miles?

73. In six days, Art drove 2100 miles. At this rate, how many days will it take him to drive 5250 miles?

74. Vanessa used 44 quarts of ice cream to make 550 milk shakes. How many quarts of ice cream are needed to make 100 milk shakes?

75. Ed used 12 quarts of sherbet to make a dessert for 100 people. How many quarts of sherbet are needed to make dessert for 350 people?

76. A basketball team won eight of its first 12 games. At this rate, how many will it win out of its first 30 games?

77. A basketball player made eight out of the first 10 free throws he attempted. At this rate, how many free throws should he make out of the first 65 he attempts?

78. There are 20 women in a class containing 32 students. At this rate, how many of the college's 2576 students would be women?

79. There are 12 men in a class containing 34 students. At this rate, how many of the college's 2737 students would be men?

80. A store sells four candy bars for 86 cents. At this rate, how much will 10 candy bars cost?

Answers to Practice Problems **6a.** $\frac{11}{13}$ **b.** $\frac{2 \text{ boys}}{3 \text{ girls}}$ **c.** $10 \frac{\$}{\text{hour}}$ **7a.** 12.4¢ per oz **b.** 16-oz jar **c.** Reggie's share = $11,250; Matt's share = $18,000 **8a.** Extremes: 7, 45; Means: 15, 21 **b.** Extremes: 5, 2; Means: 10, 1 **9a.** equal **b.** not equal **10a.** 12 **b.** 3 **11a.** 16 **b.** 22 **c.** 351

8.3 Percent

13 A *percent problem* involves three quantities.

1. The **base** (B) is the whole or original quantity.
2. The **percent** (P) is the number written with a % symbol or the word *percent*.
3. The **amount** (A) is a part of the whole or some percent of the base.

Example 15 Identify the amount, the base, and the percent.

a. What number is 80% of 1000?

b. 18 is what percent of 20?

c. 40 is 37% of what number?

d. The Johnsons have \$8000 that can be used as the down payment on a house. If the minimum down payment must be 20% of the total cost of the house, what is the highest price that they can pay for a house?

e. Mark attempted 20 shots in a game and made 70% of them. How many shots did he make?

f. Jimmy earns \$1000 a month. If he pays \$400 a month for rent, what percent of his earnings is spent on rent?

Solution To identify A, B, and P in a percent problem, rewrite the problem as the amount (A) is some percent (P) of the base (B).

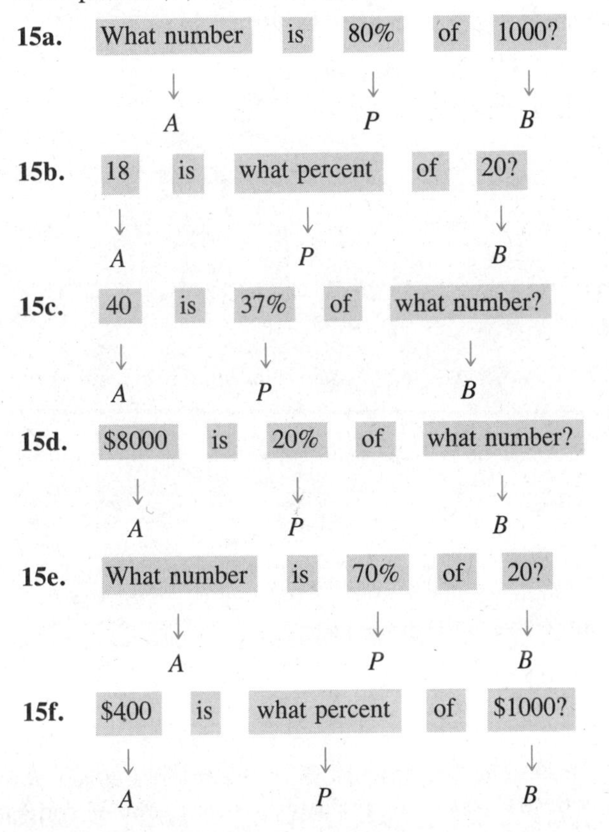

Practice Problem 12 ***Identify A, B, and P.***

a. 14 is what percent of 20?

b. What number is 30% of 50?

c. Hondo made 80% of his free throws. If he made eight free throws, how many shots did he attempt?

d. Adrienne earns a 10% commission on everything she sells. If her total sales for one week were $1200, what was her commission?

14 In a percent problem, two quantities will be given and you must find the third. To find this third quantity, use the percent formula

$$\frac{\textbf{Amount}}{\textbf{Base}} = \frac{\textbf{Percent}}{\textbf{100}} \quad \text{or} \quad \frac{A}{B} = \frac{P}{100}.$$

Example 16 Use the percent formula to find B when $A = 2$ and $P = 40\%$.

Solution To solve for a particular letter in the percent formula, replace the known quantities with their numerical values, cross-multiply, and then solve the resulting equation.

$$\frac{A}{B} = \frac{P}{100}$$

$$\frac{2}{B} = \frac{40}{100}$$

$$40B = 200 \qquad 40 \cdot B = 2 \cdot 100$$

$$B = 5$$

Practice Problem 13 ***Solve for the unknown by using the percent formula.***

a. $A = 5$, $P = ?$, and $B = 20$

b. $A = ?$, $P = 12\%$, and $B = 85$

15 The results in Ideas 13 and 14 suggest a procedure for solving percent problems.

To Solve Percent Problems:

1. Identify A, P, and B.
2. Replace the known quantities in the percent formula with their numerical values.
3. Solve for the unknown and answer the question asked.

Example 17 Solve by using the percent formula.

a. 40 is 50% of what number?

b. 20 is what percent of 80?

c. What number is 80% of 10?

d. An examination contains 20 problems. If Barbara answers 75% of the problems correctly, how many problems has she answered correctly?

e. A baseball team won 25 out of 30 games. What percent of the games did they win?

f. An investment earns 6% interest. If Penny wants to earn \$78 interest from this investment, how much money should she invest?

Solution To solve a percent problem, solve the percent formula for the unknown quantity and answer the question asked.

17a. $A = 40, B = \boxed{?}, P = 50\%$

$$\frac{A}{B} = \frac{P}{100}$$
$$\frac{40}{B} = \frac{50}{100}$$
$$50B = 4000$$
$$B = 80$$

40 is 50% of 80.

17b. $A = 20, B = 80, P = \boxed{?}$

$$\frac{A}{B} = \frac{P}{100}$$
$$\frac{20}{80} = \frac{P}{100}$$
$$80P = 2000$$
$$P = 25$$

20 is 25% of 80.

17c. $A = \boxed{?}, B = 10, P = 80\%$

$$\frac{A}{B} = \frac{P}{100}$$
$$\frac{A}{10} = \frac{80}{100}$$
$$100A = 800$$
$$A = 8$$

8 is 80% of 10.

17d. Rewrite: What number is 75% of 20?
$A = \boxed{?}, B = 20, P = 75\%$

$$\frac{A}{B} = \frac{P}{100}$$
$$\frac{A}{20} = \frac{75}{100}$$
$$100A = 1500$$
$$A = 15$$

or

$$A = P \cdot B$$
$$= (.75)(20)$$
$$= 15$$

Note: $75\% = 75(.01) = .75$

Barbara answered 15 questions correctly.

17e. Rewrite: 25 is what percent of 30?

$A = 25, B = 30, P = \boxed{?}$

$$\frac{A}{B} = \frac{P}{100}$$
$$\frac{25}{30} = \frac{P}{100}$$
$$30P = 2500$$
$$P \doteq 83 \quad \text{Round to nearest one}$$

The team won 83% of its games.

17f. Rewrite: \$78 is 6% of what number?

$$A = 78,\ B = \boxed{?},\ P = 6\%$$
$$\frac{A}{B} = \frac{P}{100}$$
$$\frac{78}{B} = \frac{6}{100}$$
$$6B = 7800$$
$$B = 1300$$

Penny should invest \$1300.

Practice Problem 14 ***Solve.***

a. 210 is 15% of what number?

b. What number is 25% of 40?

c. Jim earns \$1400 per month. If his rent is \$210 per month, what percent of his income does he spend on rent?

d. There are 200 employees in the ABT company. If 80% of the employees are students, how many employees are students?

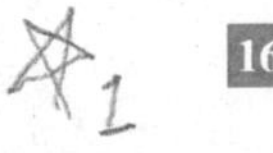

16 Sometimes the numbers given in a percent problem are not the values that you must substitute for A, P, or B.

Example 18 Solve.

a. Rich wants to buy a pair of jeans that cost \$50. If the owner gives him a 20% discount, how much should Rich pay for the jeans?

b. Susan earned \$20,000 last year. This year her salary is \$24,000. What was the percent of increase?

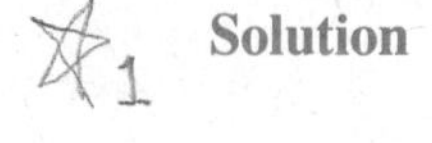

Solution **18a.** Since Rich gets a 20% discount, he must pay only 80% (100% − 20%) of the original cost. Rewrite: What number is 80% of \$50?

$$A = \boxed{?},\ B = 50,\ P = 80\%$$
$$\frac{A}{B} = \frac{P}{100}$$
$$\frac{A}{50} = \frac{80}{100}$$
$$100A = 4000$$
$$A = 40$$

The jeans will cost Rich \$40.

NOTE: When working percent problems, remember that (1) the whole or original amount is 100% and (2) $100\% - N\% = (100 - N)\%$. For example, $100\% - 70\% = (100 - 70)\% = 30\%$.

18b. To determine the percent of increase, we must find the amount of her salary increase. This amount is \$4000 (\$24,000 − \$20,000). Rewrite: \$4000 is what percent of \$20,000?

$$A = 4000,\ B = 20{,}000,\ P = \boxed{?}$$
$$\frac{4000}{20{,}000} = \frac{P}{100}$$
$$20{,}000P = 400{,}000$$
$$P = 20$$

Susan received a 20% salary increase.

Practice Problem 15 **Solve.**

a. Phil earns $12,000 a year and his wife earns $13,000 a year. If they can afford to buy a car that is 40% of their total income, what is the highest price that they can pay for a car?

b. Carver must pay a salesman a 5% commission to sell his car. If he wants to have $7600 after paying the commission, at what price must the car be sold?

8.3 Exercises

Solve. If necessary, round to the nearest hundredth.

1. 15 is 25% of what number?

2. What number is 25% of 60?

3. 15 is what percent of 60?

4. What number is 63% of 48?

5. 12.5 is 50% of what number?

6. What number is 3.2% of 300?

7. What number is 113% of 200?

8. 18 is what percent of 24?

9. 35 is what percent of 50?

10. 62 is 31% of what number?

11. 78 is 6% of what number?

12. What number is 9% of 90?

13. 16 is what percent of 24?

14. 92 is 50% of what number?

15. 210 is 15% of what number?

16. 52 is what percent of 217?

17. What number is 0.4% of 360?

18. 252 is 70% of what number?

19. 900 is what percent of 1220?

20. What number is 38% of 145?

Solve.

21. John won up 70% of his games. If he won 14 games, how many games has he played?

22. In a class of 50 students, 30 people passed a test. What percent of the students passed?

23. Joe earned $200 last week. If he always saves 15% of his weekly salary, how much money did he save last week?

24. Wayne won 12 out of 15 games. What percent of the games did he win?

25. In an algebra class, 20% of the students earned A's. If 5 students received an A, how many students are in the class?

26. Pam bought a shirt that costs $40. If she must pay a 6% sales tax, how much sales tax will she pay?

27. Bill earns $14,000 per year. If he wants to save 20% of his annual salary, how much should he save?

28. In a poll, 15 out of 60 people wanted to see a play. What percent of the people wanted to see the play?

29. Barbara earns $1000 per month. If she spends $80 per month to repay a loan, what percent of her salary does she spend to repay the loan?

30. Jane received a 10% increase in pay. If she is now earning an extra $140 per month, what was her monthly salary before the raise?

31. Vanessa changed her driving habits and increased the mpg rating on her car by four miles. If this increase represents a 20% boost in the car's mpg rating, what was the original rating?

32. Fifty out of 200 applicants for a job are women. What percent of the applicants are women?

33. In business, the selling price equals the cost plus the markup. If Gatsby's marks up all its merchandise by 60% of its original cost, what is the markup on a coat that cost Gatsby's $150? What is the selling price of the coat?

34. John earns $1600 per month. If $320 is deducted from his check, what percent of his salary is deducted from his monthly earnings?

35. John wants to buy a house that costs $80,000. If the owner wants at least a 20% down payment, how much money does John need as a down payment?

36. Buford paid $300 for a $400 stereo. What percent decrease in cost is this?

37. Ron weighed 200 pounds in college. If he now weighs 220 pounds, what percent increase in weight is this?

38. Kathy pays $6000 for a new car. Two years later the car is worth $4000. What percent decrease in value is this? (Round to the nearest percent.)

39. The value of a house increased 40% in 4 years. If the amount of the increase was $20,000, find the original cost and the new cost of the house.

40. Wayne won 12 out of 15 games. What percent of the games did he lose?

41. Bill earns \$14,000 per year. If he saves 20% of his income, how much money does that leave him for living expenses?

42. Reggie paid \$8616 for a new car. If this price is 7% more than the list price, find the list price. (Hint: The price he paid is 107% of the list price.)

43. A union contract called for a 11% increase in the hourly rate of pay. If the new hourly wage is \$11.10, what was the original hourly wage?

44. The owner of a clothing store always gives Rich a 20% discount on any item he buys. If he paid \$40 for a sweater, what was its original cost?

45. Phil wants to have \$73,600 after paying Ron an 8% commission to sell his house. What must his asking price be?

Answers to Practice Problems **12a.** $A = 14$, $B = 20$, $P = ?$ **b.** $A = ?$, $B = 50$, $P = 30\%$ **c.** $A = 8$, $B = ?$, $P = 80\%$ **d.** $A = ?$, $B = 1200$, $P = 10\%$ **13a.** 25% **b.** 10.2 **14a.** 1400 **b.** 10 **c.** 15% **d.** 160 **15a.** \$10,000 **b.** \$8000

8.4 Simple Interest

17 A *simple interest problem* is one that involves the interest on money borrowed or invested. In such problems, we use the formula $I = P \cdot R \cdot T$, where

P = **principal** = amount of money borrowed or invested;
R = **rate** = annual interest rate expressed as a percent;
T = **time** = years the principal is borrowed or invested; and
I = **interest** = amount of money charged or earned by using the principal.

To solve a simple interest problem by using the formula $I = PRT$, use the following procedure:

To Solve Simple Interest Problems:

1. Identify I, P, R, and T.
2. Replace the known quantities in the formula with their numerical values.
3. Solve for the unknown quantity and answer the question asked.

NOTE: When solving simple interest problems, you must be able to convert percents to decimals and decimals to percents.

Example 19 Solve by using the formula $I = PRT$.

a. John invested \$900 in a savings plan that pays 8% interest. At the end of six months, how much interest will his account earn?

b. How much money must Ron invest at a 6.5% interest rate in order to earn \$390 in interest in one year?

c. How many months will it take to earn \$600 in interest when \$8000 is invested at 10%?

d. Jan wants to deposit \$1000 in a savings plan that will earn her \$120 in interest in one year. At what interest rate must she invest her money?

Solution To solve a simple interest problem, solve the formula $I = PRT$ for the unknown quantity and answer the question asked in the original problem. However, remember that T should be given in years and $P\% = P(.01)$.

19a. $I = ?$, $P = \$900$, $R = 8\% = 8(.01) = .08$, $T = 6 \text{ mo} = \frac{1}{2} \text{ yr}$

$$\begin{aligned} I &= PRT \\ &= (900)(.08)\left(\frac{1}{2}\right) \\ &= 36 \end{aligned}$$

John will earn \$36 in interest.

19b. $I = \$390$, $P = ?$, $R = 6.5\% = 6.5(.01) = .065$, $T = 1$ yr.

$$\begin{aligned} I &= PRT \\ 390 &= (P)(.065)(1) \\ 390 &= .065P \\ 6000 &= P \qquad \text{Divide by .065} \end{aligned}$$

Ron must invest \$6000.

19c. $I = \$600$, $P = \$8000$, $R = 10\% = 10(.01) = .10$, $T = ?$

$$\begin{aligned} I &= PRT \\ 600 &= (8000)(.10)(T) \\ 600 &= 800T \\ \frac{3}{4} &= T \qquad \text{Divide by 800} \end{aligned}$$

It will take $\frac{3}{4}$ of a year, which is $\frac{3}{4}$(12 months) = 9 months.

19d. $I = \$120$, $P = \$1000$, $R = ?$, $T = 1$ yr.

$$\begin{aligned} I &= PRT \\ 120 &= (1000)(R)(1) \\ 120 &= 1000R \\ .12 &= R \qquad \text{Divide by 1000} \end{aligned}$$

The rate should be .12 or 12%.

NOTE: To change a decimal to a percent, move the decimal point two places to the right and write a % symbol.

Practice Problem 16 ***Solve by using the formula I = PRT.***

a. Find the amount of interest earned on \$500 at 6.25% for nine months (round the answer to the nearest hundredth).

b. How much money must Rich invest at 12% to earn \$600 in interest in one year?

c. If Pam wants to earn \$50 in interest in six months on a \$500 investment, at what rate should she invest her money?

18 A simple interest problem may involve more than one interest rate. To solve this type of problem, we will use the same basic steps presented in Section 8.1.

To Solve Simple Interest Problems Involving More Than One Interest Rate:

1. Represent unknowns.
2. Write an equation. It will be helpful to construct a table that contains all the given information. In many of these problems, the equation is based on the amount of interest.
3. Solve the equation and determine all solutions.
4. Check.

Example 20 Solve.

a. Of a total investment of \$10,000, Vanessa invested part at 9% and the rest at 10.5%. If her total interest for the year was \$990, how much was invested at each rate?

b. Rochelle has \$20,000 to invest. She invested \$10,000 in a certificate of deposit that pays 15%, and \$3000 in a savings plan that pays 8.5%. At what rate should she invest the remaining money if she wants to earn an annual income of \$3155 from these three investments?

Solution To solve a simple interest problem involving two or more interest rates, (1) represent the unknowns, (2) write an equation, (3) solve the equation and determine all solutions, and (4) check.

20a. Step 1:

$$x = \text{dollars invested at } 9\%$$
$$10{,}000 - x = \text{dollars invested at } 10.5\%$$

Step 2:

	Principal	Rate	Time	Interest*
Investment #1	x	$9\% = .09$	1 yr	$.09x$
Investment #2	$10{,}000 - x$	$10.5\% = .105$	1 yr	$.105(10{,}000 - x)$
Total	\$10,000			\$990

**Note:* Interest = (Principal)(Rate)(Time)

Interest at 9%	+	Interest at 10%	=	Total interest
$.09x$	+	$.105(10{,}000 - x)$	=	990

Step 3:

$$.09x + .105(10{,}000 - x) = 990$$
$$.09x + 1050 - .105x = 990$$
$$-.015x + 1050 = 990$$
$$-.015x = -60 \quad \text{Added } -1050 \text{ to both sides}$$
$$x = 4000 \quad \text{Divided both sides by } -0.15x$$
$$10{,}000 - x = 6000$$

Vanessa invested \$4000 at 9% and \$6000 at 10.5%.

Check: total invested = \$4000 + \$6000 = \$10,000
.09(\$4000) + .105(\$6000) = \$360 + \$630 = \$990.

20b. **Step 1:** x = rate at which $7000 must be invested
Step 2:

	Principal	Rate	Time	Interest
Savings Plan	$3,000	8.5% = .085	1 yr	$255
Certificate	$10,000	15% = .15	1 yr	$1500
3rd Investment	$7,000	x% = .01x	1 yr	70x
Total	$20,000			$3155

Interest on Savings Plan	+	Interest on Certificate	+	Interest on 3rd Investment	=	$3155
255	+	1500	+	70x	=	3155

Step 3:

$$255 + 1500 + 70x = 3155$$
$$1755 + 70x = 3155$$
$$70x = 1400$$
$$x = 20$$

The $7000 must be invested at a rate of 20%.

Check: The total interest is $255 + $1500 + .20($7000) = $255 + $1500 + $1400 = $3155.

Practice Problem 17 ***Solve.***

a. Barbara has $1500. She deposited some of the money into a savings account that pays 6.25%, and the rest into an account that pays 13%. If the total annual interest earned from the two accounts was $168, how much did she deposit into each account?

b. Carver invested $4000 in a Ford and a BMW and then he leased both cars. His earned annual interest income from leasing each car was 20% and 60%, respectively. If his income from the total investment was 50%, how much did Carver pay for each car?

8.4 Exercises

Solve by using the formula $I = PRT$.

1. Find the interest on a $1000 loan at 8% for one year.
2. Find the interest on a $1000 loan at 10% for one year.
3. Find the interest on a $2000 loan at 12% for one year.
4. Find the interest on a $500 loan at 18% for six months.
5. If $500 is deposited into a savings account that pays 8% interest, how much interest will be earned in nine months?
6. If $50 in interest is paid on a loan of $500 for one year, what is the interest rate?
7. If $500 in interest is paid on a loan of $8000 for one year, what is the interest rate?
8. If the interest on an $800 loan for six months is $40, what is the interest rate?
9. If the interest on a $2400 charge account is $30 for one month, what is the interest rate?
10. If $50 in interest is paid for a 10% loan for one year, what is the principal?
11. If $600 in interest is paid for an 8% loan for six months, what is the principal?
12. If $30 in interest is paid for an 18% loan for one month, what is the principal?
13. If the total interest paid on a $400 loan at 10% was $40, how many months was the principal used?
14. If the total interest paid on a $2000 loan at 15% was $100, how many months was the principal used?
15. If the total interest paid on a $3000 loan at 14.5% was $145, how many months was the principal used?
16. Ron wants to invest some money at 15%. If he wants to earn $4500 in interest in one year, how much money should he invest?

17. Penny's monthly interest charge on a charge card is computed on a closing balance of $1080. If the annual interest rate is 18%, how much interest is due at the end of the month?

18. Reggie deposited $3000 into a savings account that pays 9.75%. How much interest will he earn in six months?

19. The Second National Bank has a savings plan that pays interest by mail. If the interest rate is 12%, how much would you have to deposit in order to receive a monthly check of $200?

20. The local furniture store charged Mary $60 in interest for a $200 sofa. If she paid the bill over a nine-month period, what was the annual interest rate?

21. John deposited $2400 in an account paying 10.5% in interest. How many months will it take him to earn $189 in interest?

22. Don inherited $1800 from his sister. Part of the money was placed in a savings plan that earns 8% interest, and the rest was deposited in an account earning 9%. If his total annual interest was $152, how much money was deposited in each account?

23. Barbara received a check last week. She invested part of it at 15% and $1000 more than this amount at 16%. If she will earn $780 in interest in one year, how much was invested at each rate?

24. Cazzie earned $25,000 from the sale of his book. He invested $10,000 at 12.5% and $7000 at 15%. At what rate must he invest the remaining $8000 so that he can earn $3740 in interest in one year?

25. Betty invested $1000 at 11.5%. How much additional money must she deposit in a savings plan at 10% so that her annual interest income from both investments will be $200?

26. Guerin earned $25,000 from the sale of a house. He invested part at 15% and the rest at 25%. If the annual interest income from the 25% investment is $250 more than the 15% investment, how much was invested at each rate?

27. Van wants to earn an annual interest income of 12.5% from his total investments. If he has $5000 invested at 11%, how much additional money must he invest at 15%?

28. Pam invested a total of $40,000 in two business ventures. The first investment earned an annual interest income of 18%, and the second investment earned 5%. If the interest from the 18% investment is six times as much as the 5% investment, how much did she invest at each rate?

29. Dr. Johnson has $6000. He invested part of the money in a savings plan at 10% and the rest in a certificate of deposit at 12%. If he has a total of $340 in interest after six months, how much did he invest at each rate?

30. Matthew invested $4000 in two savings plans. If the first plan has an interest rate of 15% and the second plan has a rate of 9%, how much must be invested at each rate so that Matthew will earn a total of $420 in 10 months?

Answers to Practice Problems **16a.** $23.44 **b.** $5000 **c.** 20% **17a.** $400 at 6.25%, $1100 at 13% **b.** $1000 for the Ford, $3000 for the BMW

8.5 Mixtures

19 A *mixture problem* involves the mixing of two or more quantities. We will solve this type of problem in much the same way we solved simple interest problems. The only exception is that the equation is based on the fact that the amount (or value) of a substance present before mixing must equal the amount (or value) of that substance after mixing.

Example 21 Solve.

a. John needs 12 quarts of a 10% salt solution. If he has one container that is a 5% salt solution and another that is a 25% salt solution, how many quarts of each must be mixed to obtain the desired mixture?

b. How many ounces of pure acid must be added to 200 ounces of a 10% acid solution to obtain a solution that is 20% acid?

c. Lou has 10 pounds of walnuts that cost \$5 per pound and 15 pounds of cashews that cost \$6 per pound. How many pounds of pecans costing \$8 per pound should be added to these nuts to yield a mixture worth \$6 per pound?

Solution To solve a mixture problem, (1) represent the unknown(s), (2) write an equation, (3) solve the equation and determine all solutions, and (4) check.

21a. Step 1: x = quarts (qt) of 5% salt solution
$12 - x$ = quarts of 25% salt solution

Step 2:

	% of Salt	Qt of Solution	Qt of Salt*
Solution #1	$5\% = .05$	x	$(.05)(x) = .05x$
Solution #2	$25\% = .25$	$12 - x$	$.25(12 - x)$
Mixture	$10\% = .10$	12	$.10(12) = 1.2$

**Note:* Amount = (% of Substance)(Volume)

Salt in 5% solution	+	Salt in 25% solution	=	Salt in the final mixture
$.05x$	+	$.25(12 - x)$	=	1.2

Step 3:
$$.05x + .25(12 - x) = 1.2$$
$$.05x + 3 - .25x = 1.2$$
$$-.20x + 3 = 1.2$$
$$-.20x = -1.8 \quad \text{Added } -3 \text{ to both sides}$$
$$x = 9 \quad \text{Divided both sides by } -.20$$
$$12 - x = 3$$

John must mix nine quarts of the 5% salt solution with three quarts of the 25% salt solution.

Check: The answer is correct since:

a. 9 qt + 3 qt = 12 qt

b.

Salt before mixing	=	Salt after mixing
.05(9 qt) + .25(3 qt)	=	.10(12 qt)
.45 qt + .75 qt	=	1.20 qt
1.20 qt	=	1.20 qt

21b. Step 1: x = ounces (oz) of pure acid

Step 2:

It seems like they're writing everything they know. And, the question—

	% of Acid	Oz of Solution	Oz of Acid
Solution #1	$10\% = .10$	200	$.10(200) = 20$
Solution #2	$100\% = 1.00$	x	$1(x) = x$
Mixture	$20\% = .20$	$x + 200$	$.20(x + 200)$

is, "...how many ounces of pure acid...,"

Acid in 10% solution	+	Acid in 100% solution	=	Acid in final mixture
20	+	x	=	$.20(x + 200)$

Step 3:

$$\begin{aligned} 20 + x &= .20(x + 200) \\ 20 + x &= .20x + 40 \\ .80x + 20 &= 40 && \text{Added } -.20x \text{ to both sides} \\ .80x &= 20 && \text{Added } -20 \text{ to both sides} \\ x &= 25 && \text{Divided both sides by } .80 \end{aligned}$$

25 oz of pure acid must be added.

Check:

Acid before mixing	=	Acid after mixing
.10(200 oz) + 25 oz	=	.20(25 + 200) oz
20 oz + 25 oz	=	.20(225 oz)
45 oz	=	45 oz

21c. Step 1: x = pounds (lb) of pecans

Step 2:

	Cost per lb ($)	Number of lb	Total Cost ($)*
Walnuts	5	10	50
Cashews	6	15	90
Pecans	8	x	$8x$
Mixture	6	$25 + x$	$6(25 + x)$

**Note:* Total cost = (cost per unit)(number of units)

Walnuts cost	+	Cashews cost	+	Pecans cost	=	Mixture cost
50	+	90	+	$8x$	=	$6(25 + x)$

Step 3:

$$\begin{aligned} 50 + 90 + 8x &= 6(25 + x) \\ 140 + 8x &= 150 + 6x \\ 140 + 2x &= 150 && \text{Added } -6x \text{ to both sides} \\ 2x &= 10 && \text{Added } -140 \text{ to both sides} \\ x &= 5 && \text{Divided both sides by } 2 \end{aligned}$$

Lou must add five pounds of pecans.

Check:

Value before mixing	=	Value after mixing
\$5(10) + \$6(15) + \$8(5)	=	\$6(25 + 5)
\$50 + \$90 + \$40	=	\$6(30)
\$180	=	\$180

Practice Problem 18 ***Solve.***

a. A bottle contains 100 cubic centimeters (cc) of a 20% acid solution. How much of this solution must be removed and replaced with pure acid in order for the new solution to be 100 cc of a 30% acid solution?

b. Susan mixes some dried fruit that costs $0.75 per pound with roasted soybeans costing $1.50 per pound. If she wants a mixture of 90 pounds worth $1.24 per pound, how many pounds of each item must she use?

8.5 Exercises

Solve.

1. How many gallons of solution that is 15% salt must be mixed with a solution that is 75% salt to obtain 12 gallons of a 30% salt solution?

2. How many gallons of a 20% acid solution should be mixed with a 40% acid solution to obtain 10 gallons of a mixture that is 30% acid?

3. Betty needs six ounces of a medicine that is 10% alcohol. If she has one bottle of this medicine that is a 12% alcohol solution and another bottle that is a 7% alcohol solution, how many ounces of each must be mixed to obtain six ounces of a 10% alcohol solution?

4. How many pounds of a dried fruit costing $0.75 per pound should be mixed with another dried fruit costing $1.50 per pound to obtain a 50-pound mixture worth $1.00 per pound?

5. Ed needs a coolant that is 40% antifreeze. How many gallons of a solution that is 60% antifreeze should he mix with 10 gallons of a solution that is 6% antifreeze to obtain the desired coolant?

6. How much pure glycerin must be added to 30 quarts of a solution that is 40% glycerin to obtain a mixture that is 60% glycerin?

7. A bottle contains 25 ounces of a solution that is 30% acid. How much of this solution must be removed and replaced with distilled water in order to obtain 25 ounces of a solution that is 20% acid?

8. A farmer has 81 pounds of grass seed that costs $1.20 per pound. How many pounds of clover seed costing $1.60 per pound must be added to the grass seed to obtain a mixture costing $1.24 per pound?

9. How much water must be evaporated from 1000 gallons of 12% dye solution to obtain a solution that is 20% dye?

10. How much pure acid must be added to 200 cc of an 8% acid solution to obtain a solution that is 20% acid?

11. How many quarts of pure nitric acid must be added to three quarts of a 12% solution to obtain a solution that is 28% nitric acid?

12. How many liters of a cocktail that is 37% alcohol must be combined with five liters of a cocktail that is 45% alcohol in order to obtain a mixture that is 42% alcohol?

13. A janitor wishes to make a cleaning solution that is 40% disinfectant. How many gallons of water must be added to 20 gallons of a solution that is 50% disinfectant to obtain the desired solution?

14. Marcia mixed 20 quarts of a chocolate ice cream containing 20% butterfat with 10 quarts of a coffee ice cream containing 15% butterfat to make mocha ice cream. What percent of the mocha ice cream is butterfat?

15. How much cream that is 15% butterfat must be combined with 10 pints of milk that is 1.5% butterfat to obtain a cream that is 10% butterfat?

16. A chemistry student has 50 ounces of a solution that is 60% acid. How much of this solution must be removed and replaced with pure acid in order to bring the acid strength up to 70%?

17. How many ounces of a perfume costing $84 per ounce must be combined with 24 ounces of a perfume costing $60 per ounce to make a perfume worth $68 per ounce?

18. How many pounds of peanuts costing $4 per pound should be mixed with cashews costing $7 per pound to obtain a 30-pound mixture worth $6 per pound?

19. How many barrels of a type of glue costing $80 per barrel should be mixed with glue costing $100 per barrel to obtain 50 barrels of a glue worth $92 per barrel?

20. How many cases of a candy costing $40 per case must be mixed with a candy costing $60 a case to obtain 80 cases of a candy worth $48 per case?

Answers to Practice Problems **18a.** 12.5cc **b.** 31.2 pounds of fruit; 58.8 pounds of soybeans

8.6 Uniform Motion

20 A *uniform motion problem* deals with distance, rate, and time. The relationship among these quantities is represented by the formula $\boldsymbol{d = rt}$, where

d = the **distance** an object travels;
r = the **constant rate** at which an object travels; and
t = the **time** an object travels.

To solve a uniform motion problem by using the distance formula, use the following procedure.

To Solve Uniform Motion Problems:

1. Identify d, r, and t.
2. Replace the known quantities in the formula with their numerical values.
3. Solve for the unknown quantity and answer the questions asked in the original problem.

NOTE: When using the distance formula, make the units of measure consistent. That is, if the rate is expressed as miles per hour, then be sure to express the distance in miles and the time in hours.

Example 22 Solve.

a. If you drove four hours at a rate of 55 mph, how many miles would you travel?

b. Peter traveled 400 miles in six hours and 15 minutes. How fast was he traveling?

Solution To solve a uniform motion problem, solve the formula $d = rt$ for the unknown quantity and answer the question asked in the original problem. Remember that the units of measure must be consistent.

22a. $d = ?$, $r = 55$ mph, $t = 4$ hr

$$\begin{aligned} d &= rt \\ &= (55)(4) \\ &= 220 \end{aligned}$$

You would travel 220 miles.

22b. $d = 400$ miles, $r = ?$, $t = 6$ hrs 15 min $= 6\frac{1}{4}$ hr

(*Note:* 15 minutes $= 15\left(\frac{1}{60}\text{ hr}\right) = \frac{1}{4}$ hr)

$$\begin{aligned} d &= rt \\ 400 &= r\left(6\frac{1}{4}\right) \\ 400 &= \frac{25}{4}r \\ 64 &= r \quad \text{Multiplied both sides by } \frac{4}{25} \end{aligned}$$

Peter was traveling at 64 mph.

Practice Problem 19 ***Solve.***

a. If a bus travels 45 mph for four hours, how many miles will it travel?

b. A boat traveled 300 miles at a rate of 40 mph. How many hours did the boat travel?

21 Many uniform motion problems involve a relationship between two objects (or one object under two different conditions). We can solve such problems in much the same manner that we solved simple interest and mixture problems. The only exception is that the equation is usually based on how the distances are related.

Example 23 Solve.

a. Betty leaves school on a bike and travels at a rate of 15 mph. One hour later, Rochelle follows Betty in a car and travels 30 mph. How long will it take Rochelle to catch Betty?

b. Dan and Frank leave the gym and travel in opposite directions on a highway. If they are 30 miles apart in 20 minutes and Frank's car is traveling 5 mph faster than Dan's, how fast was each man driving?

c. It takes a boat two hours to travel downstream with a 6 mph current. The return trip upstream against the same current takes three hours. Find the rate of the boat in still water.

Solution To solve a uniform motion problem involving two objects, (1) represent the unknowns, (2) write an equation, (3) solve the equation and determine all solutions, and (4) check. For step 2, it may be helpful to make a sketch.

23a. Step 1: x = hours it takes Rochelle to catch Betty

Step 2:

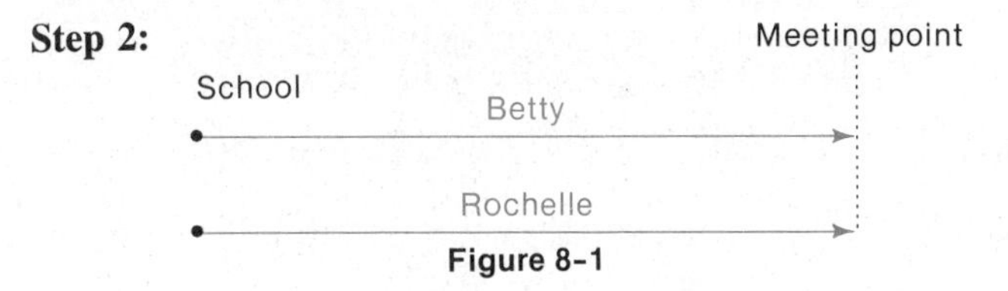

Figure 8-1

	Rate (mph)	Time (hr)	Distance (miles)*
Betty	15	$x + 1$	$15(x + 1)$
Rochelle	30	x	$30x$

**Note:* Distance = (Rate)(Time)

Figure 8–1 shows you that:

$$\text{Distance Rochelle travels} = \text{Distance Betty travels}$$
$$30x = 15(x + 1)$$

Step 3:
$30x = 15(x + 1)$
$30x = 15x + 15$
$15x = 15$ Added $-15x$ to both sides
$x = 1$ Divided both sides by 15

It will take Rochelle one hour to catch Betty.

Check: Distances traveled must be the same.
Rochelle's distance = (30 mph)(1 hr) = 30 miles
Betty's distance = (15 mph)(2 hr) = 30 miles

23b. Step 1:
x = rate (in mph) of Dan's car
$x + 5$ = rate (in mph) of Frank's car

Step 2:

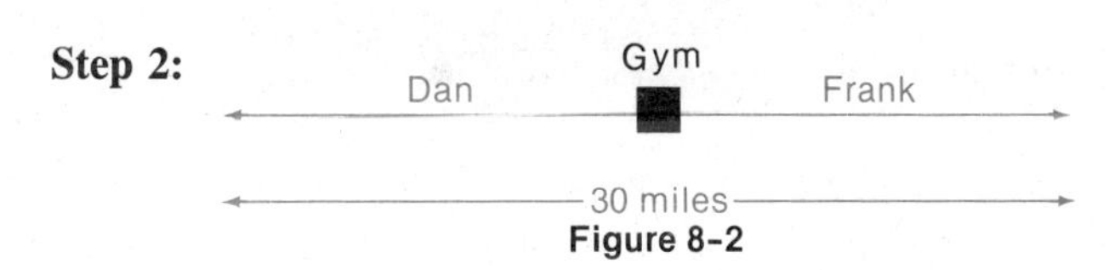

Figure 8-2

	Rate (mph)	Time (hours)	Distance (miles)
Dan	x	20 min $= \frac{1}{3}$ hr	$\frac{1}{3}x$
Frank	$x + 5$	20 min $= \frac{1}{3}$ hr	$\frac{1}{3}(x + 5)$

Figure 8–2 shows you that:

$$\text{Frank's distance} + \text{Dan's distance} = 30 \text{ miles}$$

$$\frac{1}{3}(x + 5) + \frac{1}{3}x = 30$$

Step 3: $\frac{1}{3}(x + 5) + \frac{1}{3}x = 30$

$x + 5 + x = 90$ Multiplied both sides by 3

$2x + 5 = 90$

$2x = 85$

$x = 42\frac{1}{2}$

$x + 5 = 47\frac{1}{2}$

Dan's rate is $42\frac{1}{2}$ mph and Frank's is $47\frac{1}{2}$ mph.

Check: Sum of distance traveled must be 30 miles.

Frank's distance $= \left(47\frac{1}{2} \text{ mph}\right)\left(\frac{1}{3} \text{ hr}\right) = \frac{95}{6}$ miles

Dan's distance $= \left(42\frac{1}{2} \text{ mph}\right)\left(\frac{1}{3} \text{ hr}\right) = \frac{85}{6}$ miles

Total distance $= 30$ miles

23c. **Step 1:** x = rate (mph) of the boat in still water

Step 2:

downstream $x + 6$

upstream $x - 6$

Figure 8-3

	Rate (mph)	Time (hours)	Distance (miles)
Downstream	$x + 6$	2	$2(x + 6)$
Upstream	$x - 6$	3	$3(x - 6)$

Figure 8–3 shows you that:

$$\text{Distance upstream} = \text{Distance downstream}$$

$$3(x - 6) = 2(x + 6)$$

Step 3: $3(x - 6) = 2(x + 6)$

$3x - 18 = 2x + 12$

$x - 18 = 12$ Added $-2x$ to both sides

$x = 30$ Added 18 to both sides

The boat traveled 30 mph in still water.

Check: The distances traveled must be the same.
Distance upstream = (36 mph)(2 hr) = 72 miles
Distance downstream = (24 mph)(3 hr) = 72 miles

Practice Problem 20 **Solve.**

a. Train A leaves Dallas at 10:00 A.M. and travels south at a rate of 45 mph. Train B leaves Dallas at 1:00 P.M. on a parallel track and travels south at a rate of 75 mph. How long will it take train B to catch up to train A?

b. Reggie made a trip of 360 miles to his summer home in eight hours. He traveled for six hours on an interstate highway and the rest of the time on an unpaved road. If his rate on the highway was 20 mph more than on the unpaved road, find his rate on the highway and the road.

22 In some uniform motion problems, the equation may not always be based on distance.

Example 24 Train A leaves Chicago for Boston (a distance of 1050 miles) at the same time that train B leaves Boston for Chicago. If train A travels at a rate of 60 mph and train B travels at 40 mph, how far has each train traveled when the two trains pass?

Solution **Step 1:**

x = miles train A travels before passing train B
$1050 - x$ = miles train B travels before passing train A

Step 2:

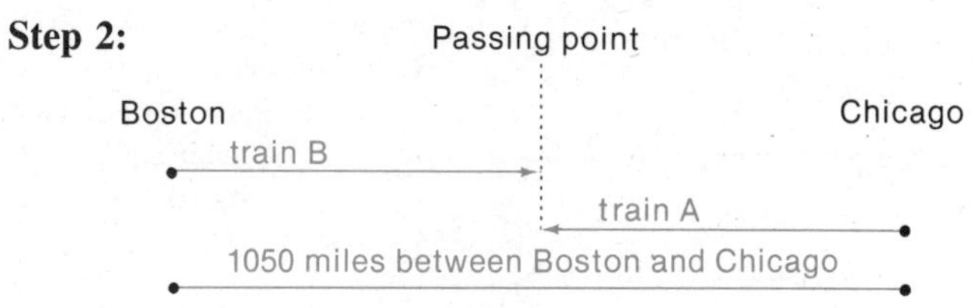

Figure 8-4

	Rate (mph)	Time (hours)*	Distance (miles)
Train A	60	$\frac{x}{60}$	x
Train B	40	$\frac{1050 - x}{40}$	$1050 - x$

**Note:* Since $d = rt$, then $t = \frac{d}{r}$.

The equation is based on the fact that:

Time train A travels = Time train B travels

$$\frac{x}{60} = \frac{1050 - x}{40}$$

Step 3:

$$\frac{x}{60} = \frac{1050 - x}{40}$$

$40x = 60(1050 - x)$ Cross-multiplied

$40x = 63{,}000 - 60x$

$100x = 63{,}000$ Added $60x$ to both sides

$x = 630$ Divided both sides by 100

$1050 - x = 420$

Train A travels 630 miles and train B travels 420 miles.

Check: The time traveled by each train must be the same.

$$\text{Time train A travels} = \frac{630 \text{ miles}}{60 \text{ mph}} = 10.5 \text{ hr}$$

$$\text{Time train B travels} = \frac{420 \text{ miles}}{40 \text{ mph}} = 10.5 \text{ hr}$$

Practice Problem 21 **Solve.**

A plane traveling 480 mph and a train traveling 64 mph leave the same point at the same time and travel north. In less than an hour, the plane has traveled 390 miles more than the train. How far has each vehicle traveled?

8.6 Exercises

Solve.

1. If you drove at a rate of 55 mph for two hours, how many miles would you travel?

2. John rode his bike at a rate of 20 mph for 45 minutes. How far did he travel?

3. Jim drove 150 miles in three hours. How fast was he driving?

4. How many hours will it take to travel 500 miles if you drive at a constant rate of 40 mph?

5. How many hours will it take a jet traveling at a rate of 550 mph to travel 1650 miles?

6. A jet traveled 3000 miles in five hours. How fast was the jet traveling?

7. Ricky completed a car race in three hours. If the race was 330 miles, how fast was Ricky traveling?

8. Denny rode his motorcycle at a constant rate of 45 mph for 40 minutes. How far did he travel?

9. June ran 28 miles in two hours and 14 minutes. What was her rate?

10. A speedboat travels at a constant rate of 90 mph for one hour and 12 minutes. How far did the boat travel?

11. Reggie leaves his house and runs at a rate of 5 mph. One hour later, his brother leaves the house on a bike and travels at a rate of 15 mph. How long will it take Reggie's brother to catch him?

12. Ed took 4 minutes to finish a race and Mike took $4\frac{1}{2}$ minutes to finish the same race. If the rate of the faster runner is three feet per second more than that of the slower runner, find each man's rate.

13. At 7:00 A.M. a boat traveling 30 mph leaves station A for station B, a distance of 150 miles. At 8:00 A.M. a boat traveling 20 mph leaves station B for station A. At what time will they pass each other?

14. A plane leaves an airport and travels 450 mph. Three hours later a second plane leaves the same airport and travels 475 mph. How long will it take the second plane to catch the first?

15. Two trains start from the same point at the same time and travel in the same direction on parallel tracks. If one train travels at 90 mph and the other at 68 mph, how many hours will it take before they will be 66 miles apart?

16. The pilot of a plane sets out to overtake the pilot of a helicopter that is 12 miles ahead. If the rate of the plane is 210 mph and the rate of the helicopter is 165 mph, how long will it take the plane to overtake the helicopter?

17. From a point on a highway, Sam and Tim ride their motorcycles in opposite directions. If they are 38 miles apart in 30 minutes and Tim's rate is 8 mph less than Sam's rate, how fast is each man traveling?

18. Ernie drove 760 miles in 14 hours. He traveled eight hours on a highway and six hours on a dirt road. If his rate on the highway was 25 mph more than on the dirt road, find the rate he traveled on both parts of his trip.

19. At 9:00 A.M., Rich leaves the train station on a bike and travels south at 8 mph. One hour later Joe leaves the same train station in a car and travels south a 20 mph. At what time will Joe overtake Rich?

20. At 11:00 A.M., Sylvia left the office to deliver a package to a restaurant. She stayed at the restaurant for one hour to eat lunch, and then she returned directly to the office, arriving at 3:00 P.M. She traveled to the restaurant at a rate of 45 mph, but her speed returning to the office was 35 mph. If she traveled the same route both ways, how far is it from her office to the restaurant?

21. Two steamboats are 50 miles apart. Both start at the same time and travel toward each other in a straight canal. If one averages 18 mph and the other averages 12 mph, how far does each travel before passing the other?

Answers to Practice Problems **19a.** 180 miles **b.** $7\frac{1}{2}$ hours **20a.** $4\frac{1}{2}$ hours **b.** 30 mph on the road, 50 mph on the highway **21.** Plane traveled 450 miles; train traveled 60 miles.

Summary

Important Terms

8.2
ratio
rate
proportion
means
extremes
cross multiplication rule

8.3
percent problem
base
percent
amount

8.4
simple interest problem
principal
rate
time
interest

8.5
mixture problem

8.6
uniform motion problem
distance
rate
time

Important Skills

8.1

Changing written statements to algebraic expressions
Representing unknown quantities in terms of the same letter
Changing written statements to linear equations
Solving word problems by using equations

8.2

Expressing written statements as a ratio
Solving word problems involving ratios
Identifying the means and extremes of a proportion
Determining if two ratios are equal
Solving proportions
Solving word problems involving proportions

8.3

Identifying A, B, and P in a percent problem
Solving word problems by using the percent formula

8.4

Solving simple interest problems by using the formula $I = PRT$
Solving simple interest problems which involve more than one rate

8.5

Solving mixture problems

8.6

Solving uniform motion problems by using the formula $d = rt$
Solving uniform motion problems which involve two objects

Review Exercises

Solve.

1. Express "five less than a number" as an algebraic expression.

2. The sum of two numbers is 50. Represent each number in terms of the same variable.

3. The sum of two consecutive integers is 263. Find the integers.

4. John won 18 out of 30 games. What percent of the games did he win?

5. What is the ratio of eight inches to three feet?

6. If $54 in interest is paid on an 18% loan for six months, what is the principal of the loan?

7. How many liters of a 40% salt solution must be added to 10 liters of a 70% salt solution to obtain a solution that is 50% salt?

8. After paying an 8% sales commission, Rich received $73,600 from the sale of his summer home. What was the actual selling price?

9. Jim earned $5200 in five months. At this rate, how much can he expect to earn in one year?

10. Two cars moving in opposite directions from the same point are 19 miles apart in 15 minutes. If one car's rate is 8 mph less than the other, find the rate of each car.

11. Jane invested $8000. Part was invested at 11% and the rest at 10%. If her total annual interest was $850, how much was invested at each rate?

12. Solve for x. $6:4\frac{1}{2} = x:30$

13. If a car travels 500 miles in 6 hours 15 minutes, how fast was the car traveling?

14. How much distilled water must be added to 50 ounces of a 12% acid solution to obtain a solution that is 6% acid?

15. A car rental company charges $19.95 per day plus 18¢ per mile to rent a compact car. If Joe rented a car for one day and his total bill was $34.44, how many miles did he travel?

16. The profit from the sale of a boat must be divided among three men in a ratio of 5:7:8. If the total profit from the sale of the boat was $22,400, how much money should each man receive?

GRAPHING

OBJECTIVES

1. Graph ordered pairs in the rectangular coordinate system.

2. Find the coordinates of a point.

3. Graph linear equations in the rectangular coordinate system.

4. Graph linear equations in the rectangular coordinate system using the slope-intercept method.

5. Graph linear inequalities in the rectangular coordinate system.

PRETEST

EXPLANATION

1. Graph.
   a. (0, 3) b. (−2, 3)
   c. (−1, −3) d. (0, 0)
   e. (−3, 0) f. (2, −3)

Section 9.1
Idea 3

2. Find the coordinates.

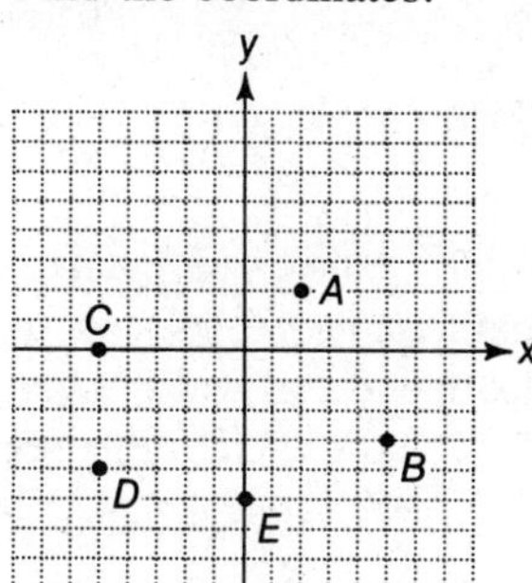

Section 9.1
Idea 4

3. Graph.
   a. $x - y = 3$ b. $2x + y = 4$
   c. $6x - 4y = 8$ d. $x - 5 = -1$

Section 9.2
Ideas 8 and 9

4. Graph. Use the slope-intercept method.
   a. $y = 2x - 2$ b. $3x - 2y = -6$

Section 9.3
Ideas 10–12

5. Graph.
   a. $x + y < 4$ b. $2x - y \geq 4$
   c. $x - 5y \leq 0$ d. $2y - 8 > -2$

Section 9.4
Ideas 15 and 16

9.1 Ordered Pairs

Idea 1 The **rectangular coordinate system** is formed by two intersecting perpendicular number lines. The point of intersection is called the **origin** (see Figure 9–1).

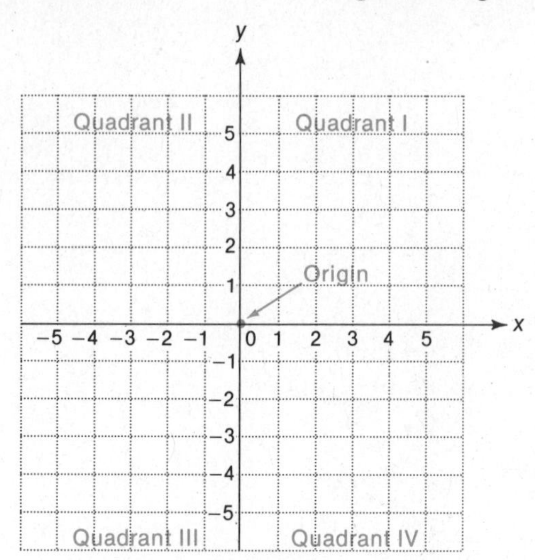

Figure 9–1

The rectangular coordinate system is divided into four regions called **quadrants.** The quadrants are numbered (with Roman numerals) counterclockwise beginning with quadrant I in the upper right region. The horizontal line is called the ***x*-axis** and the vertical line is called the ***y*-axis.** An arrow on each axis is used to show which direction is positive, and the origin is labeled "0."

2 Points in the rectangular coordinate system correspond to **ordered pairs** of numbers (x, y). For example, $(0, 2)$, $(-4, 2)$, $(-0.5, -3)$, and $(2\frac{1}{2}, 5)$ are ordered pairs that correspond to points in the coordinate system. The numbers in the ordered pair are called **coordinates** of the point. The first number is the x-coordinate or **abscissa** of the point. The second number is the y-coordinate or **ordinate** of the point.

3 Finding points in the coordinate system is called **graphing** or **plotting points.** When you are given the coordinates of a point and are asked to find the point (x, y) in the coordinate system, use the following procedure:

To Graph a Point:

1. Locate x on the x-axis and draw a vertical line through this point. This line will be parallel to the y-axis.
2. Locate y on the y-axis and draw a horizontal line through this point. This line will be parallel to the x-axis.
3. The intersection of these two lines is the graph of the ordered pair (x, y).

Example 1 Graph the following ordered pairs.

a. $(2, 4)$ **b.** $(-2, 3)$ **c.** $(-4, -2)$ **d.** $(3, -4)$

Solution Points in the coordinate system can be seen as the intersection of two lines (see Figure 9–2).

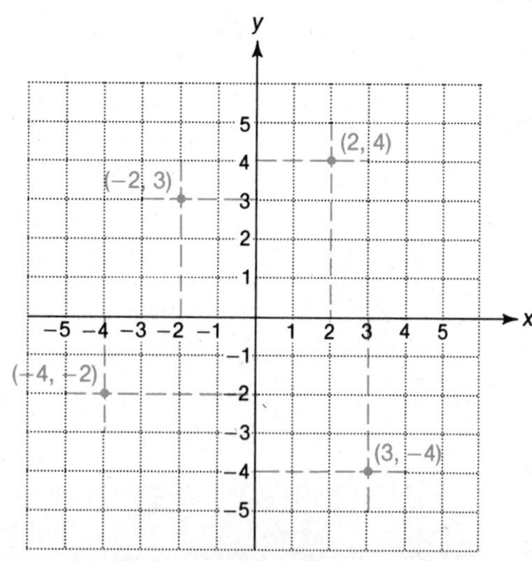

Figure 9–2

With practice you should be able to plot points without actually drawing the intersecting lines.

Example 2 Graph the following ordered pairs.

a. (0, 3) **b.** (−2, 0) **c.** (3, 0) **d.** (0, −3)

Solution All points having an x-coordinate of zero are on the y-axis. All points having a y-coordinate of zero are on the x-axis (see Figure 9–3).

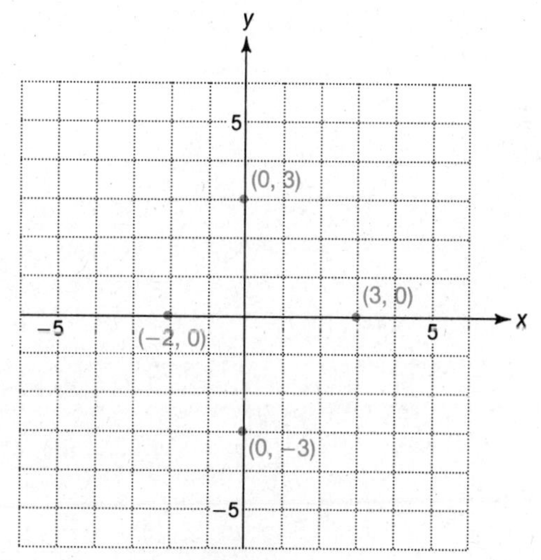

Figure 9–3

NOTE: The order in which the coordinates of a point are written is important. For example, the ordered pairs (0, 3) and (3, 0) are not the same; they represent two different points.

Practice Problem 1 ***Graph the following ordered pairs.***

a. (3, 4) **b.** (−2, 0) **c.** (3, −4) **d.** (0, 4) **e.** (−2, −3)

4 So far we have plotted points, given their coordinates. Let us now find the coordinates of a given point. To find the coordinates of a given point, use the following procedure:

To Find the Coordinates of a Given Point:

1. Draw a vertical line through the point. This line will intersect the x-axis at x (this is the first coordinate of the point).
2. Draw a horizontal line through the point. This line will intersect the y-axis at y (this is the second coordinate of the point).
3. Write the coordinates as the ordered pair (x, y).

Example 3 Find the coordinates of each point in Figure 9–4.

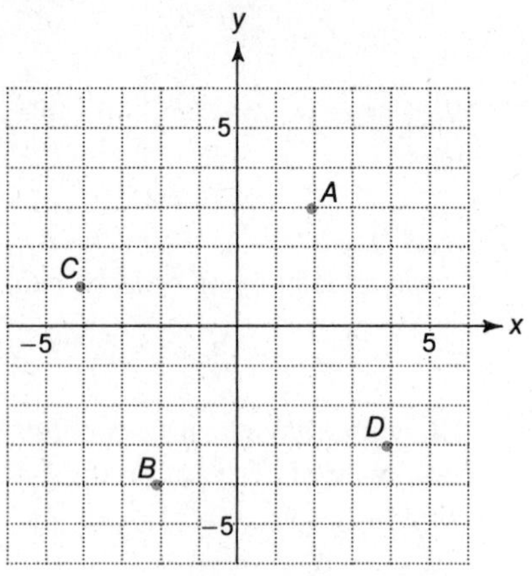

Figure 9–4

Solution

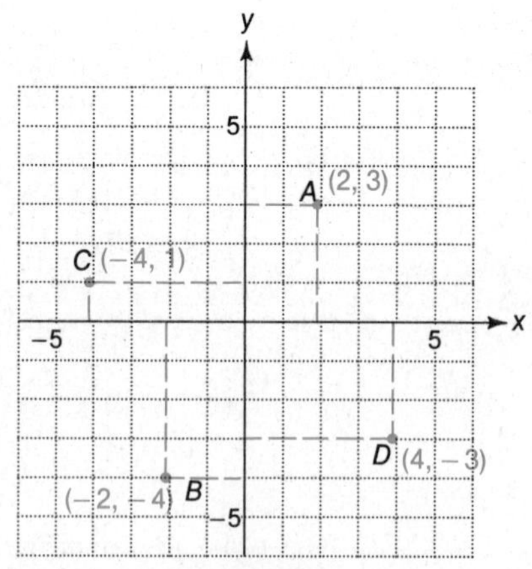

Figure 9–5

With practice you should be able to find the coordinates of a point without drawing the vertical and horizontal lines.

Example 4 Find the coordinates of each point in Figure 9–6.

Figure 9-6

Solution As shown in Figure 9–7, every point on the x-axis is of the form $(x, 0)$. Every point on the y-axis is of the form $(0, y)$.

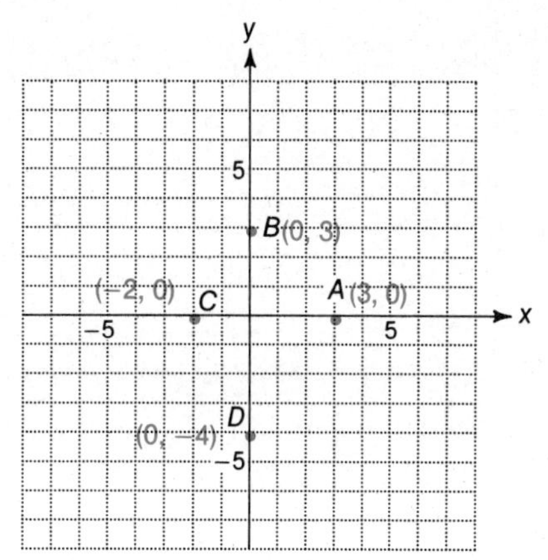

Figure 9-7

NOTE: The origin is assigned the coordinates $(0, 0)$.

Practice Problem 2 ***Find the coordinates of each point in the Figure below.***

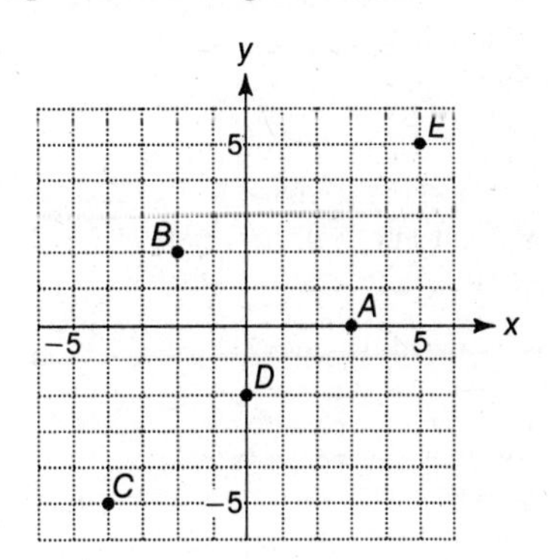

9.1 Exercises

1. Graph $(2, 3)$, $(-3, 2)$, $(-3, -3)$, and $(-2, 4)$.
2. Graph $(3, -2)$, $(-3, -1)$, $(2, 2)$, and $(-5, 1)$.
3. Graph $(0, 3)$, $(3, 0)$, $(-4, 0)$, and $\left(0, -2\frac{1}{2}\right)$.
4. Graph $(-6, 0)$, $(0, -6)$, $(2.5, 0)$, and $(0, 6)$.
5. Plot the points having the following coordinates: $(0, 4)$, $\left(-2, 3\frac{1}{2}\right)$, $(-5, 0)$, and $(-3, -4)$.
6. Plot the points having the following coordinates: $(3, -5)$, $(0, -1)$, $(0, 0)$, and $(5, -2)$.

Find the coordinates of points A, B, C, D, E, and F.

7.

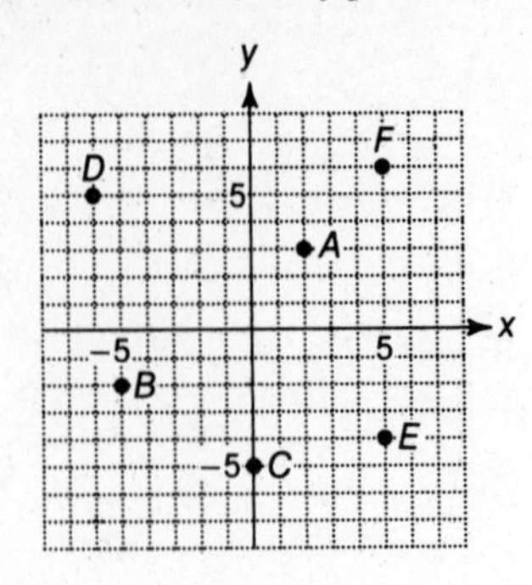

8.

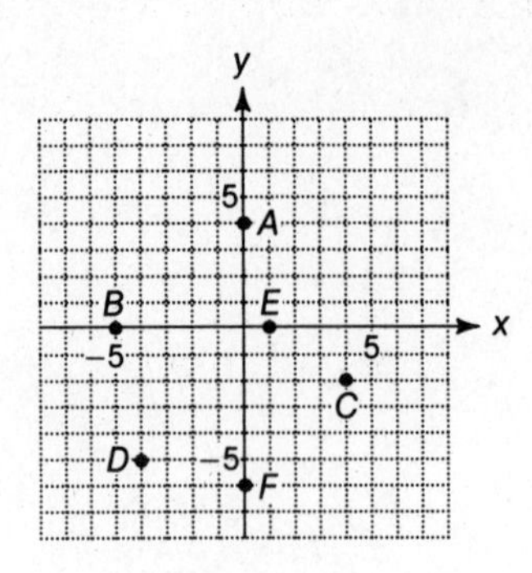

9.

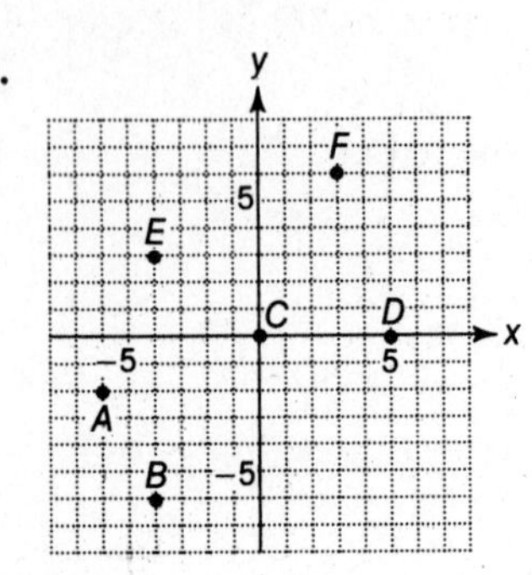

10.

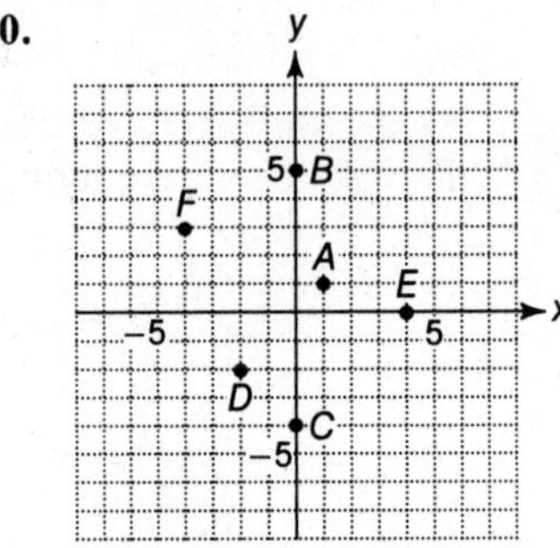

Fill in the blanks.

11. The rectangular coordinate system is divided into four regions called ________________.

12. In the rectangular coordinate system, the horizontal line is called the ________________ and the vertical line is called the ________________.

13. Points in the rectangular coordinate system correspond to ________________ of numbers.

14. The numbers in an ordered pair are called the ________________ of the point.

15. The first number in an ordered pair is the ________________ or ________________ and the second number is the ________________ or ________________ of the point.

16. To graph an ordered pair (x, y), first locate x on the x-axis and draw a ________________ line through this point. Next, locate y on the y-axis and draw a ________________ line through this point. Finally, the ________________ of these two lines is the graph of the ordered pair (x, y).

17. To find the coordinates of a given point, first draw a ________________ line through the point. This line will intersect the x-axis at ________________, which is the first ________________ of the point. Next, draw a ________________ line through the point. This line will intersect the y-axis at ________________, which is the ________________ coordinate of the point. Finally, write the ________________ as the ________________ (x, y).

18. The ________________ is assigned the coordinates (0, 0).

Answers to Practice Problems

1.

y
x
5
(0, 4) (3, 4)
(−2, 0)
−5
5
(3, −4)
(−2, −3)
−5

2. $A = (3, 0)$, $B = (-2, 2)$, $C = (-4, -5)$,

$D = (0, -2)$, $E = (5, 5)$

9.2 Linear Equations

5 An equation of the form $ax + by = c$, where x and y are variables and a, b, and c are constants (a and b are not both zero) is called a linear or first-degree equation in two variables or, simply, a **linear equation.**

Example 5 The following are linear equations in two variables.

a. $x + y = 4$ **b.** $3x - 5y = 10$

6 A **solution** for a linear equation in two variables is an ordered pair of numbers that makes the equation a true statement. For example, (1, 3) is a solution of the equation $x + y = 4$ because $x + y = 4$ is true when $x = 1$ and $y = 3$. We can also say that (1, 3) *satisfies* the equation $x + y = 4$.

$$x + y = 4$$
$$1 + 3 = 4$$
$$4 = 4 \quad \text{True}$$

NOTE: This equation has an infinite number of solutions. Some of these solutions are (0, 4), (4, 0), (2, 2), and (6, −2).

Example 6 Find and graph three ordered pairs that are solutions of $x + y = 4$.

Solution By inspection, you have found that (0, 4), (4, 0), and (2, 2) are solutions of $x + y = 4$ since these ordered pairs make the equation $x + y = 4$ a true statement. Figure 9–8 shows that the ordered pairs that are solutions of a linear equation lie on a straight line. This line is called the **graph of the equation.** All ordered pairs associated with the points on this line will satisfy the equation.

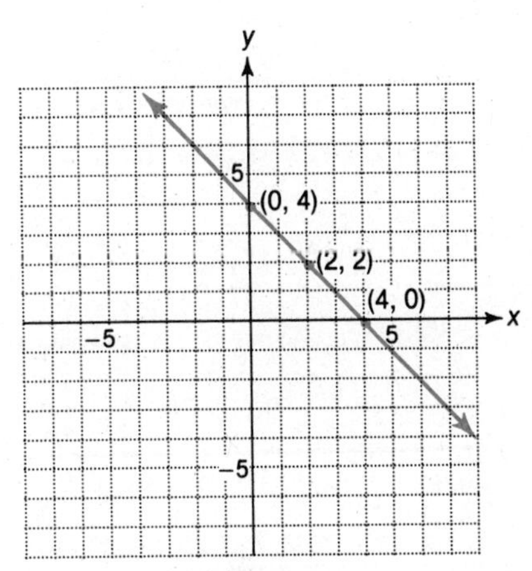

Figure 9–8

Practice Problem 3 ***Find and graph four ordered pairs that are solutions of $x - y = 5$.***

7 Quite often you will not be able to find the solution of a linear equation by inspection. When this occurs, either (1) choose a convenient value for x and find the corresponding value for y, *or* (2) choose a value for y and find the corresponding value for x.

Example 7 Find and graph three ordered pairs that are solutions of $2x - 3y = 6$.

Solution To find a solution for a linear equation, choose a value for x or y and then solve the equation for the other variable.

Let $x = 0$.

$$\begin{aligned} 2x - 3y &= 6 \\ 2(0) - 3y &= 6 \\ -3y &= 6 \\ y &= -2 \end{aligned}$$

Thus, $(0, -2)$ is a solution of $2x - 3y = 6$. The point $(0, -2)$ is called the **y-intercept** since this is the point where the graph of the equation crosses the y-axis (see Figure 9–9).

Let $y = 0$.

$$\begin{aligned} 2x - 3y &= 6 \\ 2x - 3(0) &= 6 \\ 2x &= 6 \\ x &= 3 \end{aligned}$$

Thus, $(3, 0)$ is a solution of $2x - 3y = 6$. The point $(3, 0)$ is called the ***x*-intercept** since this is the point where the graph of the equation crosses the x-axis (see Figure 9–9).

Let $x = -3$.

$$\begin{aligned} 2x - 3y &= 6 \\ 2(-3) - 3y &= 6 \\ -6 - 3y &= 6 \\ -3y &= 12 \\ y &= -4 \end{aligned}$$

Thus, $(-3, -4)$ is a solution of $2x - 3y = 6$ (see Figure 9–9).

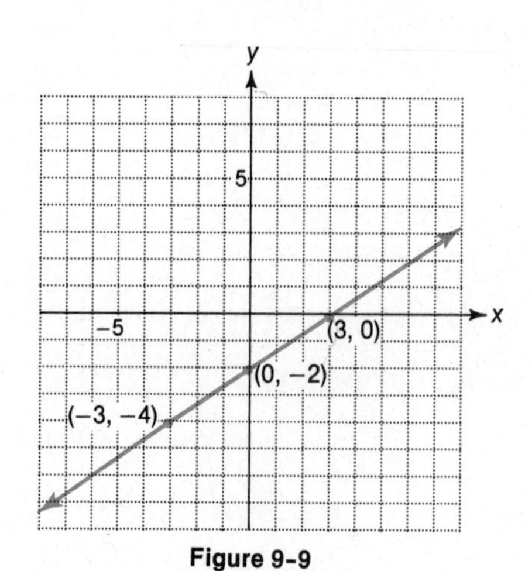

Figure 9-9

NOTE: Although you need only two points to draw a straight line, it is advisable to find at least one other point as a check.

8 Based on the results in Ideas 6 and 7, use the following procedure to graph linear equations in two variables. This is called the **intercept method.**

To Graph Linear Equations (Intercept Method):

1. Find the x-intercept by letting $y = 0$ and solving for x.
2. Find the y-intercept by letting $x = 0$ and solving for y.
3. Find a third ordered pair that satisfies the equation by letting x or y be any convenient number and then solving for the other variable.
4. Graph the three ordered pairs that satisfy the equation and connect them with a straight line. This line is the graph of the equation.

Example 8 Graph $4x - 3y = 9$.

Solution To graph a linear equation, first find at least three ordered pairs that satisfy the equation. Next, graph these ordered pairs. Finally, connect the points with a straight line.

x-intercept:

$$4x - 3y = 9; \text{ let } y = 0 \text{ and solve for } x$$
$$4x - 3(0) = 9$$
$$4x = 9$$
$$x = \frac{9}{4}$$
$$x = 2\frac{1}{4}$$

The x-intercept is $2\frac{1}{4}$ or $(2\frac{1}{4}, 0)$ (see Figure 9–10).

y-intercept:

$$4x - 3y = 9; \text{ let } x = 0 \text{ and solve for } y$$
$$4(0) - 3y = 9$$
$$-3y = 9$$
$$y = -3$$

The y-intercept is -3 or $(0, -3)$ (see Figure 9–10).

Check:

$$4x - 3y = 9; \text{ let } x = 3 \text{ and solve for } y$$
$$4(3) - 3y = 9$$
$$12 - 3y = 9$$
$$-3y = -3$$
$$y = 1$$

The checkpoint is (3, 1) (see Figure 9–10).

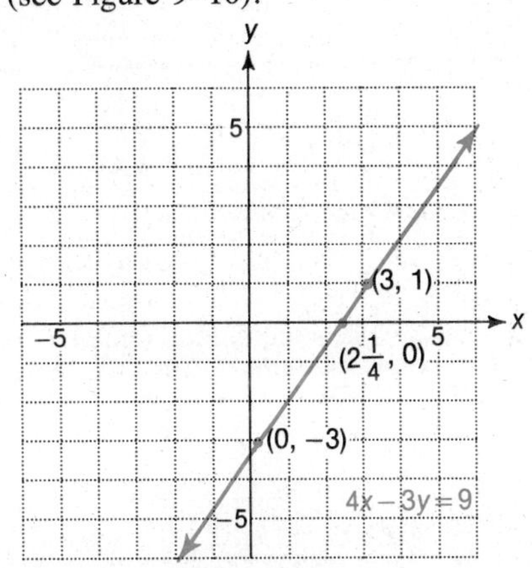

Figure 9–10

Example 9 Graph $x = 2y$.

Solution

x-intercept:

$x = 2y$; let $y = 0$ and solve for x
$x = 2(0)$
$x = 0$

The x-intercept is (0, 0) (see Figure 9–11).

y-intercept:

$x = 2y$; let $x = 0$ and solve for y
$0 = 2y$
$0 = y$

The y-intercept is (0, 0) (see Figure 9–11). Since the x and y-intercept are the same point, we must find two other ordered pairs that satisfy the equation.

Third ordered pair:

$x = 2y$; let $y = 1$ and solve for x
$x = 2(1)$
$x = 2$

The third point is (2, 1) (see Figure 9–11).

Check:

$x = 2y$; let $x = 4$ and solve for y
$4 = 2y$
$2 = y$

The checkpoint is (4, 2) (see Figure 9–11).

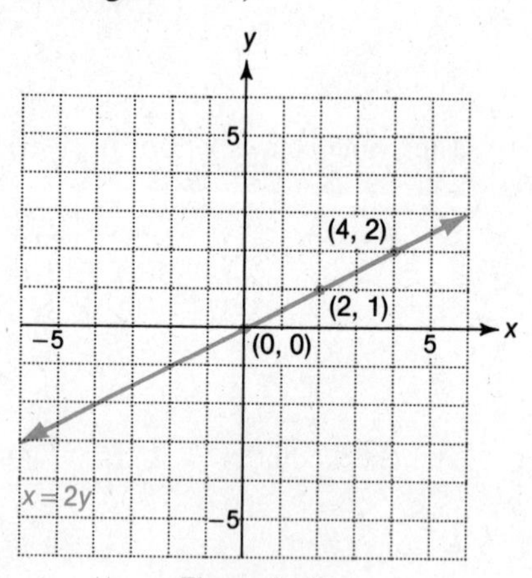

Figure 9-11

Practice Problem 4 ***Graph each equation.***

a. $2x + 3y = 9$ **b.** $2x + y = 0$

9 Some linear equations contain only one variable. The graph of such equations is either a horizontal line or a vertical line.

Example 10 Graph each equation.

a. $x = -3$ **b.** $y = 2$

Solution **10a.** Since there is no y, the equation $x = -3$ is equivalent to $x + 0y = -3$ and every ordered pair $(-3, y)$ satisfies the equation. That is, $(-3, 0)$, $(-3, 3)$, and $(-3, -2)$ satisfy the equation and correspond to points on the graph of the line $x = -3$ (see Figure 9–12).

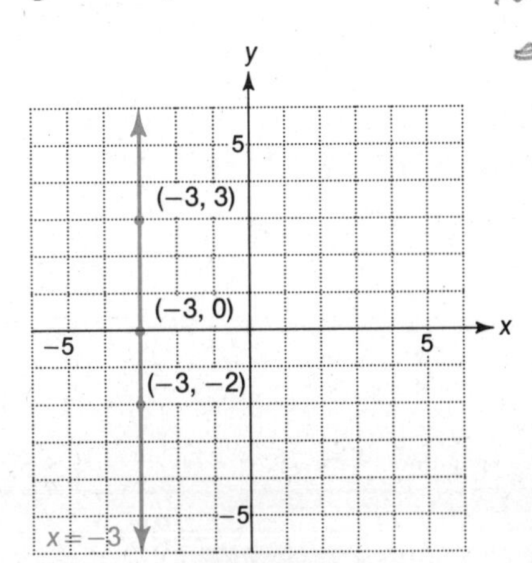

Figure 9-12

10b. Since there is no x, the equation $y = 2$ is equivalent to $0x + y = 2$, and every ordered pair $(x, 2)$ is a solution. That is, $(0, 2)$, $(-2, 2)$, and $(3, 2)$ satisfy the equation and correspond to points on the graph of the line $y = 2$ (see Figure 9–13).

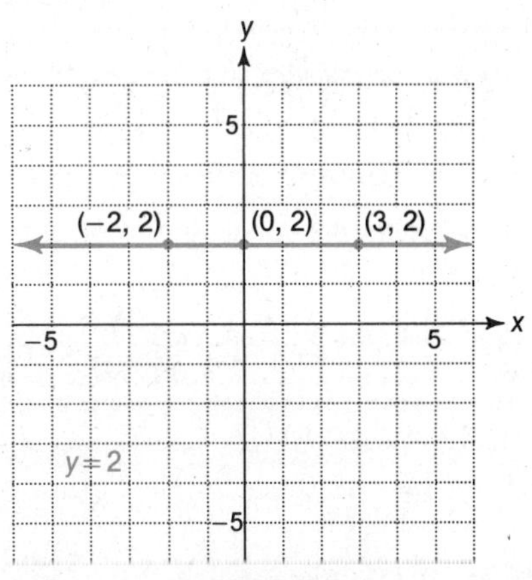

Figure 9-13

NOTE: In general, the graph of $x = a$, where a is a constant, is a vertical line that passes through the point $(a, 0)$. Similarly, the graph of $y = b$, where b is a constant, is a horizontal line that passes through the point $(0, b)$.

Practice Problem 5 ***Graph each equation.***

a. $x = 3$ **b.** $y = -\frac{7}{2}$ **c.** $x - 5 = -9$

9.2 Exercises

Find the missing coordinate by using the given equation, and then graph the ordered pairs.

1. $y = 3x + 1$ $(0,\ )$, $(-2,\ )$, $(\ , 4)$

2. $2x + y = 6$ $(\ , 0)$, $(1,\ )$, $(5,\ )$

3. $3x - 7y = 14$ $(0,\ )$, $(\ , 1)$, $(5,\ )$

4. $x = 5$ $(\ , 2)$, $(\ , -1)$, $(\ , 0)$

5. $y + 3 = 1$ $(0,\ )$, $(1,\ )$, $(-3,\ )$

Find the x-intercept and the y-intercept.

6. $x + y = 6$

7. $3x - y = 6$

8. $3x + 4y = 12$

9. $y = 3x + 6$

10. $5x - 10y = 20$

11. $\frac{x}{3} + \frac{y}{2} = 1$

12. $2x = 3y$

13. $y = \frac{x}{2} - 6$

Graph each equation using the intercept method.

14. $x + y = 3$

15. $2x + y = 4$

16. $2x - 3y = 6$

17. $x - 3y = 6$

18. $x - y = 0$

19. $2x + 5y = 10$

20. $x = -2$

21. $y = -\frac{3}{2}$

22. $x - 5 = -3$

23. $y + 4 = 6$

24. $x = -2y$

25. $x = \frac{y}{2} - 4$

26. $x = \frac{3}{4}y - 3$

27. $x = \frac{2}{3}y - 4$

28. $2x = 5(y + 2)$

29. $3x = 5(y + 3)$

30. $4y = 16 - 8x$

31. $3y = 6 - 2x$

32. $y = \frac{3}{5}x$

33. $y = \frac{3}{4}x$

34. $0.2x + 0.3y = 0.6$

35. $0.1x - 0.1y = 0.3$

36. $x - 2.5 = -5$

37. $x - 3.5 = -8$

38. $y = -3.5$

39. $y = -2.5$

Fill in the blanks.

40. A ______________ for a linear equation in two variables is an ______________ that makes the equation a true statement.

41. The graph of a linear equation consists of ordered pairs that lie on a ______________.

42. The ordered pairs associated with the points on a line which is the graph of an equation will ______________ the equation.

43. To graph a linear equation using the intercept method, find at least three ______________ that ______________ the equation. Next, plot these ______________ and connect them with a ______________.

44. The graph of $x = a$, where a is a constant, is a ______________ line that passes through the point ______________.

Solve.

45. Jim charges \$10 per hour plus \$10 for materials to repair a lawn mower.

a. Write an equation that determines Jim's total fee for a service call. Let y stand for the number of dollars he earns for working x hours.

b. Find Jim's fee for working two hours, four hours, and six hours. That is, find the value of y in the ordered pairs $(2, y)$, $(4, y)$, and $(6, y)$.

c. Graph the equation that you wrote for 45a by plotting the ordered pairs that you found for 45b on the graph at the right.

d. Determine from the graph in 45c the number of hours it will take Jim to earn \$85.

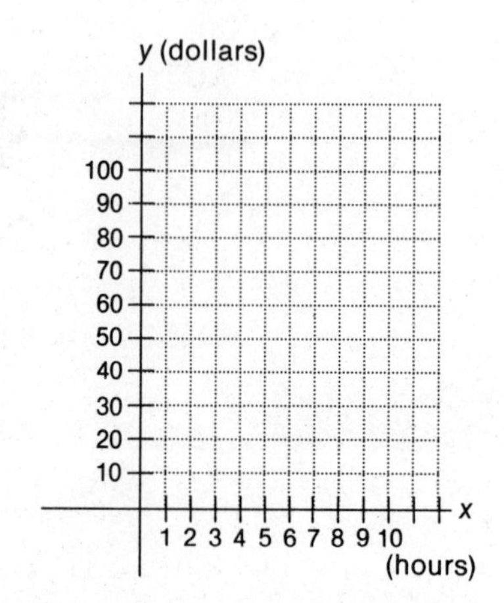

46. Using a straight-line depreciation method, a machine costing $2,000 will have a book value of $b = 2000 - 200n$ dollars n years after purchase.

a. Find the book value of the machine after one year, three years, and five years. That is, find b when $n = 1$, $n = 3$, and $n = 5$.

b. Using the fact that a solution of the equation $b = 2000 - 200n$ is the ordered pair (n, b), graph $b = 2000 - 200n$ at the right.

c. Determine from the graph in 46b the number of years it will take for the book value of the machine to be $1200.

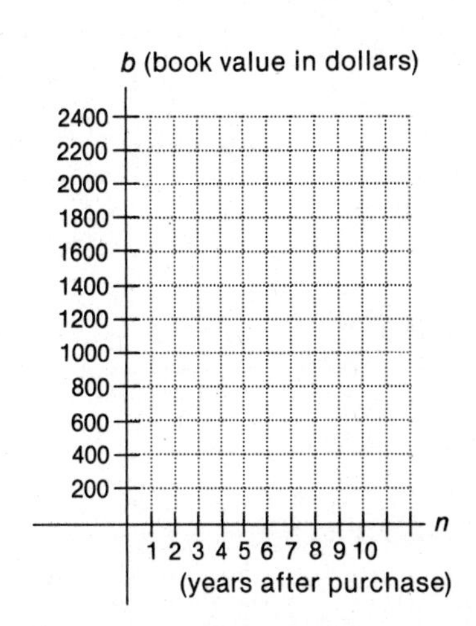

Answers to Practice Problems

3.

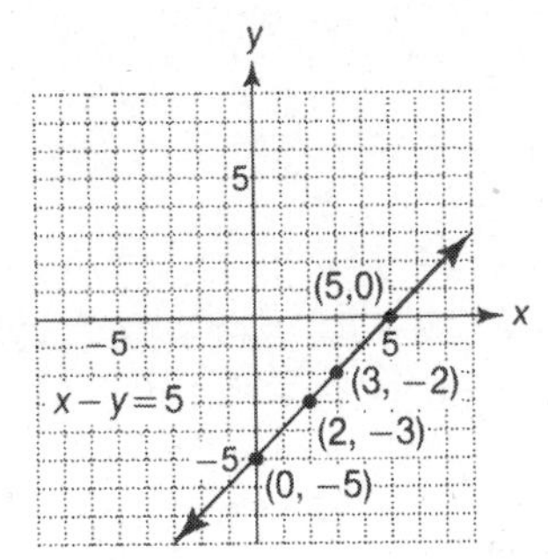

4a.

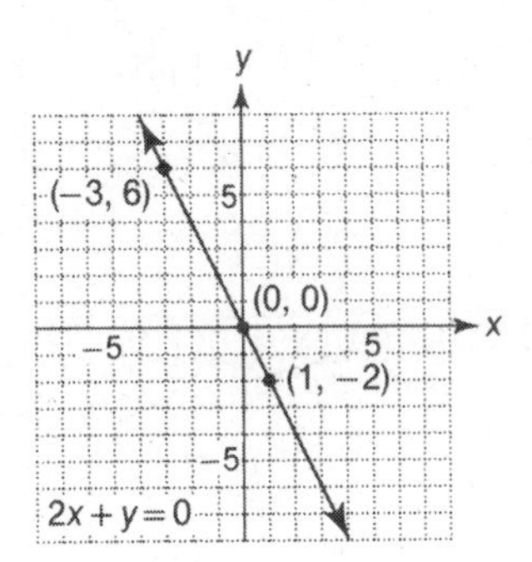

4b.

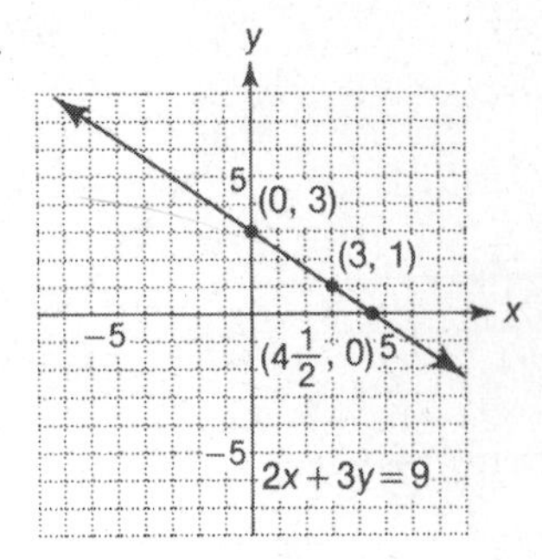

5a.

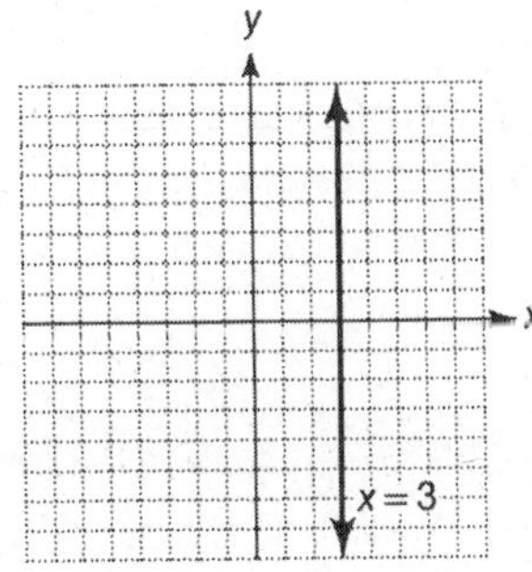

5b.

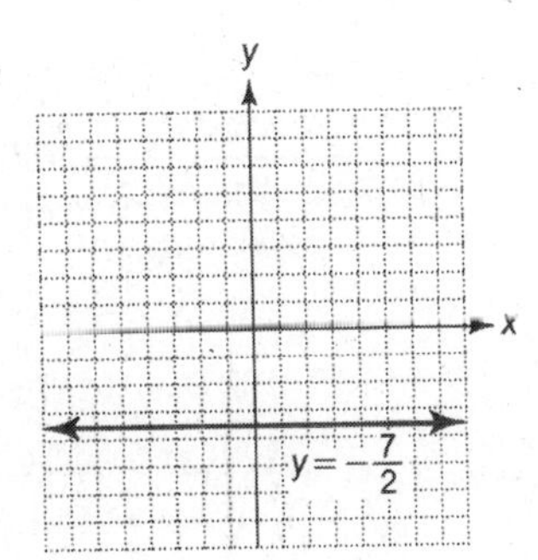

5c.

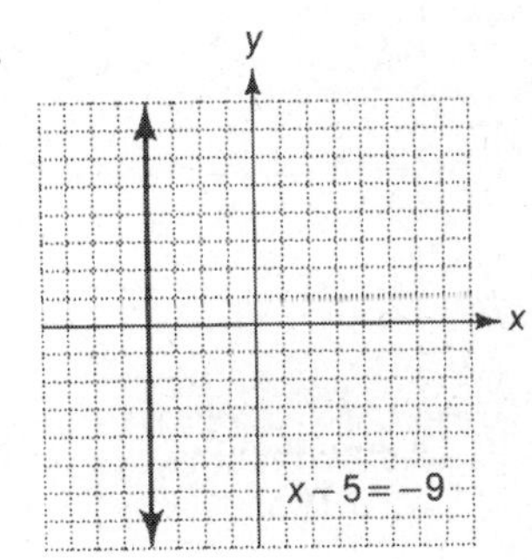

9.3 Slope-Intercept Form

10 Until now we have discussed linear equations written in the form $ax + by = c$. However, linear equations can be written in many different forms. One useful form is the **slope-intercept form** $y = mx + b$, where m is the slope of the line and b is the y-intercept. Note that the slope m of a line is the ratio of the rise (vertical rise) to the run (horizontal run).

Example 11 A slope of $\frac{4}{3}$ means that for every three units of run in a horizontal direction, there is a corresponding rise of four units. If a line slants *up* to the right (see Figure 9–14), we say that it has a positive slope.

$$\text{The slope } m = \frac{\text{rise}}{\text{run}} = \frac{4}{3}.$$

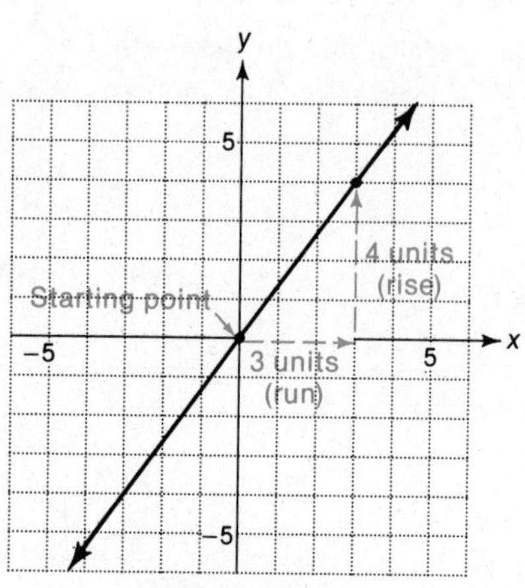

Figure 9–14

If a line slants *down* to the right (see Figure 9–15), we say that it has a negative slope.

$$\text{The slope } m = \frac{\text{rise}}{\text{run}} = \frac{-4}{2} = -2.$$

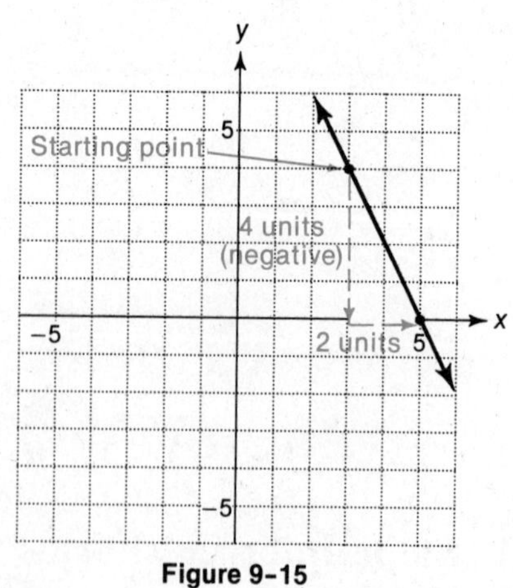

Figure 9–15

A slope of $-2\left(\frac{-2}{1}\right)$ means that for every run of 1 unit in a horizontal direction there is a corresponding fall of 2 units.

11 The slope-intercept form is useful because it lets you identify the slope m and the y-intercept $(0, b)$ directly from the equation.

Example 12 Write each equation in the slope-intercept form and identify the slope and the y-intercept.

a. $3x + y = 4$ **b.** $5x - 4y = 8$

Solution To write a linear equation in the slope-intercept form $y = mx + b$, first solve the equation for y and write the x term to the left of the constant. Then, the coefficient of x is the slope m, and the constant b is the y-intercept.

12a. $3x + y = 4$

$y = -3x + 4$ Add $-3x$ to both sides

The slope is -3 and the y-intercept is 4 or (0, 4).

12b. $5x - 4y = 8$

$-4y = -5x + 8$ Add $-5x$ to both sides

$y = \frac{5}{4}x - 2$ Divide both sides by -4

The slope is $\frac{5}{4}$ and the y-intercept is -2 or $(0, -2)$.

Practice Problem 6 ***Write each equation in slope-intercept form and identify the slope and the y-intercept.***

a. $4x + y = 4$ **b.** $5x - 3y = 8$

12 When an equation is in the slope-intercept form, the slope and y-intercept can be used to graph the equation.

To Graph Linear Equations (Slope-Intercept Method):

1. Rewrite the original equation in the slope-intercept form and identify the slope and the y-intercept.
2. Graph the y-intercept.
3. Starting at the y-intercept and remembering that the slope is the $\frac{\text{rise}}{\text{run}}$, find a second point on the graph.
4. Draw a line that passes through the y intercept and the point found in step 3.
5. Find at least one other point as a check.

Example 13 Use the slope-intercept method to graph each equation.

a. $2x + y = 3$ **b.** $3x - 4y = 10$

Solution **13a.** $2x + y = 3$

$y = -2x + 3$

The y-intercept is (0, 3). The slope is $-2\left(\frac{\text{rise}}{\text{run}} = \frac{-2}{1}\right)$. Now, graph the y-intercept (see Figure 9–16).

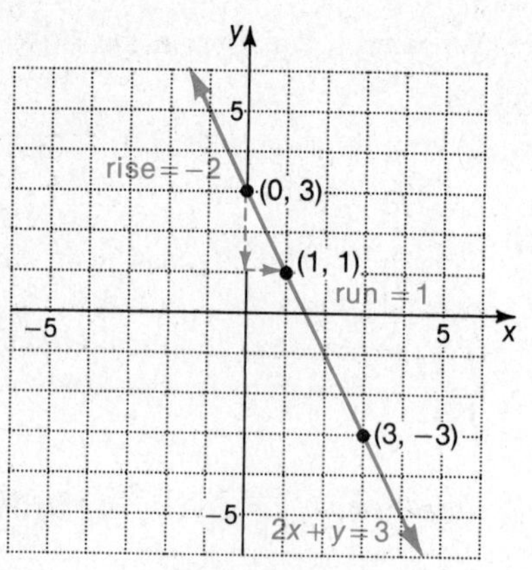

Figure 9–16

Next, starting at the y-intercept, measure two units down and then one unit to the right to obtain another point (1, 1). Finally, draw a line passing through (0, 3) and (1, 1). It is advisable to find a third point as a check. Thus, let $x = 3$ to find that

$$\begin{aligned} y &= -2x + 3 \\ &= -2(3) + 3 \\ &= -3. \qquad\qquad (3, -3) \end{aligned}$$

You can see in Figure 9–16 that $(3, -3)$ is on the line. Therefore, the graph is correct.

13b.
$$\begin{aligned} 3x - 4y &= 10 \\ -4y &= -3x + 10 \\ y &= \frac{3}{4}x - \frac{5}{2} \end{aligned}$$

The y-intercept is $(0, -2\frac{1}{2})$. The slope is $\frac{3}{4}\left(\frac{\text{rise}}{\text{run}} = \frac{3}{4}\right)$. Now, graph the y-intercept (see Figure 9–17).

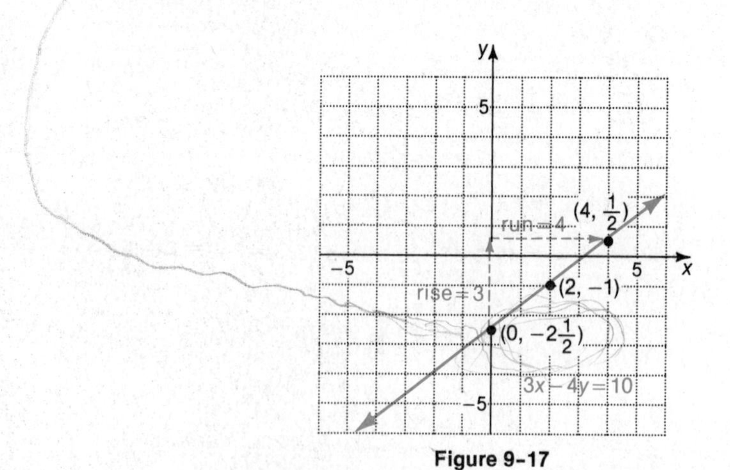

Figure 9–17

Next, starting at the y-intercept, measure three units up and then four units to the right to obtain another point $\left(4, \frac{1}{2}\right)$. Finally, draw a line passing through $\left(0, -2\frac{1}{2}\right)$ and $\left(4, \frac{1}{2}\right)$

Check:

$$y = \frac{3}{4}x - \frac{5}{2}; \text{ let } x = 2 \text{ and solve for } y$$

$$y = \frac{3}{4}(2) - \frac{5}{2}$$

$$y = \frac{3}{2} - \frac{5}{2}$$

$$y = -1 \qquad (2, -1)$$

Important: Remember

You can see in Figure 9–17 that $(2, -1)$ is on the line. Therefore, the graph is correct.

Practice Problem 7 ***Use the slope-intercept method to graph each equation.***

a. $y = 3x - 1$ **b.** $3x + 2y = 8$ **c.** $4x - 6y = 8$

9.3 Exercises

Write each equation in the slope-intercept form and identify the slope and the y-intercept.

1. $2x + y = 6$
2. $2x + 3y = 10$
3. $2x - 5y = 30$
4. $4x + 6y = 12$
5. $x - 4y = 8$
6. $3x + 5y = 15$
7. $3x + 2y = 6$
8. $3x + 4y = 12$
9. $3x + y = 6$
10. $2x + y = 4$
11. $5x - y = 8$
12. $4x - y = 9$
13. $6x + 4y = 10$
14. $4x + 6y = 14$
15. $3x - 5y = 11$
16. $2x - 3y = 5$

Use the slope-intercept method to graph each equation.

17. $6x + 2y = 4$
18. $10x - 8y = 16$
19. $y = 2x - 1$
20. $3x - 2y = 6$
21. $4x + 6y = 15$
22. $3x - y = 0$
23. $3x - 4y = -12$
24. $y = \frac{3}{2}x - 4$
25. $\frac{x}{3} + \frac{y}{2} = 1$
26. $4x - 5y = 15$
27. $5x + 2y = 10$
28. $2x + 3y = 6$
29. $4x + 3y = 12$
30. $3x - 4y = -12$
31. $2x - y = 4$
32. $3x - y = 6$
33. $\frac{x}{2} + \frac{y}{2} = 1$
34. $\frac{x}{3} + \frac{y}{3} = 1$
35. $2x = 5(y + 2)$
36. $3x = 5(y + 3)$
37. $4y = 16 - 8x$
38. $3y = 6 - 2x$
39. $y = 2x$
40. $y = 3x$
41. $x = -2y$
42. $x = -3y$

Fill in the blanks.

43. The slope-intercept form of an equation is ____________.

44. When an equation is in the slope-intercept form $y = mx + b$, the slope is ____________ and the ____________ is $(0, b)$.

45. If a line has a positive slope, it slants up to the ____________.

46. If a line has a negative slope, it slants down to the ____________.

47. The slope m of a line is the ratio of the ____________ to the ____________.

Answers to Practice Problems **6a.** $y = -4x + 4$, slope $= -4$, y-intercept $= (0,4)$ **b.** $y = \frac{5}{3}x - \frac{8}{3}$, slope $= \frac{5}{3}$, y-intercept $= \left(0, -\frac{8}{3}\right)$

7a.

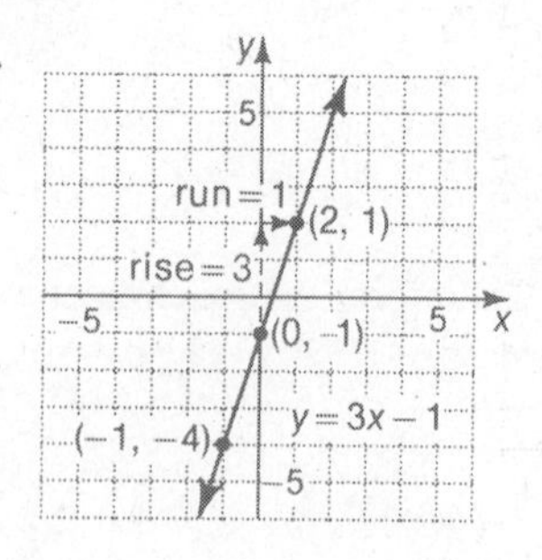

7b.

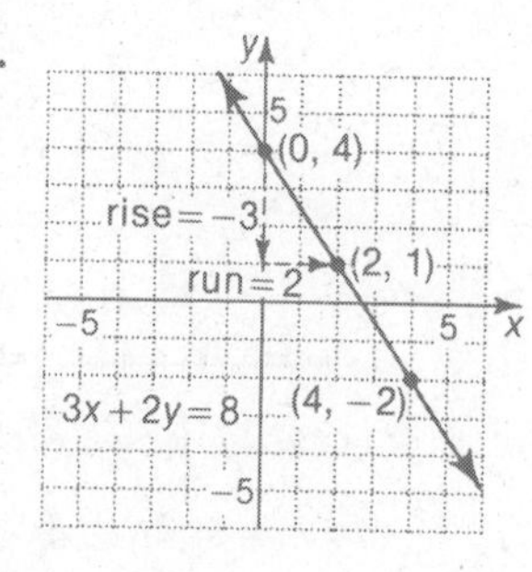

7c.

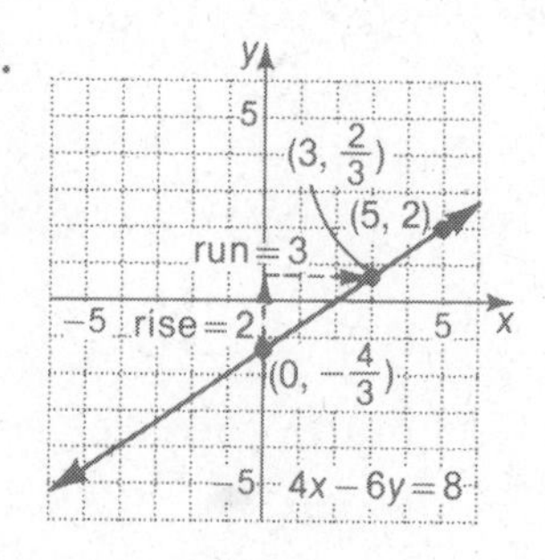

9.4 Linear Inequalities

13 You can graph linear inequalities in much the same manner that you graphed linear equations in two variables. The only distinction is that the graph of a linear inequality is a half-plane that may or may not include the boundary line. (The **boundary line** is the graph of the equation obtained when the inequality symbol is replaced with an equal sign).

Example 14 Graph the following.

a. $x + y \leq 3$ **b.** $x + y > 3$

Solution **14a.** Since the original problem contains an equality symbol, the boundary line $x + y = 3$ is part of the solution. This is indicated by a solid line (see Figure 9–18).

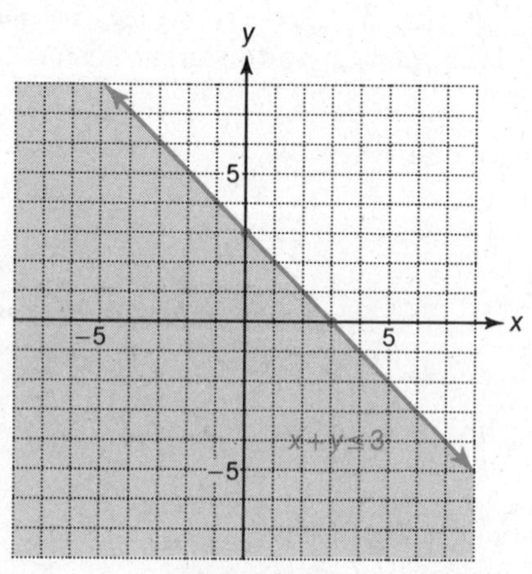

Figure 9-18

14b. Since the original problem does *not* contain an equality symbol, the boundary line $x + y = 3$ is *not* part of the solution. This is indicated by a dotted line (see Figure 9–19).

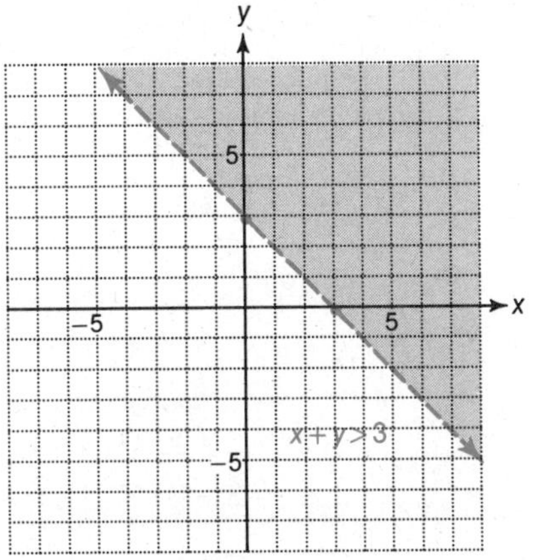

Figure 9–19

In each case, the shaded region is the solution of the inequality. That is, every point in the shaded region is a solution of the original inequality.

14 Example 14 illustrates that the boundary line divides the plane into two regions and that the points in only one of these regions are solutions of the inequality. To determine which is the correct region, choose any point that is not on the boundary line and substitute its coordinates into the inequality. If the resulting inequality is true, the solution is the half-plane containing the point selected. If the resulting inequality is false, the solution is the half-plane *not* containing the point selected. For example, in the inequality $x + y \leq 3$ (see Example 14), if you select the origin (0, 0) as a test point, you can see that

$$x + y \leq 3$$
$$0 + 0 \leq 3$$
$$0 \leq 3. \qquad \text{True}$$

Thus, the half-plane containing the point (0, 0) is the solution of the inequality.

15 Based on the results in Ideas 12 and 13, graph linear inequalities in two variables using the following procedure:

To Graph Linear Inequalities:

1. Graph the boundary line. Remember, the boundary line is a *solid* line when the original problem contains an equality symbol ($\leq$ or $\geq$) but it is a *dotted* line when the original problem does not contain an equality symbol ($<$ or $>$).
2. Select a point that is not on the line to determine which half-plane is the solution.
3. Shade the half-plane identified in step 2.

Example 15 Graph each inequality.

a. $2x + 3y < 6$ **b.** $5x - 4y \geq 10$

Solution **15a.** $2x + 3y < 6.$

Step 1: Graph the boundary line $2x + 3y = 6$ (see Figure 9–20).

x-intercept	*y*-intercept
$2x + 3y = 6$	$2x + 3y = 6$
$2x + 3(0) = 6$	$2(0) + 3y = 6$
$x = 3$; $(3, 0)$	$y = 2$; $(0, 2)$

Check:

$$2x + 3y = 6$$
$$2(-3) + 3y = 6$$
$$y = 4; (-3, 4)$$

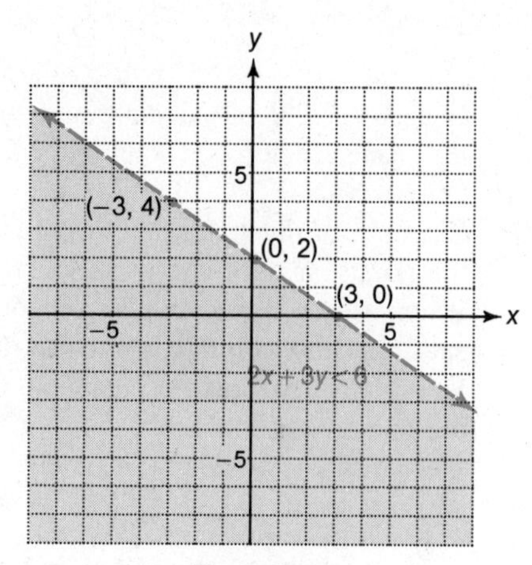

Figure 9–20

Step 2: Select the correct half-plane by using the test point (0, 0).

$$2x + 3y < 6$$
$$2(0) + 3(0) < 6$$
$$0 < 6 \quad \text{True}$$

Since this statement is true, the half-plane containing (0, 0) is the solution.

Step 3: Shade the half-plane containing the point (0, 0).

NOTE: In this case, the boundary line is not a part of the solution.

15b. $5x - 4y \geq 10$

Step 1: Graph the boundary line $5x - 4y = 10$ (see Figure 9–21).

x-intercept	*y*-intercept
$5x - 4y = 10$	$5x - 4y = 10$
$5x - 4(0) = 10$	$5(0) - 4y = 10$
$x = 2$; $(2, 0)$	$y = -\frac{5}{2}$; $\left(0, -2\frac{1}{2}\right)$

Check:

$$5x - 4y = 10$$
$$5(-2) - 4y = 10$$
$$y = -5; (-2, -5)$$

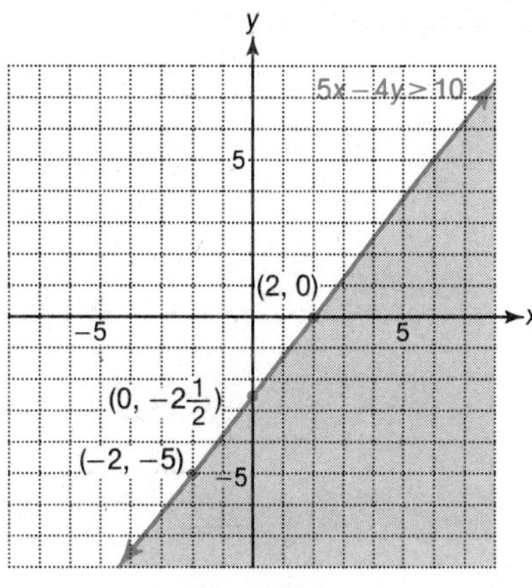

Figure 9–21

Step 2: Select the correct half-plane by using the test point (0, 0).

$$5x - 4y \geq 10$$
$$5(0) - 4(0) \geq 10$$
$$0 \geq 10 \quad \text{False}$$

Since the statement is false, the half-plane not containing (0, 0) is the solution.

Step 3: Shade the half-plane not containing (0, 0).

Example 16 Graph $x \geq 2y$.

Solution **Step 1:** Graph the boundary line $x = 2y$ (see Figure 9–22).

x-intercept	y-intercept	Third point
$x = 2y$	$x = 2y$	$x = 2y$
$x = 2(0)$	$0 = 2y$	$x = 2(-1)$
$x = 0$; (0, 0)	$0 = y$; (0, 0)	$x = -2$; $(-2, -1)$

Check:

$x = 2y$
$x = 2(2)$
$x = 4$; (4, 2)

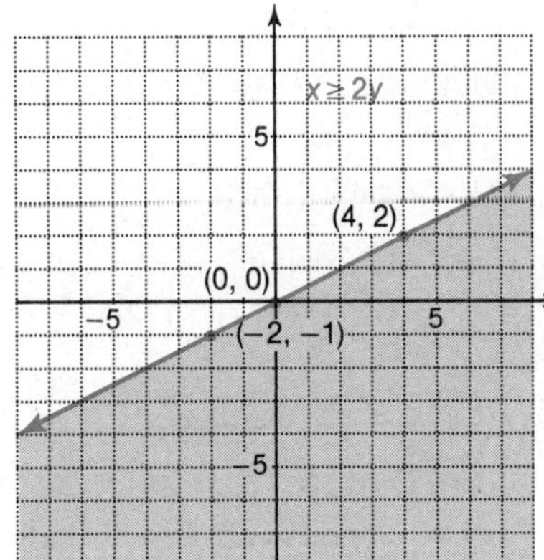

Figure 9–22

Step 2: You cannot use (0, 0) as the test point since this point is on the line. Thus, use (3, 0).

$$x \geq 2y$$
$$3 \geq 2(0)$$
$$3 \geq 0 \quad \text{True}$$

Since the statement is true, the half-plane containing (3, 0) is the solution.

Step 3: Shade the half-plane containing (3, 0).

Practice Problem 8 ***Graph each inequality.***

a. $3x + y \le 6$ **b.** $4x - 3y > 12$ **c.** $2x - 3y < 0$

16 Some inequalities contain only one variable. The boundary line of such inequalities are either a vertical or a horizontal line.

Example 17 Graph $x - 3 > -5$

Solution **Step 1:** Graph the boundary line $x - 3 = -5$

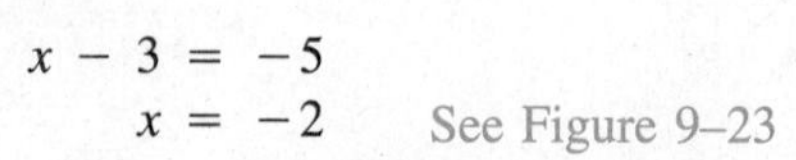

$$x - 3 = -5$$
$$x = -2 \quad \text{See Figure 9–23}$$

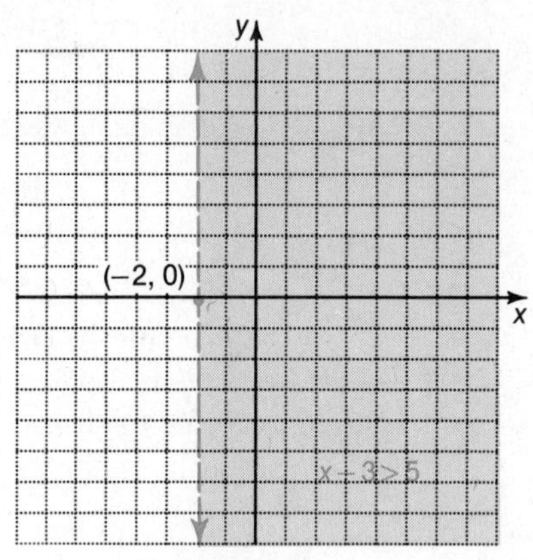

Figure 9-23

Step 2: Select the correct half-plane by using the test point (0, 0).

$$x - 3 > -5$$
$$0 - 3 > -5$$
$$-3 > -5 \quad \text{True (see Figure 9–24).}$$

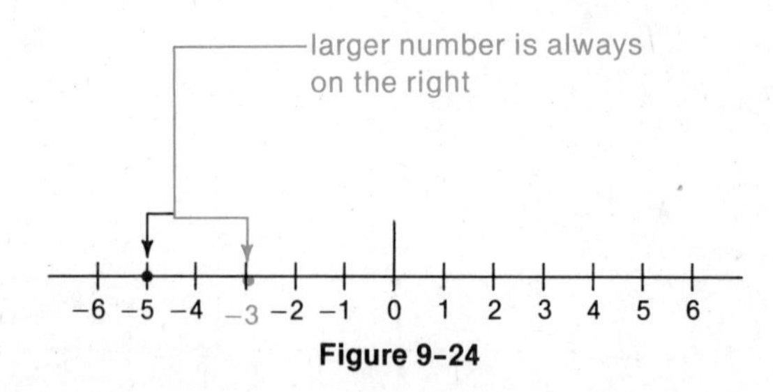

Figure 9-24

Since the statement is true, the half-plane containing (0, 0) is the solution.

Step 3: Shade the region containing the point (0, 0).

Practice Problem 9 ***Graph each inequality.***

a. $x \le 4$ **b.** $3y - 2 > 10$

9.4 Exercises

Determine if the given ordered pair is a solution of the inequality.

1. $5x + y \le 10$, (2, 0) **2.** $x < 5y$, (−3, −4) **3.** $3x + 2y > 9$, $\left(-2, \frac{3}{4}\right)$ **4.** $7x - 3y \le 8$, (1, 4)

In problems 5–10, complete the graph of the inequality by shading the correct half-plane.

5. $x - y < 3$

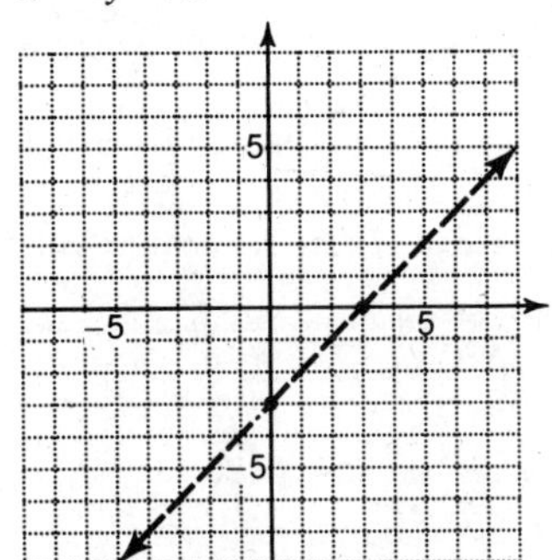

6. $2x + 5y \geq 10$

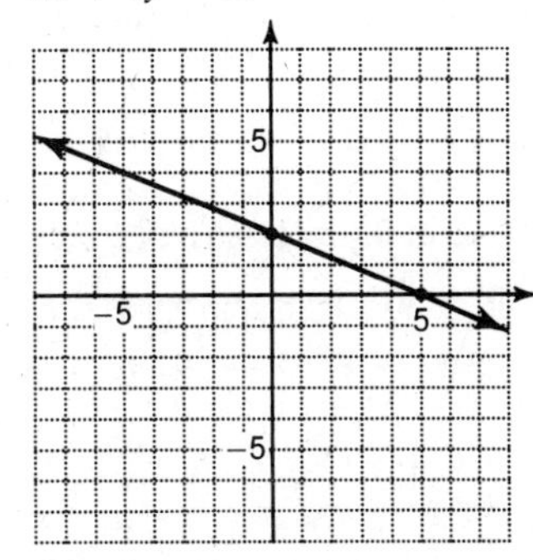

7. $3x - 4y \leq 12$

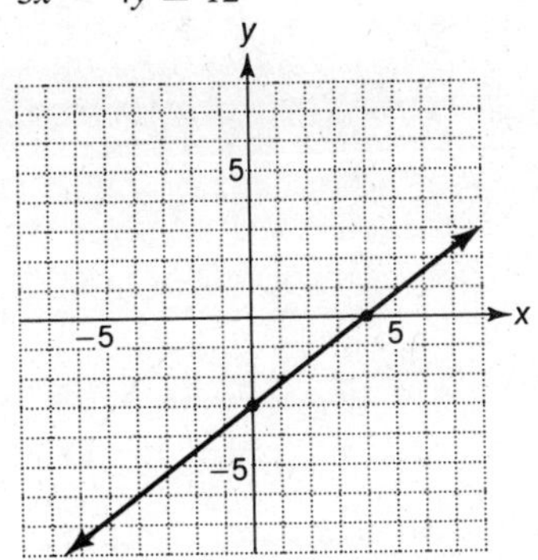

8. $x > 3y$

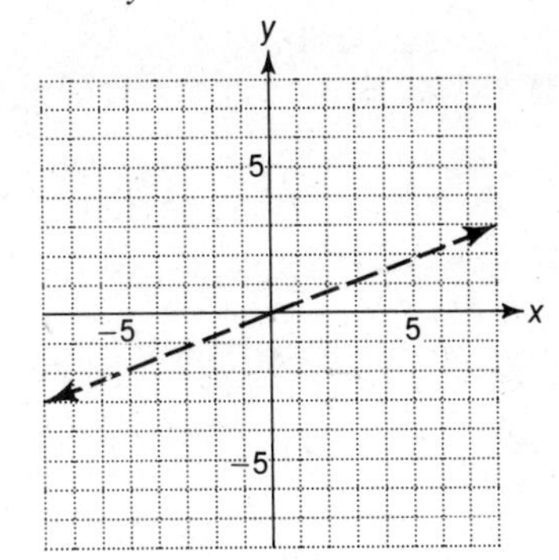

9. $x \geq -3$

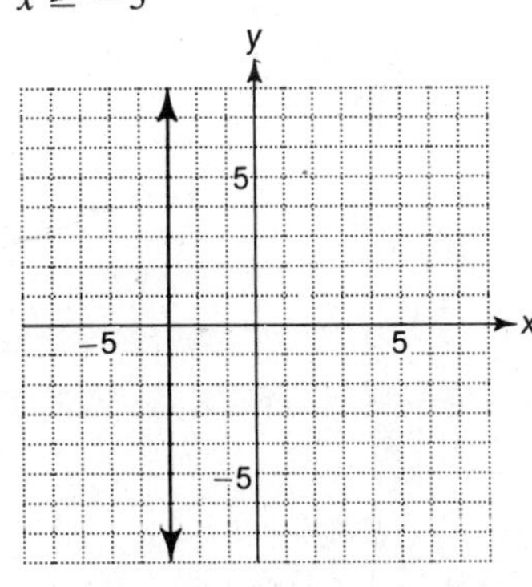

10. $y - 3 < -5$

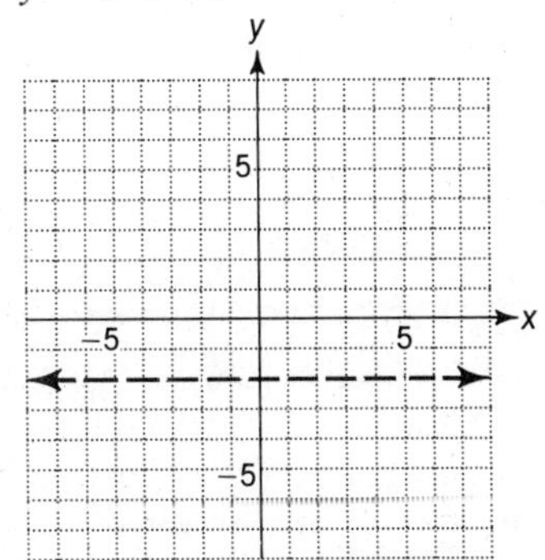

Graph each inequality.

11. $x - y < 4$
12. $x - 4y \geq 2$
13. $2x + 3y > 9$
14. $2x - 3y \leq -9$
15. $5x - 3y < 15$
16. $2x - y \geq 6$
17. $x > -4$
18. $y \leq -4$
19. $x \leq 2y + 4$
20. $x - y \leq 0$
21. $y \geq 4x + 3$
22. $2x - y < 0$
23. $5x + 2 \leq 12$
24. $4x + 5y > -10$
25. $4x + 5y > 20$
26. $3x + 5y \leq 15$
27. $\frac{x}{5} + \frac{y}{3} > 1$
28. $5x - 2y \geq 10$
29. $\frac{1}{2}x - \frac{1}{3}y > 1$
30. $2(x + 5) \leq 2$

Fill in the blanks.

31. When graphing an inequality, the boundary line is the graph of the ________________ obtained when the inequality symbol is replaced with an ________________ sign.

32. When the original inequality contains a $\leq$ or a $\geq$ symbol, the boundary line is a ________ line. In this case, the boundary line is ________ of the solution.

33. When the original inequality contains a $<$ or $>$ symbol, the boundary line is a ________ line. In this case, the boundary line is ________ of the solution.

34. To determine which half-plane is the graph of an inequality, choose any point not on the boundary line and substitute its ________ into the inequality. If the resulting inequality is true, the solution is the half–plane ________ the point selected. If the resulting inequality is false, the solution is the half-plane ________ the point selected.

35. To graph a linear inequality, first graph the ________. Next, select a point not on the ________ to determine which half-plane is the ________. Finally, ________ the half-plane identified in step 2.

Answers to Practice Problems

8a.

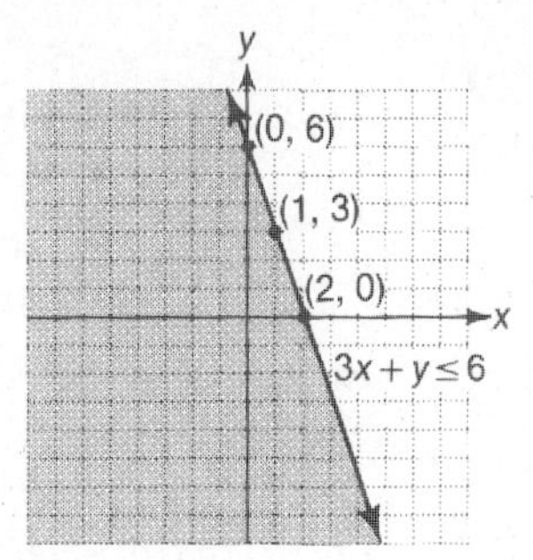

8b.

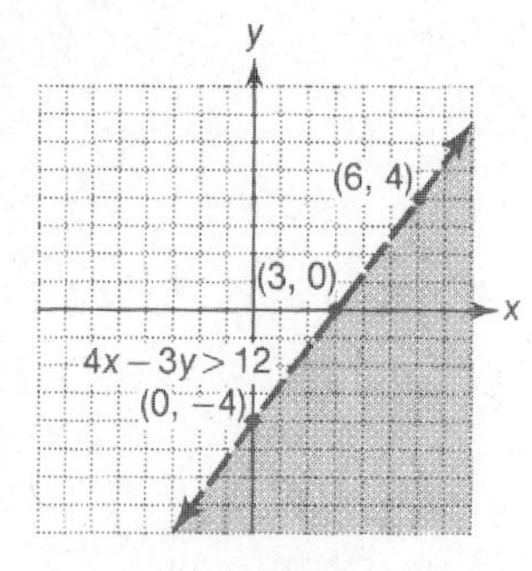

8c.

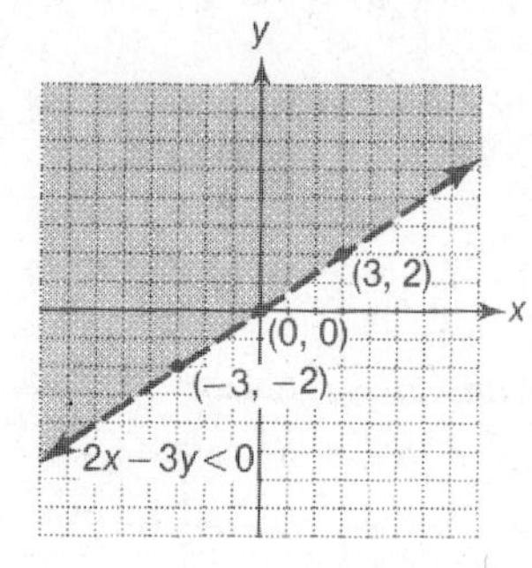

9a.

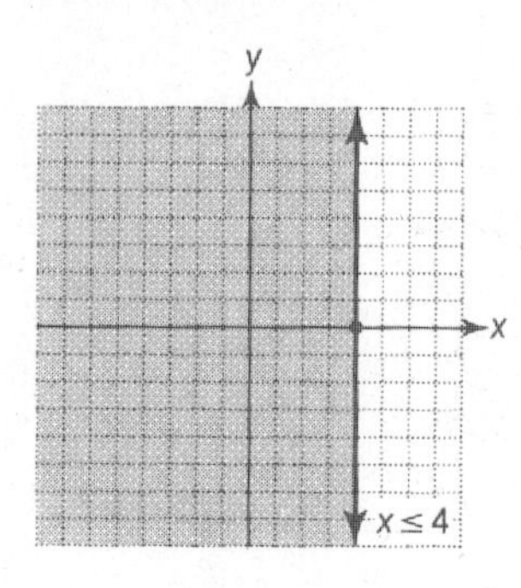

9b.

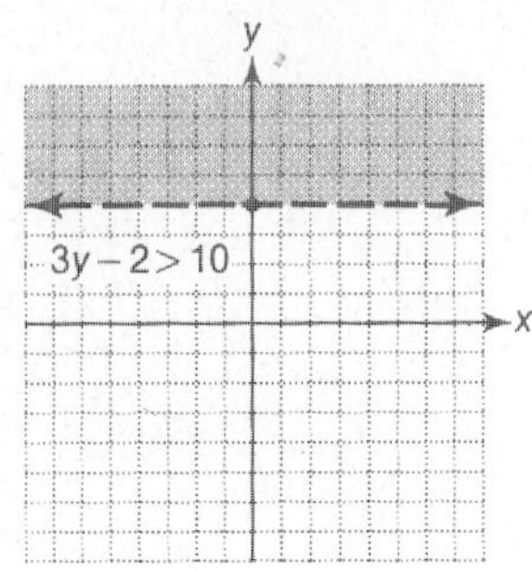

Summary Important Terms

9.1

rectangular coordinate system
quadrant
x-axis
y-axis
origin
ordered pair
point
coordinates
x-coordinate
y-coordinate
graph

9.2

linear equation
solution for a linear equation
satisfies the equation
x-intercept
y-intercept
intercept method

9.3

slope
rise
run
slope-intercept form (method)

9.4

boundary line
solution of an inequality

Important Skills

9.1

Graphing points in the coordinate system
Finding the coordinates of a given point

9.2

Graphing linear equations by using the intercept method
Writing linear equations in the slope-intercept form
Solving word problems that involve graphing equations

9.3

Graphing linear equations by using the slope-intercept method

9.4

Graphing a boundary line
Determining which half-plane is the solution of a linear inequality
Graphing linear inequalities

Review Exercises

1. Graph (1, 3), (0, 1), (−2, −3), (−2, 4).

2. Plot (0, 4), (0, 0), (3, −4), (−3, 0).

Find the coordinates of points A, B, C, D, and E.

3.

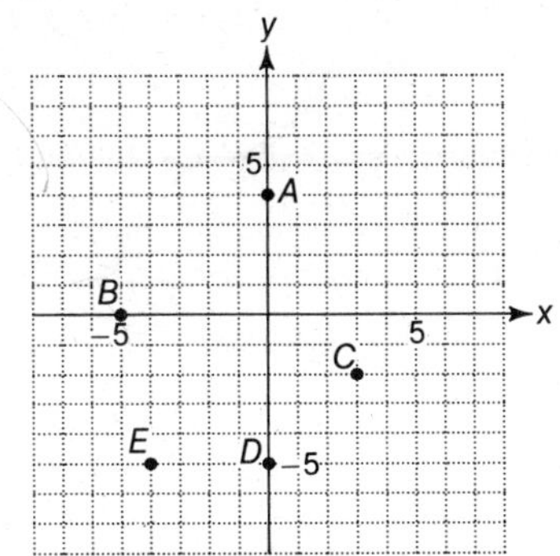

4.

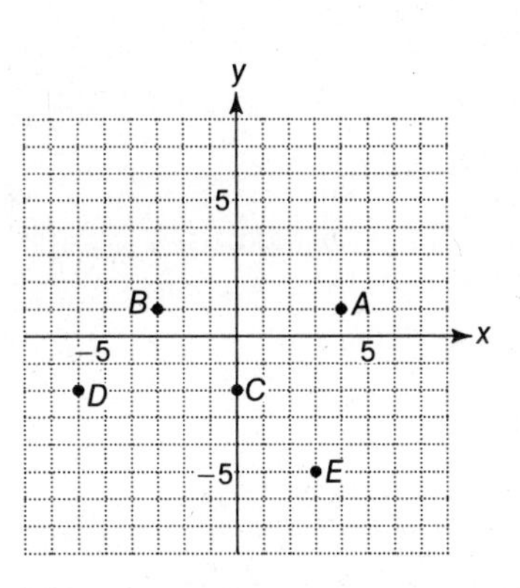

Graph.

5. $x + y = 2$

6. $3x - 2y = 6$

7. $4x - 5y < 10$

8. $3x - 4y \le 6$

9. $x - 4y = 0$

10. $y - 3 = 1$

11. $-3x < 6$

12. $8x + 6y = -24$

13. $x < -y$

14. $6x + 4y = -12$

Use the slope-intercept method to graph each equation.

15. $3x - y = 4$

16. $6x + 4y = -8$

17. $\frac{x}{2} - \frac{y}{3} = 1$

18. $y - 3x = 0$

Solve.

19. A machine costing \$4,000 will have a book value of $b = 4000 - 400n$ dollars n years after purchase.

a. Find the book value of the machine after two years, three years, and five years. That is, find b when $n = 2$, $n = 3$, and $n = 5$.

b. Graph $b = 4000 - 400n$. Let n represent the horizontal axis and let b represent the vertical axis.

c. Interpret the meaning of the ordered pair (1, 3600).

d. Using the graph in problem 19b, determine when the book value of the machine will be \$800.

LINEAR SYSTEMS

OBJECTIVES

1. Solve a system of two linear equations in two variables graphically.

2. Solve a system of two linear equations in two variables using the addition method.

3. Solve a system of two linear equations in two variables using the substitution method.

PRETEST

EXPLANATION

1. Solve each system graphically. — Section 10.1, Ideas 1–3
   a. $x + y = 5$
      $x - y = 1$
   b. $x + y = 6$
      $2x + 2y = 12$
   c. $x + 2y = 4$
      $2x + y = -1$
   d. $2x + y = 6$
      $4x + 2y = 8$

2. Solve using the addition method. — Section 10.2, Ideas 5–9
   a. $x + 2y = 1$
      $-x - 2y = 2$
   b. $2x + y = 5$
      $2x - y = -1$
   c. $4x - 3y = 12$
      $6x - 5y = 19$
   d. $\frac{x}{5} - \frac{y}{2} = 1$
      $x + y = 6$

3. Solve using the substitution method. — Section 10.3, Ideas 11–13
   a. $2x + y = 4$
      $y = -1 - 2x$
   b. $3x + y = 1$
      $9x + 3y = 3$
   c. $3x + y = 6$
      $2x + 3y = 3$
   d. $2x + 3y = -8$
      $5x + 4y = -20$
   e. Cazzie earns twice as much money as Guerin. If their combined salaries total \$615, how much does each man earn? — Idea 14

10.1 Intersecting, Parallel, and Equal Lines

Idea 1 Two equations such as

$$x + y = 4$$
$$x - y = 2$$

are called a **system of two linear equations in two variables.** A **solution** of this kind of system is an ordered pair that is a solution of both equations in the system. For example, the ordered pair (3, 1) is the solution of the above system since

$$x + y = 4 \quad \text{and} \quad x - y = 2$$
$$3 + 1 = 4 \quad\quad\quad 3 - 1 = 2$$
$$4 = 4 \quad\quad\quad 2 = 2.$$

We can also say that the ordered pair (3, 1) *satisfies* the system $x + y = 4$ and $x - y = 2$.

2 In Section 9.2 you learned that the graph of a linear equation in one variable is a straight line. What if you have a system of *two* linear equations in *two* variables? If you graph the equations in the same plane, one of three situations may occur:

1. The graphs of the equations may intersect at only one point (see Figure 10–1). In this case, the lines have one and only one point in common, and the ordered pair representing this point is the solution of the system.

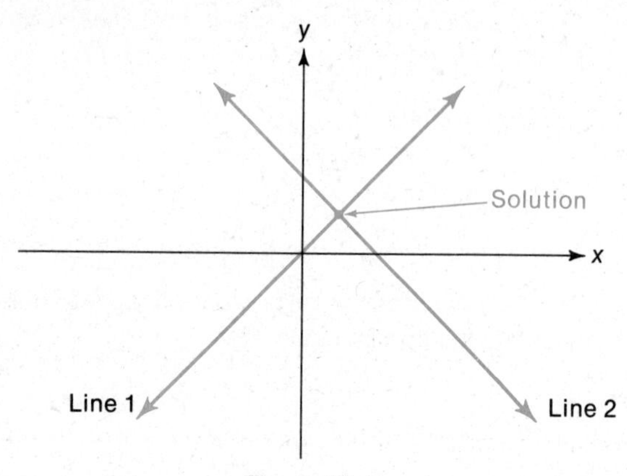

Figure 10-1

2. The graphs of the equations may be two parallel lines (see Figure 10.2). In this case, the lines have no point in common since they do not intersect. Thus, the system has no solution since there is no ordered pair that satisfies both equations.

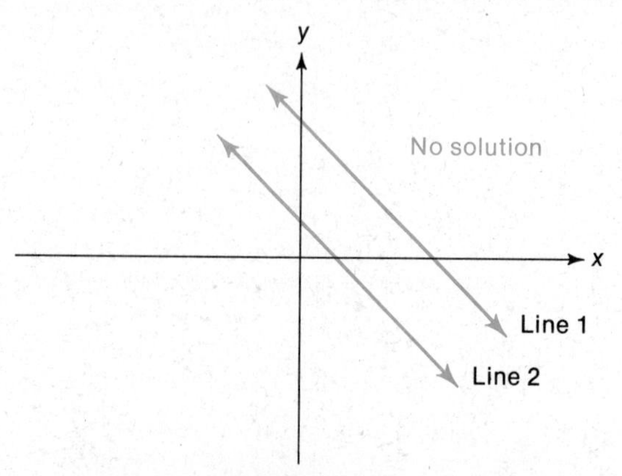

Figure 10-2

3. The graphs of the equations may be the same line (see Figure 10.3). In this case, every point on one line is also on the other line. The system has an infinite number of solutions since every ordered pair that satisfies one equation will also satisfy the other equation.

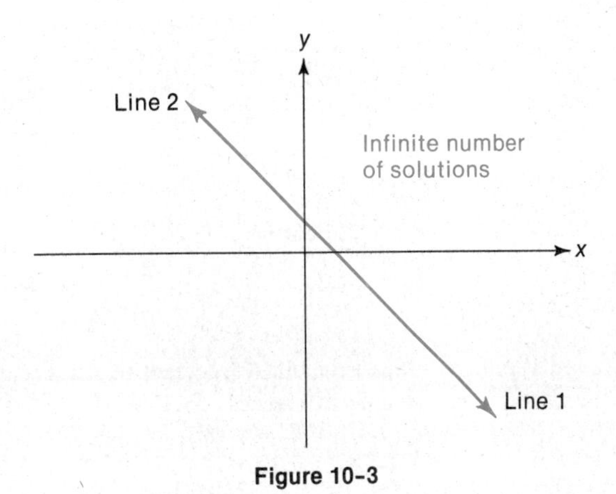

Figure 10-3

3 Based on the information provided in Idea 2, to find the solution of a system of two linear equations in two variables graphically, use the following procedure:

To Solve a System of Equations Graphically:

1. Graph each equation on the same coordinate system.

2a. If the lines *intersect,* the ordered pair representing the point of intersection is the solution to the system.

2b. If the lines are *parallel,* the system has no solution.

2c. If the graph of both equations is the *same line,* every ordered pair corresponding to a point on that line is a solution to the system.

Example 1 Solve each system graphically.

a. $x + y = 3$
$x - y = 1$

b. $2x + 3y = 6$
$4x + 6y = 12$

c. $x + y = 3$
$2x + 2y = -4$

Solution To solve a system of two linear equations in two variables graphically, first graph each equation on the same coordinate system. Then, interpret the results.

1a.

	x-intercept	y-intercept	Check
Line 1: $x + y = 3$	(3, 0)	(0, 3)	(1, 2)
Line 2: $x - y = 1$	(1, 0)	(0, −1)	(3, 2)

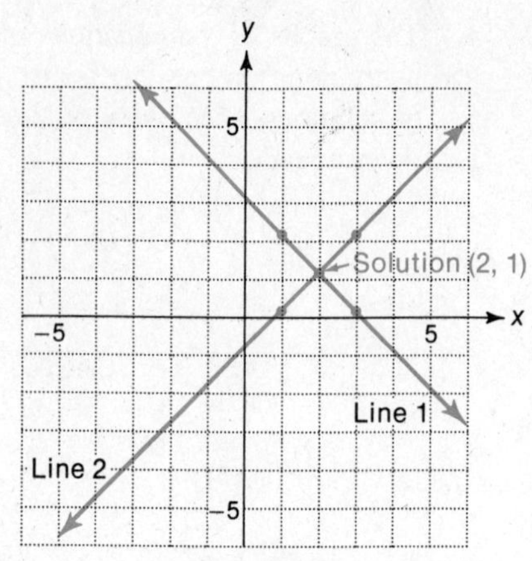

Figure 10–4

Since the lines intersect at (2, 1), this ordered pair is the system's solution (see Figure 10–4).

Check:

$$x + y = 3 \qquad x - y = 1$$
$$2 + 1 = 3 \qquad 2 - 1 = 1$$
$$3 = 3 \qquad 1 = 1$$

NOTE: When the graphs of the equations intersect at only one point, we say that the equations are **consistent.**

1b.

	x-intercept	y-intercept	Check
Line 1: $2x + 3y = 6$	(3,0)	(0,2)	(−3,4)
Line 2: $4x + 6y = 12$	(3,0)	(0,2)	(−3,4)

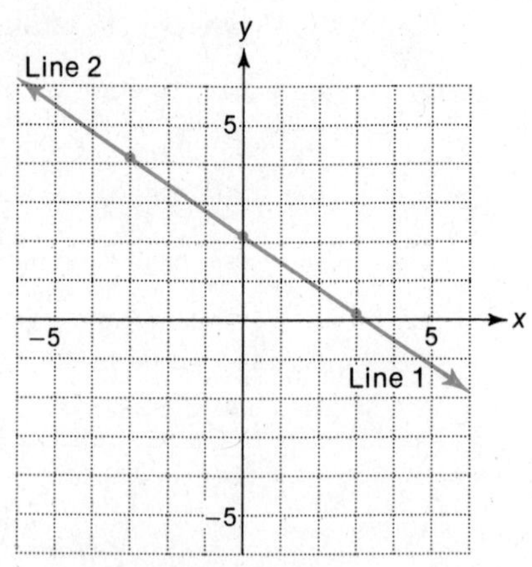

Figure 10–5

Since every point on one line also lies on the other line (see Figure 10–5), every ordered pair corresponding to a point on the line is a solution. Thus, the system has an infinite number of solutions. Since the graphs of the equations are the same line, we say that the equations are **dependent** or **equivalent.**

1c.

	x-intercept	y-intercept	Check
Line 1: $x + y = 3$	(3,0)	(0,3)	(2,1)
Line 2: $2x + 2y = -4$	(−2,0)	(0,−2)	(1,−3)

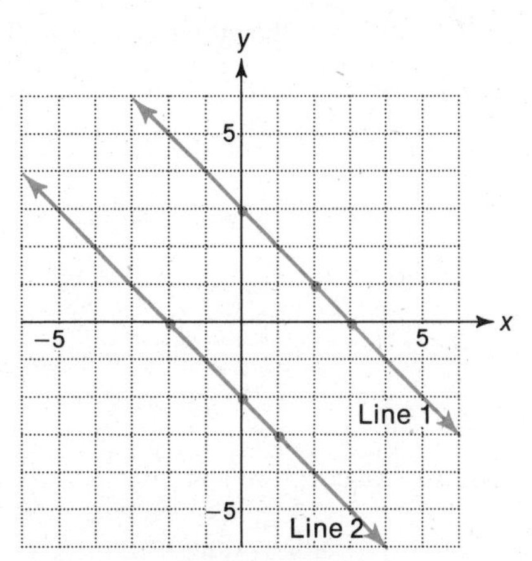

Figure 10-6

Since the lines are parallel and they will never intersect (see Figure 10–6), the system has no solution. When the graphs of the equations are distinct parallel lines, we say that the equations are **inconsistent.**

NOTE: In Example 1c, if you rewrite each equation in the slope-intercept form, the equations will have the same slope and different y-intercepts. That is,

$$x + y = 3 \quad \text{becomes} \quad y = -x + 3$$
$$2x + 2y = 4 \quad \text{becomes} \quad y = -x + 2.$$

In general, this means that the graphs of the equations are distinct parallel lines and that the equations have no common solution.

Practice Problem 1 ***Solve each system graphically.***

a. $x + y = 5$
$x - y = 1$

b. $x + 2y = 4$
$2x + 4y = -4$

c. $2x + 2y = 4$
$x + y = 2$

4 When solving a system of two linear equations in two variables, you can save time if you know whether the equations are consistent, inconsistent, or dependent. That is, given the system

$$a_1x + b_1y = c_1$$
$$a_2x + b_2y = c_2$$

where a_1, a_2, b_1, b_2, c_1, c_2 are constants, a_1 and b_1 are not both zero, and a_2 and b_2 are not both zero, the following will be true:

Case 1: If $\frac{a_1}{a_2} \neq \frac{b_1}{b_2}$ ("$\neq$" means *not equal to*), then the original equations are **consistent** and their graph is a pair of lines that intersect at one point.

Case 2: If $\frac{a_1}{a_2} = \frac{b_1}{b_2} = \frac{c_1}{c_2}$, then the original equations are **dependent** and their graph is a single straight line.

Case 3: If $\frac{a_1}{a_2} = \frac{b_1}{b_2} \neq \frac{c_1}{c_2}$, then their graph is two distinct parallel lines.

Example 2 Determine if the equations are consistent, inconsistent, or dependent.

a. $2x + y = 4$
$4x + 2y = 8$

b. $6x = 12y - 8$
$3x - 6y = 4$

c. $2x - 3y = 12$
$x + y = 1$

Solution To determine if two equations are consistent, inconsistent, or dependent, first write the equations in the form $ax + by = c$. If $\frac{a_1}{a_2} \neq \frac{b_1}{b_2}$, the equations are *consistent*. If $\frac{a_1}{a_2} = \frac{b_1}{b_2} \neq \frac{c_1}{c_2}$, the equations are *inconsistent*. If $\frac{a_1}{a_2} = \frac{b_1}{b_2} = \frac{c_1}{c_2}$, the equations are *dependent*.

2a. Since $\frac{2}{4} = \frac{1}{2} = \frac{4}{8}\left(\frac{a_1}{a_2} = \frac{b_1}{b_2} = \frac{c_1}{c_2}\right)$, the equations are dependent and their graph is a single straight line. Thus, there are an infinite number of solutions.

2b. $6x = 12y - 8$ becomes $6x - 12y = -8$
$3x - 6y = 4$ $\qquad$ $3x - 6y = 4$

Since $\frac{6}{3} = \frac{-12}{-6} \neq \frac{-8}{4}\left(\frac{a_1}{a_2} = \frac{b_1}{b_2} \neq \frac{c_1}{c_2}\right)$, the equations are inconsistent and their graph is two distinct parallel lines. Thus, the system has no solution.

2c. Since $\frac{2}{1} \neq \frac{-3}{1}\left(\frac{a_1}{a_2} \neq \frac{b_1}{b_2}\right)$, the equations are consistent and their graph is a pair of lines that intersect at one point. Thus, the system has a unique solution (x, y).

Practice Problem 2 ***Determine if the following equations are consistent, inconsistent, or dependent.***

a. $3x - 4 = 2y$
$6x - 4y = -8$

b. $2x + 5y = 10$
$3x + 4y = 1$

c. $\frac{x}{3} + \frac{y}{4} = 1$
$8x + 6y = 24$

10.1 Exercises

Determine whether or not the given ordered pair is the solution of the given system.

1. $(3, 2)$; $x + y = 5$
$x - y = 1$

2. $(4, 2)$; $x + 2y = 8$
$2x - 3y = 2$

3. $(2, -3)$; $2x + y = 4$
$4x + 2y = 8$

4. $(3, -1)$; $2x + 3y = -1$
$3x + 5y = -2$

5. $(12, 10)$; $x + y = 22$
$0.05x + 0.10y = 1.7$

6. $(1, -4)$; $3x + \frac{1}{4}y = 2$
$\frac{x}{2} + \frac{3}{4}y = -\frac{5}{2}$

7. $(-4, -8)$; $3x = y - 4$
$2x - 5y = 15$

8. $\left(\frac{5}{2}, -\frac{7}{4}\right)$; $x - 2y = 6$
$3x + 2y = 4$

Solve each system graphically. If the lines are parallel, write "no solution." Write "infinite number of solutions" if the graph is a single straight line.

9. $3x + y = 3$
$6x + 2y = 6$

10. $x + y = 6$
$x - y = 2$

11. $2x + 2y = 24$
$2x - 2y = 8$

12. $4x = y + 10$
$2x + 3y = 12$

13. $x - 3y = 6$
$2x - 6y = -18$

14. $3x + 4y = 12$
$-6x - 8y = -24$

15. $2x + 3y = -6$
$2x - 4y = 8$

16. $4x - y = 8$
$x = y + 5$

17. $x + y = 8$
$x - 2y = 2$

18. $x = 3y - 6$
$2x - 6y = 12$

19. $x = 4$
$3x - y = 6$

20. $2x - 3y = 6$
$y = -4$

21. $x - 3y = 6$
$x + 3y = -12$

22. $x + 2y = 8$
$6x + 2y = 18$

23. $\frac{x}{2} + \frac{y}{3} = 1$
$x - y = 2$

Determine if the following equations are consistent, inconsistent, or dependent (see Example 2). Also, tell whether the system has one solution, no solution, or an infinite number of solutions.

24. $x + y = 6$
$x - y = 5$

25. $8x + y = 30$
$x - 3y = 30$

26. $x + 3y = 6$
$2x + 6y = 12$

27. $3x - 4y = 8$
$6x - 8y = 8$

28. $2x + 2y = 9$
$2x - 2y = 9$

29. $-5x + y = 7$
$10x - 2y = -14$

30. $3x = y + 4$
$6x - 2y = 12$

31. $2x = 23 - 5y$
$2x + 3y = 29$

32. $x = 2y$
$3x + 2y = 29$

33. $4x - y = 3$
$-2x + \frac{y}{2} = \frac{-3}{2}$

34. $5x + 2y = -2$
$y = -2x$

35. $3x + \frac{1}{4}y = 2$
$\frac{1}{2}x + \frac{3}{4}y = \frac{5}{2}$

Fill in the blanks.

36. A solution of two linear equations in two variables is an ________ that is a ________ of ________ equations in the system.

37. If the graphs of two linear equations in two variables intersect at one and only one point, the equations are said to be ________ and the system has ________ solution.

38. If the graphs of two linear equations in two variables produce two distinct parallel lines, the equations are said to be ________ and the system has ________ solution(s).

39. If two linear equations in two variables have the same line for their graph, the equations are said to be ________ and the system has ________ solution(s).

40. Given the system $a_1x + b_1y = c_1$ and $a_2x + b_2y = c_2$, we know that if:

	Condition	Type of Equation	Number of Solutions
a.	$\frac{a_1}{a_2} \neq \frac{b_1}{b_2}$	________	________
b.	________	dependent	________
c.	________	________	no solution

41. To solve a system of two equations in two variables graphically, first graph each ________ on the same coordinate system. Then, ________ the results.

Determine if the following equations are consistent, inconsistent, or dependent. If they are consistent, find the solution graphically.

42. $x + y = 6$
$x - y = 2$

43. $x + y = 7$
$x - y = 3$

44. $2x + 3y = 8$
$4x + 6y = 5$

45. $3x + 4y = 7$
$6x + 8y = 3$

46. $x + y = 1$
$3x + 3y = 3$

47. $x + y = -2$
$2x + 2y = -4$

48. $x = 2y - 1$
$3x - 6y = 3$

49. $x = 3y + 2$
$2x - 6y = -4$

50. $x = 3 - 3y$
$2x - 3y = -12$

51. $x = 2 - 2y$
$2x - 2y = 10$

52. $x = 2y - 5$
$2x - 2y = -10$

53. $x = 3y - 2$
$2x - 6y = -4$

54. $\frac{x}{2} + \frac{y}{3} = 1$
$y = 6$

55. $\frac{x}{2} - \frac{y}{3} = 1$
$y = -3$

56. $\frac{1}{2}x + \frac{1}{2}y = 3$
$\frac{1}{3}x + \frac{1}{3}y = -3$

57. $\frac{5}{3}x + \frac{10}{3}y = \frac{10}{3}$
$\frac{1}{4}x + \frac{1}{2}y = \frac{1}{2}$

58. $\frac{2}{3}x + \frac{1}{3}y = 2$
$\frac{1}{3}x + \frac{1}{2}y = 1$

59. $\frac{3}{4}x + \frac{1}{2}y = 3$
$\frac{1}{2}x - \frac{2}{3}y = 2$

Answers to Practice Problems

1a.

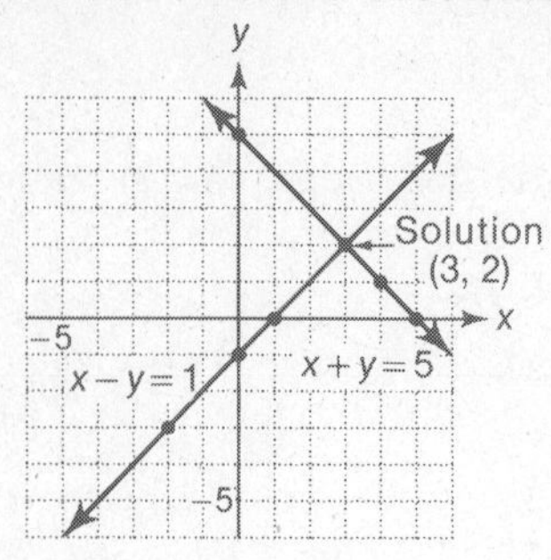

Solution is (3, 2)

1b.

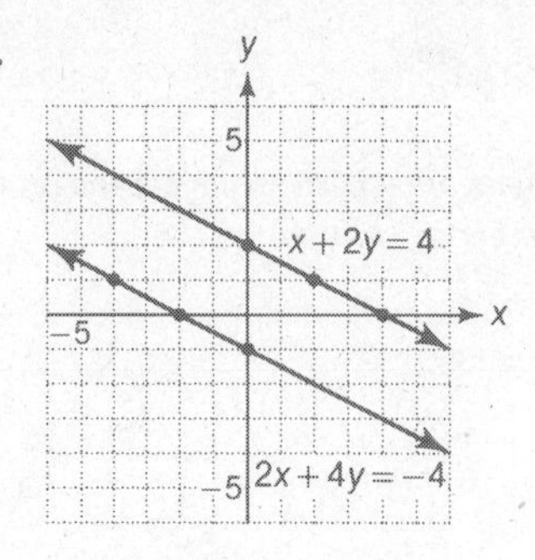

No solution

1c.

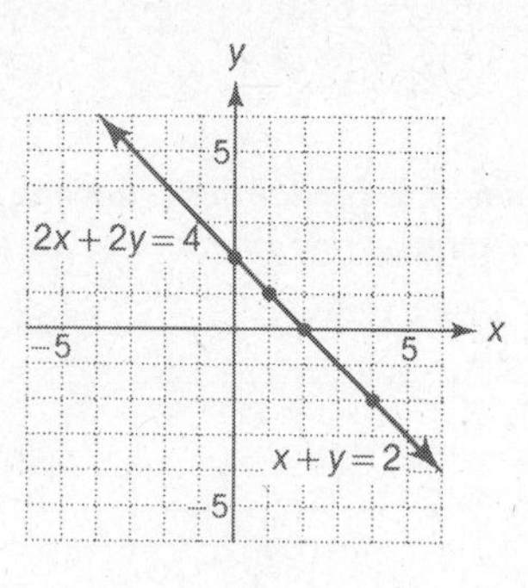

Infinite number of solutions

2a. inconsistent b. consistent c. dependent

10.2 Solving Systems of Equations by Addition

5 Solving systems of linear equations graphically can be time-consuming and inaccurate, especially when the coordinates of the point of intersection are not integers. Therefore, to obtain an exact solution, we should use an algebraic method.

6 One such method is based on the **addition property of equality.** This property states that if equals are added to equals, the results are equal. Using this property, we can eliminate one variable so that only one equation containing one variable will remain.

Example 3 Solve each system.

a. $x + y = 3$
$x - y = 1$

b. $3x - 4y = -5$
$x + 2y = 5$

Solution **3a.** Since each equation is a statement of equality, add the two equations to obtain one equation in one variable.

$$\begin{array}{r} x + y = 3 \\ \underline{x - y = 1} \\ 2x \quad\;\; = 4 \end{array}$$

Now, solve this equation.

$$2x = 4$$
$$x = 2$$

Next, find y by substituting 2 for x in one of the original equations.

$$x + y = 3$$
$$2 + y = 3$$
$$y = 1$$

Thus, the solution of the system is (2, 1).

Check:

$x + y = 3$
$2 + 1 = 3$
$3 = 3$ True

$x - y = 1$
$2 - 1 = 1$
$1 = 1$ True

3b. Adding these equations will not result in one equation containing one variable. Thus, before adding, multiply both sides of the second equation by 2 in order to make the coefficients of the y-terms additive inverses.

$$\begin{array}{rcl} 3x - 4y = -5 & \text{remains} & 3x - 4y = -5 \\ 2(x + 2y) = 2(5) & \text{becomes} & 2x + 4y = 10 \end{array}$$

Now, add the resulting equation to the first equation to obtain one equation in one variable.

$$\begin{array}{r} 3x - 4y = -5 \\ 2x + 4y = 10 \\ \hline 5x \qquad = 5 \end{array}$$

Solve this equation.

$$5x = 5$$
$$x = 1$$

Next, find y by substituting 1 for x in one of the original equations.

$$x + 2y = 5$$
$$1 + 2y = 5$$
$$2y = 4$$
$$y = 2$$

Thus, the solution of the system is the ordered pair (1, 2).

Check:

$$\begin{array}{rr} 3x - 4y = -5 & x + 2y = 5 \\ 3(1) - 4(2) = -5 & 1 + 2(2) = 5 \\ 3 - 8 = -5 & 1 + 4 = 5 \\ -5 = -5 & 5 = 5 \end{array}$$

7 The method of solving a system of linear equations in two variables by using the addition property of equality is called the **addition method** or the **elimination method.**

To Solve a System of Equations (Addition Method):

1. Write both equations in the form $ax + by = c$.
2. If the equations in step 1 are inconsistent, write "no solution." If they are dependent, write "infinite number of solutions" (see Section 10.1, Idea 4).
3. If the equations in step 1 are consistent:
   - **a.** Multiply (if necessary) one or both of the equations by a number that will make the coefficients of x (or y) additive inverses of each other.
   - **b.** Add the resulting two equations to obtain one equation in one variable.
   - **c.** Solve the equation obtained in step 3b.
   - **d.** Substitute the solution obtained in step 3c into one of the original equations and solve the resulting equation.
   - **e.** Write the values obtained in steps 3c and 3d as an ordered pair. This is your solution.
   - **f.** Check the solution in both of the original equations.

Example 4 Solve each system by the addition method.

a. $5x - y = 10$
$2x + y = 4$

b. $3x + y = 4$
$-3x - y = -4$

c. $6x + 9y = 18$
$-6x - 12y = -16$

Solution To solve a system of two linear equations in two variables by the addition method, first determine if the equations are consistent, inconsistent, or dependent.

4a. Since $\frac{5}{2} \neq \frac{-1}{1}\left(\frac{a_1}{a_2} \neq \frac{b_1}{b_2}\right)$, the equations are consistent (the graph is two intersecting lines) and there is only one solution. Now, since the coefficients of y are additive inverses, add and solve the resulting equation.

$$\begin{aligned} 5x - y &= 10 \\ 2x + y &= 4 \\ \hline 7x \quad &= 14 \\ x \quad &= 2 \end{aligned}$$

Next, find y by substituting 2 for x in one of the original equations.

$$\begin{aligned} 2x + y &= 4 \\ 2(2) + y &= 4 \\ 4 + y &= 4 \\ y &= 0 \end{aligned}$$

The solution is (2, 0).

Check:

$$\begin{aligned} 5x - y &= 10 \\ 5(2) - 0 &= 10 \\ 10 &= 10 \end{aligned} \qquad \begin{aligned} 2x + y &= 4 \\ 2(2) + 0 &= 4 \\ 4 &= 4 \end{aligned}$$

4b. Since $\frac{3}{-3} = \frac{1}{-1} = \frac{4}{-4}\left(\frac{a_1}{a_2} = \frac{b_1}{b_2} = \frac{c_1}{c_2}\right)$, the equations are dependent (graph is a single straight line). Thus, the system has an infinite number of solutions.

Alternate method
Since the coefficients of x are additive inverses, add.

$$\begin{aligned} 3x + y &= 4 \\ -3x - y &= -4 \\ \hline 0 &= 0 \end{aligned}$$

By doing this, you have eliminated both variables and the resulting equation is a true statement. In general, this means that the original equations are dependent and their graph is a straight line. Therefore, the system has an infinite number of solutions.

4c. Since $\frac{6}{-6} \neq \frac{9}{-12}\left(\frac{a_1}{a_2} \neq \frac{b_1}{b_2}\right)$, the equations are consistent and there exists one and only one solution (x, y). Next, since the coefficients of x are additive inverses, add.

$$\begin{aligned} 6x + 9y &= 18 \\ -6x - 12y &= -16 \\ \hline -3y &= 2 \\ y &= -\frac{2}{3} \end{aligned}$$

Now, find x by substituting $-\frac{2}{3}$ for y in one of the original equations.

$$6x + 9y = 18$$
$$6x + 9\left(-\frac{2}{3}\right) = 18$$
$$6x - 6 = 18$$
$$6x = 24$$
$$x = 4$$

The solution is $\left(4, -\frac{2}{3}\right)$.

Check:

$$6x + 9y = 18 \qquad -6x - 12y = -16$$
$$6(4) + 9\left(-\frac{2}{3}\right) = 18 \qquad -6(4) - 12\left(-\frac{2}{3}\right) = -16$$
$$24 - 6 = 18 \qquad -24 + 8 = -16$$
$$18 = 18 \qquad -16 = -16$$

Practice Problem 3 ***Solve each problem by the addition method.***

a. $x + y = 8$, $x - y = 4$ **b.** $-8x = -10y + 6$, $8x - 10y = 6$ **c.** $2x + 5y = 12$, $-2x - 6y = -14$

8 Sometimes you must multiply one of the equations by a number to make the coefficients of x (or y) additive inverses.

Example 5 Solve each system by the addition method.

a. $x + y = 2$, $3x + 2y = -9$ **b.** $10x - 9y = 13$, $4x + 3y = 14$ **c.** $x + 3y = 6$, $x = -2y + 5$

Solution To solve a system of two linear equations in two variables, first determine if the equations are consistent, inconsistent, or dependent.

5a. Since $\frac{1}{3} \neq \frac{1}{2}\left(\frac{a_1}{a_2} \neq \frac{b_1}{b_2}\right)$, the equations are consistent and one and only one solution exists. Now, multiply both sides of the first equation by -2. This will make the coefficients of y in the equations additive inverses. Then, add.

$$-2(x + y) = -2(2) \quad \text{becomes} \quad -2x - 2y = -4$$
$$3x + 2y = -9 \quad \text{remains} \quad 3x + 2y = -9$$
$$x = -13$$

Next, find y by substituting -13 for x in one of the original equations.

$$x + y = 2$$
$$-13 + y = 2$$
$$y = 15$$

The solution is $(-13, 15)$.

Check:

$$x + y = 2 \qquad 3x + 2y = -9$$
$$-13 + 15 = 2 \qquad 3(-13) + 2(15) = -9$$
$$2 = 2 \qquad -9 = -9$$

NOTE: By multiplying both sides of the first equation by -3, we could have obtained the same solution.

5b. Since $\frac{10}{4} \neq \frac{-9}{3} \left(\frac{a_1}{a_2} \neq \frac{b_1}{b_2}\right)$, the equations are consistent and there exists only one solution (x, y). Now, multiply both sides of the second equation by 3. This will make the coefficients of y in the equations additive inverses. Then, add.

$$\begin{array}{rcl} 10x - 9y = 13 & \text{remains} & 10x - 9y = 13 \\ 3\,(4x + 3y) = 3\,(14) & \text{becomes} & \underline{12x + 9y = 42} \\ & & 22x \qquad = 55 \\ & & x = \frac{55}{22} \\ & & x = \frac{5}{2} \end{array}$$

Next, find y by substituting $\frac{5}{2}$ for x in one of the original equations.

$$\begin{aligned} 10x - 9y &= 13 \\ 10\left(\frac{5}{2}\right) - 9y &= 13 \\ 25 - 9y &= 13 \\ -9y &= -12 \\ y &= \frac{12}{9} \\ y &= \frac{4}{3} \end{aligned}$$

The solution is $\left(\frac{5}{2}, \frac{4}{3}\right)$.

Check:

$$\begin{array}{cc} 10x - 9y = 13 & 4x + 3y = 14 \\ 10\left(\frac{5}{2}\right) - 9\left(\frac{4}{3}\right) = 13 & 4\left(\frac{5}{2}\right) + 3\left(\frac{4}{3}\right) = 14 \\ 25 - 12 = 13 & 10 + 4 = 14 \\ 13 = 13 & 14 = 14 \end{array}$$

5c. Write both equations in the form $ax + by = c$. Thus, in the second equation, add $2y$ to both sides.

$$\begin{array}{rcl} x + 3y = 6 & \text{remains} & x + 3y = 6 \\ x = -2y + 5 & \text{becomes} & x + 2y = 5 \end{array}$$

Next, since $\frac{1}{1} \neq \frac{3}{2} \left(\frac{a_1}{a_2} \neq \frac{b_1}{b_2}\right)$, the equations are consistent and there exists one and only one solution (x, y). Now, multiply both sides of the first equation by -1. This will make the coefficients of x in the equations additive inverses. Then, add.

$$\begin{array}{rcl} -1\,(x + 3y) = -1\,(6) & \text{becomes} & -x - 3y = -6 \\ x + 2y = 5 & \text{remains} & \underline{x + 2y = 5} \\ & & -y = -1 \\ & & y = 1 \end{array}$$

Next, find y by substituting 1 for y in one of the original equations.

$$\begin{aligned} x + 3y &= 6 \\ x + 3(1) &= 6 \\ x + 3 &= 6 \\ x &= 3 \end{aligned}$$

The solution is (3, 1).

Check:

$$\begin{aligned} x + 3y &= 6 \\ 3 + 3(1) &= 6 \\ 6 &= 6 \end{aligned} \qquad \begin{aligned} x &= -2y + 5 \\ 3 &= -2(1) + 5 \\ 3 &= 3 \end{aligned}$$

Practice Problem 4 ***Solve each system by the addition method.***

a. $3x - 4y = -5$, $x + 2y = 5$ **b.** $2x + 5y = 1$, $4x + 10y = 2$ **c.** $x = 16 - 3y$, $4x - 2y = 8$

9 Frequently it is necessary to multiply both equations by the appropriate numbers to make the coefficients of x (or y) additive inverses.

Example 6 Solve each system by the addition method.

a. $3x - 5y = 2$, $2x + 3y = -5$ **b.** $4x + 2y = 1$, $10x + 5y = -3$ **c.** $\frac{1}{2}x + \frac{5}{6}y = 1$, $5x + 3y = 4$

Solution **6a.** The equations are consistent and there is only one solution. Thus, multiply both sides of the first equation by 3 and multiply both sides of the second equation by 5 to make the coefficients of y additive inverses.

$$\begin{aligned} 3(3x - 5y) &= 3(2) \quad \text{becomes} \quad & 9x - 15y &= 6 \\ 5(2x + 3y) &= 5(-5) \quad \text{becomes} \quad & 10x + 15y &= -25 \\ & & 19x &= -19 \\ & & x &= -1 \end{aligned}$$

Now, find y by substituting -1 for x in one of the original equations.

$$\begin{aligned} 2x + 3y &= -5 \\ 2(-1) + 3y &= -5 \\ -2 + 3y &= -5 \\ 3y &= -3 \\ y &= -1 \end{aligned}$$

The solution is $(-1, -1)$.

Check:

$$\begin{aligned} 3x - 5y &= 2 \\ 3(-1) - 5(-1) &= 2 \\ 2 &= 2 \end{aligned} \qquad \begin{aligned} 2x + 3y &= -5 \\ 2(-1) + 3(-1) &= -5 \\ -5 &= -5 \end{aligned}$$

NOTE: By multiplying both sides of the first equation by 2 and multiplying both sides of the second equation by -3, we could have obtained the same solution.

6b. Since $\frac{4}{10} = \frac{2}{5} \neq \frac{1}{3}\left(\frac{a_1}{a_2} = \frac{b_1}{b_2} \neq \frac{c_1}{c_2}\right)$, the equations are inconsistent and their graph is two distinct parallel lines. Thus, the system has no solution.

Alternate method
Multiply both sides of the first equation by -5 and multiply both sides of the second equation by 2 to make the coefficients of y additive inverses. Then, add.

$$\begin{aligned} -5(4x + 2y) &= -5(1) \quad \text{becomes} \quad & -20x - 10y &= -5 \\ 2(10x + 5y) &= 2(-3) \quad \text{becomes} \quad & 20x + 10y &= -6 \\ & & 0 &= -11 \quad \text{False} \end{aligned}$$

We have eliminated both variables and the resulting equation is a false statement. In general, this means that the equations are inconsistent and their graph is two distinct parallel lines. Thus, the system has no solution.

6c. First, multiply both sides of the first equation by 6 (LCD) to eliminate the fractions.

$$6\left(\frac{1}{2}x + \frac{5}{6}y\right) = 6\,(1) \quad \text{becomes} \quad 3x + 5y = 6$$
$$5x + 3y = 4 \quad \text{remains} \quad 5x + 3y = 4$$

Since the equations are consistent, the system has a unique solution. Multiply both sides of the first equation by -5 and multiply both sides of the second equation by 3 to eliminate x.

$$\begin{aligned} -15x - 25y &= -30 && \text{Result after multiplying by } -5 \\ 15x + 9y &= 12 && \text{Result after multiplying by } 3 \\ \hline -16y &= -18 \\ y &= \frac{-18}{-16} \\ &= \frac{9}{8} \end{aligned}$$

To find x, substitute $\frac{9}{8}$ for y in one of the original equations.

$$\begin{aligned} 5x + 3y &= 4 \\ 5x + 3\left(\frac{9}{8}\right) &= 4 \\ 5x + \frac{27}{8} &= 4 \\ 40x + 27 &= 32 && \text{Multiply by LCD} = 8 \\ 40x &= 5 \\ x &= \frac{5}{40} \\ x &= \frac{1}{8} \end{aligned}$$

Check:

$$\begin{aligned} 5x + 3y &= 4 & \frac{1}{2}x + \frac{5}{6}y &= 1 \\ 5\left(\frac{1}{8}\right) + 3\left(\frac{9}{8}\right) &= 4 & \frac{1}{2}\cdot\frac{1}{8} + \frac{5}{6}\cdot\frac{9}{8} &= 1 \\ \frac{5}{8} + \frac{27}{8} &= 4 & \frac{1}{16} + \frac{15}{16} &= 1 \\ 4 &= 4 & 1 &= 1 \end{aligned}$$

Practice Problem 5 ***Solve each system by the addition method.***

a. $3x + 5y = -2$
$2x + 3y = -5$

b. $3x = 6y + 6$
$5x - 4y = 1$

c. $x + 2y = 4$
$-2x - 4y = -8$

10 When solving word problems involving two unknowns, it is sometimes easier to represent each unknown in terms of a different letter. However, since two variables are used, you must write two equations to find the desired solution.

Example 7 Solve.

a. The sum of two numbers is 21. Their difference is 9. Find the numbers.

b. Princess invested $10,000. She invested part at 7% and the rest at 12%. If she earned $1000 in interest in one year, how much did she invest at each rate?

Solution To solve a word problem involving two unknowns, (1) represent the unknowns, (2) write two equations, (3) solve the system obtained in step 2 and determine the solution, and (4) check your final answers.

7a. Step 1: x = the first number; y = the second number

Step 2:

first number	plus	second number	is	21
x	$+$	y	$=$	21
first number	minus	second number	is	9
x	$-$	y	$=$	9

Step 3: Now, find x.

$$\begin{aligned} x + y &= 21 \\ x - y &= 9 \\ \hline 2x &= 30 \\ x &= 15 \end{aligned}$$

Next, find y.

$$\begin{aligned} x + y &= 21 \\ 15 + y &= 21 \\ y &= 6 \end{aligned}$$

The numbers are 15 and 6.

Check: The numbers are 15 and 6, since $15 + 6 = 21$ and $15 - 6 = 9$.

7b. Step 1: x = \$ invested at 7%; y = \$ invested at 12%

Step 2:

	Principal	**Rate**	**Time**	**Interest***
Investment #1	x	7% = 0.07	1 yr	$0.07x$
Investment #2	y	12% = 0.12	1 yr	$0.12y$

*Recall that interest = (principal)(rate)(time)

$ invested at 7%	plus	$ invested at 12%	is	$10,000
x	$+$	y	$=$	10,000
Interest at 7%	plus	Interest at 12%	is	$1000
$0.07x$	$+$	$0.12y$	$=$	1000

Step 3: Now, find y. Multiply both sides of the first equation by -0.07.

$$\begin{aligned} x + y &= 10{,}000 \quad \text{becomes} \\ 0.07x + 0.12y &= 1{,}000 \quad \text{remains} \end{aligned} \qquad \begin{aligned} -0.07x - 0.07y &= -700 \\ 0.07x + 0.12y &= 1000 \\ \hline 0.05y &= 300 \\ y &= 6000 \end{aligned}$$

Next, find x.

$$\begin{aligned} x + y &= 10{,}000 \\ x + 6000 &= 10{,}000 \\ x &= 4000 \end{aligned}$$

Princess invested \$4000 at 7% and \$6000 at 12%.

Check: **a)** Total invested $= \$4000 + \$6000 = \$10{,}000$
b) Total interest $= 7\%$ of $\$4000 + 12\%$ of $\$6000$
$= \$280 + \$720$
$= \$1000$

Practice Problem 6 **Solve.**

a. The sum of two numbers is 80. Their difference is 6. Find the numbers.

b. There are 1380 people at a basketball game. A student ticket costs \$1.50 and nonstudent tickets cost \$3.00. If the total gate receipts were \$2910, how many student and how many nonstudent tickets were sold?

10.2 Exercises

Solve each system by the addition method.

1. $x - y = -5$, $x + y = 7$

2. $x - y = 3$, $x + y = -1$

3. $2x + y = 6$, $-2x - y = 8$

4. $3x + y = 7$, $6x + 2y = 14$

5. $6x - y = 2$, $8x + 3y = 7$

6. $5x - y = 10$, $x - 2y = -7$

7. $3x + 2y = 9$, $4x - 3y = 12$

8. $8x + 3y = 9$, $3x + 5y = 16$

9. $4x + 3y = -12$, $6x - 4y = 1$

10. $x - 3y = 9$, $x - 5y = 13$

11. $3x - 5y = 2$, $3x + 5y = 22$

12. $2x + y = 8$, $4x + 2y = 9$

13. $x - 7y = 1$, $2x = 2 + 14y$

14. $2x - y = -6$, $-2x + 3y = 20$

15. $5x + 7y = 3$, $2x - 3y = 7$

16. $3x + 4y = 1$, $4x + 3y = 8$

17. $2x + 2y = 4$, $2x - 3y = -8$

18. $x + 4y = 4$, $x - 2y = 10$

19. $\frac{x}{2} + \frac{y}{5} = 1$, $7x - 5y = 36$

20. $3x = 5y + 2$, $5y - 3x = 7$

21. $x + y = 8$, $2x - y = 10$

22. $x + 5y = 8$, $x - 10y = -29$

23. $3x + y = 7$, $4x - 10y = -1$

24. $8x - y = 29$, $2x + y = 11$

25. $\frac{x}{4} - \frac{y}{3} = \frac{5}{12}$, $\frac{x}{10} + \frac{y}{5} = \frac{1}{2}$

26. $3x + 5y = 2$, $2x + 3y = -5$

27. $9x - 6y = -18$, $6x - 9y = -32$

28. $0.3x + 0.2y = 0.1$, $0.4x - y = 2.6$

29. $x + y = 4$, $\frac{x}{2} + \frac{y}{2} = 2$

30. $5x + 3y = 6$, $3x + 5y = 4$

Fill in the blanks.

31. When solving a system of two linear equations in two variables by the addition method, first write both equations in the form __________. If the resulting equations are inconsistent, write __________. If they are dependent, write __________.

32. If while solving a system of two linear equations in two variables by the addition method you eliminated both variables and the resulting equation is a true statement, the original equations are __________ and the system has an __________ of solutions.

33. If while solving a system of two linear equations in two variables you eliminated both variables and the resulting equation is a false statement, the original equations are __________ and the system has __________ solution.

34. If the equations in a system of two linear equations in two variables are consistent, you can find the solution as follows:

a. **Step 1:** Multiply (if necessary) one or both of the equations by a number that will make the coefficients of x (or y) __________ of each other.

b. **Step 2:** Add the resulting equations to obtain __________ equation in __________ variable.

c. **Step 3:** __________ the equation obtained in step 2.

d. **Step 4:** Substitute the solution obtained in step 3 into ________ of the ________ equations and then ________ the resulting equation.
e. **Step 5:** Write the values obtained in steps 3 and 4 as an ________. This is our solution.
f. **Step 6:** ________ the solution in ________ equations.

Solve.

35. Determine how many gallons of a 7% solution and a 12% solution of acid should be mixed to obtain six gallons of a solution that is 10% acid.

36. Vanessa invested \$10,000. Part was invested at 9% and the rest was invested at 10.5%. If her total interest for the year was \$990, how much money was invested at each rate?

37. Carver invested \$4000 in a Ford and a BMW, and then he leased both cars. His earned annual interest income from leasing each car was 20% and 60%, respectively. If his income from the total investment was 50%, how much did Carver pay for each car?

38. The sum of two numbers is 13. Their difference is -6. What are the numbers?

39. Find two numbers such that six times the first number plus five times the second number is 34, and five times the first number plus three times the second number is 4.

40. There were 1100 people at a concert. The admission was \$2 for students and \$3 for nonstudents. If the total gate receipts were \$2800, how many student tickets and how many nonstudent tickets were sold?

41. The cost of printing 400 brochures was \$210 and the cost of printing 700 brochures was \$360. If the printer charged a fixed fee for typesetting and an additional fee for each brochure, find the typesetting fee and the fee per brochure.

42. Susan mixed a dried fruit costing \$0.75 per pound with roasted soy beans costing \$1.50 per pound. If she wants a mixture of 90 pounds worth \$1.24 per pound, how many pounds of each item must she use?

43. How many gallons of a 15% acid solution should be mixed with a 40% acid solution to obtain 16 gallons of a mixture that is 30% acid?

44. Train A leaves Chicago for Boston (a distance of 1050 miles) at the same time that train B leaves Boston for Chicago. If train A travels at a rate of 60 mph and train B travels at 40 mph, how far has each train traveled when the two trains pass?

45. The perimeter of a rectangular-shaped garden is 142 feet. If the width of the garden is 6.4 feet greater than the length, find the garden's length and width. (*Hint:* $P = 2L + 2W$)

Answers to Practice Problems **3a.** (6, 2) **b.** no solution **c.** (1, 2) **4a.** (1, 2) **b.** infinite number of solutions **c.** (4, 4) **5a.** $(-19, 11)$ **b.** $\left(-1, -\frac{3}{2}\right)$ **c.** infinite number of solutions **6a.** 43, 37 **b.** 820 student tickets, 560 nonstudent tickets

10.3 Solving Systems of Equations by Substitution

11 A system of two linear equations in two variables can also be solved by using the **substitution method.** With this method, you still eliminate one of the variables. However, you eliminate it by substitution rather than addition.

Example 8 Solve the system

$$\begin{aligned} 2x + 3y &= 40 \\ y &= 2x. \end{aligned}$$

Solution The equations are consistent. Thus, there is only one solution. Next, since $y = 2x$ in the second equation, we can substitute $2x$ for y in the first equation.

$$\begin{aligned} 2x + 3y &= 40 \\ 2x + 3(2x) &= 40 \quad \text{Replace } y \text{ with } 2x \end{aligned}$$

We now have one equation in one variable. Solve it.

$$\begin{aligned} 2x + 3(2x) &= 40 \\ 2x + 6x &= 40 \\ 8x &= 40 \\ x &= 5 \end{aligned}$$

To find y, substitute 5 for x in the second equation.

$$\begin{aligned} y &= 2x \\ y &= 2(\ 5\ ) \\ y &= 10 \end{aligned}$$

The solution is (5, 10).

Check:

$$\begin{aligned} 2x + 3y &= 40 \\ 2(5) + 3(10) &= 40 \\ 40 &= 40 \end{aligned} \qquad \begin{aligned} y &= 2x \\ 10 &= 2(5) \\ 10 &= 10 \end{aligned}$$

12 Based on the results in Example 8, to solve a system of two linear equations in two variables by the substitution method, use the following procedure:

Procedure for Solving Systems of Equations (Substitution Method):

1. Determine (mentally) if the equations are consistent, inconsistent, or dependent (see Idea 4, Section 10.1).
2. If the equations are inconsistent, write "no solution." If they are dependent, write "infinite number of solutions."
3. If the equations are consistent:
   - **a.** Solve (if necessary) one of the equations for one of the variables.
   - **b.** Substitute the expression obtained in step 3a into the other equation.
   - **c.** You now have one equation in one variable. Solve it.
   - **d.** Find the value of the other variable by substituting the value obtained in step 3c into the equation obtained in step 3a.
   - **e.** Write the values obtained in steps 3c and 3d as an ordered pair. This is the solution.
   - **f.** Check the solution in each equation.

NOTE: The substitution method is very useful when one of the equations is already solved for a variable or when one of the variables has a coefficient of 1.

Example 9 Solve each system by the substitution method.

a. $8x - y = 29$
$y = 11 - 2x$

b. $5x - 4y = 1$
$x - 2y = 2$

c. $x = 3y + 9$
$2x - 6y = -10$

Solution To solve a system of two linear equations in two variables by the substitution method, first determine (mentally) if the equations are consistent, inconsistent, or dependent.

9a. The equations are consistent $\left(\frac{8}{2} \neq \frac{-1}{1}\right)$. Therefore, a unique solution exists.

Next, since the second equation is solved for y, substitute $11 - 2x$ for y in the first equation and then solve the resulting equation.

$$\begin{aligned} 8x - y &= 29 \\ 8x - (11 - 2x) &= 29 \quad \text{Replace } y \text{ with } 11 - 2x \\ 8x - 11 + 2x &= 29 \\ 10x - 11 &= 29 \\ 10x &= 40 \\ x &= 4 \end{aligned}$$

To find y, substitute 4 for x in the equation

$$\begin{aligned} y &= 11 - 2x \\ y &= 11 - 2(4) \\ y &= 3 \end{aligned}$$

The solution is (4, 3).

Check:

$$\begin{aligned} 8x - y &= 29 \qquad & y &= 11 - 2x \\ 8(4) - 3 &= 29 \qquad & 3 &= 11 - 2(4) \\ 29 &= 29 \qquad & 3 &= 3 \end{aligned}$$

9b. The equations are consistent. Thus, solve for x in the second equation since its coefficient is 1.

$$\begin{aligned} x - 2y &= 2 \\ x &= 2 + 2y \end{aligned}$$

Now substitute $2 + 2y$ for x in the first equation and then solve the resulting equation.

$$\begin{aligned} 5x - 4y &= 1 \\ 5(2 + 2y) - 4y &= 1 \quad \text{Replace } x \text{ with } 2 + 2y \\ 10 + 10y - 4y &= 1 \\ 10 + 6y &= 1 \\ 6y &= -9 \\ y &= -\frac{9}{6} \\ y &= -\frac{3}{2} \end{aligned}$$

To find x, substitute $-\frac{3}{2}$ for y in the equation

$$\begin{aligned} x &= 2 + 2y \\ x &= 2 + 2\left(-\frac{3}{2}\right) \\ x &= -1 \end{aligned}$$

The solution is $\left(-1, -\frac{3}{2}\right)$.

Check:

$$\begin{aligned} 5x - 4y &= 1 \qquad & x - 2y &= 2 \\ 5(-1) - 4\left(-\frac{3}{2}\right) &= 1 \qquad & -1 - 2\left(-\frac{3}{2}\right) &= 2 \\ -5 + 6 &= 1 \qquad & -1 + 3 &= 2 \\ 1 &= 1 \qquad & 2 &= 2 \end{aligned}$$

9c. If we rewrite our equations in the form $ax + by = c$, then,

$x = 3y + 9$	becomes	$x - 3y = 9$
$2x - 6y = -10$	remains	$2x - 6y = -10$

which implies that the equations are inconsistent $\left(\frac{1}{2} = \frac{-3}{-6} \neq \frac{9}{-10}\right)$. Thus, the system has no solution.

Alternate method

Since the first equation is solved for x, substitute $3y + 9$ for x in the second equation and then solve the resulting equation.

$$2x - 6y = -10$$
$$2(3y + 9) - 6y = -10 \quad \text{Replace } x \text{ with } 3y + 9$$
$$6y + 18 - 6y = -10$$
$$18 = -10 \quad \text{False}$$

Since you have eliminated both variables and the resulting equation is a false statement, the two equations are inconsistent. Thus, the system has no solution.

Practice Problem 7 ***Solve each system by the substitution method.***

a. $2x + 3y = 12$, $y = -10 + 4x$ **b.** $3x - 6y = 18$, $x = 6 + 2y$ **c.** $2x + y = 5$, $4x - 2y = 10$

13 The substitution method, like the addition method, can be used to solve any system of two linear equations in two variables.

Example 10 Solve each system by the substitution method.

a. $3x + 5y = -2$, $2x + 3y = -5$ **b.** $2x + 3y = 6$, $4x + 6y = 12$ **c.** $\frac{1}{3}x + \frac{1}{4}y = 10$, $\frac{1}{3}x - \frac{1}{2}y = 4$

Solution **10a.** The equations are consistent $\left(\frac{3}{2} \neq \frac{5}{3}\right)$. Thus, solve for x in the first equation.

$$3x + 5y = -2$$
$$3x = -2 - 5y$$
$$x = \frac{-2 - 5y}{3}$$

Now, substitute $\frac{-2 - 5y}{3}$ for x in the second equation.

$$2x + 3y = -5$$
$$2\left(\frac{-2 - 5y}{3}\right) + 3y = -5 \quad \text{Replace } x \text{ with } \frac{-2 - 5y}{3}$$
$$\frac{-4 - 10y}{3} + 3y = -5 \quad \text{Multiply}$$
$$-4 - 10y + 9y = -15 \quad \text{Multiply by LCD} = 3$$
$$-4 - y = -15$$
$$-y = -11$$
$$y = 11$$

To find x, substitute 11 for y in the equation.

$$x = \frac{-2 - 5y}{3}$$
$$x = \frac{-2 - 5(11)}{3}$$
$$x = \frac{-2 - 55}{3}$$
$$x = -19$$

The solution is $(-19, 11)$.

Check:

$$3x + 5y = -2 \qquad 2x + 3y = -5$$
$$3(-19) + 5(11) = -2 \qquad 2(-19) + 3(11) = -5$$
$$-2 = -2 \qquad -5 = -5$$

10b. Since $\frac{2}{4} = \frac{3}{6} = \frac{6}{12}$, the equations are dependent. Thus, the system has an infinite number of solutions. That is, every ordered pair that satisfies one equation will satisfy the other.

Alternate method
First, solve for y in the first equation.

$$2x + 3y = 6$$
$$3y = 6 - 2x$$
$$y = \frac{6 - 2x}{3}$$

Next, substitute $\frac{6 - 2x}{3}$ for y in the second equation.

$$4x + 6y = 12$$
$$4x + \overset{2}{\cancel{6}}\left(\frac{6 - 2x}{\underset{1}{\cancel{3}}}\right) = 12 \quad \text{Replace } y \text{ with } \frac{6 - 2x}{3}$$
$$4x + 2(6 - 2x) = 12$$
$$4x + 12 - 4x = 12$$
$$12 = 12 \quad \text{True}$$

Since you eliminated both variables and the resulting equation is a true statement, the equations are dependent. Thus, the system has an infinite number of solutions.

10c. First, eliminate the fractions from both equations by multiplying both equations by their LCD.

$$12\left(\frac{1}{3}x + \frac{1}{4}y\right) = 12(10) \quad \text{becomes} \quad 4x + 3y = 120$$
$$6\left(\frac{1}{3}x - \frac{1}{2}y\right) = 6(4) \quad \text{becomes} \quad 2x - 3y = 24$$

The equations are consistent. Thus, solve for y in the first equation.

$$4x + 3y = 120$$
$$3y = 120 - 4x$$
$$y = \frac{120 - 4x}{3}$$

Now, substitute $\frac{120 - 4x}{3}$ for y in the second equation.

$$\begin{aligned} 2x - 3y &= 24 \\ 2x - \cancel{3}\left(\frac{120 - 4x}{\cancel{3}}\right) &= 24 && \text{Replace } y \text{ with } \frac{120 - 4x}{3} \\ 2x - (120 - 4x) &= 24 && \text{Multiply} \\ 2x - 120 + 4x &= 24 \\ 6x - 120 &= 24 \\ 6x &= 144 \\ x &= 24 \end{aligned}$$

To find y, substitute 24 for x in the operation.

$$\begin{aligned} y &= \frac{120 - 4x}{3} \\ y &= \frac{120 - 4(24)}{3} \\ y &= \frac{120 - 96}{3} \\ y &= 8 \end{aligned}$$

The solution is (24, 8).

Check:

$$\begin{aligned} \frac{1}{3}x + \frac{1}{4}y &= 10 & \frac{1}{3}x - \frac{1}{2}y &= 4 \\ \frac{1}{3}(24) + \frac{1}{4}(8) &= 10 & \frac{1}{3}(24) - \frac{1}{2}(8) &= 4 \\ 8 + 2 &= 10 & 8 - 4 &= 4 \\ 10 &= 10 & 4 &= 4 \end{aligned}$$

Practice Problem 8 ***Solve each system by the substitution method.***

a. $x - y = 4$
$2x - 2y = 8$

b. $2x + 3y = 3$
$3x + 4y = 3$

c. $\frac{3}{10}x + \frac{1}{2}y = -\frac{1}{5}$
$\frac{x}{6} + \frac{y}{4} = -\frac{5}{12}$

14 When you solved word problems involving systems of linear equations in Section 10.2, you used the addition method. You can also solve this type of problem by using the substitution method.

Example 11 Solve.

a. Vanessa needs 12 quarts of a 10% salt solution. If she has one container that holds a 5% salt solution and another that holds a 25% salt solution, how many quarts of each solution must be mixed to obtain the desired mixture?

b. Cazzie and Guerin leave their house and travel in opposite directions on a highway. If they are 30 miles apart in 20 minutes and Cazzie's car is traveling 5 mph faster than Guerin's, how fast was each man driving?

Solution To solve a word problem involving two unknowns, (1) represent the unknowns, (2) write two equations, (3) solve the system derived in step 2 and determine the solutions, and (4) check the final answer.

11a. Step 1: x = quarts (qt) of 5% salt solution
y = quarts of 25% salt solution

Step 2:

	% of Salt	Qt of Solution	Qt of Salt*
Solution 1	5% = 0.05	x	$0.05x$
Solution 2	25% = 0.25	y	$0.25y$
Mixture	10% = 0.10	12	0.10(12) = 1.2

*Recall that amount = (% of substance) (volume)

Qt of 5% solution	plus	Qt of 25% solution	=	12 qt
x	+	y	=	12

Salt in 5% solution	plus	Salt in 25% solution	=	Salt in mixture
$0.05x$	+	$0.25y$	=	1.2

Step 3:

$$x + y = 12$$
$$0.05x + 0.25y = 1.2$$

Solve for x in the first equation.

$$x + y = 12$$
$$x = 12 - y$$

Next, substitute $12 - y$ for x in the second equation.

$$0.05x + 0.25y = 1.2$$
$$0.05\,(12 - y) + 0.25y = 1.2 \quad \text{Replace } x \text{ with } 12 - y$$
$$0.6 - 0.05y + 0.25y = 1.2$$
$$0.20y = 0.6$$
$$y = 3$$

To find x, substitute 3 for y in the equation.

$$x = 12 - y$$
$$x = 12 - 3$$
$$x = 9$$

Vanessa must mix nine quarts of the 5% salt solution with three quarts of the 25% salt solution.

Check: The solution is correct since:

a) 9 qt + 3 qt = 12 qt

b) *Salt before mixing* = *Salt after mixing*

$$0.05(9 \text{ qt}) + 0.25(3 \text{ qt}) = 0.10(12 \text{ qt})$$
$$0.45 \text{ qt} + 0.75 \text{ qt} = 1.20 \text{ qt}$$
$$1.20 \text{ qt} = 1.20 \text{ qt}$$

11b. Step 1: x = rate (in mph) of Cazzie's car
y = rate (in mph) of Guerin's car

Step 2:

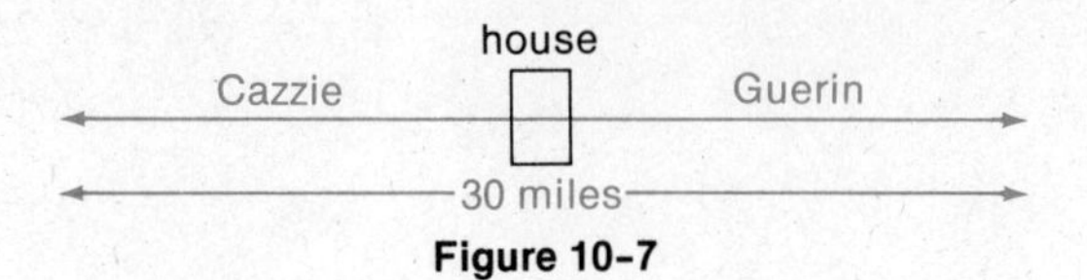

Figure 10-7

	Rate (mph)	Time (hours)	Distance (miles)*
Cazzie	x	20 min $= \frac{1}{3}$ hr	$\frac{1}{3}x$
Guerin	y	20 min $= \frac{1}{3}$ hr	$\frac{1}{3}y$

*Recall that distance = (rate)(time)

Cazzie's distance	plus	Guerin's distance	is	30 miles
$\frac{1}{3}x$	$+$	$\frac{1}{3}y$	$=$	30

Cazzie's rate	equals	Guerin's rate plus 5 mph
x	$=$	$y + 5$

Step 3:

$$\frac{1}{3}x + \frac{1}{3}y = 30$$
$$x = y + 5$$

Substitute $y + 5$ for x in the first equation.

$$\frac{1}{3}x + \frac{1}{3}y = 30$$
$$\frac{1}{3}(y + 5) + \frac{1}{3}y = 30 \quad \text{Replace } x \text{ with } y + 5$$
$$y + 5 + y = 90 \quad \text{Multiply by LCD} = 3$$
$$2y + 5 = 90$$
$$2y = 85$$
$$y = 42\frac{1}{2}$$

To find x, substitute $42\frac{1}{2}$ for y in the equation.

$$x = y + 5$$
$$x = 42\frac{1}{2} + 5$$
$$x = 47\frac{1}{2}$$

Cazzie's rate is $47\frac{1}{2}$ mph and Guerin's is $42\frac{1}{2}$ mph.

Check: **a)** Cazzie's rate $\left(47\frac{1}{2}\text{ mph}\right)$ is 5 mph more than Guerin's rate $\left(42\frac{1}{2}\text{ mph}\right)$.

b)

Cazzie's distance	=	$\left(47\frac{1}{2} \text{ mph}\right)\left(\frac{1}{3} \text{ hr}\right)$	$= 15\frac{5}{6}$ miles
Guerin's distance	=	$\left(42\frac{1}{2} \text{ mph}\right)\left(\frac{1}{3} \text{ hr}\right)$	$= 14\frac{1}{6}$ miles
Total distance	=		30 miles

Practice Problem 9 ***Solve by using the substitution method.***

a. Kay needs six ounces of a medicine that is 10% alcohol. If she has one bottle of this medicine that is a 12% alcohol solution and another bottle that is a 7% alcohol solution, how many ounces of each must be mixed to obtain the desired mixture?

b. Truck A leaves Chicago for Boston (a distance of 1050 miles) at the same time that truck B leaves Boston for Chicago. If truck A travels at a rate of 60 mph and truck B travels 40 mph, how far has each truck traveled when they pass each other?

10.3 Exercises

Solve each system by substitution.

1. $x + y = 6$; $x = 6 + y$

2. $x = 6 - 3y$; $x + 2y = 2$

3. $y = 6 - 2x$; $3x + y = 6$

4. $y = 9 - 3x$; $x + 2y = 8$

5. $y = 11 - 2x$; $x = 13 - 2y$

6. $x + 2y = 4$; $x + 2y = 10$

7. $2x + y = 10$; $6x - 2y = 10$

8. $3x + y = 7$; $2x + 4y = 8$

9. $5x - y = 10$; $x + y = 7$

10. $2x + 4y = 14$; $x + y = 2$

11. $3x + 2y = 8$; $-6x - 4y = -16$

12. $5x + 4y = 12$; $2x - 5y = 8$

13. $4x + 6y = -8$; $5x + 4y = -20$

14. $2x + 6y = 14$; $2x + 5y = 12$

15. $2x + 5y = 10$; $4x + 10y = 10$

16. $3x - 6y = 8$; $6x - 12y = 16$

17. $3x + y = -2$; $-6x + 3y = -14$

18. $5x - 2y = 10$; $2x + 5y = 8$

19. $5x - 6y = 9$; $10x - 3y = 6$

20. $y = 2x + 9$; $y = 3x - 5$

21. $x = 8y + 7$; $x = 2y - 1$

22. $9x + 3y = 8$; $y = 3x - 4$

23. $x = -18 - 4y$; $3x + 5y = -19$

24. $4x = 3y$; $x + y = 7$

25. $-2x + y = 8$; $5x + 2y = -20$

26. $5x + 3y = -9$; $-7x - y = -3$

27. $2x - y = -4$; $-x - y = 2$

28. $\frac{x}{2} + \frac{y}{5} = \frac{18}{5}$; $y = 36 - 4x$

29. $\frac{3x}{4} + \frac{y}{2} = \frac{-11}{4}$; $\frac{3x}{2} = 7 + \frac{y}{4}$

30. $\frac{x}{3} + \frac{y}{3} = \frac{4}{3}$; $\frac{x}{2} + \frac{y}{2} = \frac{3}{2}$

31. $0.3x + 0.2y = 0.1$; $-0.4 + y = -2.6$

32. $0.1x - y = -13.95$; $0.12x - y = -12.95$

Fill in the blanks.

33. When using the substitution method to solve a system of linear equations, you still __________ one of the variables. However, you __________ the variable by __________ rather than __________.

34. When solving a system of two linear equations in two variables by substitution, first determine if the __________ are consistent, inconsistent, or dependent. If the __________ are inconsistent, write __________. If they are dependent, write __________.

35. If the equations in a system of two linear equations in two variables are consistent, you can find the solution by the substitution method as follows:

a. Step 1: Solve (if necessary) __________ of the equations for __________ of the variables.
b. Step 2: __________ the expression obtained in step 1 into the __________.
c. Step 3: You now have __________ equation in __________ variable. Solve it.
d. Step 4: Find the value of the other variable by __________ the value obtained in step 3 into the __________ obtained in step 1.
e. Step 5: Write the values obtained in steps 3 and 4 as an __________. This is the solution.
f. Step 6: __________ the solution in both equations.

Solve by substitution.

36. The sum of two consecutive even integers is 70. What are the integers?

37. In a school election, Howard received 140 more votes than Gene. If the total number of votes cast was 384, how many votes did each candidate receive?

38. Charles earned twice as much money as Lee. If their combined salaries total $615, how much did each man earn?

39. A house and lot cost $55,000. If the lot costs $15,000 less than the house, how much does each cost?

40. The daily payroll of the Pace Educational Center is $775. The teachers earn $50 per day, and the tutors earn $25 per day. If the center employs 21 people, find the number of teachers and the number of tutors employed.

41. Barbara has $1500. She deposited some of the money into a savings account that pays 6.25% interest and the rest into an account that pays 13% interest. If the total annual interest earned from the two accounts is $168, how much money did she deposit into each account?

42. Matthew invested a total of $4000 in two savings plans. If the first plan has an interest rate of 15% and the second plan has a rate of 9%, how much must be invested at each rate in order for Matthew to earn a total of $420 in 10 months?

43. How many gallons of a solution that is 15% salt must be mixed with a solution that is 75% salt to obtain 12 gallons of a 30% salt solution?

44. How many pounds of caramels costing $1.90 per pound and how many pounds of Hershey bars costing $2.40 per pound must be mixed to obtain 20 pounds of a candy worth $46?

45. Jerry traveled 360 miles to his summer home in eight hours. He drove for six hours on an interstate highway and the rest of the time on an unpaved road. If his rate on the highway was 20 mph more than his rate on the unpaved road, find his rate on the highway and on the road.

46. Tom and Jerry leave school at the same time and travel in opposite directions on the same highway. Tom drives at a rate of 40 mph and Jerry drives at 50 mph. How far has each man traveled when they are exactly 270 miles apart?

47. Two boats are 50 miles apart. Both start at 10:00 A.M. and travel toward each other in a straight canal. If one boat averages 18 mph and the other averages 12 mph, how far has each boat traveled when they pass each other?

Answers to Practice Problems **7a.** (3, 2) **b.** infinite numbers of solutions **c.** $\left(\frac{5}{2}, 0\right)$ **8a.** infinite number of solutions **b.** $(-3, 3)$ **c.** $(-19, 11)$ **9a.** 3.6 oz. of the 12% solution; 2.4 oz. of the 7% solution **b.** Truck A traveled 630 miles; Truck B traveled 420 miles

Summary

Important Terms

10.1
system of equations
solution of a system of equations
consistent equations
dependent equations
inconsistent equations
no solution
infinite number of solutions

10.2
addition method

10.3
substitution method

Important Skills

10.1
Solving systems of equations graphically
Determining if equations are consistent, inconsistent, or dependent

10.2
Solving systems of equations by the addition method
Solving word problems by using the addition method

10.3
Solving systems of equations by the substitution method
Solving word problems by using the substitution method

Review Exercises

Solve each system graphically.

1. $x + y = 5$
$x - y = 3$

2. $x + 2y = 4$
$2x + 4y = 8$

3. $4x + 3y = 12$
$y - 3 = -7$

Solve each system using the addition method.

4. $2x + 2y = 9$
$2x - 2y = -7$

5. $3x + y = 2$
$6x + 2y = 1$

6. $5x + 7y = 14$
$3x - 4y = -8$

Solve each system by substitution.

7. $3x - 2y = -5$
$x = -2y + 9$

8. $6x + 8y = 14$
$3x + 5y = 8$

9. $4x + 6y = -9$
$x - 3y = 6$

Solve each system by any method.

10. $3x + y = 6$
$6x + 2y = -7$

11. $5x + 2y = 10$
$x - 2y = -4$

12. $x + 4y = 2$
$x - 2y = -4$

13. $3x + 5y = 10$
$x = 3y + 8$

14. $0.4x - 0.3y = 1.2$
$0.6x - 0.5y = 1.9$

15. $\frac{x}{3} + \frac{y}{2} = 1$
$\frac{x}{5} + \frac{y}{2} = -\frac{7}{10}$

Solve.

16. Find two numbers whose sum is 70 and whose difference is 6.

17. How many pounds of pumpkin seeds costing \$0.60 per pound must be mixed with sunflower seeds costing \$0.90 per pound to obtain 30 pounds of a mixture worth \$24?

18. Jody earned \$25,000 from the sale of a house. He invested part at 15% and the rest at 25%. If the annual interest income from the 25% investment is \$250 more than the 15% investment, how much did he invest at each rate?

19. Ed took four minutes to finish a race and Mike took $4\frac{1}{2}$ minutes to finish the same race. If the rate of the faster runner is three feet per second more than the slow runner, find each man's rate.

20. Richard received an annual income of \$2100 from \$10,000 invested in real estate and \$6000 invested in stocks. His son received \$930 from an investment of \$4000 in the same real estate and \$3000 in the same stocks. Find the interest rate they received from each investment.

SPECIAL PRODUCTS AND FACTORING

OBJECTIVES

1. Multiply two binomials using the FOIL method.
2. Square a binomial.
3. Multiply the sum and the difference of the same terms.
4. Factor polynomials in which the terms have a common factor.
5. Factor trinomials of the form $x^2 + px + q$.
6. Factor trinomials of the form $ax^2 + bx + c$.
7. Factor the difference of two squares.
8. Factor a perfect square trinomial.
9. Factor polynomials by grouping.

PRETEST

1. Multiply.
   a. $(x + 3)(x - 5)$ **b.** $(3x + 2)(x - 5)$

   Section 11.1 Idea 2

2. Square each binomial.
   a. $(x - 3)^2$ **b.** $(3x + 5)^2$

   Section 11.1 Idea 4

3. Multiply.
   a. $(x - 4)(x + 4)$ **b.** $(3x - 1)(3x + 1)$

   Section 11.1 Idea 6

4. Factor.
   a. $16x^{13} - 18x^5$ **b.** $15x^6 - 18x^5 + 10x^2$

   Section 11.1 Ideas 8 and 9

5. Factor.
   a. $x^2 - 6x + 8$ **b.** $x^2 + 5x - 24$

   Section 11.3 Ideas 10 and 11

6. Factor.
   a. $6x^2 - x - 7$ **b.** $6x^3 + 26x^2 - 20x$

   Section 11.3 Ideas 12–14

7. Factor.
   a. $x^2 - 16$ **b.** $16x^2 - 25$

   Section 11.4 Idea 16

8. Factor.
   a. $x^2 + 6x + 9$ **b.** $9x^2 - 30x + 25$

   Section 11.4 Idea 18

9. Factor.
   a. $ax - ay + bx - by$
   b. $x^2 - 2x + 1 - 81y^2$

   Section 11.4 Idea 19

EXPLANATION

11.1 Special Products

Idea 1 Quite often in mathematics you must find the product of two binomials. In Section 6.5, you did this by using the distributive property.

Example 1

a. $(x^2 + 2)(x + 3) = x^2(x + 3) + 2(x + 3)$
$= x^3 + 3x^2 + 2x + 6$

b. $(x - 6)(x + 3) = x(x + 3) + (-6)(x + 3)$
$= x^2 + 3x - 6x - 18$
$= x^2 - 3x - 18$

2 To multiply two binomials, use the following procedure:

To Multiply Two Binomials:

1. Write the product of the first terms of the binomials.
2. Write the product of the outside terms of the binomials.
3. Write the product of the inside terms of the binomials.
4. Write the product of the last terms of the binomials.
5. If possible, combine like terms.

NOTE: The word FOIL will be useful in helping you to remember how to multiply binomials. That is, *F* means to multiply the first terms, *O* means to multiply the outer terms, *I* means to multiply the inner terms, and *L* means to multiply the last terms. This procedure is sometimes called the **FOIL method** and the outer (or inner) products are called **cross-products.**

Example 2 Multiply.

a. $(x + 3)(x + 5)$ **b.** $(x - 6)(x + 7)$ **c.** $\left(x - \frac{1}{2}\right)\left(x + \frac{3}{4}\right)$

d. $(x^2 - 3)(x^2 - 4)$ **e.** $(6x + 5y)(3x - 7y)$

Solution To multiply two binomials, use the FOIL method.

First term, Last term; First term, Last term

2a. $(x + 3)(x + 5)$ (F O I L) $= x \cdot x + 5 \cdot x + 3 \cdot x + 5 \cdot 3$
$= x^2 + 5x + 3x + 15$
$= x^2 + 8x + 15$

2b. $(x - 6)(x + 7)$
$= x^2 + 7x - 6x - 42$
$= x^2 + x - 42$

2c. $\left(x - \frac{1}{2}\right)\left(x + \frac{3}{4}\right)$
$= x^2 + \frac{3}{4}x - \frac{1}{2}x - \frac{3}{8}$
$= x^2 + \frac{1}{4}x - \frac{3}{8}$

2d. $(x^2 - 3)(x^2 - 4)$
$= x^4 - 4x^2 - 3x^2 + 12$
$= x^4 - 7x^2 + 12$

2e. $(6x + 5y)(3x - 7y)$
$= 18x^2 - 42xy + 15xy - 35y^2$
$= 18x^2 - 27xy - 35y^2$

With practice, you should be able to multiply two binomials mentally.

Practice Problem 1 ***Multiply using the FOIL method.***

a. $(x + 2)(x + 6)$ **b.** $(x - 4)(x + 6)$ **c.** $(x + 0.3)(x - 0.2)$

d. $(3x - 2y)(5x + 7y)$ **e.** $\left(3x + \frac{1}{3}\right)\left(4x + \frac{2}{3}\right)$ **f.** $(5x - 2)(5x + 2)$

Practice Problem 2 ***Multiply mentally. Just write the answer.***

a. $(x + 3)(x + 4)$ **b.** $(2x - 3)(3x - 2)$ **c.** $(x + 8)(x - 3)$

3 Two special binomial products occur so often that it will be helpful to be able to recognize these products and to memorize the form of the answer. The first product is the square of a binomial.

Example 3
$(x + 3)^2 = (x + 3)(x + 3)$ Definition of a^n
$= x^2 + 3x + 3x + 9$ FOIL method
$= x^2 + 6x + 9$ Combine like terms

4 The results in Example 3 suggest that to square a binomial, use the following procedure:

To Square a Binomial:

1. Square the first term of the binomial.
2. Add twice the product of the first and last term of the binomial.
3. Add the square of the last term of the binomial.

This procedure is sometimes written $(a + b)^2 = a^2 + 2ab + b^2$, and the answer $a^2 + 2ab + b^2$ is called a **perfect square trinomial.**

Example 4 Square each binomial.

a. $(x + 5)^2$ **b.** $(x - 5)^2$ **c.** $(4x + 3)^2$

d. $(3x^3 - 2y)^2$ **e.** $(x + 0.1)^2$ **f.** $\left(3x - \frac{1}{4}\right)^2$

Solution To square a binomial, use the formula $(a + b)^2 = a^2 + 2ab + b^2$.

4a. $(x + 5)^2 = x^2 + 2(x)(5) + 5^2$
$= x^2 + 10x + 25$

4b. $(x - 5)^2 = x^2 + 2(x)(-5) + 5^2$
$= x^2 - 10x + 25$

4c. $(4x + 3)^2 = (4x)^2 + 2(4x)(3) + 3^2$
$= 16x^2 + 24x + 9$ *Think:* $(4x)^2 = 4^2 \cdot x^2 = 16x^2$

4d. $(3x^3 - 2y)^2 = (3x^3)^2 + 2(3x^3)(-2y) + (-2y)^2$
$= 9x^6 - 12x^3y + 4y^2$

4e. $(x + 0.1)^2 = x^2 + 2(x)(0.1) + 0.1^2$
$= x^2 + 0.2x + 0.01$

4f. $\left(3x - \frac{1}{4}\right)^2 = (3x)^2 + 2(3x)\left(-\frac{1}{4}\right) + \left(-\frac{1}{4}\right)^2$
$= 9x^2 - \frac{3}{2}x + \frac{1}{16}$

With practice you should be able to square a binomial mentally.

Practice Problem 3 ***Square each binomial.***

a. $(x - 6)^2$ **b.** $(2x + y)^2$ **c.** $\left(5x - \frac{3}{4}\right)^2$

Practice Problem 4 ***Square each binomial mentally. Just write the answer.***

a. $(y + 5)^2$ **b.** $(2x - 3)^2$ **c.** $\left(x - \frac{1}{2}y\right)^2$

5 The second special binomial product that occurs frequently in mathematics involves multiplying the sum and the difference of the same two terms.

Example 5 $(x + 3)(x - 3) = x^2 - 3x + 3x - 9$ FOIL method
$= x^2 - 9$ Combine like terms

6 In Example 5, note that the final product contains only two terms since the sum of the outer and inner products is zero. Thus, in general, to multiply the sum and the difference of the same terms, use the following procedure:

To Multiply a Sum and a Difference of the Same Terms:

1. Square the first term of the binomial.
2. Write a minus sign and then square the last term of the binomial.

This procedure is sometimes written as $(a + b)(a - b) = a^2 - b^2$, and the answer $a^2 - b^2$ is called the **difference of two squares.**

Example 6 Multiply.

a. $(x + 5)(x - 5)$ **b.** $(7x + 3)(7x - 3)$ **c.** $(2x - 5y)(2x + 5y)$

Solution To multiply the sum and the difference of the same terms, use the formula $(a + b)(a - b) = a^2 - b^2$.

6a. $(x + 5)(x - 5) = x^2 - 5^2$
$= x^2 - 25$

6b. $(7x + 3)(7x - 3) = (7x)^2 - 3^2$
$= 49x^2 - 9$

6c. $(2x - 5y)(2x + 5y) = (2x)^2 - (5y)^2$
$= 4x^2 - 25y^2$

With practice you should be able to multiply the sum and the difference of the same terms mentally.

Practice Problem 5 **Multiply.**

a. $(x + 7)(x - 7)$ **b.** $\left(3x + \frac{1}{2}\right)\left(3x - \frac{1}{2}\right)$ **c.** $(9x - y)(9x + y)$

Practice Problem 6 ***Multiply mentally. Just write the answer.***

a. $(3x + 4)(3x - 4)$ **b.** $(x - 1)(x + 1)$ **c.** $(y + 0.3)(y - 0.3)$

11.1 Exercises

Multiply using the FOIL method.

1. $(2x + 3)(3x + 2)$
2. $(x - 5)(x + 7)$
3. $(4x - 1)(x + 5)$
4. $(x - 7)(x - 8)$
5. $(5x + 1)(x - 9)$
6. $\left(x + \frac{1}{2}\right)\left(x - \frac{1}{3}\right)$
7. $(x + 0.2)(x - 0.3)$
8. $(3x + y)(2x - 5y)$
9. $(xz - 5)(2xz + 2)$
10. $(x - 3y)(x + 4y)$
11. $(x - 2y)(x + 3y)$
12. $(3x - 5)(2x - 1)$
13. $(2x - 3)(3x - 1)$
14. $(x + 3)(x + 3)$
15. $(x + 4)(x + 4)$
16. $(2x - 7)(3x - 6)$
17. $(3x - 5)(2x - 5)$
18. $(x + 2)(x - 2)$
19. $(x + 5)(x - 5)$
20. $(x - y)(x + 2y)$
21. $(x - y)(x + 3y)$
22. $(5x + 7)(5x - 7)$
23. $(3x + 4)(3x - 4)$
24. $(3x - 2y)(2x - 3y)$
25. $(2x - 5y)(3x - y)$
26. $(x + 5y)(2x - y)$
27. $(5x - y)(x + 3y)$
28. $(7x + 5y)(5x - 7y)$
29. $(6x + 1)(3x - 2)$
30. $(11x - 2)(11x + 5)$
31. $(x - 0.2)(x - 0.2)$
32. $(x - 0.3)(x - 0.3)$
33. $(x + 0.5y)(x - 0.5y)$
34. $\left(x + \frac{3}{4}\right)\left(x - \frac{2}{3}\right)$
35. $\left(x + \frac{2}{3}\right)\left(x - \frac{5}{2}\right)$
36. $\left(\frac{1}{2}x + \frac{1}{3}y\right)\left(\frac{1}{2}x - \frac{1}{3}y\right)$

Square each binomial (see Example 4).

37. $(x + 2)^2$
38. $(x + 5)^2$
39. $(x - 7)^2$
40. $(x - 2y)^2$
41. $(2x + y)^2$
42. $(8x - 3y)^2$
43. $\left(2x + \frac{1}{4}\right)^2$
44. $(x - 0.5)^2$
45. $(5x + 2y)^2$
46. $(x + 8)^2$
47. $(x + 9)^2$
48. $(x - 4)^2$
49. $(x - 3)^2$
50. $(x + 6)^2$
51. $(x + 7)^2$
52. $(x - 2)^2$
53. $(x - 6)^2$
54. $(x + 2y)^2$
55. $(x + 3y)^2$
56. $(x - 7y)^2$
57. $(x - 5y)^2$
58. $(3x - 5y)^2$
59. $(2x - 3y)^2$
60. $(x + 0.1)^2$
61. $(x + 0.2)^2$
62. $\left(3x + \frac{1}{3}\right)^2$
63. $\left(2x + \frac{1}{2}\right)^2$
64. $(2x + 0.1)^2$
65. $(3x + 0.1)^2$
66. $(3x + 6y)^2$
67. $(2x + 7y)^2$
68. $(x^2 + 11)^2$
69. $(x^2 + 13)^2$

Multiply (see Example 6).

70. $(x + 5)(x - 5)$
71. $(x + 3)(x - 3)$
72. $(2x - 3)(2x + 3)$
73. $(5x + 1)(5x - 1)$
74. $(a + b)(a - b)$
75. $(3x^2 - 10)(3x^2 + 10)$
76. $(xy - 1)(xy + 1)$
77. $(ab + 1)(ab - 1)$
78. $(2x + y)(2x - y)$
79. $\left(x + \frac{3}{4}\right)\left(x - \frac{3}{4}\right)$
80. $\left(x + \frac{1}{3}\right)\left(x - \frac{1}{3}\right)$
81. $\left(\frac{x}{2} + 5\right)\left(\frac{x}{2} - 5\right)$

Fill in the blanks.

82. The word FOIL is useful in helping us to multiply two ______________. The F means multiply the ______________, O means multiply ______________, I means multiply the ______________, and L means multiply the ______________.

83. To square a binomial, we ______________ the first term of the binomial, add ______________ the product of the first and last term of the binomial, and then add the ______________ of the last term of the binomial.

84. To multiply the sum and the difference of the same two terms, we ______________ the first term of the binomial, write a ______________ sign, and then ______________ the last term.

Multiply mentally. Just write the answer.

85. $(4x - 3)(5x - 1)$
86. $(x + 8)(x - 6)$
87. $(x + 9)^2$
88. $(x + y)(x - y)$
89. $(x + 2)(x + 4)$
90. $(x - 2y)(x + 2y)$
91. $(3x - 2)^2$
92. $(3x - 1)(3x - 5)$
93. $(x + 2)(x - 2)$
94. $(x - 10)^2$
95. $\left(x - \frac{3}{4}\right)^2$
96. $(2x + 9y)(3x - y)$
97. $\left(x + \frac{1}{2}\right)\left(x - \frac{1}{2}\right)$
98. $(x + 0.2)^2$
99. $(3x + 5y)(3x - 5y)$
100. $(7x - y)(3x + 2y)$
101. $(x + 5)(2x - 7)$
102. $(3x - 4)(3x + 5)$
103. $(x - 8)^2$
104. $\left(\frac{1}{4}x - y\right)\left(\frac{1}{4}x + y\right)$
105. $(3x^3 - y^2)(2x^3 + y^2)$

Answers to Practice Problems **1a.** $x^2 + 8x + 12$ **b.** $x^2 + 2x - 24$ **c.** $x^2 + 0.1x - 0.06$ **d.** $15x^2 + 11xy - 14y^2$ **e.** $12x^2 + \frac{10}{3}x + \frac{2}{9}$ **f.** $25x^2 - 4$ **2a.** $x^2 + 7x + 12$ **b.** $6x^2 - 13x + 6$ **c.** $x^2 + 5x - 24$ **3a.** $x^2 - 12x + 36$ **b.** $4x^2 + 4xy + y^2$ **c.** $25x^2 - \frac{15}{2}x + \frac{9}{16}$ **4a.** $y^2 + 10y + 25$ **b.** $4x^2 - 12x + 9$ **c.** $x^2 - xy + \frac{1}{4}y^2$ **5a.** $x^2 - 49$ **b.** $9x^2 - \frac{1}{4}$ **c.** $81x^2 - y^2$ **6a.** $9x^2 - 16$ **b.** $x^2 - 1$ **c.** $y^2 - 0.09$

11.2 Common Monomial Factors

7 In Section 6.5, the distributive law was used to multiply a monomial by a polynomial. The distributive law can also be used to write a polynomial as a product of its factors. This process of writing a polynomial as a product is called **factoring.**

Example 7

	Multiplication	Factoring
		Common factor is 2
a.	$2(x + 7) = 2 \cdot x + 2 \cdot 7$	$2x + 14 = 2 \cdot x + 2 \cdot 7$
	$= 2x + 14$	$= 2(x + 7)$
		Common factor is x^3
b.	$x^3(2x^2 + 5) = x^3 \cdot 2x^2 + x^3 \cdot 5$	$2x^5 + 5x^3 = x^3 \cdot 2x^2 + x^3 \cdot 5$
	$= 2x^5 + 5x^3$	$= x^3(2x^2 + 5)$

Clearly, the relationship between multiplication and factoring is as follows:

Multiplication

$2(x + 7) = 2x + 14$

Factors Product

Factoring

NOTE: The **greatest common factor (GCF)** of a polynomial is the largest factor that will divide each term of a polynomial. For example, 2 is the greatest common factor of $2x + 14$ since it is the largest factor that will divide both $2x$ and 14 evenly.

8 Therefore, to factor a polynomial in which the terms have a common monomial factor, use the following procedure:

1. Find the greatest common monomial factor (GCF) of the polynomial. The greatest common monomial factor of a polynomial is the product of (a) the largest common factor of the numerical coefficients of the polynomial, and (b) the smallest power of every variable that appears in each term of the polynomial.
2. Divide each term of the original polynomial by its greatest common factor (do this mentally).
3. Write the expressions obtained in steps 1 and 2 as a product.

Example 8 Factor.

a. $10x^3 + 15x^2$ **b.** $8x^5y + 10x^4y^2 - 14x^3y^3$

c. $12x^3y^2z - 18xy^2$ **d.** $-6x^5 - 12x^4 + 42x^3$

Solution **8a.** $10x^3 + 15x^2 = $ (first factor)(second factor)

First factor (GCF of $10x^3 + 15x^2$): The GCF of 10 and 15 is 5 while x^2 is the smallest power of the variable which appears in each term. Therefore, the GCF of $10x^3 + 15x^2$ is $5 \cdot x^2 = 5x^2$.

Second factor ($10x^3 + 15x^2 \div$ GCF): $\dfrac{10x^3}{5x^2} + \dfrac{15x^2}{5x^2} = 2x + 3$

Finally, $10x^3 + 15x^2 = 5x^2(2x + 3)$

8b. $8x^5y + 10x^4y^2 - 14x^3y^3 = $ (first factor)(second factor)

First factor: GCF of polynomial is $2x^3y$

Second factor: $\dfrac{8x^5y}{2x^3y} + \dfrac{10x^4y^2}{2x^3y} - \dfrac{14x^3y^3}{2x^3y} = 4x^2 + 5xy - 7y^2$

Therefore, $8x^5y + 10x^4y^2 - 14x^3y^3 = 2x^3y(4x^2 + 5xy - 7y^2)$

8c. $12x^3y^2z - 18xy^2 = $ (first factor)(second factor)

First factor: GCF of the polynomial is $6xy^2$

Second factor: $\dfrac{12x^3y^2z}{6xy^2} - \dfrac{18xy^2}{6xy^2} = 2x^2z - 3$

Therefore, $12x^3y^2z - 18xy^2 = 6xy^2(2x^2z - 3)$

8d. $-6x^5 - 12x^4 + 42x^3 =$ (first factor)(second factor)

First factor: Since the first term of the polynomial is negative, it will be helpful to include a minus sign in the GCF. Thus, the GCF is $-6x^3$.

Second factor: $\dfrac{-6x^5}{-6x^3} - \dfrac{12x^4}{-6x^3} + \dfrac{42x^3}{-6x^3} = x^2 + 2x - 7$

Therefore, $-6x^5 - 12x^4 + 42x^3 = -6x^3(x^2 + 2x - 7)$

NOTE: With practice you should be able to mentally factor a polynomial in which the terms contain a common monomial factor.

Practice Problem 7 ***Factor.***

a. $8x^2 - 28x$ **b.** $36x^4y + 24x^3y^2 - 18x^2y$

c. $24x^5y - 35x^4y^2$ **d.** $-8x^6 - 10x^5 - 12x^3$

9 Sometimes you may not be able to recognize the GCF of the numerical coefficients of a polynomial by inspection.

To Find the GCF of the Coefficients of a Polynomial:

1. Write the prime factorization of each coefficient.
2. The GCF is the product of the common prime factors of the coefficients. Each common factor is used the minimum number of times it occurs in any factorization.

Example 9 Find the GCF of the numerical coefficients.

a. $64x^2y + 96xy^2$ **b.** $30x^2 + 75x - 90$

Solution **9a.** $64 = 2 \cdot 2 \cdot 2 \cdot 2 \cdot 2 \cdot 2 = 2^6$ $96 = 2 \cdot 2 \cdot 2 \cdot 2 \cdot 2 \cdot 3 = 2^5 \cdot 3$

Since the only common prime factor is 2 and the minimum number of times it occurs in any factorization is 5, the GCF of 64 and 96 is $2 \cdot 2 \cdot 2 \cdot 2 \cdot 2 = 32$.

9b. $30 = 2 \cdot 3 \cdot 5$ $75 = 3 \cdot 5 \cdot 5 = 3 \cdot 5^2$ $-90 = -2 \cdot 3 \cdot 3 \cdot 5 = -2 \cdot 3^2 \cdot 5$

Since the only common prime factors are 3 and 5 and the minimum number of times each occurs in any factorization is 1, the GCF of 30, 75, and -90 is $3 \cdot 5 = 15$.

Example 10 Factor: $96x^5 - 144x^4 + 72x^3 - 48x^2$

Solution $96x^5 - 144x^4 + 72x^3 - 48x^2 =$ (first factor)(second factor)

First factor:
The GCF of the coefficients is 24* and x^2 is the smallest power of the variable appearing in each term. Thus, the GCF of the polynomial is $24 \cdot x^2 = 24x^2$

*Note:
$96 = 2 \cdot 2 \cdot 2 \cdot 2 \cdot 2 \cdot 3 = 2^5 \cdot 3$
$144 = 2 \cdot 2 \cdot 2 \cdot 2 \cdot 3 \cdot 3 = 2^4 \cdot 3^2$
$72 = 2 \cdot 2 \cdot 2 \cdot 3 \cdot 3 = 2^3 \cdot 3^2$
$48 = 2 \cdot 2 \cdot 2 \cdot 2 \cdot 3 = 2^4 \cdot 3$
$\text{GCF} = 2 \cdot 2 \cdot 2 \cdot 3 = 24 = 2^3 \cdot 3$

Second factor:

$$\frac{96x^5}{24x^2} - \frac{144x^4}{24x^2} + \frac{72x^3}{24x^2} - \frac{48x^2}{24x^2} = 4x^3 - 6x^2 + 3x - 2$$

Therefore, $96x^5 - 144x^4 + 72x^3 - 48x^2 = 24x^2(4x^3 - 6x^2 + 3x - 2)$

Practice Problem 8 **Factor.**

$24a^4b - 60a^3b^2 + 84a^2b^3 + 108ab^4$

11.2 Exercises

Factor.

1. $2x + 8$

2. $6x^2 - 8x$

3. $9x^2y + 15xy$

4. $2x^3 - 10$

5. $25xy - 35y^2$

6. $12x^4y^2 - 16x^2y^4$

7. $10x^3 - 25x^2 + 20$

8. $16y^4 - 24y^3 + 32y^2$

9. $50x^5 + 100x^3 - 150x^2$

10. $5x^6 + 25x^4 - 20x^2$

11. $16x^3y + 8x^2y + 24xy$

12. $45a^4b^5 - 36a^2b^6 + 81ab^7$

13. $19x^3 + 38x^2y^2$

14. $12x^5 + 14x^4 + 16x^3 + 18$

15. $12a^3d^2 - 18a^2d^3$

16. $45x^3 - 15b^2 + 30$

17. $-27b^4 - 18b^3 + 36b$

18. $-10a^3b^6 - 15a^6b^3$

19. $14x^5y^6 - 42x^3y^7 + 28xy^8$

20. $15x^2yz - 16xy^2$

21. $12a^3b^2 - 18a^2b^3$

22. $15xy^3 - 45x^2y^4$

23. $44a^{14}b^7 - 33a^{10}b^5$

24. $26x^8y^6 - 39x^{12}y^5$

25. $12x^8y^9 + 18x^5y^4 - 20xy^3$

26. $9x^7y^8 - 12xy^5 + 15x^2y^5$

27. $15a^{12}b - 8ab^{12} + 9ab$

28. $10xy^6 + 5x^6y - 4xy$

29. $-30a^3b^4 - 45a^8b^7 - 15a^2b^2$

30. $-18a^3b^4 + 14x^{10}y^6 + 24x^2y^2$

31. $18x^3 - 12y^2 - 48x^4$

32. $32x^5 - 24x^8 + 40y^5$

33. $16x^3y^2 + 24xy^3 - 40x^4y^2$

34. $20x^2y - 30xy^3 - 40xy$

35. $-20x^5 - 10x^3 + 5x^2$

36. $-30x^8 - 10x^5 + 5x^4$

37. $44x^3y^4z - 55x^2y^2z$

38. $66a^6b^4c^2 - 77a^4b^3c^2$

39. $40a^2b^3c^4 - 80a^3b^2c^3$

40. $30x^6y^5z^4 - 60x^3y^2z^2$

Find the GCF (see Example 9).

41. 24, 36, 96

42. 16, 24, 32, 64

43. 26, 39, 65, 91

44. 140, 210, 350

45. 63, 42, 105

46. 48, 60, 84

47. 111, 93, 123

48. 240, 300, 420

49. 75, 90, 120

50. 90, 126, 198

51. 78, 130, 182

52. 66, 110, 132

53. 1001, 1309

Factor.

54. $88x^2 + 121$

55. $65x^3 - 91x^2$

56. $150x^3 - 200x^2 + 250x$

57. $70x^3y - 105x^2y^2 + 140xy^3$

58. $-42a^3b^2c - 66a^2b^3c$

59. $121x^4 - 143x^3 + 187$

60. $24xy^2z + 36xyz + 96xy$

61. $-48x^3y + 60x^2 - 84xy$

Fill in the blanks.

62. The process of writing a polynomial as a product of two or more factors is called ______________.

63. The ______________ of a polynomial is the largest factor that will divide each term of the polynomial.

64. The greatest common monomial factor of a polynomial is the ______________ of the greatest common divisor of the ______________ of the polynomial and the ______________ power of every variable that appears in each term of the polynomial.

65. To factor a polynomial in which the terms have a common monomial factor, write the ________________ of the polynomial and the expression obtained when each ________________ of the polynomial is divided by the ________________ of the polynomial as a product.

66. To find the GCF of the numerical coefficients, write the ________________ of each coefficient. The GCF is the ________________ of the common prime factors of the coefficients. Each common factor is used the ________________ number of times it occurs in any factorization.

Answers to Practice Problems **7a.** $4x(2x - 7)$ **b.** $6x^2y(6x^2 + 4xy - 3)$ **c.** $x^4y(24x - 35y)$ **d.** $-2x^3(4x^3 + 5x^2 + 6)$ **8.** $12ab(2a^3 - 5a^2b + 7ab^2 + 9b^3)$

11.3 Factoring Trinomials

10 In Section 11.1, you multiplied two binomials and obtained a trinomial.

$$(x + a)(x + b) = x^2 + (a + b)x + ab$$

In this section, you will do the reverse by factoring trinomials as the product of two binomials.

$$x^2 + (a + b)x + ab = (x + a)(x + b)$$

Example 11 Factor.

a. $x^2 + 7x + 10$ **b.** $x^2 - 7x + 10$ **c.** $x^2 - 2x - 15$

Solution

11a. To factor $x^2 + 7x + 10$ as $(x + a)(x + b)$, you must find two integers a and b whose product is 10 and whose sum is 7. The numbers are 2 and 5. Therefore, $x^2 + 7x + 10 = (x + 2)(x + 5)$.

11b. To factor $x^2 - 7x + 10$ as $(x + a)(x + b)$, you must find two integers a and b whose product is 10 and whose sum is -7. If the product of the two integers must be positive and their sum must be negative, both integers must be negative. Therefore, the numbers are -2 and -5, and $x^2 - 7x + 10 = (x - 2)(x - 5)$.

11c. To factor $x^2 - 2x - 15$ as $(x + a)(x + b)$, you must find two integers whose product is -15 and whose sum is -2. Since the product of the two integers must be negative, one integer is positive and the other is negative. In this case, it may be helpful to list all pairs of integers whose product is -15.

Factors of -15	Sum of factors
$-3, 5$	2
$3, -5$	-2 ← Correct sum
$-1, 15$	14
$1, -15$	-14

The numbers are 3 and -5. Therefore, $x^2 - 2x - 15 = (x + 3)(x - 5)$

NOTE: When factoring a trinomial $x^2 + px + q$ as a product of two binomials, the following information may be helpful.

Sign of the Last Term (q)	Coefficient of the Middle Term (p)	Signs of the Last Terms of Binomials
Positive	Positive	Both positive
Positive	Negative	Both negative
Negative	Positive or negative	One positive, one negative

11 Based on the results in Example 11, you can factor a trinomial $x^2 + px + q$ as a product of two binomials using the following procedure:

To Factor $x^2 + px + q$:

1. Make sure that the first terms of each binomial are the same and that their product equals the first term of the trinomial.
2. Find two integers (if they exist) whose product is the same as the constant term (q) of the trinomial and whose sum is the same as the coefficient of the middle term (p) of the trinomial.
3. The last terms of the binomial factors are the integers found in step 2.
4. Check the answer mentally. The product of the binomial factors must be the same as the original trinomial.

NOTE: In step 2, if the integers do not exist, the trinomial is not factorable. In this case, we say that the trinomial is *prime with respect to the integers*.

Example 12 Factor.

a. $x^2 - 9x + 14$ **b.** $x^2 + 4x - 21$ **c.** $x^2 + 7x + 6$

d. $y^2 - y - 12$ **e.** $a^2 + 7a + 8$ **f.** $x^4 - 7x^2 - 44$

Solution To factor a trinomial $x^2 + px + q$ as $(x + a)(x + b)$, you must find two integers a and b such that $ab = q$ and $a + b = p$.

12a. Step 1: First term of each binomial must be the same and product must be x^2

$$x^2 - 9x + 14 = (x \quad)(x \quad)$$

Step 2: Find two integers whose product is 14 and whose sum is -9.

14	Sum
$-1, -14$	-15
$-2, -7$	-9 ← Correct sum

Step 3: Since -9 is the correct sum, the last term of the binomials are -2 and -7.

$$x^2 - 9x + 14 = (x - 2)(x - 7)$$

Check: $(x - 2)(x - 7) = x^2 - 9x + 14$ This is done mentally

12b. $x^2 + 4x - 21 = (x \quad)(x \quad)$

-21	Sum
$3, -7$	-4
$-3, 7$	4 ← Correct sum
$1, -21$	
$-1, 21$	

Therefore, $x^2 + 4x - 21 = (x - 3)(x + 7)$.

12c. $x^2 + 7x + 6 = (x \quad)(x \quad)$

$\downarrow$

6	Sum	
$2 \cdot 3$	5	
$6 \cdot 1$	7	← Correct sum

Therefore, $x^2 + 7x + 6 = (x + 6)(x + 1)$

12d. $y^2 - y - 12 = (y \quad)(y \quad)$

$\downarrow$

−12	Sum	
2, −6	−4	
−2, 6	4	
3, −4	−1	← Correct sum
−3, 4		
1, −12		
−1, 12		

Therefore, $y^2 - y - 12 = (y + 3)(y - 4)$

12e. $a^2 + 7a + 8 = (a \quad)(a \quad)$

$\downarrow$

8	Sum
2, 4	6
1, 8	9
−2, −4	−6
−1, −8	−9

Since there are no two integers whose product is 8 and whose sum is 7, $a^2 + 7a + 8$ is not factorable (with respect to the integers).

12f. $x^4 - 7x^2 - 44 = (x^2 \quad)(x^2 \quad)$

$\downarrow$

−44	Sum	
2, −22	−20	
−2, 22	20	
4, −11	−7	← Correct sum
−4, 11		
1, −44		
−1, 44		

Therefore, $x^4 - 7x^2 - 44 = (x^2 + 4)(x^2 - 11)$.

NOTE: With practice you should be able to factor trinomials of the form $x^2 + px + q$ mentally.

Practice Problem 9 ***Factor.***

a. $x^2 - 8x + 15$ **b.** $y^2 - 3y - 10$ **c.** $x^2 + 10x - 75$

Practice Problem 10 ***Factor mentally. Just write the answer.***

a. $x^2 + 8x - 20$ **b.** $y^2 - 9y + 20$ **c.** $x^2 - 7x - 30$

12 Sometimes it will be necessary to factor trinomials in which the coefficient of the squared term is not equal to 1. When factoring this type of trinomial, we will use the trial-and-error method.

Example 13 Factor: $3x^2 - 7x - 6$

Solution The product of the first terms of the binomial factors must be $3x^2$ and the product of the last terms must be -6. To obtain $3x^2$, the factors must be $3x$ and x. Therefore, you have $3x^2 - 7x - 6 = (x \quad)(3x \quad)$.

Now, identify the possible factors of -6 and list the possible binomial factors of $3x^2 - 7x - 6$.

Possible binomial factors	Resulting middle term	
$(x + 2)(3x - 3)$	$3x$	
$(x + 3)(3x - 2)$	$7x$	
$(x - 2)(3x + 3)$	$-3x$	
$(x - 3)(3x + 2)$	$-7x$	← Correct middle term
$(x + 6)(3x - 1)$	$17x$	
$(x + 1)(3x - 6)$	$-3x$	
$(x - 6)(3x + 1)$	$-17x$	
$(x - 1)(3x + 6)$	$3x$	

Finally, since $-7x$ is the same as the middle term of the trinomial, the correct factors are $(x - 3)(3x + 2)$.

NOTE: It may be helpful to remember that *if the original trinomial has no common factor, neither of its factors will, either.* For example, $(x + 2)(3x - 3)$ could not be the factors of $3x^2 - 7x - 6$ since $3x - 3$ has a common factor and the original trinomial $3x^2 - 7x - 6$ has no common factor.

13 Based on the results in Example 13, you can factor a trinomial of the form $ax^2 + bx + c$ as a product of two binomials in the following way:

To Factor $ax^2 + bx + c$:

1. Write two parentheses and fill in any obvious information.

2. Identify all possible factors for the coefficients of the first term and the last term of the trinomial ($ax^2 + bx + c$). Also, list all possible binomial factors of the trinomial and their resulting middle term.

3a. Select the correct factors. These are the factors whose resulting middle term is the same as the middle term of the trinomial.

3b. If the factors do not exist, write "not factorable."

NOTE: In step 2, you need not always list *all* the possible binomial products. Instead, start listing the possible factors along with their resulting middle terms and then stop when you obtain the correct pair.

Example 14 Factor.

a. $5x^2 - 23x - 10$ **b.** $7x^2 - 31x + 12$ **c.** $2x^2 + 3x + 3$

d. $8y^2 + 14y - 9$ **e.** $6x^2 - 35x + 50$ **f.** $4z^2 + 12yz - 7y^2$

Solution **14a. Step 1:** The only possible factors of $5x^2$ for this problem are $5x$ and x.

$$5x^2 - 23x - 10 = (5x \quad)(x \quad)$$

Step 2: The factors of -10 are -2 and 5, 2 and -5, 1 and -10, and -1 and 10.

Possible binomial factors	Resulting middle term	
$(5x - 2)(x + 5)$	$23x$	Sign is wrong

NOTE: Since you want a middle term of $-23x$ (not $23x$), interchange the signs of the last terms of the binomials to obtain the correct factors:

$(5x + 2)(x - 5)$ $-23x \leftarrow$ Correct middle term

Step 3: $5x^2 - 23x - 10 = (5x + 2)(x - 5)$

14b. Step 1: The only possible factors of $7x^2$ for this problem are $7x$ and x.

$$7x^2 - 31x + 12 = (7x \quad)(x \quad)$$

Step 2: Since the last term is positive and the middle term is negative, both factors of the last terms of the binomials must be negative. Therefore, the possible factors of 12 are -2 and -6, -3 and -4, and -1 and -12.

Possible binomial factors	Resulting middle term	
$(7x - 2)(x - 6)$	$-44x$	
$(7x - 6)(x - 2)$	$-20x$	
$(7x - 3)(x - 4)$	$-31x$	$\leftarrow$ Correct middle term

Step 3: $7x^2 - 31x + 12 = (7x - 3)(x - 4)$

14c. Step 1: The only possible factors of $2x^2$ for this problem are $2x$ and x.

$$2x^2 + 3x + 3 = (2x \quad)(x \quad)$$

Step 2: Since the last term is positive and the middle term is positive, the signs of both factors of the last term must be positive. Therefore, the possible factors of 3 are 1 and 3.

Possible binomial factors	Resulting middle term
$(2x + 3)(x + 1)$	$5x$
$(2x + 1)(x + 3)$	$7x$

Step 3: Since none of the possible factors yields a middle term of $3x$, the trinomial is prime with respect to the integers and you should write "not factorable."

14d. Step 1: Since both $8x^2$ and -9 have several possible factors, start by writing

$$8y^2 + 14y - 9 = (\quad)(\quad).$$

Step 2: The factors of $8x^2$ are $2x$ and $4x$, and $8x$ and x. The factors of -9 are 3 and -3, 1 and -9, and -1 and 9.

Possible binomial factors	Resulting middle term
$(2y + 3)(4y - 3)$	$6y$
$(2y + 9)(4y - 1)$	$34y$
$(2y - 1)(4y + 9)$	$14y \leftarrow$ Correct middle term

Step 3: $8y^2 + 14y - 9 = (2y - 1)(4y + 9)$

14e. Step 1: Since $6x^2$ and 50 have several factors, begin by writing

$$6x^2 - 35x + 50 = (\quad)(\quad).$$

Step 2: Since the last term of the trinomial is positive and the coefficient of the middle term is negative, both factors of the last term of the trinomial must be negative. Therefore, the possible factors of 50 are -2 and -25; -5 and -10; and -1 and -50. The factors of $6x^2$ are $3x$ and $2x$; and x and $6x$.

Possible binomial factors	Resulting middle term
$(3x - 5)(2x - 10)$	*Note:* This could not be correct since the second binomial has a common factor, while the trinomial does not.
$(3x - 10)(2x - 5)$	$-35x \leftarrow$ Correct middle term

Step 3: $6x^2 - 35x + 50 = (3x - 10)(2x - 5)$

14f. Step 1: Although this trinomial is not of the form $ax^2 + bx + c$, you can still factor it by using the techniques presented in this section.

$$4z^2 + 12yz - 7y^2 = (\quad)(\quad)$$

Step 2: The possible factors of $4z^2$ are $2z$ and $2z$, or $4z$ and z. The possible factors of $-7y^2$ are y and $-7y$, or $-y$ and $7y$.

Possible binomial factors	Resulting middle term
$(2z - y)(2z + 7y)$	$12yz \leftarrow$ Correct middle term

Step 3: $4z^2 + 12yz - 7y^2 = (2z - y)(2z + 7y)$

NOTE: When factoring trinomials of the form $ax^2 + bx + c$, you must use the trial-and-error method. However, with practice you can make intelligent guesses and eliminate the possibilities in a systematic manner. Also, remember to do as many of the steps as possible mentally.

Practice Problem 11 ***Factor.***

a. $3x^2 + 16x + 5$ **b.** $6x^2 - 13x - 5$ **c.** $6x^2 + 5x - 6$

d. $8x^2 - 6x - 9$ **e.** $2x^2 - 17x + 5$ **f.** $12x^2 + xy - 20y^2$

14 Sometimes a trinomial will have a common monomial factor.

Example 15 Factor.

a. $2x^2 - 20x + 48$ **b.** $24x^3 - 54x^2 + 30x$

Solution When factoring trinomials, first factor out the GCF (if it exists) and then factor the trinomial obtained when you divide each term of the original trinomial by the GCF. The complete factorization will be the product of the GCF and the binomial factors.

15a. The GCF of $2x^2 - 20x + 48$ is 2.

$$2x^2 - 20x + 48 = 2(x^2 - 10x + 24)$$

Now, factor $x^2 - 10x + 24$.

$$x^2 - 10x + 24 = (x - 6)(x - 4)$$

Finally, $2x^2 - 20x + 48 = 2(x - 6)(x - 4)$.

15b. The GCF of $24x^3 - 54x^2 + 30x$ is $6x$.

$$24x^3 - 54x^2 + 30x = 6x(4x^2 - 9x + 5)$$

Now, factor $4x^2 - 9x + 5$ by the trial-and-error method.

$$4x^2 - 9x + 5 = (4x - 5)(x - 1)$$

Finally, $24x^3 - 54x^2 + 30x = 6x(4x - 5)(x - 1)$

Practice Problem 12 ***Factor.***

a. $8x^5 - 2x^4 - 6x^3$ **b.** $2x^2 + 30x + 112$

11.3 Exercises

Factor.

1. $x^2 + 12x + 35$
2. $x^2 + 6x + 8$
3. $x^2 - 6x + 5$
4. $x^2 + 4x - 21$
5. $x^2 - x - 30$
6. $x^2 - 9x + 14$
7. $x^2 - 11x + 24$
8. $x^2 + 5x + 8$
9. $x^2 - 9x - 36$
10. $x^2 + 7x + 12$
11. $x^2 + 2x - 35$
12. $x^2 - 3x + 2$
13. $x^2 - 11x + 28$
14. $x^2 + 17x - 60$
15. $x^2 + 17x - 16$
16. $3x^2 + 10x + 3$
17. $5x^2 - 13x - 6$
18. $10x^2 + 37x + 7$
19. $6x^2 - x - 5$
20. $6x^2 + 7x - 3$
21. $2x^2 - 13x + 20$
22. $12x^2 - 7x - 10$
23. $3x^2 - 2x - 8$
24. $10x^2 + 11x - 6$
25. $20x^2 - 19x + 3$
26. $6x^2 + 23x + 20$
27. $12x^2 - 16x - 35$
28. $15x^2 + 2x - 8$
29. $36x^2 - 13x + 1$
30. $4x^2 - 24x + 35$
31. $y^2 + 2y - 15$
32. $2x^2 + 12x + 16$
33. $x^2 + x - 6$
34. $x^2 - 7x + 12$
35. $a^2 - 3a - 18$
36. $6x^2 - 13x - 5$
37. $21x^2 - 29x + 10$
38. $x^2 - 9x + 12$
39. $3x^2 - 48x - 108$
40. $2x^2 - 22x + 20$
41. $27x^2 + 54x + 24$
42. $6x^2 - 13x + 6$
43. $7x^4 + 69x^2 - 10$
44. $12x^2 - 49x + 4$
45. $x^2 - 7xy - 18y^2$
46. $5x^2 - 3x - 14$
47. $3x^2 + 12x + 10$
48. $3x^2 + 10x + 8$
49. $12x^2 - 22x - 20$
50. $3x^3 - 30x^2 + 72x$
51. $x^7 - 5x^6 - 14x^5$
52. $6x^2 - 48x - 120$
53. $x^2 + x - 42$
54. $6y^5 - 36y^4 + 30y^3$
55. $8x^4 - 6x^2 - 9$
56. $28x^4 - 58x^3 - 30x^2$
57. $8x^2 + 13x + 5$
58. $12x^2 - 7x - 12$
59. $6x^2 - 7x - 20$
60. $8x^3 + 20x^2 - 12x$
61. $15x^2 - x - 6$
62. $40x^2 + x - 6$
63. $10x^2 - x - 24$
64. $4y^2 - 5y - 6$
65. $15x^4 - 22x^2 + 8$
66. $15a^2 - 7a - 4$

Fill in the blanks.

67. To factor a trinomial of the form $x^2 + px + q$ as a product of two binomials, do the following:

a. Step 1: Make sure that the first terms of each binomial are the ______________ and that their product equals the ______________ of the trinomial.

b. Step 2: Find two integers (if they exist) whose ______________ is the same as the constant term (q) and whose ______________ is the same as the ______________ of the middle term (p) of the trinomial.

c. Step 3: The ______________ of the binomial factors are the integers found in step 2.

d. Check the answer mentally. The ______________ of the binomial factors must be the ______________ as the original trinomial.

68. To factor a trinomial of the form $ax^2 + bx + c$ as a product of two binomials, do the following:

a. Step 1: Write two parentheses and ______________ in any obvious information.

b. Step 2: Identify all possible factors for the ______________ of the first term and the ______________ of the trinomial. Also, list all possible binomial factors of the trinomial and their resulting ______________.

c. Step 3a: Select the correct factors. These are the factors whose resulting ______________ is the same as the ______________ of the ______________.

d. Step 3b: If the factors do not exist, write ______________.

69. When factoring any trinomial, you should always look first for a ______________ factor.

Factor.

70. $6x^2 + 5xy - 6y^2$ **71.** $25x^2 - 5xy - 2y^2$ **72.** $9x^2 + 18xy + 8y^2$

73. $4s^2 - 5s - 6$ **74.** $x^6 + 3x^3 - 7$ **75.** $16x^2 - 30x + 9$

76. $a^2 + 3ab - 10b^2$ **77.** $2a^2 + 4ab - 30b^2$ **78.** $6a^2 - 19ab + 15b^2$

79. $12a^4 - 7a^2b^2 - 10b^4$ **80.** $16a^4b^2 - 80a^3b^2 + 384a^2b^2$ **81.** $63c^2 - 31c - 10$

Answers to Practice Problems **9a.** $(x - 5)(x - 3)$ **b.** $(y - 5)(y + 2)$ **c.** $(x + 15)(x - 5)$ **10a.** $(x + 10)(x - 2)$ **b.** $(y - 5)(y - 4)$ **c.** $(x - 10)(x + 3)$ **11a.** $(3x + 1)(x + 5)$ **b.** $(3x + 1)(2x - 5)$ **c.** $(3x - 2)(2x + 3)$ **d.** $(4x + 3)(2x - 3)$ **e.** not factorable **f.** $(3x + 4y)(4x - 5y)$ **12a.** $2x^3(4x + 3)(x - 1)$ **b.** $2(x + 8)(x + 7)$

11.4 Special Factorization

15 In Section 11.1, we discussed the following special products:

$$(a + b)(a - b) = a^2 - b^2$$
$$(a + b)^2 = a^2 + 2ab + b^2$$

Since factoring is the reverse of multiplication, we will now consider the following special factorizations:

$$a^2 - b^2 = (a + b)(a - b)$$
$$a^2 + 2ab + b^2 = (a + b)^2$$

Example 16 Factor: $x^2 - 25$

Solution This binomial is the difference of two squares.

$$x^2 - 25 = (x)^2 - (5)^2$$

Therefore, it can be factored as the product of the sum and difference of the same terms.

$$x^2 - 25 = (x)^2 - (5)^2 = (x + 5)(x - 5)$$

In other words, $a^2 - b^2 = (a + b)(a - b)$.

NOTE: In the binomial $x^2 - 25$, x^2 and 25 are called **perfect squares** since they can be expressed as a quantity squared. Also, x and 5 are called the **square roots** of x^2 and 25 since they are the quantities that were squared to obtain x^2 and 25. Similarly, $16x^2 = (4x)^2$ is a perfect square and its square root is $4x$.

16 The results in Example 16 show that you can factor the difference of two squares by using the following formula:

$$a^2 - b^2 = (a + b)(a - b),$$

where a and b are the square roots of a^2 and b^2.

Example 17 Factor.

a. $x^2 - 36$ **b.** $4x^2 - 49$ **c.** $50x^2 - 2$ **d.** $x^4 - 16$

Solution To factor the difference of two squares, use the formula $a^2 - b^2 = (a + b)(a - b)$.

17a. $x^2 - 36 = x^2 - 6^2$
$= (x + 6)(x - 6)$

17b. $4x^2 - 49 = (2x)^2 - 7^2$
$= (2x + 7)(2x - 7)$

17c. $50x^2 - 2 = 2(25x^2 - 1)$ *Note:* 2 is a common factor.
$= 2(5x + 1)(5x - 1)$

17d. $x^4 - 16 = (x^2)^2 - 4^2$
$= (x^2 + 4)(x^2 - 4)$ *Note:* $x^2 - 4$ can be factored.
$= (x^2 + 4)(x + 2)(x - 2)$

NOTE: With practice you should be able to factor the difference of two squares mentally. Also, when one of the factors can still be factored, you should factor it. This process of factoring until none of the factors can be factored further is sometimes called **factoring completely.**

Practice Problem 13 ***Factor.***

a. $x^2 - 100$ **b.** $64x^2 - 1$ **c.** $2x^2 - 162$

17 Another special factorization involves a perfect square trinomial which is the square of a binomial. To recognize a perfect square trinomial, remember that:

1. The first and last terms must be perfect squares.
2. The middle term is twice the product of the square roots of the perfect squares ($2ab$) or its additive inverse ($-2ab$).

Example 18 Factor: $4x^2 + 20x + 25$

Solution This trinomial could be factored by trial and error. However, since $4x^2$ and 25 are perfect squares and the middle term $20x$ $(2 \cdot 2x \cdot 5)$ is twice the product of the square roots of $4x^2$ and 25, $4x^2 + 20x + 25$ is a perfect square trinomial. Therefore, it can be factored as a binomial squared.

$$4x^2 + 20x + 25 = (2x)^2 + 2(2x \cdot 5) + 5^2$$
$$= (2x + 5)^2$$

In other words, $a^2 + 2ab + b^2 = (a + b)^2$

18 Based on the results in Example 18, to factor a perfect square trinomial $a^2 + 2ab + b^2$ as a binomial squared use the following procedure:

To Factor a Perfect Square Trinomial:

1. Write the square root of the first term.
2. Write the sign of the middle term of the trinomial.
3. Write the square root of the last term and then square the resulting binomial.

Example 19 Factor each perfect square trinomial.

a. $x^2 + 18x + 81$ **b.** $16x^2 - 8x + 1$ **c.** $4x^2 + 10x + 25$
d. $25x^2 - 30x + 9$ **e.** $x^2 - 14x - 49$

Solution To factor a perfect square trinomial, use the formula $a^2 + 2ab + b^2 = (a + b)^2$.

19a. $x^2 + 18x + 81 = x^2 + 2(x \cdot 9) + 9^2$ is a perfect square trinomial. Therefore,

$$x^2 + 18x + 81 = (x + 9)^2.$$

19b. $16x^2 - 8x + 1 = (4x)^2 - 2(4x \cdot 1) + 1^2$ is a perfect square trinomial. Therefore,

$$16x^2 - 8x + 1 = (4x - 1)^2.$$

19c. $4x^2 + 10x + 25$ is not a perfect square trinomial since the middle term $10x$ is not twice the product of the square roots of $4x^2$ and 25. Therefore, it cannot be factored as a binomial squared.

19d. $25x^2 - 30x + 9 = (5x)^2 - 2(5x \cdot 3) + 3^2$ is a perfect square trinomial. Therefore,

$$25x^2 - 30x + 9 = (5x - 3)^2.$$

19e. $x^2 + 14x - 49$ is not a perfect square trinomial since the last term is negative. Remember that in a perfect square trinomial, the coefficients of the first and last terms must be positive.

NOTE: With practice you should be able to recognize and factor perfect square trinomials mentally.

Practice Problem 14 ***Factor each perfect square trinomial.***

a. $x^2 - 6x + 9$ **b.** $9x^2 + 15x + 25$ **c.** $4x^2 + 12x + 9$

19 Until now we have discussed special factorizations of binomials and trinomials. Let us now consider a special factorization technique called **factoring by grouping** which involves polynomials containing four terms. In this technique, the terms of a polynomial are rearranged into smaller groups that are factorable.

Example 20 Factor.

a $ax - ay + bx - by$ **b.** $6x^2 - 2xy - 9x + 3y$

c. $ab + b + a + 1$ **d.** $x^2 + 10x + 25 - 4y^2$

Solution To factor by grouping, first try to group together terms that have a common factor or groups that are special forms. Then, factor the resulting expressions by using the techniques discussed in this chapter.

20a. Although the four terms have no common factor, notice that the first two terms and the last two terms have a common factor.

$ax - ay + bx - by$
$= (ax - ay) + (bx - by)$ Rewrite as two groups
$= a(x - y) + b(x - y)$ Factor each group
$= (x - y)(a + b)$ Factor out GCF of $x - y$

20b. $6x^2 - 2xy - 9x + 3y$
$= (6x^2 - 2xy) - (9x - 3y)$ Rewrite as two groups
$= 2x(3x - y) - 3(3x - y)$ Factor each group
$= (3x - y)(2x - 3)$ Factor out GCF of $3x - y$

20c. $ab + b + a + 1$
$= (ab + b) + (a + 1)$ Rewrite as two groups
$= b(a + 1) + (a + 1)$ Factor each group
$= (a + 1)(b + 1)$ Factor out GCF of $a + 1$

20d. The first three terms are a perfect square trinomial.

$x^2 + 10x + 25 - 4y^2$
$= (x^2 + 10x + 25) - 4y^2$ Rewrite as two groups
$= (x + 5)^2 - (2y)^2$ Rewrite as difference of squares
$= [(x + 5) + 2y][(x + 5) - 2y]$ Factor difference of squares
$= (x + 2y + 5)(x - 2y + 5)$ Rearrange terms of factors

Practice Problem 15 ***Factor.***

a. $ax + bx + ay + by$ **b.** $ay + a - y - 1$ **c.** $z^2 - x^2 + 4xy - 4y^2$

11.4 Exercises

Factor each binomial that is the difference of squares.

1. $x^2 - 9$ **2.** $x^2 - 16$ **3.** $9x^2 - 4$ **4.** $x^2 - 36$
5. $x^2 - 64$ **6.** $4y^2 - 100$ **7.** $9s^2 - 1$ **8.** $4x^4 - 9$
9. $x^4 - 16$ **10.** $x^2 - 12$ **11.** $x^2 + 9$ **12.** $4x^2 - 81$
13. $81x^2 - 1$ **14.** $s^4 - 25$ **15.** $25y^2 - 4$ **16.** $36x^2 - 25$

Factor each perfect square trinomial.

17. $x^2 + 4x + 4$ **18.** $x^2 + 2x + 1$ **19.** $a^2 - 10a + 25$
20. $x^2 - 14x + 49$ **21.** $y^2 - 8y + 16$ **22.** $x^2 - 20x + 100$
23. $x^2 + 8x - 16$ **24.** $9x^2 - 14x + 25$ **25.** $4x^2 + 20x - 25$
26. $x^2 - 6x + 9$ **27.** $9x^2 - 15x + 25$ **28.** $9x^2 + 30x + 25$
29. $x^2 + 16x + 64$ **30.** $16x^2 - 8x + 1$ **31.** $64x^2 - 24x + 9$

Factor.

32. $ap + bp + aq + bq$ **33.** $cx + cy + dx + dy$ **34.** $ax - cx - ay + cy$
35. $bx - b + ax - a$ **36.** $x^2 - y^2 + 2x + 2y$ **37.** $y^3 + 3y^2 + 4y + 12$
38. $b^2 + ab + b + a$ **39.** $y^3 - 3y^2 - 9y + 27$ **40.** $x^2 - 14x + 49 - 9y^2$

41. $z^2 - x^2 + 4xy - 4y^2$

42. $2xy + 3xz - 10y - 15z$

43. $6b^2 - 3bc - 14b + 7c$

44. $x^3 + 5x^2 + 3x + 15$

45. $x^2 + 8x + 16 - 25y^2$

46. $ax + ay + bx + by$

47. $bx + by + cx + cy$

48. $aw - bw - az + bz$

49. $cw - dw - cz + dz$

50. $ay + y - a - 1$

51. $by + y - b - 1$

52. $x^2 - y^2 + 4x - 4y$

53. $x^2 - y^2 + 5x - 5y$

54. $2x^2 - 6xy + 5x - 15y$

55. $4s^2 - 12st + 7s - 21t$

56. $x^2 - 4x + 4 - 25y^2$

57. $x^2 - 6x + 9 - 36y^2$

58. $x^2 - y^2 + 10y - 25$

59. $y^2 - x^2 - 12x - 36$

60. $a^2 - b^2 + a - b$

61. $x^2 - y^2 + x + y$

Fill in the blanks.

62. The difference of two squares $a^2 - b^2$ is factored as $(a + b)(a - b)$, where a and b are the ______________ of ______________ and ______________.

63. To recognize a perfect square trinomial, remember that the first and last terms are ______________. Also, the middle term is ______________ the product of the square roots of the ______________ or its ______________.

64. To factor a perfect square trinomial, first write the ______________ of the first term. Then, write the sign of the ______________ of the trinomial. Finally, write the ______________ of the last term and ______________ the resulting binomial.

65. To factor by grouping, first try to group together terms that have a ______________ factor or groups which are special forms. Then, ______________ the resulting expression.

Factor.

66. $2x^3 - 32x$

67. $9x^2 + 48x + 64$

68. $x^4 - 81$

69. $xy - ay + xp - ap$

70. $6x^2 + 3xy + 2x + y$

71. $8x^2 - 24x + 18$

72. $18x^2 - 2y^2$

73. $32x^2 + 16x + 2$

74. $x^2 - z^2 + 10z - 25$

75. $3x^2 - 75$

Answers to Practice Problems **13a.** $(x + 10)(x - 10)$ **b.** $(8x + 1)(8x - 1)$ **c.** $2(x + 9)(x - 9)$ **14a.** $(x - 3)^2$ **b.** not a perfect square trinomial **c.** $(2x + 3)^2$ **15a.** $(x + y)(a + b)$ **b.** $(y + 1)$ $(a - 1)$ **c.** $(x - 2y + z)(-x + 2y + z)$

Summary

Important Terms

11.1

FOIL method
inner product
outer product
perfect square trinomial
difference of two squares

11.4

perfect square
square root
factoring completely
factoring by grouping

11.2

greatest common factor (GCF)

Important Skills

11.1

Multiplying two binomials by using the FOIL method
Squaring binomials
Multiplying the sum and the difference of the same two terms

11.2

Finding the greatest common monomial factor of a polynomial
Factoring a polynomial in which the terms have a common monomial factor

11.3

Factoring trinomials of the form $x^2 + px + q$
Factoring trinomials of the form $ax^2 + bx + c$

11.4

Factoring the difference of two squares
Factoring perfect square trinomials
Factoring polynomials by grouping

Review Exercises

Factor completely.

1. $4x^2 + 12x + 9$
2. $x^2 - 3x - 10$
3. $x^2 - 11x + 30$
4. $x^4 - 81$
5. $2x^2 + 12x - 14$
6. $25x^2 - 10x + 1$
7. $xy - y - x - 1$
8. $b^3 - 2b^2 - 4b + 8$
9. $x^2 - 10$
10. $32x^2 - 2$
11. $4x^2 + 4x - 3$
12. $18x^6 - 24x^4 + 36x^3$
13. $60x^9 + 84x^8 - 96x^7 + 24x^4$
14. $4x^2 - 12x + 9 - 25y^2$
15. $2x^2 + 5x - 3$
16. $12x^2 - 7x - 12$
17. $6x^2 - 7x - 10$
18. $2x^2 + 22x + 28$
19. $64x^4 - 1$
20. $24x^2 - 10xy - 21y^2$
21. $9x^2 - 42x + 21$
22. $15x^4 - 27x^3 - 9$
23. $x^2 - 12xy + y^2$
24. $x^6 - 4$
25. $18x^2 - 3x - 10$

Perform the indicated operation.

26. $(3x - 2)(3x - 5)$
27. $(2x + 5)^2$
28. $(x + 9)(x - 9)$
29. $(x - 0.2)^2$
30. $(x + 8)(x - 9)$
31. $\left(\frac{1}{4}x - 2\right)\left(\frac{1}{4}x + 2\right)$
32. $(x + 0.3)(x - 0.5)$
33. $\left(x + \frac{1}{3}\right)\left(x - \frac{3}{4}\right)$

SQUARE ROOTS AND QUADRATIC EQUATIONS

OBJECTIVES

1. Find the square root of a number.
2. Find the decimal approximation of a square root.
3. Multiply square roots.
4. Simplify square roots.
5. Divide one square root by another.
6. Rationalize the denominator of a fraction.
7. Add and subtract square roots.
8. Solve quadratic equations using the square root method.
9. Solve quadratic equations by factoring.
10. Solve quadratic equations by using the quadratic formula.

PRETEST	EXPLANATION
1. Find the square root. **a.** $\sqrt{16}$ **b.** $-\sqrt{y^8}$	Section 12.1 Ideas 1–3
2. Find the decimal approximation. **a.** $\sqrt{13}$ **b.** $\sqrt{97}$	Section 12.1 Idea 4
3. Multiply. **a.** $\sqrt{2}\sqrt{11}$ **b.** $\sqrt{3}\sqrt{27}$	Section 12.2 Idea 5
4. Simplify. **a.** $\sqrt{20x^3}$ **b.** $\sqrt{539}$	Section 12.2 Ideas 6 and 7
5. Divide. **a.** $\dfrac{\sqrt{32x}}{\sqrt{2}}$ **b.** $\dfrac{\sqrt{64x^7}}{\sqrt{x}}$	Section 12.2 Idea 9
6. Rationalize the denominator. **a.** $\dfrac{1}{\sqrt{3}}$ **b.** $\dfrac{\sqrt{3}}{\sqrt{20}}$	Section 12.2 Idea 12
7. Perform the indicated operations. **a.** $\sqrt{3} - 4\sqrt{3}$ **b.** $2\sqrt{18} + 3\sqrt{50} - \sqrt{8}$	Section 12.3 Idea 14 Idea 14
8. Solve. **a.** $(x + 3)^2 = 16$ **b.** $6x^2 - 8 = 1$	Section 12.4 Ideas 16 and 17
9. Solve. **a.** $3x^2 = 15x$ **b.** $6x^2 + 3x = 3$	Section 12.5 Ideas 19 and 20
10. Solve. **a.** $x^2 + 6x = -9$ **b.** $2x^2 + 4x = 1$	Section 12.6 Ideas 22 and 23

12.1 Square Roots

Idea 1 To determine the square root of a given number, you must find a number whose square is that given number. For example, one square root of 25 is 5 since $5^2 = 25$. Another square root of 25 is -5 since $(-5)^2 = 25$. Therefore, 25 has two square roots. In general, every positive number has two square roots (one positive and one negative).

2 The **positive** or **principal square root** of a number is written with the symbol $\sqrt{}$. The negative square root of a number is written $-\sqrt{}$. The $\sqrt{}$ is called a **radical symbol** and it is also used to represent the square root of zero ($\sqrt{0} = 0$).

Example 1 Find the square root of each number.

a. $\sqrt{16}$ b. $\sqrt{100}$ c. $-\sqrt{36}$ d. $-\sqrt{81}$ e. $\sqrt{1}$

Solution To find the square root of a number x, remember that K is the square root of x if $K^2 = x$.

1a. $\sqrt{16} = 4$ since $4^2 = 16$ **1b.** $\sqrt{100} = 10$ since $10^2 = 100$

1c. $-\sqrt{36} = -6$ **1d.** $-\sqrt{81} = -9$

1e. $\sqrt{1} = 1$

NOTE: The expression under the radical symbol is called the **radicand.** The entire expression (radical symbol and radicand) is called a **radical.** For example, $\sqrt{36}$ is a radical and 36 is the radicand.

Practice Problem 1 ***Find the square root of each number.***

a. $\sqrt{49}$ b. $-\sqrt{9}$ c. $\sqrt{64}$ d. $\sqrt{25}$ e. $-\sqrt{4}$

3 Sometimes the radicand is a number raised to a power or a variable. (In this book, always assume that the variables represent positive numbers.)

Example 2 Find the square root of each number.

a. $\sqrt{x^6}$ b. $-\sqrt{y^8}$ c. $\sqrt{y^4}$ d. $\sqrt{5^{30}}$ e. $-\sqrt{3^{60}}$

Solution To find the square root of a variable or a number raised to an even power, keep the same base and divide the exponent by 2.

2a. $\sqrt{x^6} = x^{\frac{6}{2}} = x^3$ since $(x^3)^2 = x^6$

2b. $-\sqrt{y^8} = -y^4$

2c. $\sqrt{y^4} = y^2$

2d. $\sqrt{5^{30}} = 5^{15}$

2e. $-\sqrt{3^{60}} = -3^{30}$

NOTE: In Example 2, each radicand is a perfect square.

Practice Problem 2 ***Find the square root of each number.***

a. $\sqrt{x^{10}}$ b. $-\sqrt{y^6}$ c. $\sqrt{7^{20}}$ d. $-\sqrt{11^{50}}$

4 You can also find the square root of a number by using a table (see Appendix Table A–4). However, if the number is not a perfect square, the square root given in the table is a decimal approximation.

Example 3 Find the decimal approximation of each square root.

a. $\sqrt{37}$ **b.** $\sqrt{10}$ **c.** $-\sqrt{96}$

Solution To find the square root of a number by using a table (see Appendix), locate the number in the column labeled n and then find its square root in the column labeled $\sqrt{n}$.

3a. Locate 37 in the column labeled n. Move your finger across this row to the column headed $\sqrt{n}$ and you should see 6.083. Therefore, $\sqrt{37} \doteq 6.083$.

3b. $\sqrt{10} \doteq 3.162$ **3c.** $-\sqrt{96} \doteq -9.798$

NOTE: Remember that every point on the number line is associated with a real number. Until now, the only real numbers we have considered have been **rational numbers.** Rational numbers such as $\sqrt{9}$ and $\sqrt{25}$ are numbers that can be expressed in the form $\frac{a}{b}$, where a and b are integers and $b \neq 0$. Numbers such as $\sqrt{37}$ and $\sqrt{96}$ are also real numbers, but they are called **irrational numbers** (see Figure 12–1). Irrational numbers are numbers that cannot be expressed as a fraction $\frac{a}{b}$, where a and b are integers and $b \neq 0$.

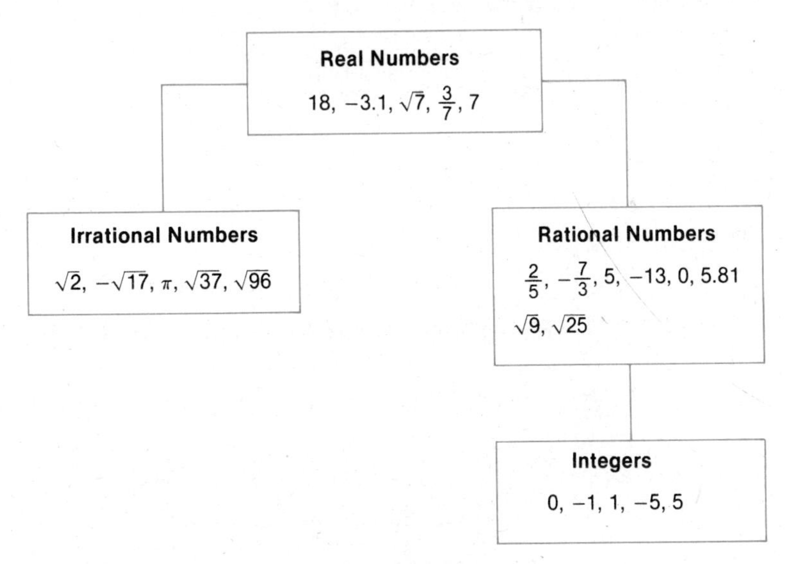

Figure 12-1

Practice Problem 3 ***Find the decimal approximation of each square root.***

a. $\sqrt{79}$ **b.** $-\sqrt{68}$ **c.** $\sqrt{87}$

12.1 Exercises

Find the square root of each number.

1. 4 **2.** 9 **3.** 16 **4.** 25 **5.** 81

6. 100 **7.** 1 **8.** 0 **9.** 49 **10.** 36

Find the square root of each number.

11. $-\sqrt{25}$ **12.** $\sqrt{121}$ **13.** $-\sqrt{1}$ **14.** $-\sqrt{49}$ **15.** $\sqrt{x^{16}}$

16. $\sqrt{y^{10}}$ **17.** $-\sqrt{2^{10}}$ **18.** $\sqrt{3^6}$ **19.** $-\sqrt{x^{100}}$ **20.** $\sqrt{z^8}$

21. $-\sqrt{7^{70}}$ **22.** $\sqrt{x^{10}}$ **23.** $-\sqrt{a^{20}}$ **24.** $-\sqrt{b^{30}}$ **25.** $\sqrt{5^{70}}$

26. $-\sqrt{c^{66}}$

Find the decimal approximation of each square root.

27. $\sqrt{32}$ **28.** $\sqrt{92}$ **29.** $\sqrt{47}$ **30.** $\sqrt{84}$ **31.** $\sqrt{14}$

32. $\sqrt{83}$ **33.** $\sqrt{80}$ **34.** $-\sqrt{87}$ **35.** $-\sqrt{13}$ **36.** $-\sqrt{18}$

37. $-\sqrt{7}$ **38.** $-\sqrt{15}$

Fill in the blanks.

39. Every positive number has one ________________ square root and one ________________ square root.

40. The symbol $-\sqrt{}$ is used to indicate the ________________ of a number.

41. In the expression $\sqrt{18}$, 18 is called the ________________ and $\sqrt{}$ is a ________________ symbol.

42. The $\sqrt{-9}$ is not a real number since there is no real number that you can ________________ to obtain -9.

43. To determine the square root of a given number, you must find a number whose ________________ is that ________________.

44. To find the square root of a number raised to an even power, keep the same ________________ and divide the ________________ by 2.

45. To find the square root of a number by using a table, locate the ________________ in a column labeled n and then find its principal square root in the column labeled ________________.

Answers to Practice Problems **1a.** 7 **b.** -3 **c.** 8 **d.** 5 **e.** -2 **2a.** x^5 **b.** $-y^3$ **c.** 7^{10} **d.** -11^{25} **3a.** 8.888 **b.** -8.246 **c.** 9.327

12.2 Products, Simplification, and Division

5 You can multiply square roots by using the **multiplication rule for square roots.**

Rule for Multiplying Square Roots:

$$\sqrt{a} \cdot \sqrt{b} = \sqrt{a \cdot b},$$

where a and b are not negative.

In other words, the product of two or more square roots is the square root of the product of the radicands.

Example 4 Multiply.

a. $\sqrt{2}\,\sqrt{7}$ **b.** $\sqrt{6}\,\sqrt{x}$ **c.** $\sqrt{8}\,\sqrt{8}$

d. $\sqrt{3}\sqrt{12}$ **e.** $\sqrt{3}\sqrt{5}\sqrt{7}$ **f.** $\sqrt{x}\sqrt{y}$

Solution To multiply square roots having positive radicands, remember that $\sqrt{a}\sqrt{b} = \sqrt{a \cdot b}$.

4a. $\sqrt{2}\sqrt{7} = \sqrt{2 \cdot 7}$
$= \sqrt{14}$

4b. $\sqrt{6}\sqrt{x} = \sqrt{6 \cdot x}$
$= \sqrt{6x}$

4c. $\sqrt{8}\sqrt{8} = \sqrt{8 \cdot 8}$
$= \sqrt{64}$
$= 8$

4d. $\sqrt{3}\sqrt{12} = \sqrt{3 \cdot 12}$
$= \sqrt{36}$
$= 6$

4e. $\sqrt{3}\sqrt{5}\sqrt{7} = \sqrt{3 \cdot 5 \cdot 7}$
$= \sqrt{105}$

4f. $\sqrt{x}\sqrt{y} = \sqrt{x \cdot y}$
$= \sqrt{xy}$

Practice Problem 4 ***Multiply.***

a. $\sqrt{3}\sqrt{2}$ **b.** $\sqrt{2}\sqrt{8}$ **c.** $\sqrt{19}\sqrt{x}$

6 According to the multiplication rule for square roots, $\sqrt{a}\sqrt{b} = \sqrt{a \cdot b}$. However, if you interchange the left and right members of the equation, you can see that

$$\sqrt{a \cdot b} = \sqrt{a}\sqrt{b},$$

which is a formula that can be used to simplify square roots.

Example 5 Simplify each square root.

a. $\sqrt{75}$ **b.** $\sqrt{48x}$ **c.** $\sqrt{x^7}$ **d.** $\sqrt{12x^{19}}$ **e.** $\sqrt{45x^2}$

Solution To simplify a square root, use the formula $\sqrt{a \cdot b} = \sqrt{a}\sqrt{b}$, where a is the largest perfect square factor of the radicand.

5a. $\sqrt{75} = \sqrt{25 \cdot 3}$ 25 is the largest perfect square factor
$= \sqrt{25}\sqrt{3}$
$= 5\sqrt{3}$ *Note:* $5\sqrt{3} = 5 \cdot \sqrt{3}$

5b. $\sqrt{48x} = \sqrt{16 \cdot 3x}$
$= \sqrt{16}\sqrt{3x}$
$= 4\sqrt{3x}$

5c. $\sqrt{x^7} = \sqrt{x^6 \cdot x}$
$= \sqrt{x^6}\sqrt{x}$
$= x^3\sqrt{x}$

5d. $\sqrt{12x^{19}} = \sqrt{4x^{18} \cdot 3x}$
$= \sqrt{4x^{18}}\sqrt{3x}$
$= 2x^9\sqrt{3x}$

5e. $\sqrt{45x^2} = \sqrt{9x^2 \cdot 5}$
$= \sqrt{9x^2}\sqrt{5}$
$= 3x\sqrt{5}$

NOTE: In general, a square root is in its simplest form when 1 is the only perfect square factor of the radicand.

Practice Problem 5 ***Simplify each square root.***

a. $\sqrt{32}$ **b.** $\sqrt{y^{15}}$ **c.** $\sqrt{200x^2}$ **d.** $\sqrt{5x^6}$

7 When you cannot find the largest perfect square factor of the radicand by inspection (trial-and-error method), it may be helpful to factor.

Example 6 Simplify each square root.

a. $\sqrt{567}$ **b.** $\sqrt{450}$ **c.** $\sqrt{875x^2}$

Solution To simplify the square root of a large number, write the prime factorization of the radicand and then use the formula $\sqrt{a \cdot b} = \sqrt{a}\sqrt{b}$, where a is the largest perfect square factor of the radicand.

6a.
$$\begin{aligned}\sqrt{567} &= \sqrt{3^4 \cdot 7}\\ &= \sqrt{3^4}\sqrt{7}\\ &= 3^2\sqrt{7}\\ &= 9\sqrt{7}\end{aligned}$$

$567 = 3 \cdot 3 \cdot 3 \cdot 3 \cdot 7 = 3^4 \cdot 7$

Think: $\sqrt{3^4} = 3^{\frac{4}{2}} = 3^2$

6b.
$$\begin{aligned}\sqrt{450} &= \sqrt{3^2 \cdot 5^2 \cdot 2}\\ &= \sqrt{3^2}\sqrt{5^2}\sqrt{2}\\ &= 3 \cdot 5 \cdot \sqrt{2}\\ &= 15\sqrt{2}\end{aligned}$$

$450 = 2 \cdot 5 \cdot 5 \cdot 3 \cdot 3 = 2 \cdot 5^2 \cdot 3^2$

6c.
$$\begin{aligned}\sqrt{875x^2} &= \sqrt{5^2 \cdot x^2 \cdot 5 \cdot 7}\\ &= \sqrt{5^2}\sqrt{x^2}\sqrt{5}\sqrt{7}\\ &= 5 \cdot x \cdot \sqrt{5} \cdot \sqrt{7}\\ &= 5x\sqrt{35}\end{aligned}$$

$875x^2 = 5^3 \cdot 7 \cdot x^2$

NOTE: Both the formula for simplifying square roots and Appendix Table A–4 can be used to find the square root of large numbers. For instance, in Example 6a,

$$\sqrt{567} = 9\sqrt{7} \doteq 9(2.646) \doteq 23.814.$$

Practice Problem 6 ***Simplify each square root.***

a. $\sqrt{441}$ **b.** $\sqrt{360x^6}$ **c.** $\sqrt{1764}$

8 Sometimes you can simplify a square root *after* you have multiplied.

Example 7 Multiply and simplify.

a. $\sqrt{2}\sqrt{6}$ **b.** $\sqrt{8x}\sqrt{10x^5}$

Solution **7a.**
$$\begin{aligned}\sqrt{2}\sqrt{6} &= \sqrt{2 \cdot 6}\\ &= \sqrt{12}\\ &= \sqrt{4 \cdot 3}\\ &= \sqrt{4}\sqrt{3}\\ &= 2\sqrt{3}\end{aligned}$$

7b.
$$\begin{aligned}\sqrt{8x}\sqrt{10x^5} &= \sqrt{8x \cdot 10x^5}\\ &= \sqrt{80x^6}\\ &= \sqrt{16x^6 \cdot 5}\\ &= \sqrt{16x^6}\sqrt{5}\\ &= 4x^3\sqrt{5}\end{aligned}$$

Practice Problem 7 ***Multiply and simplify.***

a. $\sqrt{2}\sqrt{10}$ **b.** $\sqrt{3x}\sqrt{15x^3}$ **c.** $\sqrt{11}\sqrt{33}$

9 The **division rule for square roots** is similar to the multiplication rule.

Rule for Dividing Two Square Roots:

$$\frac{\sqrt{a}}{\sqrt{b}} = \sqrt{\frac{a}{b}},$$

where a and b are not negative and $b \neq 0$.

In other words, the quotient of two square roots equals the square root of the quotient of the radicands.

Example 8 Divide and simplify.

a. $\frac{\sqrt{30}}{\sqrt{2}}$ **b.** $\frac{\sqrt{48}}{\sqrt{3}}$ **c.** $\frac{\sqrt{49x^9}}{\sqrt{x}}$ **d.** $\frac{\sqrt{160x^6}}{\sqrt{20x}}$

Solution To find the quotient of two square roots, use the formula $\frac{\sqrt{a}}{\sqrt{b}} = \sqrt{\frac{a}{b}}$.

8a. $\frac{\sqrt{30}}{\sqrt{2}} = \sqrt{\frac{30}{2}} = \sqrt{15}$

8b. $\frac{\sqrt{48}}{\sqrt{3}} = \sqrt{\frac{48}{3}} = \sqrt{16} = 4$

8c. $\frac{\sqrt{49x^9}}{\sqrt{x}} = \sqrt{\frac{49x^9}{x}} = \sqrt{49x^8} = 7x^4$

8d. $\frac{\sqrt{160x^6}}{\sqrt{20x}} = \sqrt{\frac{160x^6}{20x}} = \sqrt{8x^5} = \sqrt{4x^4 \cdot 2x} = 2x^2\sqrt{2x}$

Practice Problem 8 ***Divide and simplify.***

a. $\frac{\sqrt{28}}{\sqrt{4}}$ **b.** $\frac{\sqrt{80}}{\sqrt{5}}$ **c.** $\frac{\sqrt{75x^3}}{\sqrt{3x}}$

10 The division rule for square roots can also be used to determine the square root of a fraction.

Rule for Finding the Square Root of a Fraction:

$$\sqrt{\frac{a}{b}} = \frac{\sqrt{a}}{\sqrt{b}},$$

where a and b are not negative and $b \neq 0$.

Example 9 Simplify.

a. $\sqrt{\frac{25}{4}}$ **b.** $\sqrt{\frac{3}{16}}$ **c.** $\sqrt{\frac{8}{x^4}}$ **d.** $\sqrt{\frac{x^8}{y^6}}$

Solution To find the square root of a fraction, write the square root of the numerator over the square root of the denominator.

9a. $\sqrt{\frac{25}{4}} = \frac{\sqrt{25}}{\sqrt{4}} = \frac{5}{2}$

9b. $\sqrt{\frac{3}{16}} = \frac{\sqrt{3}}{\sqrt{16}} = \frac{\sqrt{3}}{4}$

9c. $\sqrt{\frac{8}{x^4}} = \frac{\sqrt{8}}{\sqrt{x^4}} = \frac{\sqrt{4}\,\sqrt{2}}{x^2} = \frac{2\sqrt{2}}{x^2}$

9d. $\sqrt{\frac{x^8}{y^6}} = \frac{\sqrt{x^8}}{\sqrt{y^6}} = \frac{x^4}{y^3}$

NOTE: If the radicand is a fraction, the square root is not in its simplest form.

Practice Problem 9 ***Simplify.***

a. $\sqrt{\frac{16}{9}}$ b. $\sqrt{\frac{15}{81}}$ c. $\sqrt{\frac{32x^5}{2x}}$

11 Sometimes when you find the square root of a fraction, there will be a square root left in the denominator. When this occurs, you can eliminate it by using the multiplication rule for fractions.

Example 10

$$\sqrt{\frac{1}{3}} = \frac{\sqrt{1}}{\sqrt{3}} = \frac{1}{\sqrt{3}} = \frac{1 \cdot \sqrt{3}}{\sqrt{3} \cdot \sqrt{3}} = \frac{\sqrt{3}}{\sqrt{9}} = \frac{\sqrt{3}}{3}$$

Multiplying by $\sqrt{3}$ will make the radicand in the denominator a perfect square.

12 Based on the results in Example 10, to eliminate a square root from the denominator of a fraction, do the following:

1. Multiply the numerator and denominator of the fraction by the square root appearing in the denominator, or choose a multiplier that will make the radicand in the denominator a perfect square.
2. Simplify the square roots.

This process of eliminating a square root from the denominator of a fraction is called **rationalizing the denominator.**

NOTE: When a radical appears in the denominator of a fraction, the expression is not in its simplest form.

Example 11 Simplify.

a. $\frac{3}{\sqrt{5}}$ b. $\frac{\sqrt{2}}{\sqrt{7}}$ c. $\sqrt{\frac{9}{6}}$ d. $\frac{1}{\sqrt{20}}$

Solution To rationalize the denominator of a fraction, first multiply the numerator and denominator by a square root that will make the radicand of the denominator a perfect square. Then, simplify the square roots.

11a. $\frac{3}{\sqrt{5}} = \frac{3 \cdot \sqrt{5}}{\sqrt{5} \cdot \sqrt{5}} = \frac{3\sqrt{5}}{\sqrt{25}} = \frac{3\sqrt{5}}{5}$

11b. $\frac{\sqrt{2}}{\sqrt{7}} = \frac{\sqrt{2} \cdot \sqrt{7}}{\sqrt{7} \cdot \sqrt{7}} = \frac{\sqrt{14}}{\sqrt{49}} = \frac{\sqrt{14}}{7}$

11c. $\sqrt{\frac{9}{6}} = \frac{\sqrt{9}}{\sqrt{6}} = \frac{\sqrt{9}\sqrt{6}}{\sqrt{6}\sqrt{6}} = \frac{\overset{1}{\cancel{3}}\sqrt{6}}{\underset{2}{\cancel{6}}} = \frac{\sqrt{6}}{2}$

Alternate method

$\sqrt{\frac{9}{6}} = \sqrt{\frac{3}{2}}$ Reduce

$= \frac{\sqrt{3}}{\sqrt{2}} = \frac{\sqrt{3}\sqrt{2}}{\sqrt{2}\sqrt{2}} = \frac{\sqrt{6}}{2}$

11d. $\frac{1}{\sqrt{20}} = \frac{1 \cdot \sqrt{5}^*}{\sqrt{20} \cdot \sqrt{5}} = \frac{\sqrt{5}}{\sqrt{100}} = \frac{\sqrt{5}}{10}$

*You could have multiplied by $\sqrt{20}$. However, multiplying by $\sqrt{5}$ will make your computations easier.

NOTE: You can find the square root of a fraction by using the quotient rule for square roots and Appendix Table A–4. For instance, in Example 11c,

$$\sqrt{\frac{9}{6}} = \frac{\sqrt{6}}{2} \doteq \frac{2.449}{2} \doteq 1.2245.$$

Practice Problem 10 ***Simplify.***

a. $\frac{2}{\sqrt{3}}$ b. $\frac{\sqrt{2}}{\sqrt{5}}$ c. $\sqrt{\frac{2}{12}}$ d. $\frac{\sqrt{8}}{\sqrt{5}}$

13 At this point it may be helpful to review what is meant when we say that a square root is in its simplest form. An expression containing radicals is in its simplest form:

1. if the number 1 is the only perfect square factor of the radicand;

2. if the radicand does not contain a fraction; or
3. if the denominator does not contain a radical.

Example 12 Simplify.

a. $\sqrt{18x^3}$ b. $\sqrt{0.12}$ c. $\dfrac{\sqrt{20}}{\sqrt{3}}$

Solution 12a.

$$\sqrt{18x^3} = \sqrt{9x^2 \cdot 2x} = \sqrt{9x^2}\sqrt{2x} = 3x\sqrt{2x}$$

12b.

$$\sqrt{0.12} = \sqrt{\frac{12}{100}} \quad \text{Express 0.12 as a fraction}$$

$$= \frac{\sqrt{12}}{\sqrt{100}}$$

$$= \frac{2\sqrt{3}}{10} \quad \sqrt{12} = \sqrt{4}\sqrt{3} = 2\sqrt{3}$$

$$= \frac{\sqrt{3}}{5}$$

12c.

$$\frac{\sqrt{20}}{\sqrt{3}} = \frac{\sqrt{20}\sqrt{3}}{\sqrt{3}\sqrt{3}}$$

$$= \frac{\sqrt{60}}{3}$$

$$= \frac{2\sqrt{15}}{3} \quad \sqrt{60} = \sqrt{4}\sqrt{15} = 2\sqrt{15}$$

Practice Problem 11 ***Simplify.***

a. $\sqrt{150x^6}$ b. $\sqrt{\dfrac{1}{12}}$ c. $\dfrac{\sqrt{12}}{\sqrt{7}}$

12.2 Exercises

Simplify.

1. $\sqrt{8}$
2. $-\sqrt{32}$
3. $\sqrt{40}$
4. $\sqrt{20}$
5. $-\sqrt{125}$
6. $\sqrt{300}$
7. $\sqrt{28}$
8. $-\sqrt{75}$
9. $\sqrt{x^5}$
10. $-\sqrt{y^7}$
11. $\sqrt{x^{33}}$
12. $-\sqrt{x^{17}}$
13. $\sqrt{54x^4}$
14. $\sqrt{x^{11}}$
15. $\sqrt{48}$
16. $-\sqrt{27x}$
17. $\sqrt{50}$
18. $\sqrt{80}$
19. $\sqrt{72x^4}$
20. $\sqrt{32x^6}$
21. $\sqrt{98}$
22. $\sqrt{60}$
23. $\sqrt{x^{19}}$
24. $\sqrt{x^{21}}$
25. $-\sqrt{x^7}$
26. $-\sqrt{x^{13}}$
27. $-\sqrt{x^{31}}$
28. $-\sqrt{x^{29}}$
29. $\sqrt{25x^5}$
30. $\sqrt{36x^3}$
31. $\sqrt{16x}$
32. $\sqrt{9x}$
33. $\sqrt{c^7}$
34. $\sqrt{x^9}$
35. $-\sqrt{90}$
36. $-\sqrt{20}$
37. $\sqrt{100}$
38. $\sqrt{486}$
39. $-\sqrt{490}$
40. $-\sqrt{425}$
41. $\sqrt{588x}$
42. $\sqrt{720x}$
43. $-\sqrt{2700x^8}$
44. $-\sqrt{7500x^6}$

Multiply and simplify.

45. $\sqrt{3}\sqrt{5}$
46. $\sqrt{5}\sqrt{6}$
47. $\sqrt{x}\sqrt{13}$
48. $\sqrt{3}\sqrt{6}$
49. $\sqrt{19}\sqrt{19}$
50. $\sqrt{2}\sqrt{8}$
51. $\sqrt{7}\sqrt{112}$
52. $\sqrt{8x}\sqrt{6x}$
53. $\sqrt{10}\sqrt{10}$
54. $\sqrt{xy}\sqrt{x^3y}$
55. $\sqrt{2}\sqrt{27x^3}$
56. $\sqrt{6x}\sqrt{6x^9}$
57. $\sqrt{20xy}\sqrt{10x^{11}}$
58. $\sqrt{2xy}\sqrt{250x^2y^5}$

Divide and simplify.

59. $\frac{\sqrt{12}}{\sqrt{3}}$
60. $\frac{\sqrt{27}}{\sqrt{3}}$
61. $\frac{\sqrt{128}}{\sqrt{2}}$
62. $\frac{\sqrt{80x}}{\sqrt{5x}}$
63. $\frac{\sqrt{75x^7}}{\sqrt{3x^2}}$
64. $\frac{\sqrt{64x^{15}}}{\sqrt{16x^{11}}}$
65. $\frac{\sqrt{75x^{13}}}{\sqrt{x^4}}$
66. $\frac{\sqrt{x^8y^{11}}}{\sqrt{x^2y}}$
67. $\frac{\sqrt{250x^{11}}}{\sqrt{5x}}$
68. $\frac{\sqrt{350x^{13}}}{\sqrt{2x}}$

Simplify.

69. $\sqrt{\frac{4}{9}}$
70. $\sqrt{\frac{16}{81}}$
71. $\sqrt{\frac{36}{49}}$
72. $\sqrt{\frac{1}{100}}$
73. $\sqrt{\frac{5}{9}}$
74. $\sqrt{\frac{1}{16}}$
75. $\sqrt{\frac{11}{25}}$
76. $\sqrt{\frac{3}{49}}$

Rationalize the denominator.

77. $\frac{5}{\sqrt{2}}$
78. $\frac{5}{\sqrt{3}}$
79. $\frac{1}{\sqrt{6}}$
80. $\frac{1}{\sqrt{14}}$
81. $\frac{\sqrt{2}}{\sqrt{3}}$
82. $\frac{\sqrt{5}}{\sqrt{2}}$
83. $\frac{\sqrt{3}}{\sqrt{5}}$
84. $\frac{\sqrt{5}}{\sqrt{3}}$

Rationalize the denominator.

85. $\frac{1}{\sqrt{2}}$
86. $\frac{13}{\sqrt{5}}$
87. $\frac{\sqrt{2}}{\sqrt{8}}$
88. $\frac{1}{\sqrt{7}}$
89. $\frac{1}{\sqrt{x}}$
90. $\frac{\sqrt{5}}{\sqrt{18}}$
91. $\frac{8x}{\sqrt{6}}$
92. $\frac{\sqrt{5}}{\sqrt{20}}$
93. $\frac{\sqrt{5}}{\sqrt{8x}}$
94. $\frac{\sqrt{3}}{\sqrt{21}}$
95. $\frac{\sqrt{8x^5}}{\sqrt{3x}}$
96. $\frac{\sqrt{x^5}}{\sqrt{32x}}$

Fill in the blanks.

97. The product of two or more square roots is the _______________ of the _______________ of the radicands.

98. To simplify a square root, use the formula $\sqrt{a \cdot b} = \sqrt{a}\sqrt{b}$, where a is the largest _______________ of the radicand.

99. To simplify the square root of a large number, first write the prime factorization of the _______________. Then, use the formula $\sqrt{a \cdot b} = \sqrt{a}\sqrt{b}$, where a is the largest _______________ of the radicand.

100. The quotient of two square roots is the _______________ of the quotient of the _______________.

101. To find the square root of a fraction, write the square root of the _______________ over the square root of the _______________.

102. To rationalize the denominator of a fraction, first _______________ the numerator and the denominator by a square root that will make the _______________ in the denominator a _______________. Then, simplify the square roots.

Simplify.

103. $\sqrt{\frac{4}{25}}$ **104.** $\sqrt{\frac{9}{16}}$ **105.** $\sqrt{\frac{3}{4}}$ **106.** $\sqrt{\frac{81}{100}}$

107. $\sqrt{\frac{9}{12}}$ **108.** $\sqrt{\frac{3}{5}}$ **109.** $\sqrt{0.48}$ **110.** $\sqrt{90x^8y^6}$

111. $\frac{\sqrt{10}}{\sqrt{80}}$ **112.** $\frac{\sqrt{15}}{\sqrt{75}}$ **113.** $\frac{\sqrt{8}}{\sqrt{32}}$ **114.** $\sqrt{726x^4y^2z^5}$

115. $\sqrt{10}\sqrt{\frac{7}{40}}$ **116.** $\sqrt{\frac{9}{50}}\sqrt{\frac{2}{3}}$ **117.** $\sqrt{x^{15}y^{20}z^3}$

Answers to Practice Problems **4a.** $\sqrt{6}$ **b.** 4 **c.** $\sqrt{19x}$ **5a.** $4\sqrt{2}$ **b.** $y^7\sqrt{y}$ **c.** $10x\sqrt{2}$ **d.** $x^3\sqrt{5}$ **6a.** 21 **b.** $6x^3\sqrt{10}$ **c.** 42 **7a.** $2\sqrt{5}$ **b.** $3x^2\sqrt{5}$ **c.** $11\sqrt{3}$ **8a.** $\sqrt{7}$ **b.** 4 **c.** $5x$ **9a.** $\frac{4}{3}$ **b.** $\frac{\sqrt{15}}{9}$ **c.** $4x^2$ **10a.** $\frac{2\sqrt{3}}{3}$ **b.** $\frac{\sqrt{10}}{5}$ **c.** $\frac{\sqrt{6}}{6}$ **d.** $\frac{2\sqrt{10}}{5}$ **11a.** $5x^3\sqrt{6}$ **b.** $\frac{\sqrt{3}}{6}$ **c.** $\frac{2\sqrt{21}}{7}$

12.3 Addition and Subtraction

14 Square roots are added and subtracted in such the same manner that polynomials are added and subtracted.

> **To Add or Subtract Square Roots:**
>
> **1.** Simplify each square root.
>
> **2.** Use the distributive law to combine like square roots.

Remember, you can only add and subtract like square roots. **Like square roots** are square roots having the same radicand.

Example 13 Perform the indicated operation.

a. $2\sqrt{3} + 5\sqrt{3}$ **b.** $7\sqrt{5} - 9\sqrt{5}$

c. $\sqrt{2} + \sqrt{3}$ **d.** $5\sqrt{7} - 8\sqrt{7} + 14\sqrt{7}$

Solution To add or subtract like radicals, use the distributive law $ab + cd = (a + c)b$.

13a. $2\sqrt{3} + 5\sqrt{3} = (2 + 5)\sqrt{3}$
$= 7\sqrt{3}$

13b. $7\sqrt{5} - 9\sqrt{5} = (7 - 9)\sqrt{5}$
$= -2\sqrt{5}$

13c. $\sqrt{2} + \sqrt{3}$ cannot be simplified since $\sqrt{2}$ and $\sqrt{3}$ are not like square roots.

13d. $5\sqrt{7} - 8\sqrt{7} + 14\sqrt{7} = (5 - 8 + 14)\sqrt{7}$
$= 11\sqrt{7}$

Practice Problem 12 ***Perform the indicated operation.***

a. $3\sqrt{11} + 5\sqrt{11}$ **b.** $\sqrt{13} - 6\sqrt{13}$ **c.** $\sqrt{2} - 3\sqrt{2} + 2\sqrt{2}$

Example 14 Perform the indicated operation.

a. $5\sqrt{2} + \sqrt{8}$ **b.** $\sqrt{20} - \sqrt{125}$ **c.** $\sqrt{7} + \sqrt{\frac{1}{7}}$

d. $\sqrt{12} + 4\sqrt{75}$ **e.** $\sqrt{27} - \sqrt{48} + 2\sqrt{3} + \sqrt{6}$

Solution To add or subtract unlike square roots, first simplify each square root. Then, use the distributive law to combine like square roots.

14a.

$$\begin{aligned} 5\sqrt{2} + \sqrt{8} &= 5\sqrt{2} + \sqrt{4}\,\sqrt{2} && \text{Multiplication rule} \\ &= 5\sqrt{2} + 2\sqrt{2} \\ &= (5 + 2)\sqrt{2} \\ &= 7\sqrt{2} \end{aligned}$$

14b.

$$\begin{aligned} \sqrt{20} - \sqrt{125} &= \sqrt{4}\,\sqrt{5} - \sqrt{25}\,\sqrt{5} \\ &= 2\sqrt{5} - 5\sqrt{5} \\ &= (2 - 5)\sqrt{5} \\ &= -3\sqrt{5} \end{aligned}$$

14c.

$$\begin{aligned} \sqrt{7} + \sqrt{\frac{1}{7}} &= \sqrt{7} + \frac{1}{\sqrt{7}} && \text{Division rule} \\ &= \sqrt{7} + \frac{1 \cdot \sqrt{7}}{\sqrt{7} \cdot \sqrt{7}} && \text{Rationalize denominator} \\ &= \sqrt{7} + \frac{\sqrt{7}}{7} \\ &= \sqrt{7} + \frac{1}{7}\sqrt{7} \\ &= \left(1 + \frac{1}{7}\right)\sqrt{7} \\ &= \frac{8}{7}\sqrt{7} \end{aligned}$$

14d.

$$\begin{aligned} \sqrt{12} + 4\sqrt{75} &= \sqrt{4}\,\sqrt{3} + 4 \cdot \sqrt{25}\,\sqrt{3} \\ &= 2\sqrt{3} + 4 \cdot 5 \cdot \sqrt{3} \\ &= 2\sqrt{3} + 20\sqrt{3} \\ &= (2 + 20)\sqrt{3} \\ &= 22\sqrt{3} \end{aligned}$$

14e.

$$\begin{aligned} \sqrt{27} - \sqrt{48} + 2\sqrt{3} + \sqrt{6} &= \sqrt{9}\,\sqrt{3} - \sqrt{16}\,\sqrt{3} + 2\sqrt{3} + \sqrt{6} \\ &= 3\sqrt{3} - 4\sqrt{3} + 2\sqrt{3} + \sqrt{6} \\ &= (3 - 4 + 2)\sqrt{3} + \sqrt{6} \\ &= \sqrt{3} + \sqrt{6} \end{aligned}$$

Practice Problem 13 ***Perform the indicated operations.***

a. $5\sqrt{12} - \sqrt{27}$ **b.** $\sqrt{98} + \sqrt{50}$ **c.** $\sqrt{\frac{5}{3}} + \sqrt{\frac{3}{5}}$ **d.** $\sqrt{18} - \sqrt{27} + 3\sqrt{32}$

12.3 Exercises

Perform the indicated operations.

1. $4\sqrt{2} + 5\sqrt{2}$
2. $3\sqrt{5} + 4\sqrt{5}$
3. $6\sqrt{7} - 8\sqrt{7}$
4. $7\sqrt{11} - \sqrt{11}$
5. $7\sqrt{x} + 5\sqrt{x}$
6. $-\sqrt{3} - 5\sqrt{3}$
7. $8\sqrt{13} + 3\sqrt{13}$
8. $\sqrt{7} + \sqrt{11}$
9. $4\sqrt{2} - 8\sqrt{2}$
10. $\sqrt{18} + \sqrt{32}$
11. $\sqrt{27} - \sqrt{3}$
12. $\sqrt{20} + \sqrt{45}$
13. $\sqrt{72} - \sqrt{98}$
14. $\sqrt{24} - \sqrt{54}$
15. $-\sqrt{3} + \sqrt{12}$
16. $\sqrt{48} + 2\sqrt{27}$
17. $3\sqrt{80} - \sqrt{45}$
18. $4\sqrt{50} - \sqrt{32}$
19. $\sqrt{18} + 3\sqrt{48} - 7\sqrt{28}$
20. $-\sqrt{18} - 3\sqrt{32} - \sqrt{50}$

Fill in the blanks.

21. Like square roots are square roots having the same __________.
22. To add or subtract square roots, first __________ each square root. Then, use the __________ to combine the like square roots.
23. You cannot add $\sqrt{5} + \sqrt{7}$ since $\sqrt{5}$ and $\sqrt{7}$ are not __________.

Perform the indicated operations.

24. $\sqrt{8} + 3\sqrt{18} - 5\sqrt{2}$
25. $\sqrt{2} + \sqrt{3} + \sqrt{5}$
26. $3\sqrt{48} - 2\sqrt{27} + \sqrt{12}$
27. $\sqrt{12} + \sqrt{27} + \sqrt{48}$
28. $\sqrt{\frac{1}{3}} + \sqrt{27}$
29. $\sqrt{\frac{1}{2}} - \sqrt{2}$
30. $\sqrt{\frac{1}{12}} + \sqrt{\frac{1}{27}}$
31. $\sqrt{\frac{3}{7}} + \sqrt{\frac{7}{3}}$
32. $\frac{3}{4}\sqrt{6} + \frac{1}{4}\sqrt{6}$
33. $3\sqrt{\frac{1}{6}} + 5\sqrt{\frac{3}{2}}$
34. $\sqrt{48} - \sqrt{\frac{1}{3}}$
35. $\sqrt{\frac{25}{2}} - \sqrt{\frac{9}{2}} + \sqrt{12}$
36. $\sqrt{x^3} - 3\sqrt{4x^3}$
37. $\sqrt{16a} - \sqrt{9a} + \sqrt{36a}$
38. $\sqrt{8x^2} - 4x\sqrt{2}$
39. $\sqrt{40x^4} + x^2\sqrt{90}$
40. $\sqrt{112} + \sqrt{252}$
41. $3\sqrt{275} - 4\sqrt{396} + \sqrt{99}$

Answers to Practice Problems **12a.** $8\sqrt{11}$ **b.** $-5\sqrt{13}$ **c.** 0 **13a.** $7\sqrt{3}$ **b.** $12\sqrt{2}$ **c.** $\frac{8}{15}\sqrt{15}$ **d.** $15\sqrt{2} - 3\sqrt{3}$

12.4 Quadratic Equations

15

A **quadratic** or **second-degree equation** in one variable is an equation that can be written in the form

$$ax^2 + bx + c = 0$$

where a, b, and c are real numbers and $a \neq 0$. A quadratic equation written in this form is said to be in **standard form.**

Example 15 Write each equation in standard form.

a. $3x^2 + 5x = 4$ **b.** $7x = 8 - 2x^2$

Solution

To write a quadratic equation in standard form, use the rules for solving linear equations to make the right side of the equation equal to zero.

15a. $3x^2 + 5x = 4$

$3x^2 + 5x - 4 = 0$ Add -4 to both sides

$a = 3$, $b = 5$, and $c = -4$

15b. $7x = 8 - 2x^2$

$2x^2 + 7x = 8$ Add $2x^2$ to both sides

$2x^2 + 7x - 8 = 0$ Add -8 to both sides

$a = 2$, $b = 7$, and $c = -8$

Practice Problem 14

Write each equation in standard form.

a. $5x^2 = 6x + 4$ **b.** $-2x = -6x^2 + 2$ **c.** $6x + 7x^2 = x^2 - 9 + 2x$

16

When $b = 0$, the equation $ax^2 + bx + c = 0$ becomes $ax^2 + 0x + c = 0$ or $ax^2 + c = 0$. To solve this type of equation, use the following procedure:

To Solve the Equation $ax^2 + c = 0$:

1. Rewrite the equation (if necessary) so that x^2 is on one side of the equation and a constant is on the opposite side.
2. Take the square root of both sides. Remember that every positive real number has both a positive and a negative square root.
3. Check both solutions.

This method of solving a quadratic equation is called the **square root method.**

Example 16

Solve.

a. $x^2 = 9$ **b.** $x^2 = 20$ **c.** $x^2 + 16 = 0$

d. $4x^2 - 9 = 0$ **e.** $9x^2 - 7 = 25$ **f.** $2x^2 - 3.5 = -0.5$

Solution

To solve a quadratic equation containing no x term, first write the equation in the form $x^2 = K$. Then, take the square root of both sides and simplify.

16a. $x^2 = 9$

$x = \pm\sqrt{9}$ *Note:* $x = \pm\sqrt{9}$ is shorthand for the statement

$x = \pm 3$ $x = \sqrt{9}$ or $x = -\sqrt{9}$

The solutions are 3 and -3

Check:

For 3	*For −3*
$x^2 = 9$	$x^2 = 9$
$3^2 = 9$	$(-3)^2 = 9$
$9 = 9$	$9 = 9$

16b. $x^2 = 20$

$x = \pm\sqrt{20}$

$x = \pm 2\sqrt{5}$ *Think:* $\sqrt{20} = \sqrt{4}\sqrt{5} = 2\sqrt{5}$

The solutions are $2\sqrt{5}$ and $-2\sqrt{5}$

Check:

For $2\sqrt{5}$

$x^2 = 20$

$(2\sqrt{5})^2 = 20$

$(2)^2(\sqrt{5})^2 = 20$

$20 = 20$

For $-2\sqrt{5}$

$x^2 = 20$

$(-2\sqrt{5})^2 = 20$

$(-2)^2(\sqrt{5})^2 = 20$

$20 = 20$

16c. $x^2 + 16 = 0$

$x^2 = -16$ Add -16 to both sides

$x = \pm\sqrt{-16}$

This equation has no real solution since there is no real number that we can square and obtain -16.

16d. $4x^2 - 9 = 0$

$4x^2 = 9$ Add 9 to both sides

$x^2 = \frac{9}{4}$ Divide both sides by 4

$x = \pm\sqrt{\frac{9}{4}}$

$x = \pm\frac{3}{2}$

The solutions are $\frac{3}{2}$ and $-\frac{3}{2}$

Check:

For $\frac{3}{2}$

$4x^2 - 9 = 0$

$4\left(\frac{3}{2}\right)^2 - 9 = 0$

$4\left(\frac{9}{4}\right) - 9 = 0$

$9 - 9 = 0$

$0 = 0$

For $-\frac{3}{2}$

$4x^2 - 9 = 0$

$4\left(-\frac{3}{2}\right)^2 - 9 = 0$

$4\left(\frac{9}{4}\right) - 9 = 0$

$9 - 9 = 0$

$0 = 0$

16e. $9x^2 - 7 = 25$

$9x^2 = 32$ Add 7 to both sides

$x^2 = \frac{32}{9}$ Divide both sides by 9

$x = \pm\sqrt{\frac{32}{9}}$

$x = \pm\frac{4\sqrt{2}}{3}$ Simplify the square root

The solutions are $\frac{4\sqrt{2}}{3}$ and $-\frac{4\sqrt{2}}{3}$.

Check:

For $\frac{4\sqrt{2}}{3}$

$$\begin{aligned} 9x^2 - 7 &= 25 \\ 9\left(\frac{4\sqrt{2}}{3}\right)^2 - 7 &= 25 \\ 9\left(\frac{32}{9}\right) - 7 &= 25 \\ 32 - 7 &= 25 \\ 25 &= 25 \end{aligned}$$

For $-\frac{4\sqrt{2}}{3}$

$$\begin{aligned} 9x^2 - 7 &= 25 \\ 9\left(-\frac{4\sqrt{2}}{3}\right)^2 - 7 &= 25 \\ 9\left(\frac{32}{9}\right) - 7 &= 25 \\ 32 - 7 &= 25 \\ 25 &= 25 \end{aligned}$$

16f.

$$\begin{aligned} 2x^2 - 3.5 &= -0.5 \\ 2x^2 &= 3 && \text{Add 3.5 to both sides} \\ x^2 &= \frac{3}{2} && \text{Divide both sides by 2} \\ x &= \pm\sqrt{\frac{3}{2}} \\ x &= \pm\frac{\sqrt{6}}{2} && \text{Simplify the square root} \end{aligned}$$

The solutions are $\frac{\sqrt{6}}{2}$ and $-\frac{\sqrt{6}}{3}$.

Check:

For $\frac{\sqrt{6}}{2}$

$$\begin{aligned} 2x^2 - 3.5 &= -0.5 \\ 2\left(\frac{\sqrt{6}}{2}\right)^2 - 3.5 &= -0.5 \\ 2\left(\frac{6}{4}\right) - 3.5 &= -0.5 \\ 3 - 3.5 &= -0.5 \\ -0.5 &= -0.5 \end{aligned}$$

For $-\frac{\sqrt{6}}{2}$

$$\begin{aligned} 2x^2 - 3.5 &= -0.5 \\ 2\left(-\frac{\sqrt{6}}{2}\right)^2 - 3.5 &= -0.5 \\ 2\left(\frac{6}{4}\right) - 3.5 &= -0.5 \\ 3 - 3.5 &= -0.5 \\ -0.5 &= -0.5 \end{aligned}$$

NOTE: The equations in Example 16 are called **pure quadratic equations** since the x term is missing.

Practice Problem 15 ***Solve and check.***

a. $x^2 = 25$ **b.** $2x^2 = 16$ **c.** $5x^2 - 7 = 5$

17 The square root method for solving pure quadratic equations can also be used to solve equations in which one side of the equation is a binomial squared and the other side is a constant.

Example 17 Solve and check.

a. $(x + 4)^2 = 5$ **b.** $(3x + 2)^2 = 25$

Solution **17a.**

$$\begin{aligned} (x + 4)^2 &= 5 \\ x + 4 &= \pm\sqrt{5} && \text{Take the square root of both sides} \end{aligned}$$

$x + 4 = \sqrt{5}$ or $x + 4 = -\sqrt{5}$

$x = -4 + \sqrt{5}$ $\qquad$ $x = -4 - \sqrt{5}$

The solutions are $-4 + \sqrt{5}$ and $-4 - \sqrt{5}$.

Check:

For $-4 + \sqrt{5}$

$$\begin{aligned} (x + 4)^2 &= 5 \\ (-4 + \sqrt{5} + 4)^2 &= 5 \\ (\sqrt{5})^2 &= 5 \\ 5 &= 5 \end{aligned}$$

For $-4 - \sqrt{5}$

$$\begin{aligned} (x + 4)^2 &= 5 \\ (-4 - \sqrt{5} + 4)^2 &= 5 \\ (-\sqrt{5})^2 &= 5 \\ 5 &= 5 \end{aligned}$$

17b. $(3x + 2)^2 = 25$

$3x + 2 = \pm 5$ Take the square root of both sides

$$\begin{aligned} 3x + 2 &= 5 \quad \text{or} \quad & 3x + 2 &= -5 \\ 3x &= 3 & 3x &= -7 \\ x &= 1 & x &= -\frac{7}{3} \end{aligned}$$

Check:

For 1

$$\begin{aligned} (3x + 2)^2 &= 25 \\ (3 \cdot 1 + 2)^2 &= 25 \\ (5)^2 &= 25 \\ 25 &= 25 \end{aligned}$$

For $-\frac{7}{3}$

$$\begin{aligned} (3x + 2)^2 &= 25 \\ \left(3\left[-\frac{7}{3}\right] + 2\right)^2 &= 25 \\ (-7 + 2)^2 &= 25 \\ (-5)^2 &= 25 \\ 25 &= 25 \end{aligned}$$

Practice Problem 16 **Solve and check.**

a. $(x + 3)^2 = 36$ **b.** $(x + 1)^2 = 6$

18 One interesting application of quadratic equations involves right triangles. A **right triangle** is a triangle that has a right angle (square corner) (see Figure 12–2). In a right triangle, the side opposite the right angle is called the **hypotenuse,** while the other two sides (a and b) are called the **legs.** It will be helpful to remember that *the sum of the squares of the legs of a right triangle is equal to the square of the hypotenuse* ($a^2 + b^2 = c^2$). This relationship concerning the sides of a right triangle is called the **Pythagorean Theorem.** This theorem is used to find the length of one side of a right triangle when the lengths of the other two sides are known.

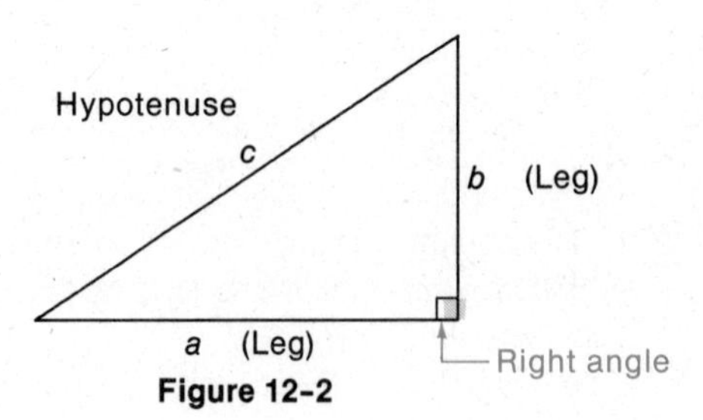

Figure 12-2

Example 18 Solve.

a. Find the length of the hypotenuse of a right triangle whose legs are 3 meters and 4 meters.

b. Find the length of the leg b in the right triangle shown in Figure 12–3.

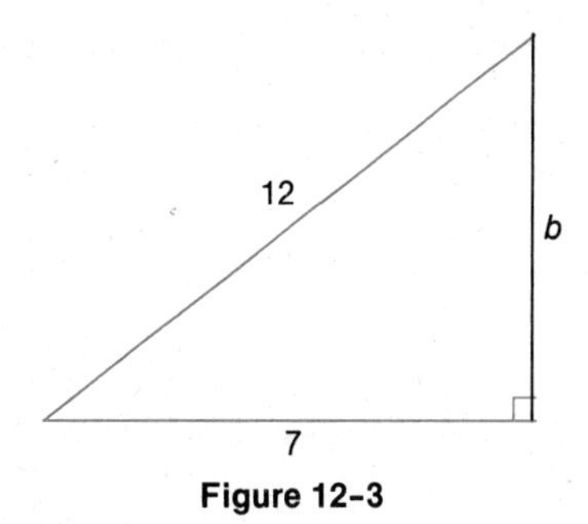

Figure 12-3

c. A ladder that is 15 feet long leans against a house and reaches a point on the house that is 10 feet above the ground. How far is the foot of the ladder from the side of the house?

Solution **18a.** Make a sketch (see Figure 12–4) and then use the Pythagorean Theorem to solve.

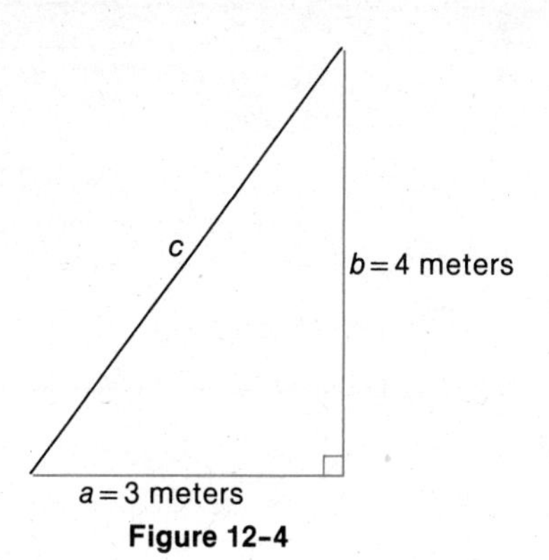

Figure 12-4

$$a^2 + b^2 = c^2$$
$$3^2 + 4^2 = c^2$$
$$9 + 16 = c^2$$
$$25 = c^2 \quad \text{Take the square root of both sides}$$
$$c = 5$$

Only the positive square root is considered since the length of a side of a triangle cannot be negative. Therefore, the hypotenuse is 5 meters.

18b.

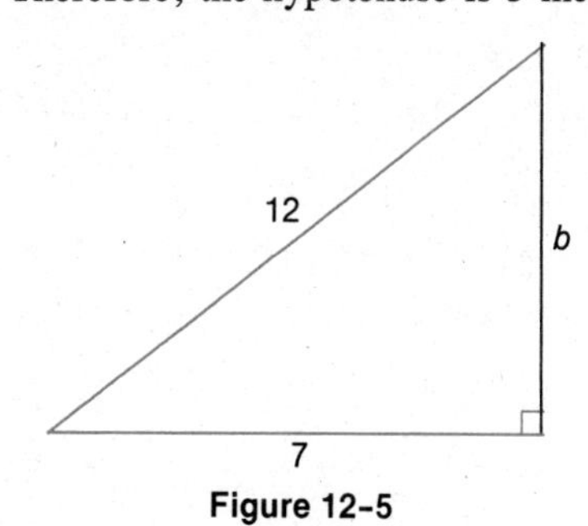

Figure 12-5

$$a^2 + b^2 = c^2$$
$$7^2 + b^2 = 12^2$$
$$49 + b^2 = 144$$
$$b^2 = 95$$
$$b \doteq 9.747 \quad \text{See Appendix Table A–4}$$

18c. Sketch the house and ladder (see Figure 12–6).

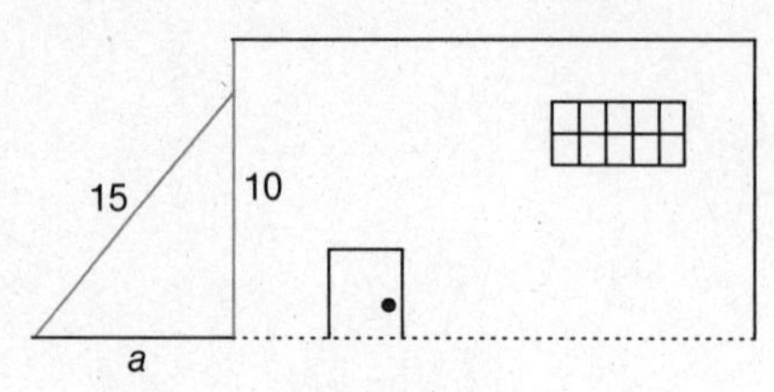

Figure 12-6

Step 1: a = distance (in feet) from foot of the ladder to the house

Step 2: $a^2 + b^2 = c^2$

$a^2 + 10^2 = 15^2$

Step 3: $a^2 + 10^2 = 15^2$

$a^2 + 100 = 225$

$a^2 = 125$

$a = \sqrt{125}$

$a = 5\sqrt{5}$ *Think:* $\sqrt{125} = \sqrt{25}\,\sqrt{5} = 5\sqrt{5}$

$a \doteq 5(2.236)$ See Appendix Table A–4

$a \doteq 11.18$

The foot of the ladder is approximately 11.18 feet from the house.

Practice Problem 17 **Solve.**

a. Find the length of the leg a in the right triangle in the figure shown below.

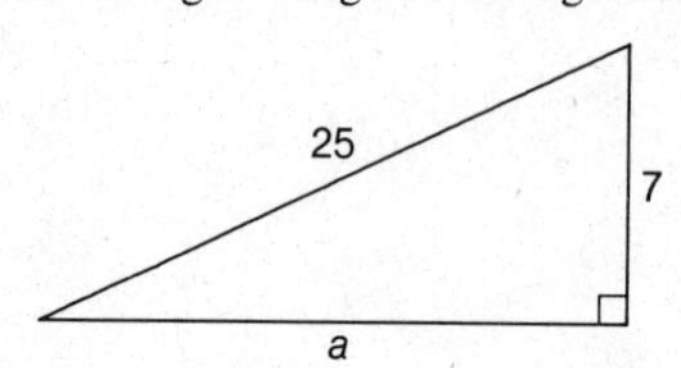

b. A brace is needed to reinforce a rectangular shaped gate. If the gate is 8 feet long and 6 feet high, how long is the brace?

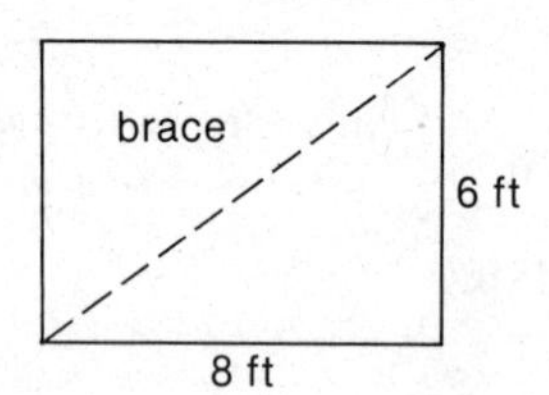

12.4 Exercises

Write each equation in standard form.

1. $4x^2 - 7x = 9$
2. $x^2 = 5x - 9$
3. $3x - 1 = 7 - 3x^2$
4. $(2x + 3)(3x - 7) = 7x + 2$
5. $5x^2 - 9x + 2 = 2x^2 + 8x$
6. $2x^2 = 9$

Solve.

7. $x^2 = 9$
8. $x^2 = 16$
9. $x^2 = 36$
10. $x^2 = 81$
11. $x^2 = 12$
12. $x^2 = 18$
13. $x^2 = 7$
14. $x^2 = 10$
15. $x^2 + 8 = 0$
16. $9x^2 = 25$
17. $4x^2 = 1$
18. $x^2 - 80 = 0$

19. $9x^2 - 4 = 0$
20. $4x^2 - 6 = 19$
21. $9x^2 - 4 = 6$
22. $2x^2 - 9 = 40$
23. $8x^2 - 16 = -15$
24. $20x^2 = 3$
25. $16x^2 - 101 = -1$
26. $25x^2 - 7 = 13$
27. $5x^2 - 3 = 5$
28. $81x^2 = 18$
29. $4x^2 - 1 = -7$
30. $32x^2 - 5 = 10$
31. $4x^2 - 15 = 60$
32. $x^2 - 1 = 97$
33. $9x^2 - 4 = 28$
34. $(x + 2)^2 = 7$
35. $(3x + 2)^2 = 36$
36. $(x - 3)^2 = 1$
37. $(x - 3)^2 = 4$
38. $(2x - 1)^2 = 9$
39. $(3x - 7)^2 = 4$
40. $(3x + 5)^2 = 9$
41. $(6x - 2)^2 = 121$
42. $(7x - 10)^2 = 144$
43. $(3x - 2)^2 = 27$
44. $(5x - 3)^2 = 50$
45. $(2x - 5)^2 = 98$

Find the length of the third side of each right triangle.

46.

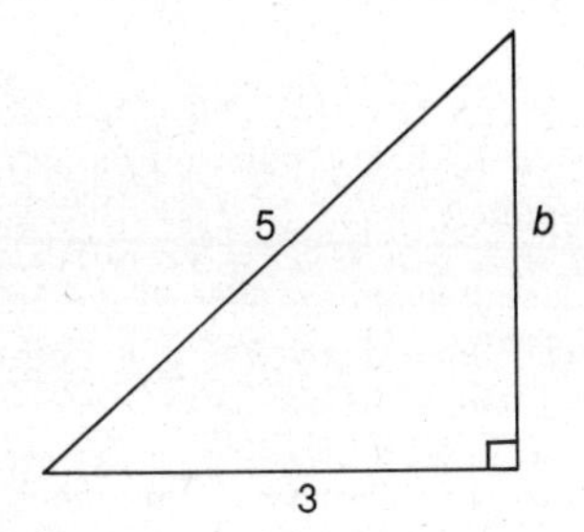

47.

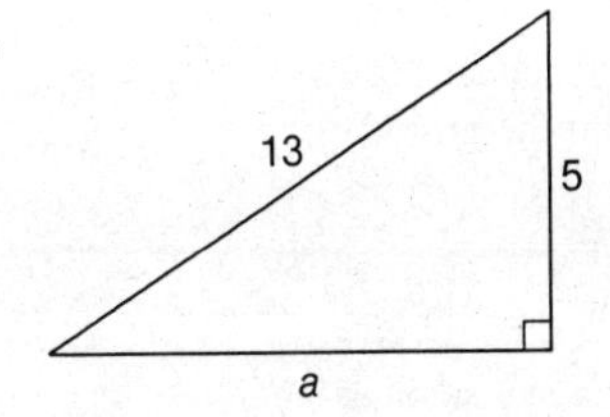

48.

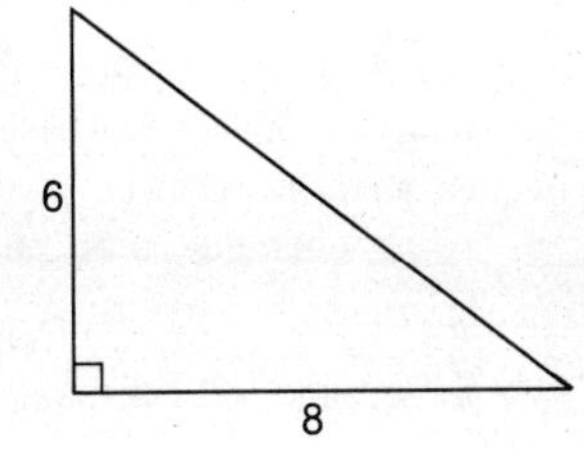

49.

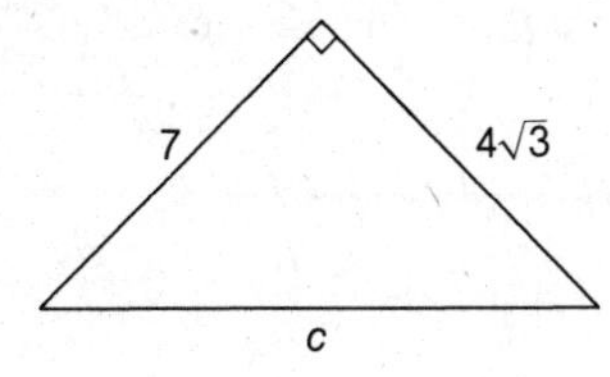

50.

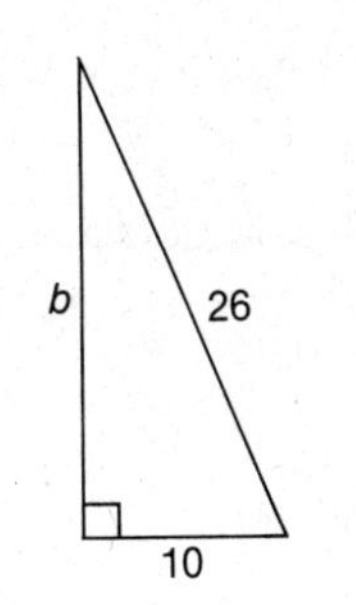

51.

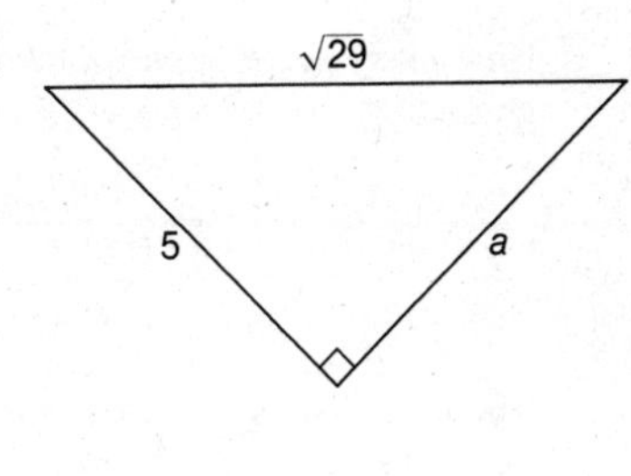

52.

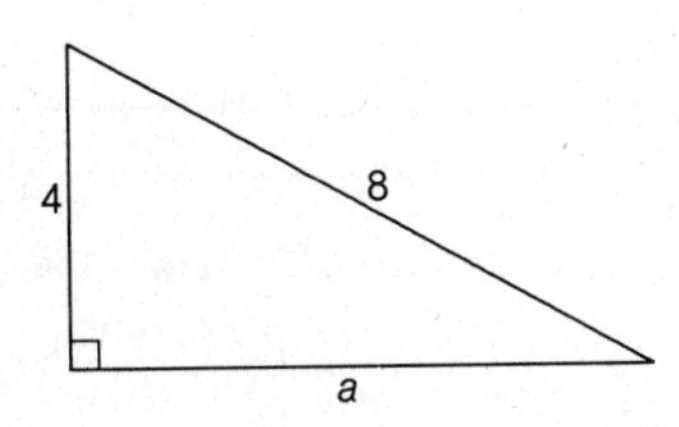

53.

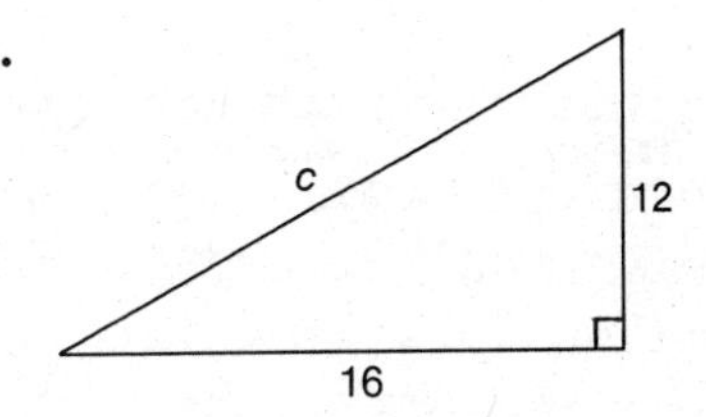

54.

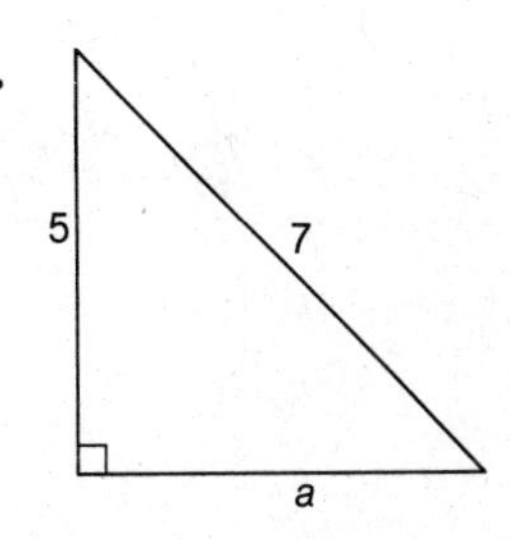

Fill in the blanks.

55. An equation of the form $ax^2 + bx + c = 0$ is called a ______________ equation.

56. To solve a pure quadratic equation, first write the equation in the form $x^2 = K$. Then, take the ______________ of both sides and ______________.

57. In a right triangle, the side opposite the right angle is called the ______________, and the other two sides are called ______________ of the triangle.

58. The Pythagorean Theorem states that the sum of the squares of the ______________ of a right triangle is equal to the square of the ______________.

Solve.

59. A ball is dropped from a cliff that is 320 feet above the ground. The distance S (in feet) it falls in t seconds is given by the formula

$$S = 16t^2.$$

How many seconds will it take for the ball to hit the ground? (Round the answer to the nearest tenth of a second.)

60. A rock falls from a tower that is 640 feet high. As it is falling, its height h (in feet) is given by the formula

$$h = 640 - 16t^2$$

where the falling time t is measured in seconds. How many seconds will it take for the rock to hit the ground? (Round the answer to the nearest tenth of a second. *Hint:* when the object hits the ground, $h = 0$.)

61. An object is dropped from a tower that is 400 feet above the ground. The distance S (in feet) it falls in t seconds is given by the formula

$$S = 16t^2.$$

How many seconds will it take for the object to hit the ground?

62. An object falls from a building 576 feet high. While it is falling, its height h (in feet) is given by the formula

$$h = 576 - 16t^2$$

where the falling time t is measured in seconds. How many seconds will it take for the object to hit the ground?

63. A slow-pitch softball diamond (see figure below) is a square whose sides are each 60 feet long. What is the distance from first base to third base?

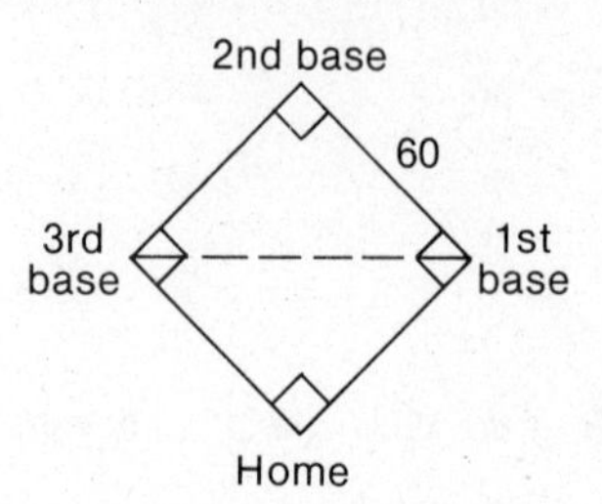

64. How long must a piece of board be to reach from the top of a building 15 feet high to a point on the ground eight feet from the building?

65. A 17-foot ladder is leaning against a wall. If the ladder is 15 feet from the wall, how many feet is it from the top of the ladder to the bottom of the wall?

66. While trying to "square" the corner of a tree house, a carpenter takes the measurements shown in the Figure below. Are the walls of the tree house square?

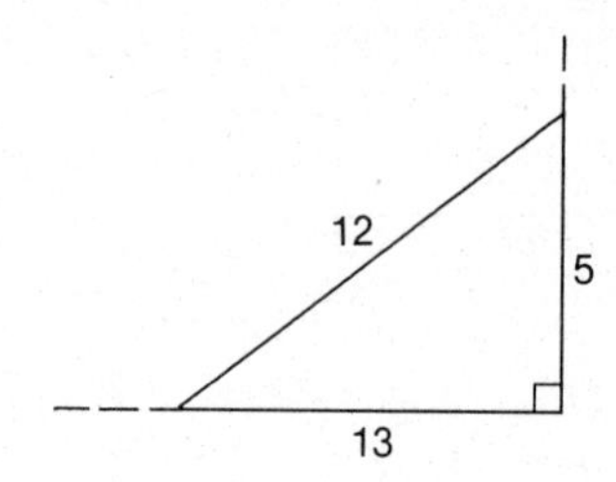

Answers to Practice Problems **14a.** $5x^2 - 6x - 4 = 0$ **b.** $6x^2 - 2x - 2 = 0$ **c.** $6x^2 + 4x + 9 = 0$ **15a.** $5, -5$ **b.** $2\sqrt{2}, -2\sqrt{2}$ **c.** $\frac{2\sqrt{15}}{5}, -\frac{2\sqrt{15}}{5}$ **16a.** $3, -9$ **b.** $-1 + \sqrt{6}, -1 - \sqrt{6}$ **17a.** 24 **b.** 10 ft

12.5 Solving Quadratic Equations by Factoring

19 The square root method cannot be used to solve the equation $x^2 - 3x - 10 = 0$ since it is not a pure quadratic equation. To solve this type of equation, you must use the factoring techniques discussed in Chapter 11 and a rule called the **zero-factor property.** The zero-factor property states that if the product of two numbers is zero, then at least one of the numbers must be zero. In other words, if $a \cdot b = 0$, then $a = 0$ or $b = 0$, or both a and b are zero.

Example 19 Solve: $x^2 - 3x - 10 = 0$

Solution

$x^2 - 3x - 10 = 0$ — Factor left side

$(x - 5)(x + 2) = 0$ — Zero-factor property

$x - 5 = 0$ or $x + 2 = 0$

$x = 5 \quad x = -2$

The solutions are 5 and -2.

Check: *For 5*

$$\begin{aligned} x^2 - 3x - 10 &= 0 \\ 5^2 - 3(5) - 10 &= 0 \\ 25 - 15 - 10 &= 0 \\ 0 &= 0 \end{aligned}$$

For −2

$$\begin{aligned} x^2 - 3x - 10 &= 0 \\ (-2)^2 - 3(-2) - 10 &= 0 \\ 4 + 6 - 10 &= 0 \\ 0 &= 0 \end{aligned}$$

20 The results in Example 19 tell us that to solve a quadratic equation by factoring, we must use the following procedure:

To Solve Quadratic Equations by Factoring:

1. Write the equation in standard form.
2. Factor the polynomial, if possible.
3. Make each factor equal to zero and solve the resulting first-degree equations.
4. The values found in step 3 are the solutions of the equation.
5. Check the solutions in the original equation.

Example 20 Solve by factoring.

a. $x^2 + 15x = -36$ **b.** $2x^2 - 13x + 20 = 0$ **c.** $3x^2 = 21x$

d. $4x^2 - 25 = 0$ **e.** $5(4x - 5) = 4x^2$

Solution To solve a quadratic equation by factoring, first write the equation in standard form. Next, factor the polynomial and then use the zero-factor property. Finally, solve the resulting first-degree equations.

20a.

$$\begin{aligned} x^2 + 15x &= -36 \\ x^2 + 15x + 36 &= 0 && \text{Standard form} \\ (x + 12)(x + 3) &= 0 && \text{Factor} \end{aligned}$$

$$x + 12 = 0 \qquad x + 3 = 0 \qquad \text{Zero-factor property}$$

$$x = -12 \qquad x = -3$$

The solutions are -12 and -3.

Check: *For −12*

$$\begin{aligned} x^2 + 15x &= -36 \\ (-12)^2 + 15(-12) &= -36 \\ 144 + (-180) &= -36 \\ -36 &= -36 \end{aligned}$$

For −3

$$\begin{aligned} x^2 + 15x &= -36 \\ (-3)^2 + 15(-3) &= -36 \\ 9 + (-45) &= -36 \\ -36 &= -36 \end{aligned}$$

20b.

$$\begin{aligned} 2x^2 - 13x + 20 &= 0 \\ (2x - 5)(x - 4) &= 0 \end{aligned}$$

$$2x - 5 = 0 \qquad x - 4 = 0$$

$$2x = 5 \qquad x = 4$$

$$x = \frac{5}{2}$$

Check: *For* $\frac{5}{2}$

$$\begin{aligned} 2x^2 - 13x + 20 &= 0 \\ 2\left(\frac{5}{2}\right)^2 - 13\left(\frac{5}{2}\right) + 20 &= 0 \\ \frac{25}{2} - \frac{65}{2} + \frac{40}{2} &= 0 \\ 0 &= 0 \end{aligned}$$

For 4

$$\begin{aligned} 2x^2 - 13x + 20 &= 0 \\ 2(4)^2 - 13(4) + 20 &= 0 \\ 32 - 52 + 20 &= 0 \\ 0 &= 0 \end{aligned}$$

20c.

$$3x^2 = 21x$$
$$3x^2 - 21x = 0$$
$$3x(x - 7) = 0$$
$$3x = 0 \qquad x - 7 = 0$$
$$x = 0 \qquad x = 7$$

The solutions are 0 and 7.

Check:

For 0

$$3x^2 = 21x$$
$$3(0)^2 = 21(0)$$
$$0 = 0$$

For 7

$$3x^2 = 21x$$
$$3(7)^2 = 21(7)$$
$$147 = 147$$

20d.

$$4x^2 - 25 = 0$$
$$(2x + 5)(2x - 5) = 0$$
$$2x + 5 = 0 \qquad 2x - 5 = 0$$
$$2x = -5 \qquad 2x = 5$$
$$x = -\frac{5}{2} \qquad x = \frac{5}{2}$$

The solutions are $-\frac{5}{2}$ and $\frac{5}{2}$.

Check:

For $-\frac{5}{2}$

$$4x^2 - 25 = 0$$
$$4\left(-\frac{5}{2}\right)^2 - 25 = 0$$
$$25 - 25 = 0$$
$$0 = 0$$

For $\frac{5}{2}$

$$4x^2 - 25 = 0$$
$$4\left(\frac{5}{2}\right)^2 - 25 = 0$$
$$25 - 25 = 0$$
$$0 = 0$$

20e.

$$5(4x - 5) = 4x^2$$
$$20x - 25 = 4x^2$$
$$4x^2 - 20x + 25 = 0 \qquad \text{Add } -4x^2\text{. Then, multiply by } -1.$$
$$(2x - 5)(2x - 5) = 0$$
$$2x - 5 = 0 \qquad 2x - 5 = 0$$
$$2x = 5 \qquad 2x = 5$$
$$x = \frac{5}{2} \qquad x = \frac{5}{2}$$

The solution is $\frac{5}{2}$. Also, since the two solutions are equal, $\frac{5}{2}$ is called a **double root.**

Check:

$$5(4x - 5) = 4x^2$$
$$5\left(4\left[\frac{5}{2}\right] - 5\right) = 4\left(\frac{5}{2}\right)^2$$
$$5(10 - 5) = 4\left(\frac{25}{4}\right)$$
$$25 = 25$$

Practice Problem 18 **Solve by factoring.**

a. $x^2 + 2x - 8 = 0$ **b.** $2x^2 = 5 - 3x$

c. $9x^2 - 100 = 0$ **d.** $5x^2 = 15x$

21 Knowing how to solve quadratic equations by factoring will help you in solving some word problems.

Example 21 Solve.

a. The product of two consecutive even integers is 48. What are the integers?

b. The length of a rectangle is five inches more than twice its width. If the area of the rectangle is 33 square inches, find the length and width of the rectangle.

c. Part of a rectangular-shaped field that is six miles by 12 miles is to be used for a recreational area. The City Council decides that the recreational area will cover 40 square miles of the field (see Figure 12–7). The remainder of the field will be used as a picnic area. If the picnic area will have a uniform width, find the width of the picnic area.

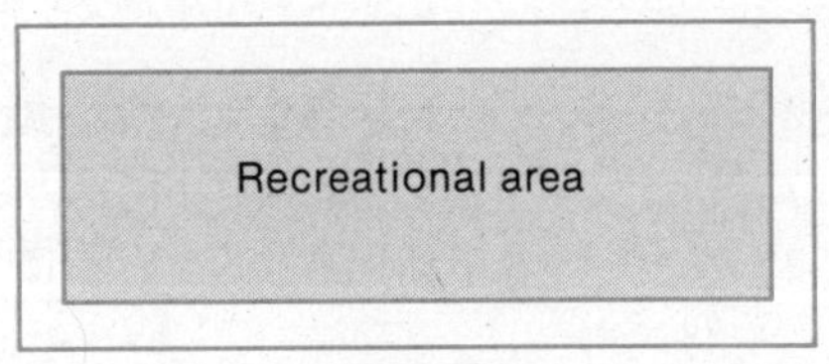

Figure 12-7

Solution To solve a word problem, (1) represent the unknown(s), (2) write an equation, (3) solve the equation written in step 2 and determine the solution(s), and (4) check.

21a. Step 1:

$x =$ an even integer

$x + 2 =$ the next consecutive even integer

Step 2: Product of two numbers is 48

$x(x + 2) = 48$

Step 3:

$$x(x + 2) = 48$$
$$x^2 + 2x = 48$$
$$x^2 + 2x - 48 = 0$$
$$(x + 8)(x - 6) = 0$$
$$x + 8 = 0 \qquad x - 6 = 0$$
$$x = -8 \qquad x = 6$$
$$x + 2 = -6 \qquad x + 2 = 8$$

The solutions are -8 and -6, and 6 and 8.

Check:

For -8 and -6

a. -8 and -6 are consecutive even integers

b. $(-8)(-6) = 48$

For 6 and 8

a. 6 and 8 are consecutive even integers

b. $(6)(8) = 48$

21b. Step 1:

$x =$ width of rectangle (in inches)

$2x + 5 =$ length of rectangle (in inches)

Step 2: Draw the rectangle (see Figure 12–8).

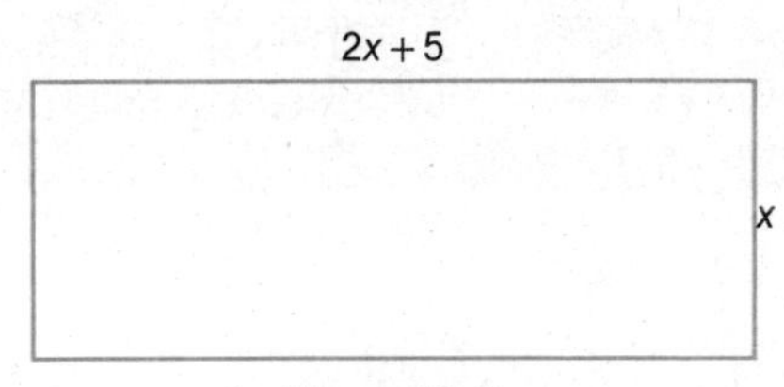

Figure 12-8

$$\text{Area of rectangle} = (\text{length})(\text{width})$$
$$33 = (2x + 5)(x)$$

Step 3:

$$33 = (2x + 5)(x)$$
$$33 = 2x^2 + 5x$$
$$2x^2 + 5x - 33 = 0$$
$$(2x + 11)(x - 3) = 0$$
$$2x + 11 = 0 \qquad x - 3 = 0$$
$$2x = -11 \qquad x = 3 \quad \text{(width)}$$
$$x = -\frac{11}{2} \qquad 2x + 5 = 11 \quad \text{(length)}$$

The solution $-\frac{11}{2}$ is meaningless since a rectangle cannot have a width that is negative. Therefore, the width of the rectangle is 3 inches and the length is 11 inches.

Check: The rectangle is 3 inches by 11 inches since:

a. The length (11 inches) is 5 inches more than twice the width (3 inches).

b. Area = (11 inches)(3 inches) = 33 square inches.

21c. **Step 1:** x = width of picnic area

Step 2: Draw the rectangular-shaped field (see Figure 12–9).

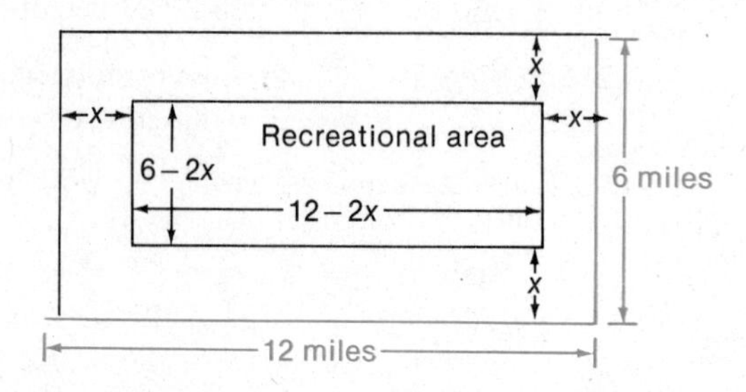

From this figure, the recreational area is $6 - 2x$ by $12 - 2x$. Therefore,

$$\text{Area} = (\text{length})(\text{width})$$
$$40 = (12 - 2x)(6 - 2x).$$

Step 3:

$$40 = (12 - 2x)(6 - 2x)$$
$$(12 - 2x)(6 - 2x) = 40$$
$$72 - 36x + 4x^2 = 40$$
$$4x^2 - 36x + 32 = 0$$
$$x^2 - 9x + 8 = 0 \quad \text{Divide both sides by 4}$$
$$(x - 8)(x - 1) = 0$$
$$x - 8 = 0 \qquad x - 1 = 0$$
$$x = 8 \qquad x = 1$$

The solution cannot be 8 miles since one of the dimensions of the field is only 6 miles. Therefore, the width of the picnic area is 1 mile.

Check: The solution checks since if the width of the picnic area is 1 mile, then the area of the recreational area is

$$A = (10 \text{ mi})(4 \text{ mi}) = 40 \text{ mi.}^2$$

Practice Problem 19 **Solve.**

a. The product of two consecutive positive integers is 90. What are the integers?

b. The area of a square is 2.5 times its perimeter. Find the length of each side. (*Hint*: Perimeter $= 4s$.)

12.5 Exercises

Solve by factoring.

1. $x^2 + 2x - 8 = 0$

2. $2x^2 - 3x = 5$

3. $x^2 + 12x + 35 = 0$

4. $x^2 - 8x - 48 = 0$

5. $x^2 = 5x - 6$

6. $2x^2 + 5x - 3 = 0$

7. $x^2 + 3x - 10 = 0$

8. $3x^2 + 16x = 12$

9. $x^2 - 16 = 0$

10. $9x^2 - 25 = 0$

11. $x^2 - 14x + 49 = 0$

12. $9x^2 = 30x - 25$

13. $10x^2 - 29x + 21 = 0$

14. $6x^2 + 17x - 45 = 0$

15. $x^2 - 6x + 8 = 0$

16. $x^2 + x = 6$

17. $4x^2 + x = 0$

18. $8x^2 - 6x = 0$

19. $x^2 - 9x = 0$

20. $x^2 + 5x = 0$

21. $3x^2 = 2x$

22. $7x^2 = -5x$

23. $8x^2 + 8x = 0$

24. $5x^2 - 6x = 0$

25. $x^2 + 11x = -24$

26. $2x^2 - x - 15 = 0$

27. $(x - 3)(x - 5) = 3$

28. $2(x^2 + 10) = 13x$

29. $x(x - 6) + 9 = 0$

30. $-11x = 3(2x^2 + 1)$

31. $12x^2 - x - 20 = 0$

32. $14x^2 + 29x - 15 = 0$

33. $25x^2 = 36$

34. $6x^2 - 23x = -20$

35. $4x^2 - 5x - 6 = 0$

36. $3x(3x - 4) = -4$

Fill in the blanks.

37. The zero-factor property states that if the ____________ of two numbers is zero, then at least one of the numbers must be ____________.

38. To solve a quadratic equation by factoring, first write the equation in ____________. Next, factor the polynomial and then use the ____________. Finally, solve the resulting first degree equations and check your solutions.

39. When the two solutions of a quadratic equation are equal, the solution is called a ____________.

Solve.

40. $(3p + 4)(p - 1) = -2$

41. $\dfrac{b^2}{2} + b = -b - 2$

42. $(b - 1)(b + 1) = 12(b - 3)$

43. $\dfrac{a^2}{2} = \dfrac{a}{4} + \dfrac{5}{2}$

44. $6(2x + 1) = x(5x - 1)$

45. $(2y + 9)^2 = 9(2y + 9)$

Solve.

46. The product of two consecutive odd integers is 63. What are the integers?

47. The product of two consecutive positive integers is 72. What are the integers?

48. The sum of two numbers is 17. Their product is 66. Find the numbers.

49. One number is five more than another. Their product is 24. Find the two numbers.

50. Two times a positive number is 15 less than its square. What is the number?

51. The length of a rectangle is four meters more than its width. If the area of the rectangle is 32 square meters, find the length and the width of the rectangle.

52. A rectangular-shaped lot is 20 yards by 14 yards. Jim decides to plant a garden on this lot with a uniform strip of lawn around it. If the area of the garden is 160 square yards, find the width of the strip of lawn.

53. The height of a triangle is six inches more than the base. If the area of the triangle is 20 square inches, find the base and the height of the triangle. (*Hint:* $A = \frac{1}{2}bh$.)

54. A rectangular-shaped floor is 20 feet by 12 feet. Betty decides to place a rug on the floor so that the wooden border surrounding the rug has a uniform width. If the area of the rug is 84 square feet, find the width of the wooden border.

55. The length of a rectangle is twice its width. If the area of the rectangle is numerically 20 more than the perimeter, find the length and the width of the rectangle.

Answers to Practice Problems **18a.** $-4, 2$ **b.** $-\frac{5}{2}, 1$ **c.** $\frac{10}{3}, -\frac{10}{3}$ **d.** $0, 3$ **19a.** $9, 10$ **b.** 10

12.6 The Quadratic Formula

22 The methods that we have discussed so far for solving quadratic equations can be used to solve only some quadratic equations. However, if a quadratic equation is written in standard form, you can always find the solution by using the quadratic formula.

Quadratic Formula:

$$x = \frac{-b \pm \sqrt{b^2 - 4ac}}{2a}$$

where a is the coefficient of the x^2 term; b is the coefficient of the x term; and c is the constant.

NOTE: The radicand $b^2 - 4ac$ is called the **discriminant.** When $b^2 - 4ac = 0$, the equation has a solution that is a double root. When $b^2 - 4ac$ is positive, the equation has two real solutions. When $b^2 - 4ac$ is negative, the equation has no real solutions.

Example 22 Solve $x^2 - 8x + 7 = 0$ using the quadratic formula.

Solution Since the equation is in standard form, $a = 1$, $b = -8$, and $c = 7$. Now, substitute these values into the formula and simplify.

$$\begin{aligned} x &= \frac{-b \pm \sqrt{b^2 - 4ac}}{2a} \\ &= \frac{-(-8) \pm \sqrt{(-8)^2 - 4(1)(7)}}{2(1)} \\ &= \frac{8 \pm \sqrt{64 - 28}}{2} \\ &= \frac{8 \pm \sqrt{36}}{2} \\ &= \frac{8 \pm 6}{2} \end{aligned}$$

Finally, write the two solutions separately by first using the plus sign and then the minus sign.

$$x = \frac{8 + 6}{2} \quad \text{or} \quad x = \frac{8 - 6}{2}$$

$$x = \frac{14}{2} \qquad\qquad x = \frac{2}{2}$$

$$x = 7 \qquad\qquad x = 1$$

The solutions are 7 and 1.

NOTE: The equation in Example 22 could have been solved by factoring. Therefore, when you are asked to solve a quadratic equation, first check to see if the equation can be solved by factoring since this method is usually less time-consuming. If it cannot be solved by factoring, then use the quadratic formula.

23 The results in Example 22 suggest that the following procedure can be used to solve quadratic equations by the quadratic formula.

To Solve Quadratic Equations Using the Quadratic Formula:

1. Write the equation in standard form.
2. Determine the values for a, b, and c.
3. Substitute the values of a, b, and c into the quadratic formula.
4. Simplify the resulting expression.

Example 23 Solve by using the quadratic formula.

a. $3x^2 - 4x - 2 = 0$ **b.** $4x^2 - 20x + 25 = 0$

c. $x^2 - 2x + 3 = 0$ **d.** $\frac{1}{3}x^2 + \frac{5}{6}x = \frac{1}{2}$

e. $4x(x - 3) = 3x^2 - 6x - 1$

Solution To solve a quadratic equation by the quadratic formula, first write the equation in standard form. Then, substitute the values for a, b, and c into the quadratic formula and simplify.

23a. Since $3x^2 - 4x - 2 = 0$ is in standard form, use the values $a = 3$, $b = -4$, and $c = -2$.

$$x = \frac{-b \pm \sqrt{b^2 - 4ac}}{2a}$$

$$x = \frac{-(-4) \pm \sqrt{(-4)^2 - 4(3)(-2)}}{2(3)}$$

$$x = \frac{4 \pm \sqrt{16 + 24}}{6}$$

$$x = \frac{4 \pm \sqrt{40}}{6}$$

$$x = \frac{4 \pm 2\sqrt{10}}{6} \qquad \textit{Think: } \sqrt{40} = \sqrt{4}\sqrt{10} = 2\sqrt{10}$$

$$x = \frac{\overset{1}{\cancel{2}}(2 \pm \sqrt{10})}{\underset{3}{\cancel{6}}} \qquad \text{Factor}$$

$$x = \frac{2 \pm \sqrt{10}}{3}$$

The solutions are $\frac{2 + \sqrt{10}}{3}$ and $\frac{2 - \sqrt{10}}{3}$.

NOTE: Use the square root table to find an approximate solution. From Appendix Table A–4, $\sqrt{10} \doteq 3.162$.

$$\frac{2 + \sqrt{10}}{3} \doteq \frac{2 + 3.162}{3} \doteq \frac{5.162}{3} \doteq 1.7$$

$$\frac{2 - \sqrt{10}}{3} \doteq \frac{2 - 3.162}{3} \doteq \frac{-1.162}{3} \doteq -0.4$$

23b. Since $4x^2 - 20x + 25 = 0$, $a = 4$, $b = -20$, and $c = 25$.

$$x = \frac{-b \pm \sqrt{b^2 - 4ac}}{2a}$$
$$x = \frac{-(-20) \pm \sqrt{(-20)^2 - 4(4)(25)}}{2(4)}$$
$$x = \frac{20 \pm \sqrt{400 - 400}}{8}$$
$$x = \frac{20 \pm \sqrt{0}}{8}$$
$$x = \frac{20 \pm 0}{8}$$
$$x = \frac{20 + 0}{8} \qquad x = \frac{20 - 0}{8}$$
$$x = \frac{5}{2} \qquad x = \frac{5}{2}$$

The solution is $\frac{5}{2}$, which is a double root.

23c. Since $x^2 - 2x + 3 = 0$, $a = 1$, $b = -2$, and $c = 3$.

$$x = \frac{-b \pm \sqrt{b^2 - 4ac}}{2a}$$
$$x = \frac{-(-2) \pm \sqrt{(-2)^2 - 4(1)(3)}}{2(1)}$$
$$x = \frac{2 \pm \sqrt{4 - 12}}{2}$$
$$x = \frac{2 \pm \sqrt{-8}}{2}$$

The radical $\sqrt{-8}$ is not a real number. Therefore, the equation has no real solution.

Alternate method
The discriminant of $x^2 - 2x + 3 = 0$ is

$$\begin{aligned} b^2 - 4ac &= (-2)^2 - 4(1)(3) \\ &= 4 - 12 \\ &= -8. \end{aligned}$$

Since the discriminant is negative, $x^2 - 2x + 3 = 0$ has no real solution.

NOTE: When solving quadratic equations by the formula, it may be helpful to first determine if the discriminant is positive or negative. If it is negative, the equation has no real solution.

23d. To eliminate the fractions, multiply both sides of the equation by 6 (LCD).

$$6\left(\frac{1}{3}x^2 + \frac{5}{6}x\right) = 6\left(\frac{1}{2}\right)$$
$$2x^2 + 5x = 3$$

Now, write the equation in standard form.

$$2x^2 + 5x = 3$$
$$2x^2 + 5x - 3 = 0$$

In standard form, $a = 2$, $b = 5$, and $c = -3$

$$x = \frac{-b \pm \sqrt{b^2 - 4ac}}{2a}$$
$$x = \frac{-5 \pm \sqrt{5^2 - 4(2)(-3)}}{2(2)}$$
$$x = \frac{-5 \pm \sqrt{25 + 24}}{4}$$
$$x = \frac{-5 \pm \sqrt{49}}{4}$$
$$x = \frac{-5 \pm 7}{4}$$
$$x = \frac{-5 + 7}{4} \qquad x = \frac{-5 - 7}{4}$$
$$x = \frac{1}{2} \qquad x = -3$$

The solutions are $\frac{1}{2}$ and -3.

23e. Write the equation in standard form.

$$4x(x - 3) = 3x^2 - 6x - 1$$
$$4x^2 - 12x = 3x^2 - 6x - 1$$
$$x^2 - 6x + 1 = 0 \quad \text{Add } -3x^2, 6x, \text{ and } 1.$$

In standard form, $a = 1$, $b = -6$, and $c = 1$.

$$x = \frac{-b \pm \sqrt{b^2 - 4ac}}{2a}$$
$$x = \frac{-(-6) \pm \sqrt{(-6)^2 - 4(1)(1)}}{2(1)}$$
$$x = \frac{6 \pm \sqrt{36 - 4}}{2}$$
$$x = \frac{6 \pm \sqrt{32}}{2}$$
$$x = \frac{6 \pm 4\sqrt{2}}{2}$$
$$x = \frac{\cancel{2}(3 \pm 2\sqrt{2})}{\cancel{2}}$$
$$x = 3 \pm 2\sqrt{2}$$

The solutions are $3 + 2\sqrt{2}$ and $3 - 2\sqrt{2}$.

Practice Problem 20 ***Solve using the quadratic formula.***

a. $x^2 - 5x + 4 = 0$ **b.** $5x^2 - 2 = 3x$

c. $x^2 - 10x = -23$ **d.** $4x(x + 3) = 7$

24 Sometimes when working a word problem you will need to use the quadratic formula.

Example 24 One leg of a right triangle is two meters longer than the other leg. If the hypotenuse is six meters long, what are the lengths of the two legs?

Solution **Step 1:** x = length of one leg (in meters)
$x + 2$ = length of the other leg (in meters)

Step 2: Draw a right triangle (see Figure 12–10).

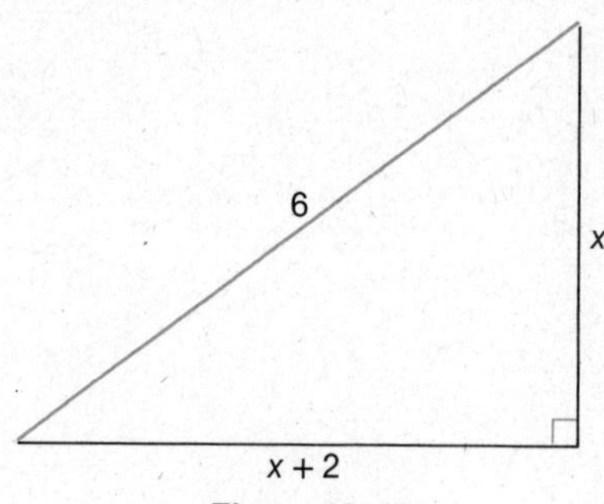

Figure 12-10

$$a^2 + b^2 = c^2$$
$$(x + 2)^2 + x^2 = 6^2$$

Step 3:
$$(x + 2)^2 + x^2 = 6^2$$
$$x^2 + 4x + 4 + x^2 = 36$$
$$2x^2 + 4x - 32 = 0$$
$$x^2 + 2x - 16 = 0 \quad \text{Divided by 2}$$

Since you cannot solve this equation by factoring, use the quadratic formula.
$a = 1, b = 2, c = -16$

$$x = \frac{-b \pm \sqrt{b^2 - 4ac}}{2a}$$
$$x = \frac{-2 \pm \sqrt{2^2 - 4(1)(-16)}}{2(1)}$$
$$x = \frac{-2 \pm \sqrt{4 + 64}}{2}$$
$$x = \frac{-2 \pm \sqrt{68}}{2}$$
$$x \doteq \frac{-2 \pm 8.246}{2} \qquad \sqrt{68} \doteq 8.246$$

$$x \doteq \frac{-2 + 8.246}{2} \qquad x \doteq \frac{-2 - 8.246}{2}$$
$$x \doteq \frac{6.246}{2} \qquad x \doteq \frac{-10.246}{2}$$
$$x \doteq 3.123 \qquad x \doteq -5.123$$
$$x + 2 \doteq 5.123$$

A length of -5.123 is meaningless. Therefore, one leg is approximately 3.123 meters and the other is approximately 5.123 meters.

Practice Problem 21 ***Solve.***

One leg of a right triangle is three inches longer than the other leg. If the hypotenuse is five inches, find the length of the two legs.

12.6 Exercises

Solve using the quadratic formula.

1. $x^2 - 6x + 8 = 0$
2. $x^2 - 4x + 3 = 0$
3. $5x^2 - 3x + 2 = 0$
4. $9x^2 - 12x + 4 = 0$
5. $2x^2 - 13x = -20$
6. $6x^2 - 7x - 5 = 0$
7. $x^2 - 2x + 1 = 0$
8. $x^2 = 2 - 2x$
9. $2x^2 - 3x + 5 = 0$
10. $x^2 + 2x - 5 = 0$
11. $3x^2 - 2x - 8 = 0$
12. $x^2 = 25$
13. $9x^2 - 16 = 0$
14. $3x^2 + 6x = 0$
15. $9x^2 = 15x$
16. $x^2 = 10x - 22$
17. $3x^2 - 7x = 1$
18. $5x^2 - 8x = 3$
19. $3x^2 + 2x + 7 = 0$
20. $2x^2 = 7x - 3$
21. $x^2 = 6x + 3$
22. $4x^2 - 5x - 2 = 0$
23. $2x^2 = 12x - 2$
24. $2x^2 - 5x = 3$
25. $4 + 4x - x^2 = 0$
26. $x^2 + 2x = 12 - 3x - x$
27. $x^2 = (2x - 2)(x + 2)$
28. $\frac{2}{3}x^2 = \frac{11}{3}x + 2$
29. $\frac{1}{8}x^2 = \frac{1}{2} - \frac{x}{2}$
30. $(2x + 1)^2 - 8x = 0$

Fill in the blanks.

31. In the quadratic formula, a is the ______________ of the x^2-term, b is the ______________ of the x-term, and c is the ______________ term.

32. The radicand $b^2 - 4ac$ is called the ______________. When $b^2 - 4ac$ is negative, the quadratic equation has ______________ solutions.

33. To solve a quadratic equation by the quadratic formula, first write the equation in ______________ form. Then substitute the values for ______________, ______________, and ______________ into the quadratic formula and simplify the resulting expression.

Solve and approximate the solutions to the nearest tenth.

34. $x^2 - 10x + 23 = 0$
35. $3x^2 + 6x - 2 = 0$
36. $4x^2 - 8x - 2 = 3x^2 - 6x$
37. $4x^2 - 2 = 5x$
38. $x^2 = 8$
39. $4x^2 - 3x + 5 = 0$

Solve.

40. One leg of a right triangle is 2 cm longer than the other. If the hypotenuse is 3 cm, find the length of the two legs.

41. One leg of a right triangle is four meters shorter than the other. If the hypotenuse is five meters, find the length of the two legs.

42. The length of a rectangle is two feet longer than the width. If the area is 18 ft^2, find the length and the width of the rectangle.

43. The length of a rectangle is three meters less than the width. If the area is 15 square meters, find the length and the width of the rectangle.

44. The area of a triangle is 14 square inches. If the height of the triangle is three inches more than the base, find the height and the base of the triangle. (*Hint:* $A = \frac{1}{2}bh$.)

45. The length of one leg of a right triangle is one foot longer than the other leg. If the hypotenuse is three feet, find the length of the two legs.

Answers to Practice Problems **20a.** 4, 1 **b.** $-\frac{2}{5}$, 1 **c.** $5 + \sqrt{2}, 5 - \sqrt{2}$ **d.** $\frac{1}{2}, -\frac{7}{2}$ **21.** 1.7015 inches, 4.7015 inches

Summary

Important Terms

12.1

principal square root, $\sqrt{\ }$
negative square root, $-\sqrt{\ }$
radical
radicand
rational numbers
irrational numbers

12.2

product rule for square roots
quotient rule for square roots
rationalizing the denominator

12.3

like square roots

12.4

quadratic equation
standard form
square root method
pure quadratic equation
right triangle
legs
hypotenuse

12.5

zero-factor property
double root

12.6

quadratic formula
discriminant

Important Skills

12.1

Finding the square root of a number
Finding the decimal approximation of a square root

12.2

Multiplying square roots
Simplifying square roots
Finding the quotient of two square roots
Finding the square root of a fraction
Rationalizing the denominator of a fraction

12.3

Adding and subtracting like square roots
Adding and subtracting unlike square roots

12.4

Writing quadratic equations in standard form
Solving quadratic equations by using the square root method
Solving word problems by using the Pythagorean theorem

12.5

Solving quadratic equations by factoring
Solving word problems by factoring

12.6

Solving quadratic equations by using the quadratic formula
Solving word problems by using the quadratic formula

Review Exercises

Perform the indicated operations and simplify.

1. $\sqrt{36}$
2. $-\sqrt{81}$
3. $\sqrt{2}\sqrt{14}$
4. $\sqrt{40x^2}$
5. $\dfrac{\sqrt{80}}{\sqrt{5}}$
6. $\sqrt{\dfrac{12}{49}}$
7. $\sqrt{3} + 4\sqrt{3}$
8. $\sqrt{45} - \sqrt{20}$
9. $\dfrac{1}{\sqrt{8}}$
10. $\sqrt{\dfrac{3}{5}}$
11. $\sqrt{891}$
12. $\sqrt{\dfrac{3}{20}}\sqrt{\dfrac{8}{15}}$
13. $\sqrt{5} + \sqrt{\dfrac{1}{5}}$
14. $\sqrt{5x}\sqrt{15x^5}$
15. $\sqrt{18} - 3\sqrt{32} + 5\sqrt{50}$

Find the decimal approximation of each square root.

16. $\sqrt{43}$

17. $\sqrt{252}$

18. $\sqrt{\frac{5}{4}}$

19. Solve by factoring: $6x^2 - 11x = 10$

20. Solve using the quadratic formula: $x^2 = 4x + 1$

21. Solve using the square root method: $4x^2 - 7 = 59$

Solve.

22. $x^2 + 2x - 8 = 0$

23. $x^2 = 49$

24. $8x^2 = 10x$

25. $3x^2 - 4x + 2 = 0$

26. $4x^2 = 7 - 12x$

27. $\frac{1}{2}x^2 = \frac{x}{6} + \frac{1}{3}$

28. $(3x - 8)^2 = 16$

29. $x(x + 16) = 16$

Solve.

30. Find the length and width of a square-shaped room if its area is 196 square yards.

31. The product of two consecutive positive integers is 90. What are the integers?

32. The base of a right triangle is three yards shorter than the height. If the hypotenuse is five yards, find the height and the base.

33. The perimeter of a rectangle is 48 feet. The area of this rectangle is 140 square feet. Find the length and the width of the rectangle. (*Hint:* The length plus the width is 24 feet.)

GEOMETRY

OBJECTIVES

1. Find the perimeter of a polygon.

2. Find the circumference of a circle.

3. Find the area of a geometric figure.

4. Find the volume of a three-dimensional figure.

PRETEST

1. Find the perimeter.

a.

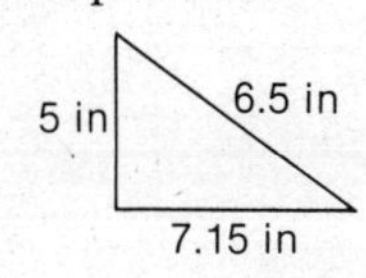

b.

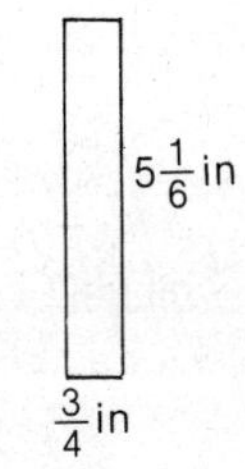

2. Find the circumference.

a.

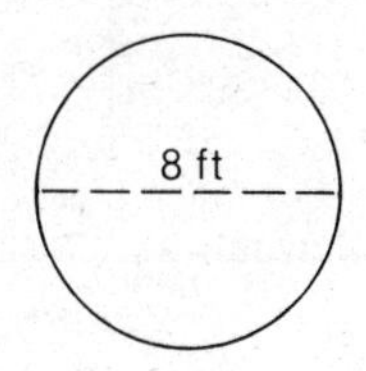

b.

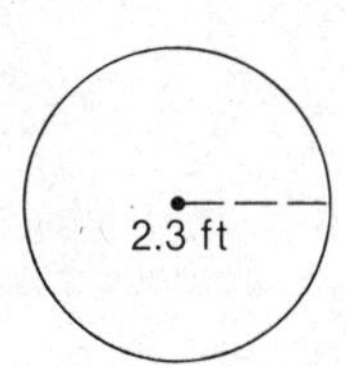

3. Find the area.

a.

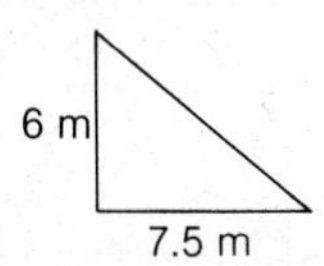

b.

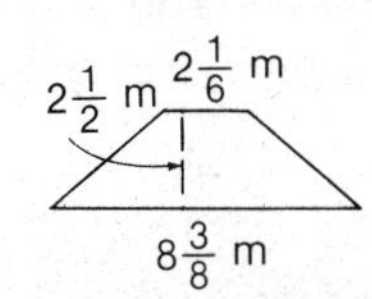

c.

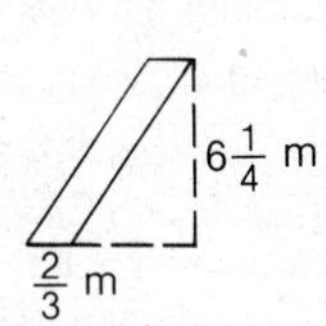

d.

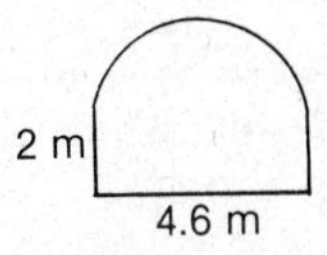

4. Find the volume.

a.

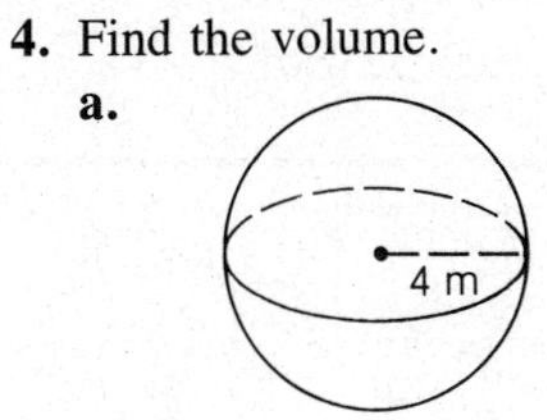

b.

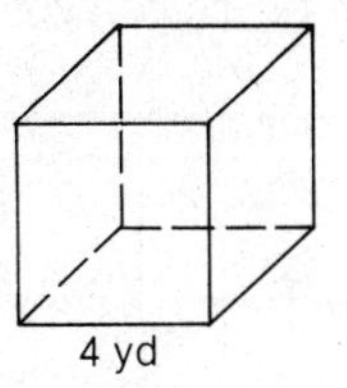

c.

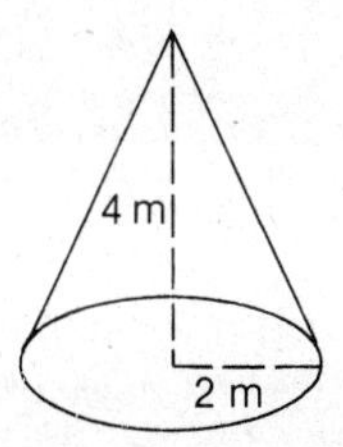

d.

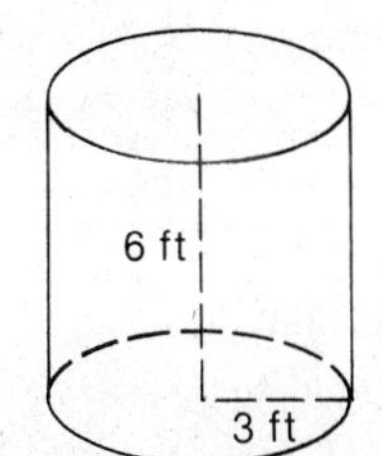

EXPLANATION

Section 13.1
Idea 3

Section 13.1
Idea 4

Section 13.2
Ideas 8–16

Section 13.3
Idea 19

13.1 Perimeter

Idea 1 A **polygon** is a closed figure whose sides are straight lines.

Example 1 The figures shown in Figure 13–1 are polygons.

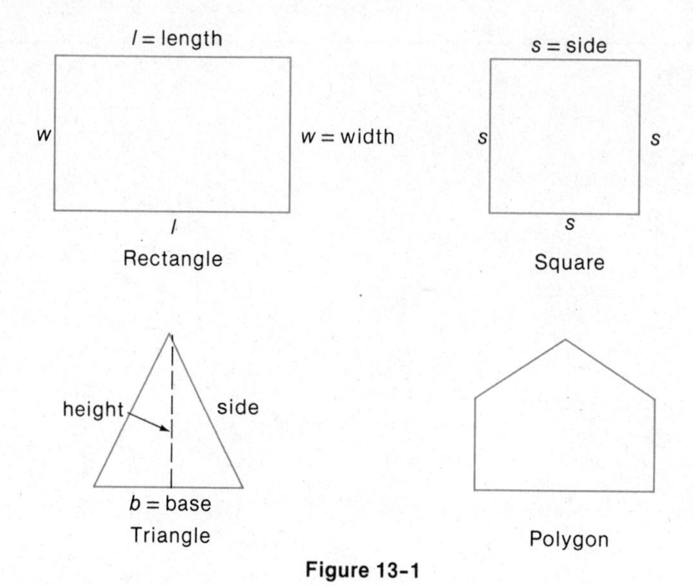

Figure 13-1

2 The **perimeter** of a polygon is the total distance around the figure. You can find the perimeter of a polygon by using the following procedure:

> **To Find the Perimeter of a Polygon:**
>
> **1.** Determine the lengths of all sides of the polygon.
>
> **2.** Compute the sum of the lengths of the sides.

Example 2 Find the perimeter of the polygons shown in Figures 13–2 through 13–5.

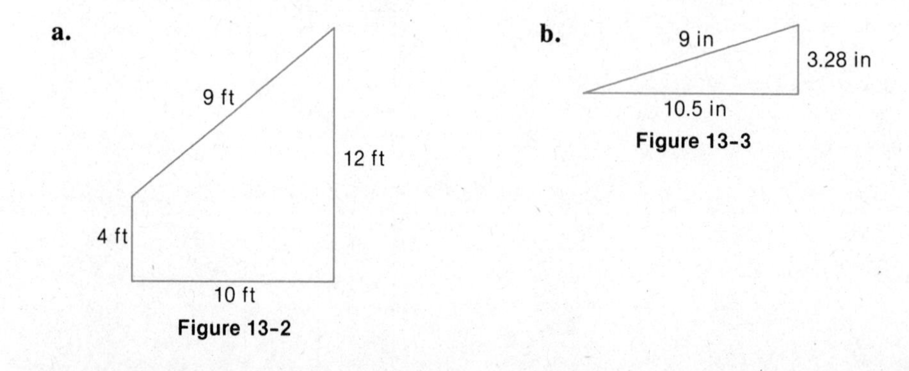

Figure 13-2

Figure 13-3

c.

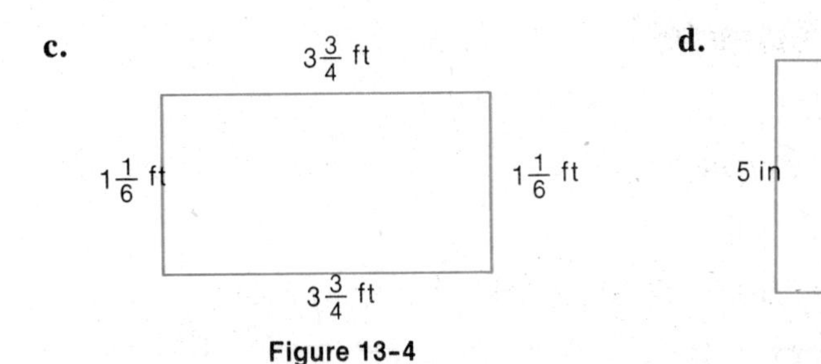

Figure 13-4

d.

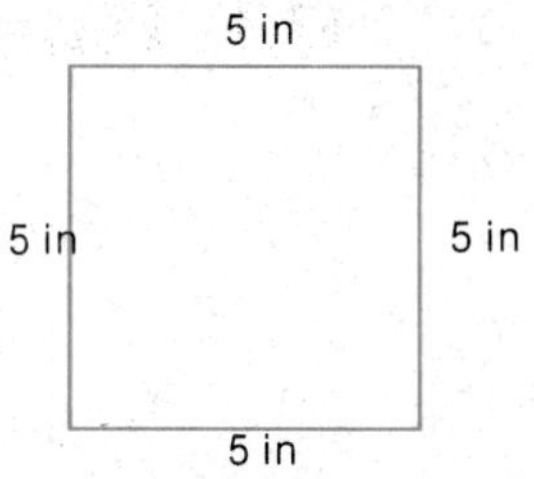

5 in

5 in 5 in

5 in

Figure 13-5

Solution To determine the perimeter (P) of a polygon, find the sum of the lengths of the sides.

2a.
$$\begin{aligned} P &= 4\text{ ft} + 9\text{ ft} + 12\text{ ft} + 10\text{ ft} \\ &= (4 + 9 + 12 + 10)\text{ ft} \\ &= 35\text{ ft} \end{aligned}$$
Read as "35 feet"

2b.
$$\begin{aligned} P &= 9\text{ in} + 3.28\text{ in} + 10.5\text{ in} \\ &= 22.78\text{ in} \end{aligned}$$
Read as "22.78 inches"

2c.
$$\begin{aligned} P &= 1\frac{1}{6}\text{ in} + 3\frac{3}{4}\text{ in} + 1\frac{1}{6}\text{ in} + 3\frac{3}{4}\text{ in} \\ &= 1\frac{2}{12}\text{ in} + 3\frac{9}{12}\text{ in} + 1\frac{2}{12}\text{ in} + 3\frac{9}{12}\text{ in} \\ &= 8\frac{22}{12}\text{ in} \\ &= 9\frac{5}{6}\text{ in} \end{aligned}$$

Alternate method

Figure 13–4 is a rectangle. Its perimeter is given by the formula $P = 2l + 2w$, where l is the length of the rectangle and w is the width. Therefore,

$$\begin{aligned} P &= 2l + 2w \\ &= 2\left(3\frac{3}{4}\text{ in}\right) + 2\left(1\frac{1}{6}\text{ in}\right) \\ &= 2\left(\frac{15}{4}\text{ in}\right) + 2\left(\frac{7}{6}\text{ in}\right) \\ &= \frac{15}{2}\text{ in} + \frac{7}{3}\text{ in} \\ &= \frac{45}{6}\text{ in} + \frac{14}{6}\text{ in} \\ &= \frac{59}{6}\text{ in} \quad\text{or}\quad 9\frac{5}{6}\text{ in.} \end{aligned}$$

2d.
$$\begin{aligned} P &= 5\text{ in} + 5\text{ in} + 5\text{ in} + 5\text{ in} \\ &= 20\text{ in} \end{aligned}$$

Alternate method

Figure 13–5 is a square. Its perimeter is given by the formula $P = 4s$, where s is the length of a side. Therefore,

$$\begin{aligned} P &= 4s \\ &= 4(5\text{ in}) \\ &= 20\text{ in.} \end{aligned}$$

Example 3 Find the perimeter of the Figure 13–6.

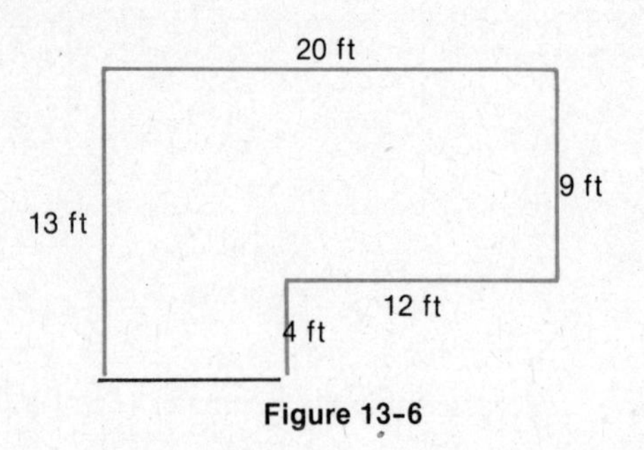

Figure 13-6

Solution To find the perimeter, first find the length of the missing side. The missing side is 20 ft − 12 ft = 8 ft (see Figure 13–7).

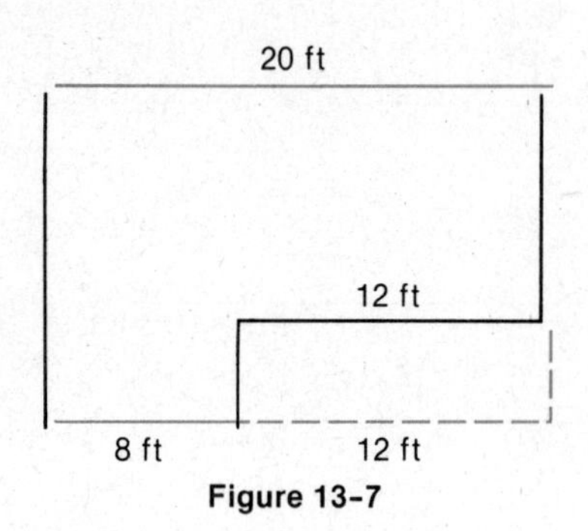

Figure 13-7

$$P = 13 \text{ ft} + 20 \text{ ft} + 9 \text{ ft} + 12 \text{ ft} + 4 \text{ ft} + 8 \text{ ft}$$
$$= 66 \text{ ft}$$

Practice Problem 1 ***Find the perimeter of each figure.***

a.

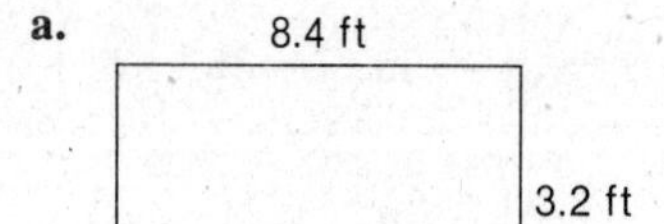

b.

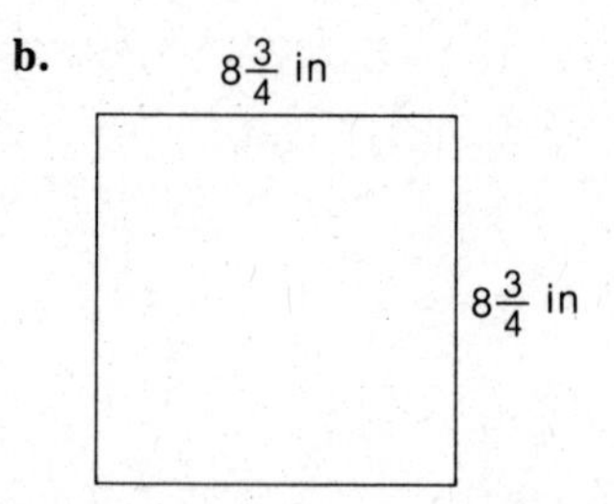

c.

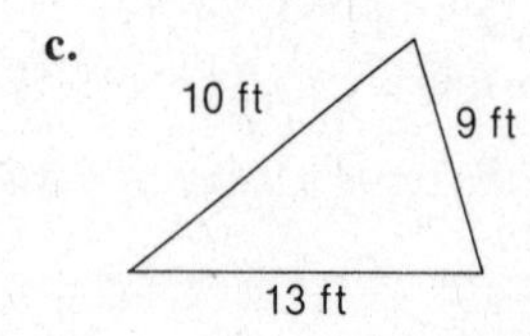

d.

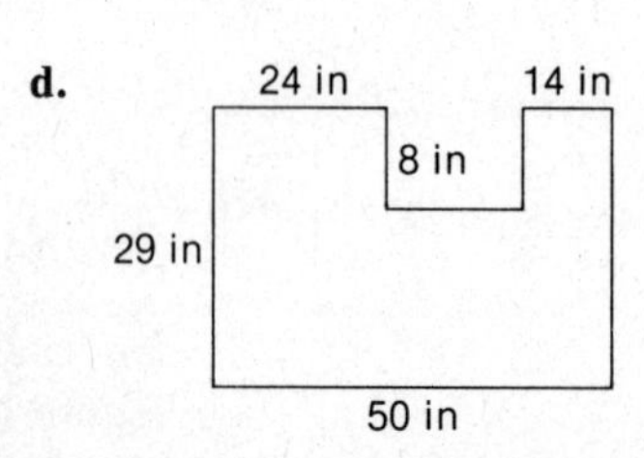

3 We know that the distance around a polygon or any closed figure is called the perimeter. The perimeter of a circle is called the **circumference** (see Figure 13–8).

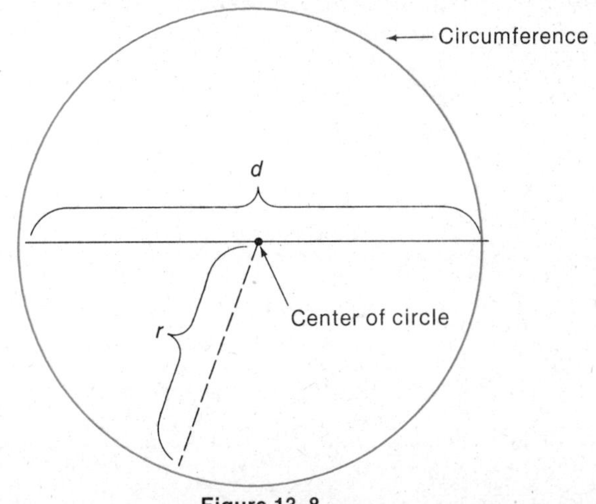

Figure 13-8

4

You can find the circumference (C) of a circle by using the formula

$$C = \pi d,$$

where π (read as "pi") is approximately 3.14.

NOTE: The **radius** (r) of a circle is the distance from the center to any point on the circumference. The **diameter** (d) is twice the radius. Therefore, $d = 2r$ or $r = \frac{d}{2}$.

Example 4 Find the circumference of the circles shown in Figures 13–9 and 13–10. The units are given in inches.

a.

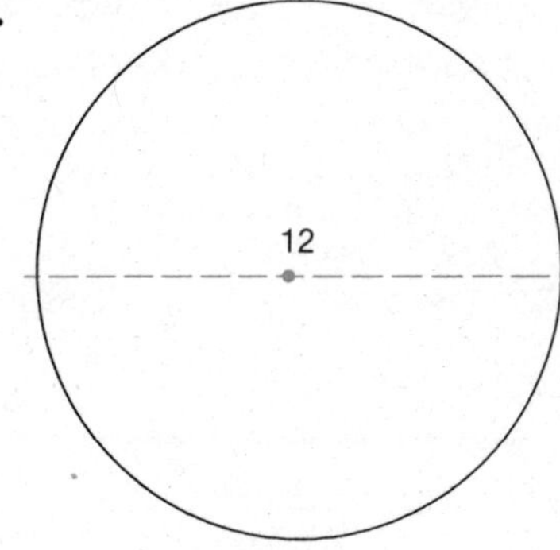

Figure 13-9

b.

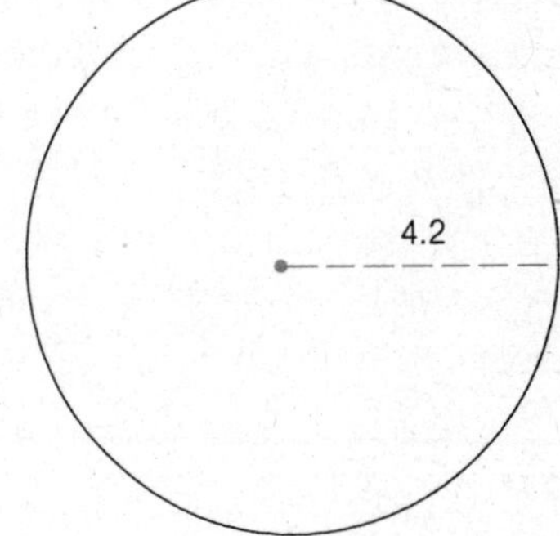

Figure 13-10

Solution To find the circumference of a circle, use the formula $C = \pi d$. Remember that $\pi \doteq 3.14$ and that $d = 2r$.

4a. $C = \pi d$
$\doteq (3.14)(12 \text{ in})$
$\doteq 37.68 \text{ in}$

4b. $C = \pi d$
$\doteq (3.14)(8.4 \text{ in})$
$\doteq 26.376 \text{ in}$

Example 5 Find the length of the semicircles shown in Figures 13–11 and 13–12. The units are given in feet.

a.

3

Figure 13-11

b.

5.85

Figure 13-12

Solution Since a semicircle is half a circle, the length (l) is half the circumference of a circle. Thus,

$$l = \frac{1}{2} \cdot \pi \cdot d = \pi \cdot \frac{d}{2} = \pi r,$$

where r is the radius of the circle.

5a. $l = \pi r$
$\doteq (3.14)(3 \text{ ft})$
$\doteq 9.42 \text{ ft}$

5b. $l = \pi r$
$\doteq (3.14)(2.925 \text{ ft})$
$\doteq 9.1845 \text{ ft}$

NOTE: The radius $r = \frac{d}{2} = \frac{5.85 \text{ ft}}{2} = 2.925$ ft.

Practice Problem 2 ***Find the circumference of the circles (Figures a and b) and the length of the semicircle (Figure c). The units are given in inches.***

a.

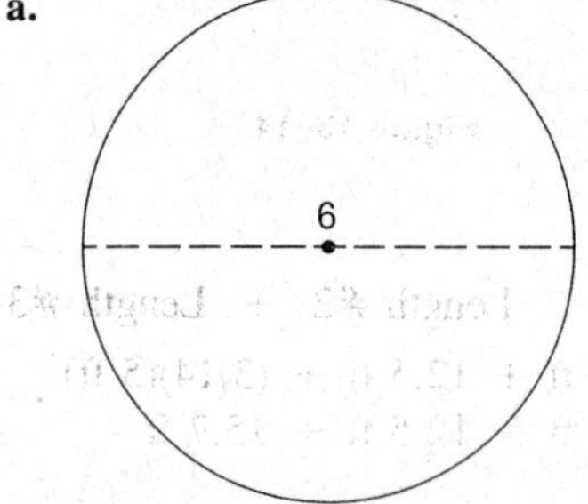

b.

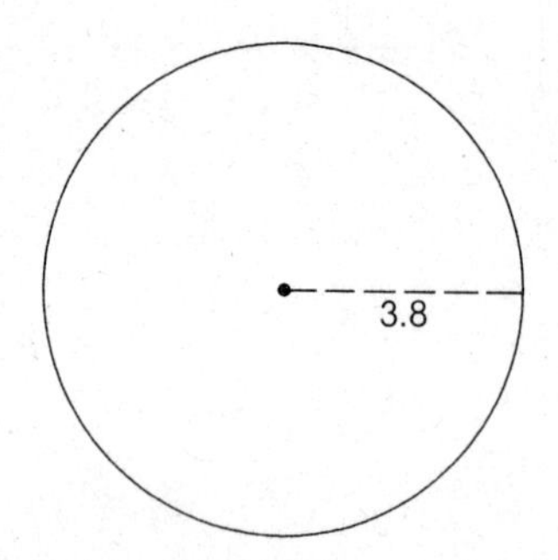

c.

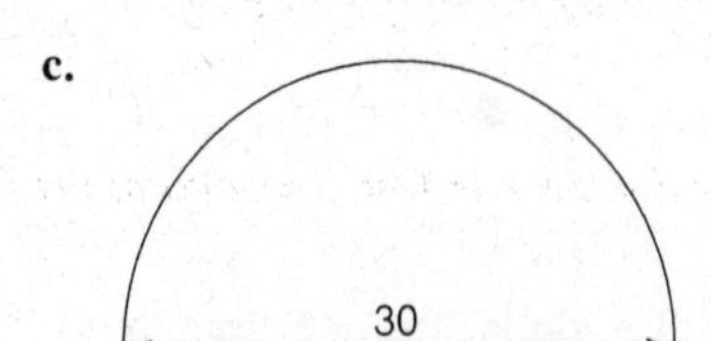

5 Sometimes you will need to find the perimeter of some unfamiliar geometric figures.

Example 6 Find the perimeter of Figure 13–13. The units are given in feet.

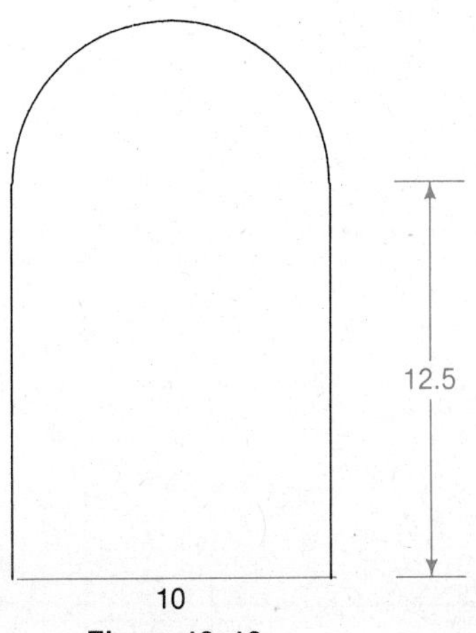

Figure 13-13

Solution To find the perimeter of an unfamiliar figure, first divide the figure into parts that you can recognize such as line segments, rectangles, circles or semicircles. Then find the sum of the lengths of these parts.

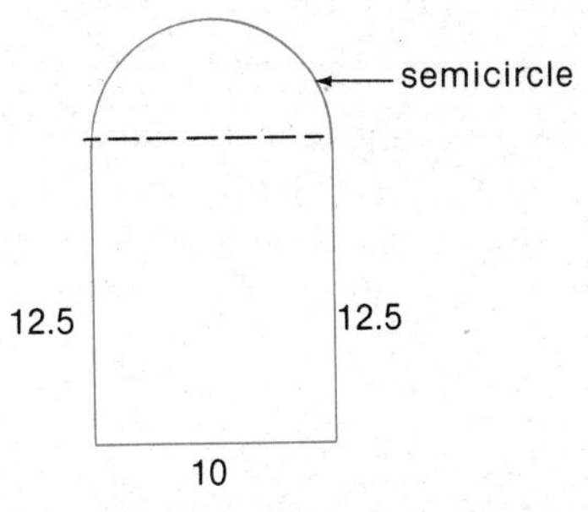

Figure 13-14

$$\begin{aligned} P &= \text{Length \#1} + \text{Length \#2} + \text{Length \#3} + \text{Length of semicircle} \\ &= 12.5 \text{ ft} + 10 \text{ ft} + 12.5 \text{ ft} + (3.14)(5 \text{ ft}) \\ &= 12.5 \text{ ft} + 10 \text{ ft} + 12.5 \text{ ft} + 15.7 \text{ ft} \\ &= 50.7 \text{ ft} \end{aligned}$$

NOTE: The diameter of the semicircle is 10 ft.

Practice Problem 3 ***Find the perimeter. The units are given in feet. The ends of the figure are semicircles.***

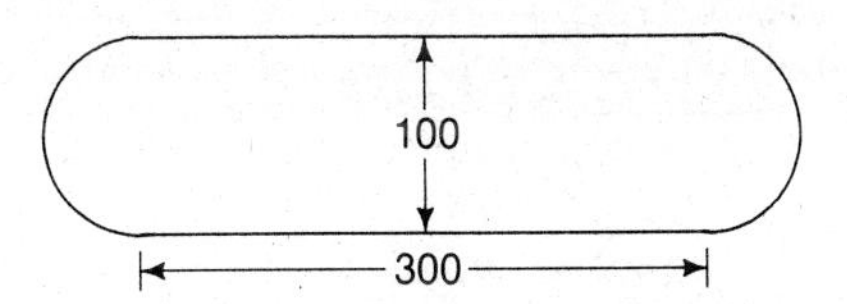

13.1 Exercises

Find the perimeter. The units are given in inches.

1.

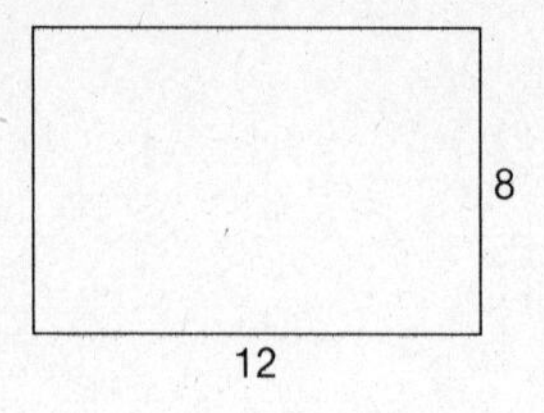

2.

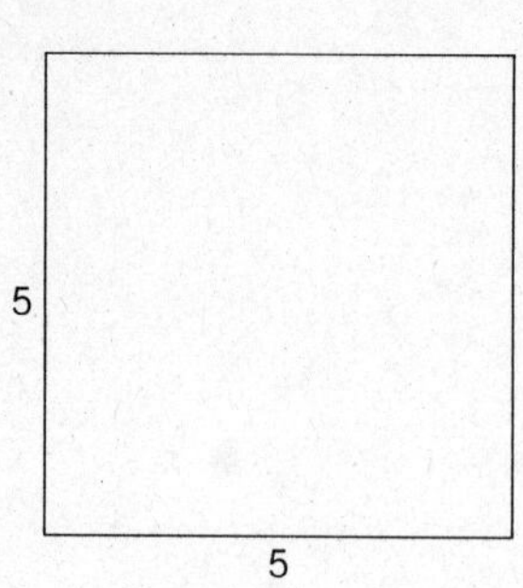

3.

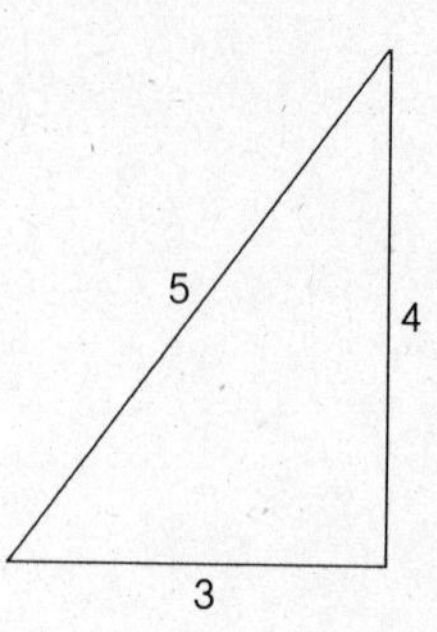

4.

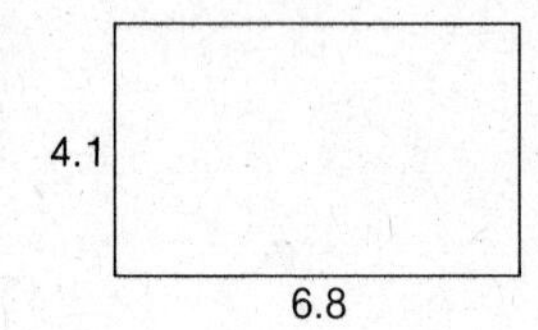

5.

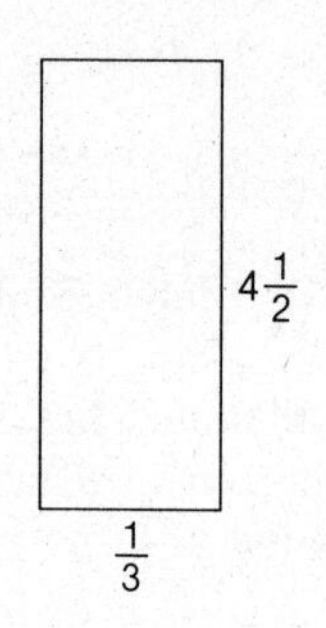

6.

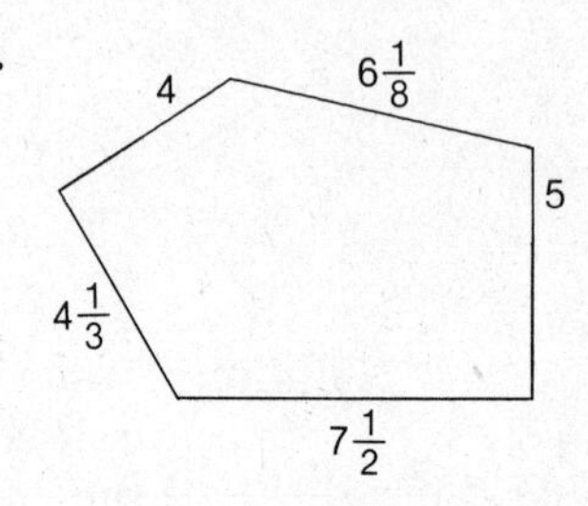

Find the perimeter. The units are given in feet.

7.

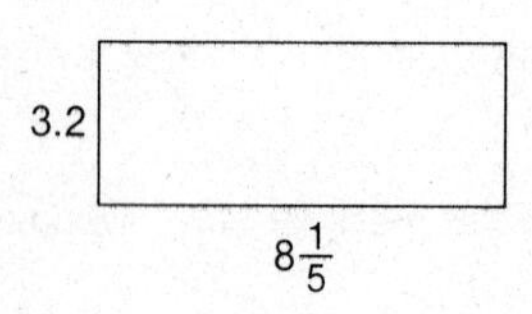

8.

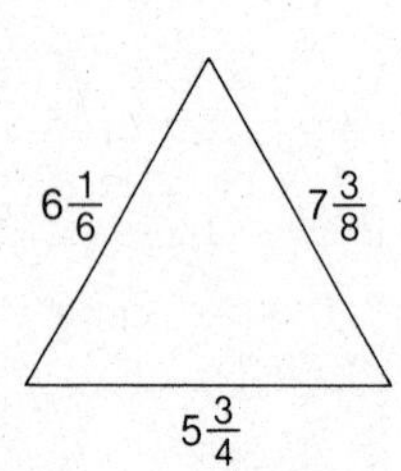

9.

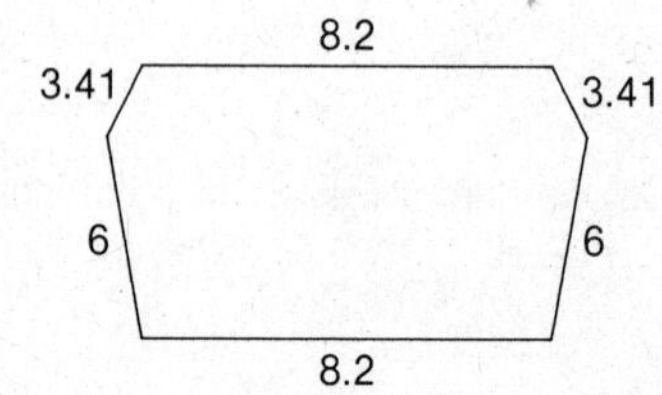

Find the circumference of the circles and the lengths of the semicircles. The units are given in feet.

10.

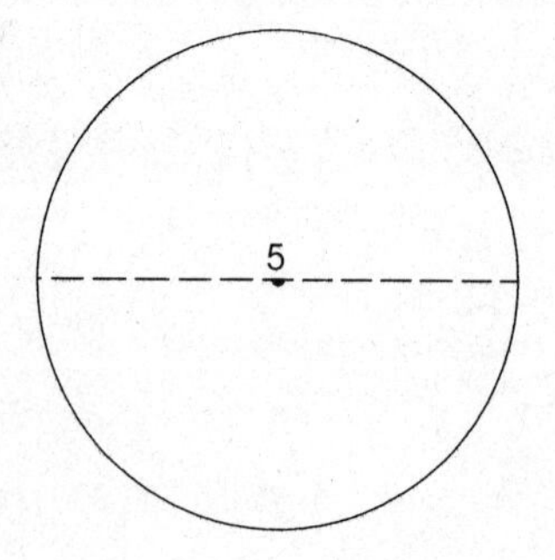

11.

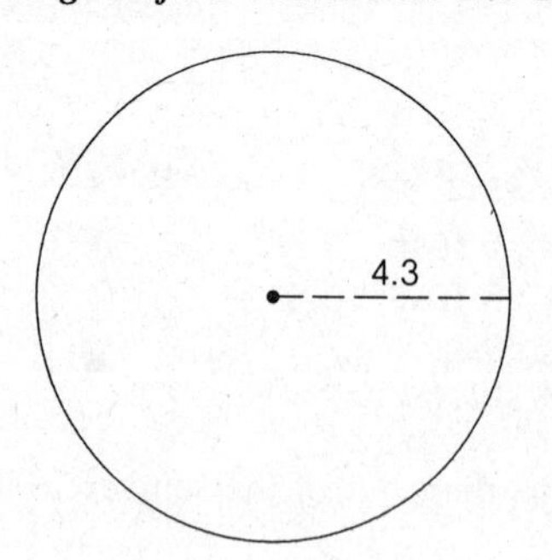

12.

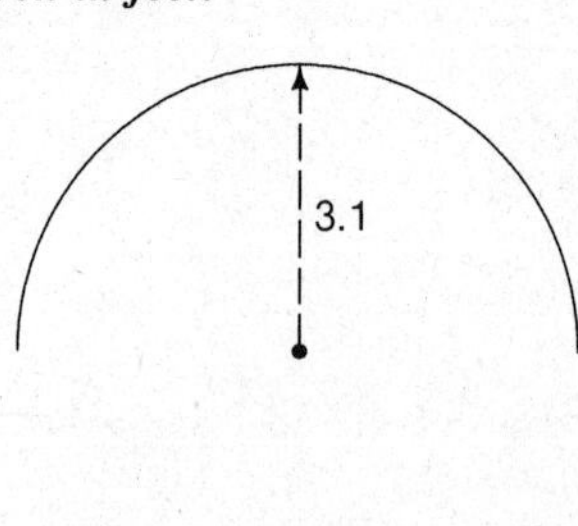

13.

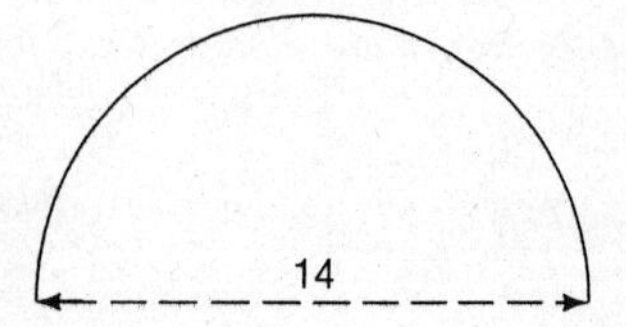

14.

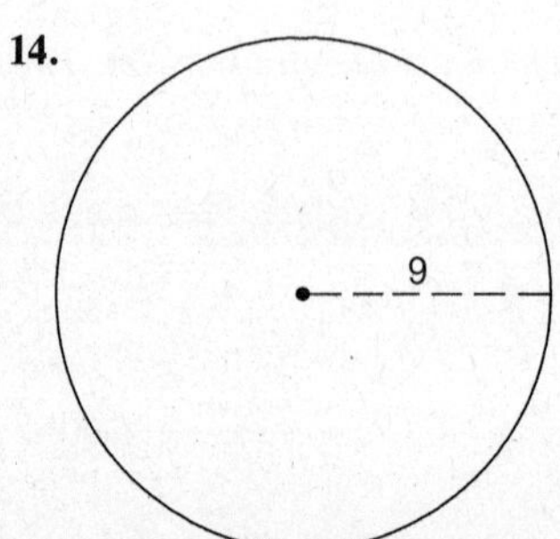

15.

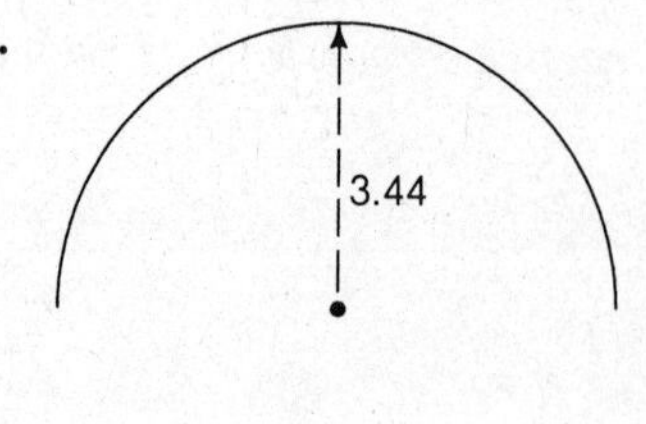

Find the perimeter. The units are given in yards.

16.

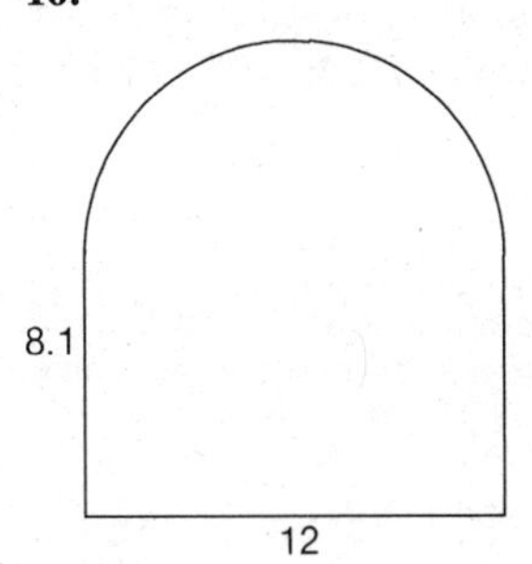

17.

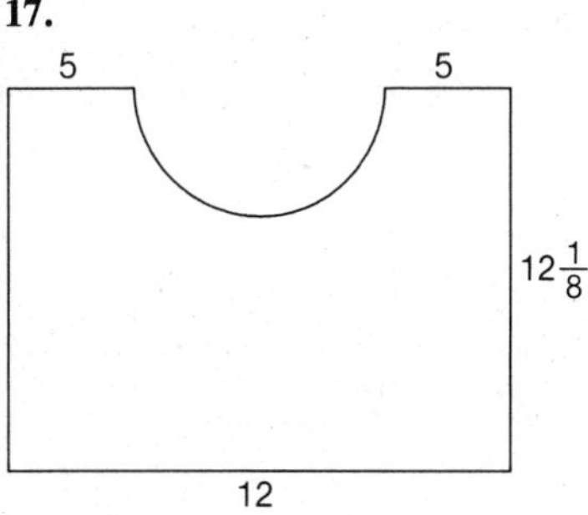

18.

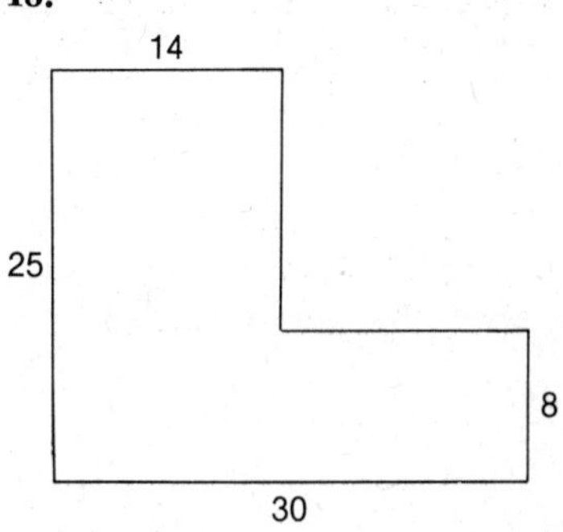

19.

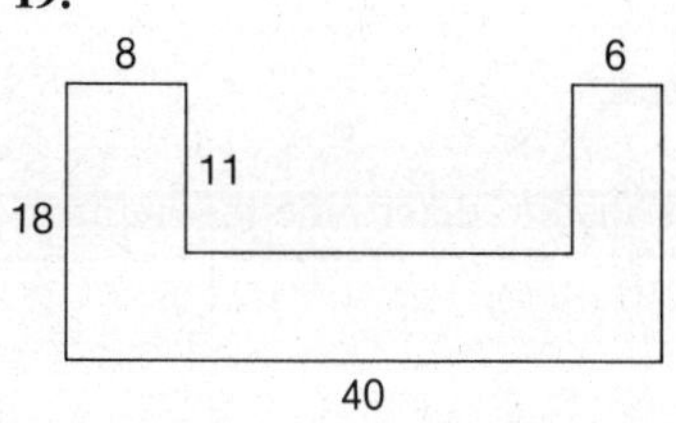

20.

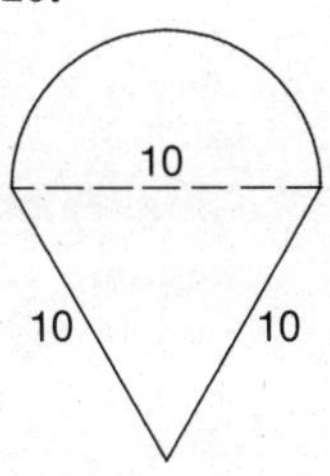

21.

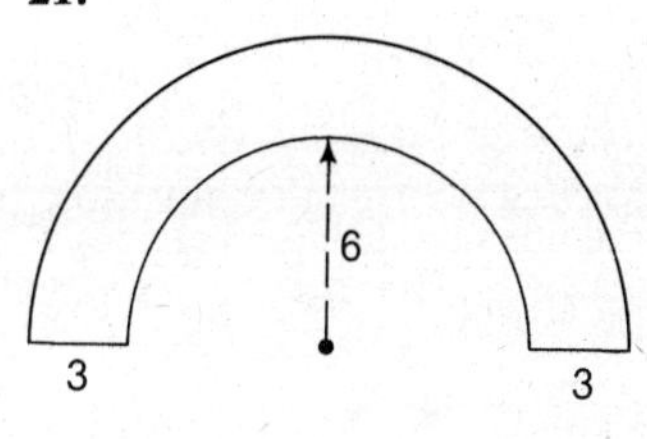

Fill in the blanks.

22. A ______________ is a closed figure whose sides are straight lines.

23. The ______________ of a polygon is the total distance around the figure.

24. The perimeter of a circle is called the ______________.

25. To find the perimeter of a polygon, first determine the lengths of all ______________ of the polygon. Next, compute the ______________ of the lengths of the ______________.

26. You can find the perimeter of a rectangle by using the formula ______________.

27. You can find the circumference of a circle by using the formula ______________.

Solve.

28. A rectangular-shaped garden is 30 feet by 19 feet. If fencing costs $3.50 a foot and Pete wants to build a fence around this garden, what is the cost of the fencing?

29. If Joe can put one foot of binding on a rug in $1\frac{1}{2}$ minutes, how long will it take him to put binding on a rug that is 16 feet by 12 feet?

30. Betty is planting a circular garden. How many feet of fencing will she need to enclose this garden if the diameter of the garden is 9.2 feet?

31. Cazzie wants to make a magnet by wrapping wire around a metal pole with a three-inch diameter. If he plans to wrap the wire around the pole 11 times, how much wire will he need?

32. The Johnson's den is sketched below. How much floor molding will they need to buy to go around the walls?

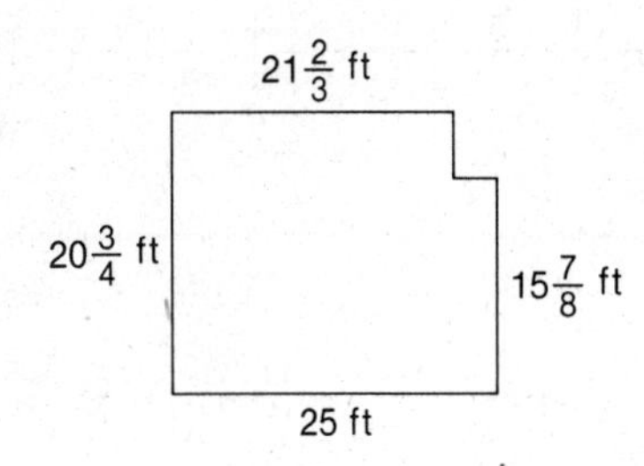

33. Fencing costs $6 per yard. Find the total cost of fencing in a diamond-shaped play area whose sides are six yards each.

Answers to Practice Problems **1a.** 23.2 ft **b.** 35 in **c.** 32 ft **d.** 174 in **2a.** 18.84 in **b.** 23.864 in **c.** 47.1 in **3.** 914 ft

13.2 Area

6 **Area** is a measure of a surface, and it is expressed in square units. Two such units—a square inch (sq in) and a square centimeter (sq cm)—are shown in Figures 13–15 and 13–16.

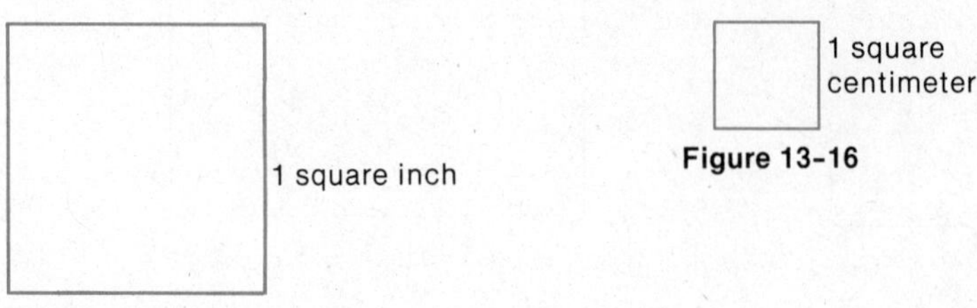

Figure 13-15

Figure 13-16

NOTE: 1 inch is equal to approximately 2.54 centimeters.

7 When you are asked to find the area of a region, you will need to determine the number of square units contained within the region.

Example 7 Consider a rectangle with a length of 3 centimeters and a width of 2 centimeters (see Figure 13–17). Since this rectangle contains six square centimeters, its area is 6 sq cm or 6 cm^2).

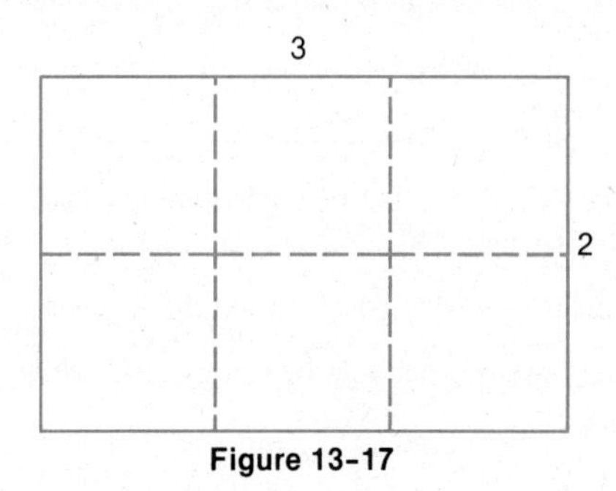

Figure 13-17

8 Examine the rectangle in Figure 13–17 carefully. If you multiply its length (3 cm) by its width (2 cm), the result will be 6 cm^2.

In general, you can find the area of a rectangle by using the formula

$$A = l \cdot w,$$

where l is the length and w is the width.

NOTE: When using this formula, you must express the length and width in the same units.

Example 8 Find the area of Figures 13–18 and 13–19. The units are given in feet.

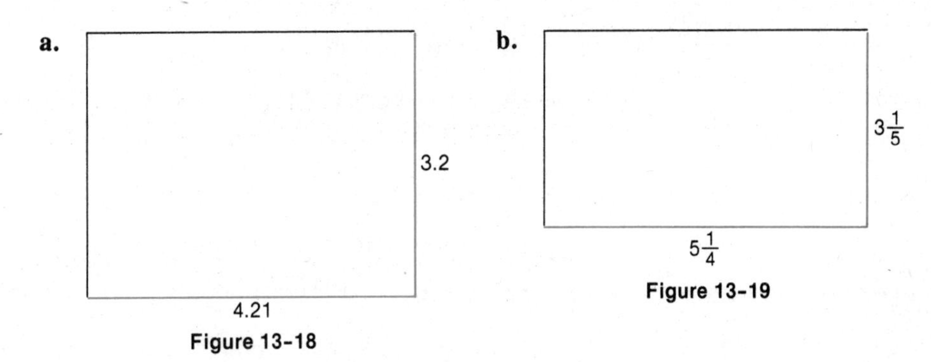

Figure 13-18

Figure 13-19

Solution To find the area of a rectangle, use the formula $A = lw$.

8a. $A = lw$
$= (4.21 \text{ ft})(3.2 \text{ ft})$
$= 13.472 \text{ ft}^2$ Read as "13.472 square feet"

8b. $A = lw$

$$= \left(3\frac{1}{5} \text{ ft}\right)\left(5\frac{1}{4} \text{ ft}\right)$$

$$= \left(\frac{16}{5} \text{ ft}\right)\left(\frac{21}{4} \text{ ft}\right)$$

$$= \frac{\overset{4}{\cancel{16}} \cdot 21}{5 \cdot \underset{1}{\cancel{4}}} \text{ ft}^2$$

$$= \frac{84}{5} \text{ ft}^2$$

$$= 16\frac{4}{5} \text{ ft}^2$$

NOTE: Since a square is a rectangle in which the length and width are equal, we can use the formula $A = lw$ to find the area of a square.

Practice Problem 4 ***Find the area. The units are given in inches.***

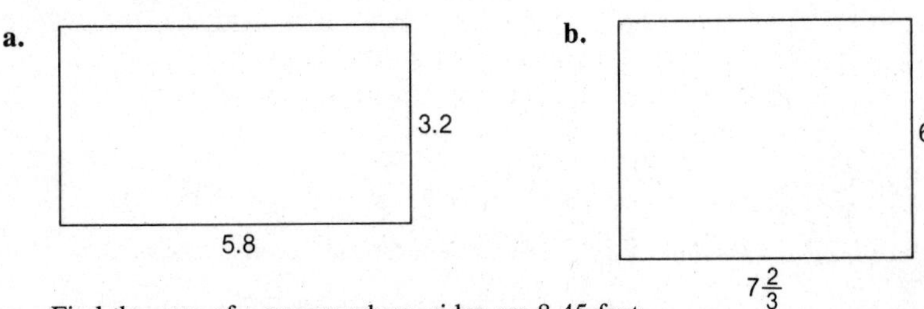

c. Find the area of a square whose sides are 8.45 feet.

9 Let us now consider how to find the area of some other geometric figures. The first figure is a parallelogram (see Figure 13–20). A **parallelogram** is a four-sided polygon whose opposite sides are equal and parallel.

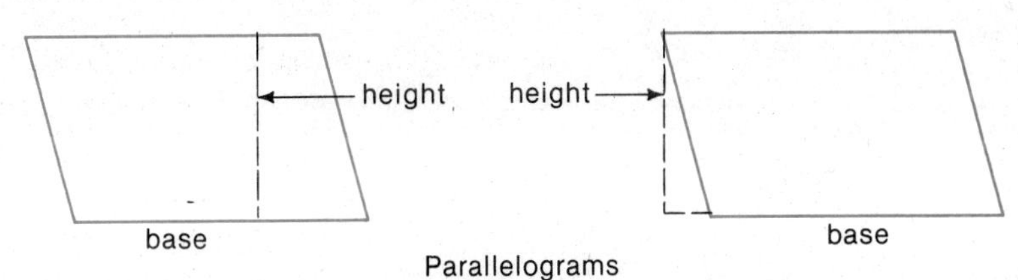

Parallelograms

Figure 13-20

10 You can find the area of a parallelogram by using the formula

$$A = b \cdot h,$$

where b is the length of the base and h is the height.

Example 9 Find the area of the parallelograms in Figures 13–21 and 13–22.

a.

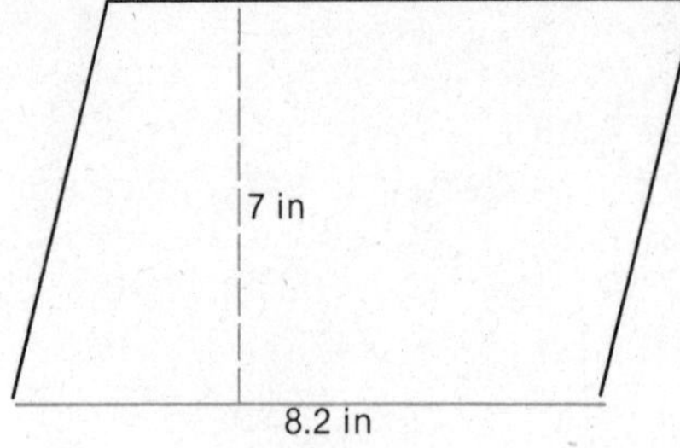

Figure 13-21

b.

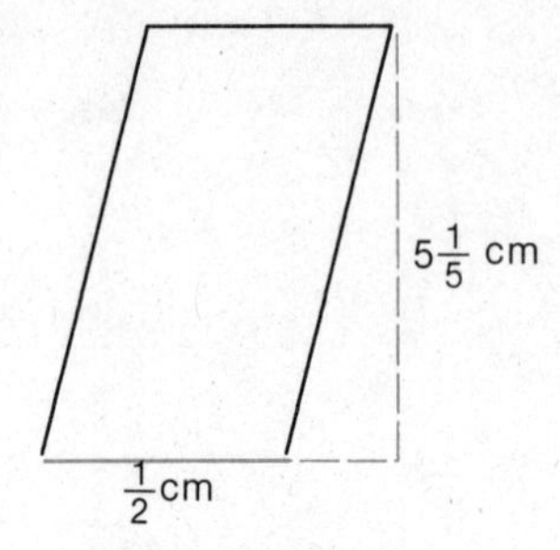

Figure 13-22

Solution To find the area of a parallelogram, use the formula $A = bh$.

9a.
$$\begin{aligned} A &= bh \\ &= (8.2 \text{ in})(7 \text{ in}) \\ &= 57.4 \text{ in}^2 \end{aligned}$$

9b.
$$\begin{aligned} A &= bh \\ &= \left(\frac{1}{2} \text{ cm}\right)\left(5\frac{1}{5} \text{ cm}\right) \\ &= \left(\frac{1}{2} \text{ cm}\right)\left(\frac{26}{5} \text{ cm}\right) \\ &= \frac{1 \cdot \overset{13}{\cancel{26}}}{\underset{1}{\cancel{2}} \cdot 5} \text{ cm}^2 \\ &= \frac{13}{5} \text{ cm}^2 \\ &= 2\frac{3}{5} \text{ cm}^2 \end{aligned}$$

Practice Problem 5 ***Find the area. The units are given in centimeters.***

a.

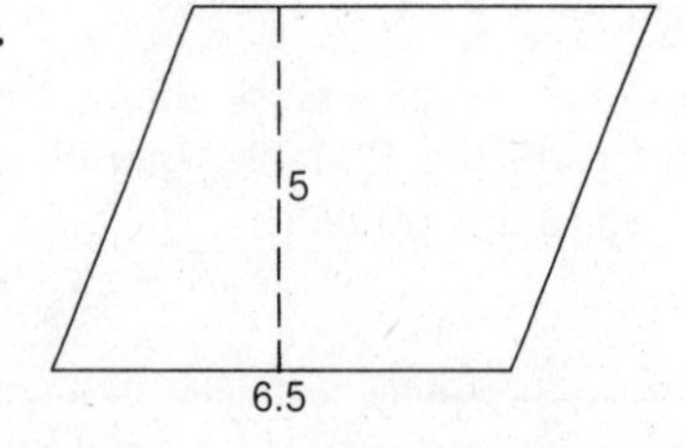

b.

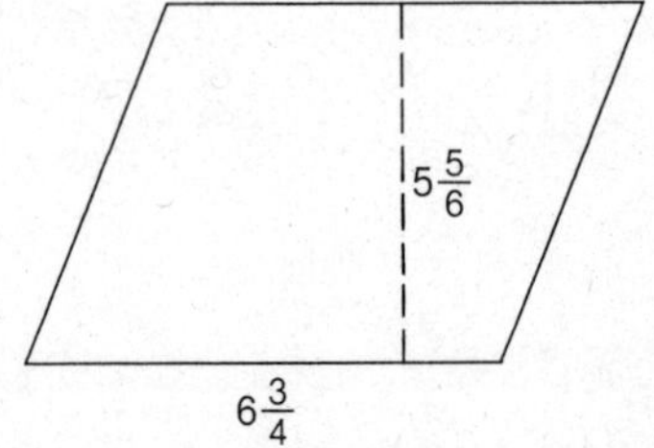

11 The second geometric figure we wish to find the area of is a triangle (see Figure 13–23). Recall that a **triangle** is a three-sided polygon.

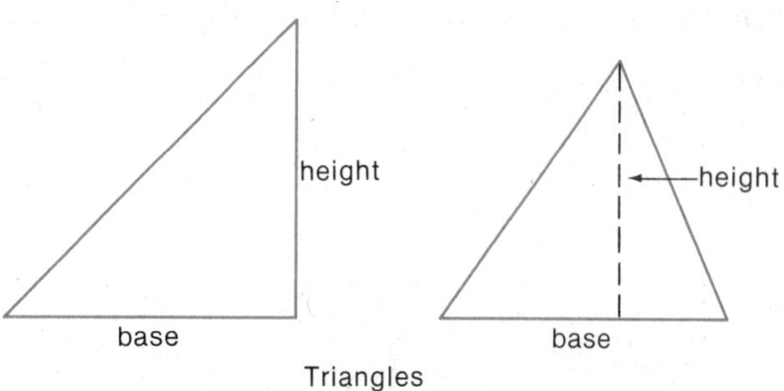

Triangles
Figure 13-23

> **12** To find the area of a triangle, use the formula
>
> $$A = \frac{1}{2}bh,$$
>
> where b is the length of the base and h is the height.

Example 10 Find the area of the triangles in Figures 13–24 and 13–25. The units are given in feet.

a.

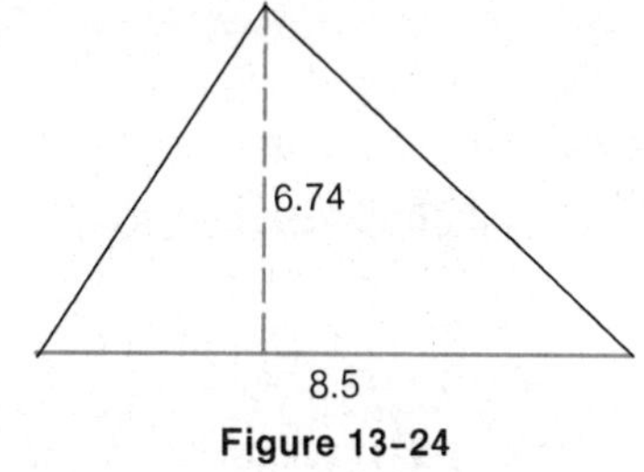

Figure 13-24

b.

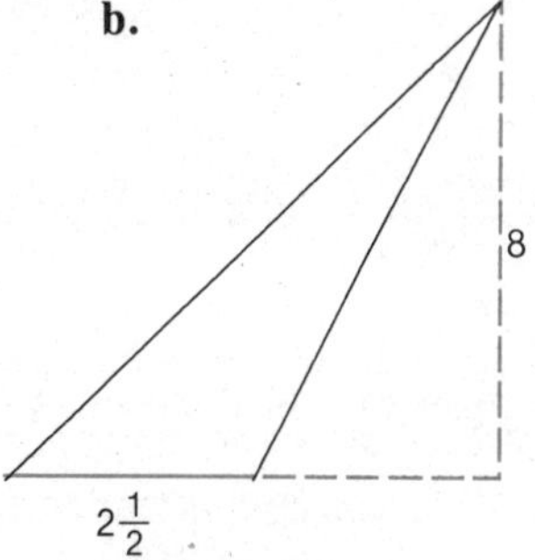

Figure 13-25

Solution To find the area of a triangle, use the formula $A = \frac{1}{2}bh$.

10a.
$$\begin{aligned} A &= \frac{1}{2}\,bh \\ &= \frac{1}{2}\,(8.5 \text{ ft})(6.74 \text{ ft}) \\ &= (0.5)(8.5 \text{ ft})(6.74 \text{ ft}) \\ &= 28.645 \text{ ft}^2 \end{aligned}$$

10b.
$$\begin{aligned} A &= \frac{1}{2}\,bh \\ &= \frac{1}{2}\left(2\frac{1}{2} \text{ ft}\right)(8 \text{ ft}) \\ &= \left(\frac{1}{2}\right)\left(\frac{5}{2} \text{ ft}\right)\left(\frac{8}{1} \text{ ft}\right) \\ &= \frac{1 \cdot 5 \cdot \overset{\not{4}\ 2}{\not{8}}}{\underset{1}{\not{2}} \cdot \underset{1}{\not{2}} \cdot 1} \text{ ft}^2 \\ &= 10 \text{ ft}^2 \end{aligned}$$

Practice Problem 6 ***Find the area. The units are given in centimeters.***

a.

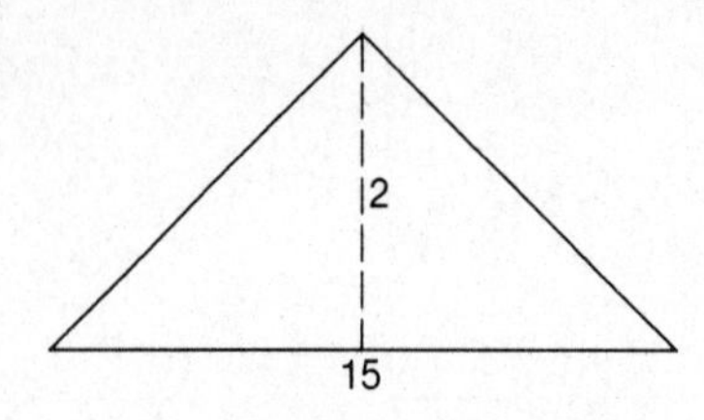

b.

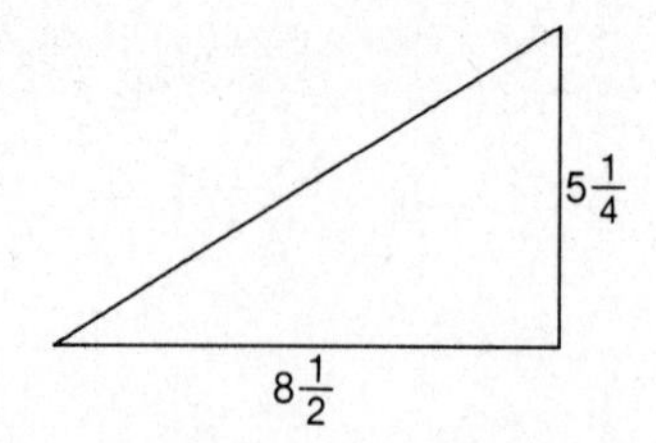

13 The third figure you should be able to find the area of is a trapezoid (see Figure 13–26). A **trapezoid** is a four-sided polygon having only two parallel sides.

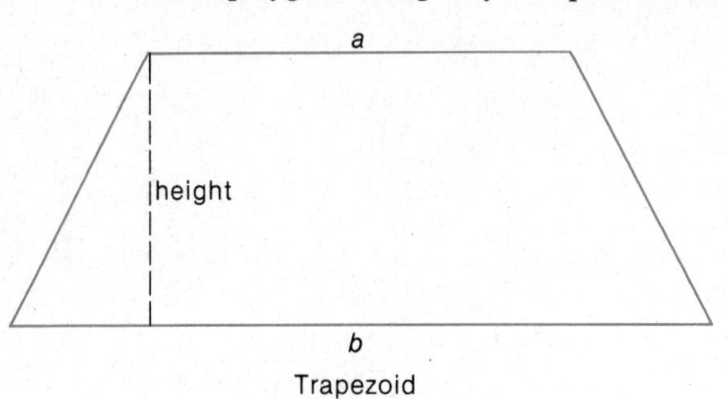

Trapezoid
Figure 13-26

14 You can find the area of a trapezoid by using the formula

$$A = \frac{1}{2}(a + b) \cdot h,$$

where h is the height and a and b are the lengths of the parallel sides.

Example 11 Find the area of the trapezoids in Figures 13–27 and 13–28. The units are given in feet.

a.

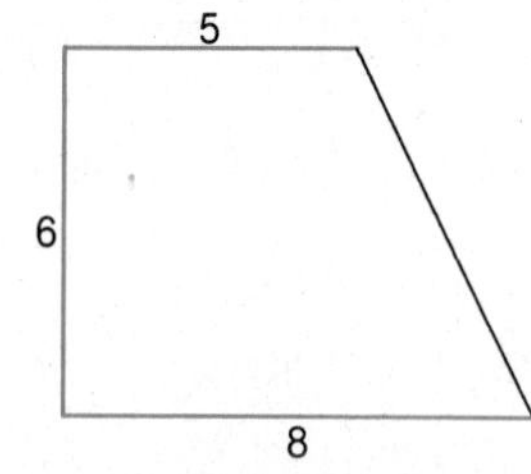

Figure 13-27

b.

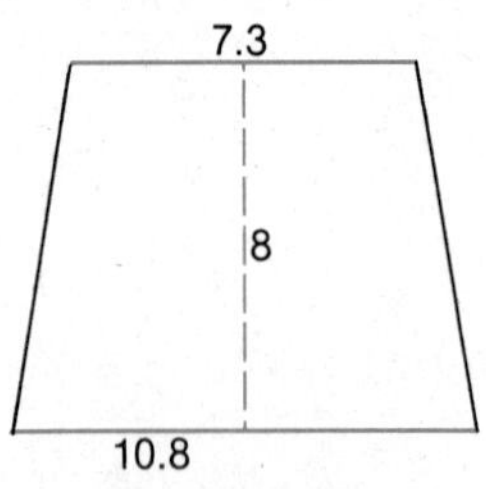

Figure 13-28

Solution To find the area of a trapezoid, use the formula $A = \frac{1}{2}(a + b) \cdot h$.

11a.

$$\begin{aligned} A &= \frac{1}{2}(a + b) \cdot h \\ &= \frac{1}{2}(5 \text{ ft} + 8 \text{ ft})(6 \text{ ft}) \\ &= \frac{(13 \text{ ft})(\overset{3}{\cancel{6}} \text{ ft})}{\underset{1}{\cancel{2}}} \\ &= 39 \text{ ft}^2 \end{aligned}$$

11b.

$$\begin{aligned} A &= \frac{1}{2}(a + b) \cdot h \\ &= \frac{1}{2}(7.3 \text{ ft} + 10.8 \text{ ft})(8 \text{ ft}) \\ &= \frac{(18.1 \text{ ft})(\overset{4}{\cancel{8}} \text{ ft})}{\underset{1}{\cancel{2}}} \\ &= 72.4 \text{ ft}^2 \end{aligned}$$

Practice Problem 7 ***Find the area. The units are given in centimeters.***

a.

b.

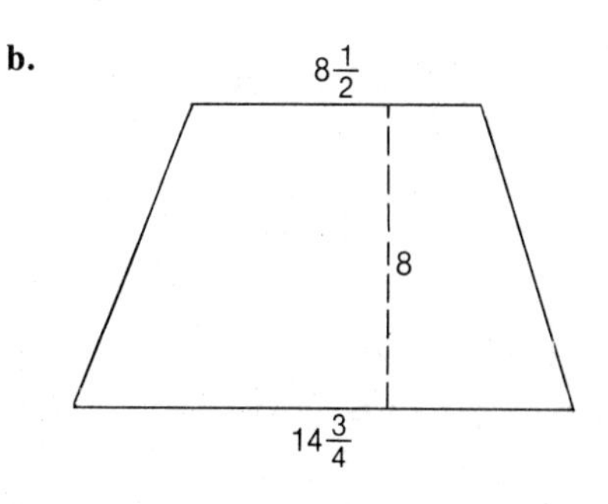

15 In Section 13.1 you learned that a circle has a radius r and a diameter d (see Figure 13–29).

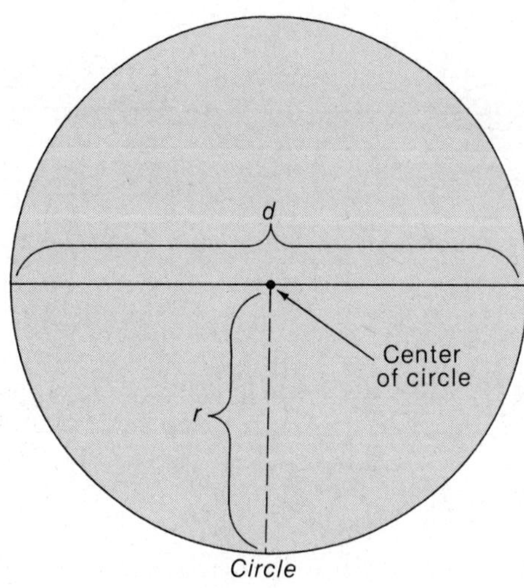

Circle

Figure 13-29

> To find the area of a circle, use the formula
>
> $$A = \pi r^2,$$
>
> where r is the radius of the circle and $\pi \doteq 3.14$.

Example 12 Find the area of the circles in Figures 13–30 and 13–31. The units are given in meters (m).

a.

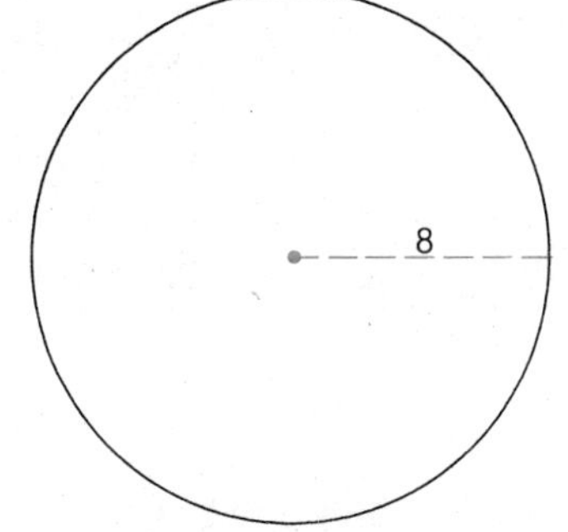

Figure 13-30

b.

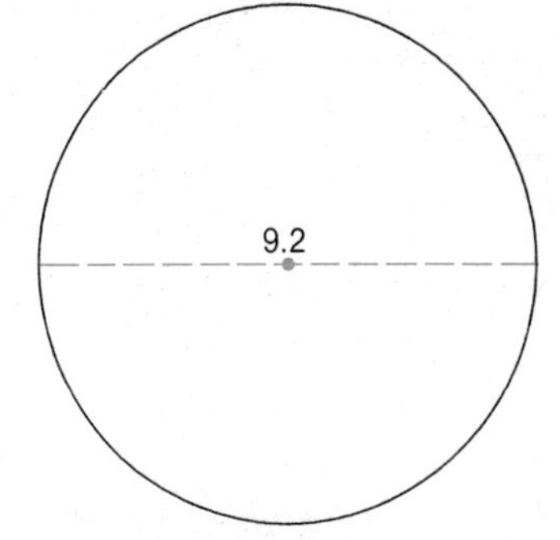

Figure 13-31

NOTE: A meter is approximately 3.3 feet.

Solution To find the area of a circle, use the formula $A = \pi r^2$.

12a. $A = \pi r^2$
$\doteq (3.14)(8 \text{ m})^2$
$\doteq (3.14)(64 \text{ m}^2)$
$\doteq 200.96 \text{ m}^2$

12b. $A = \pi r^2$
$\doteq (3.14)(4.6 \text{ m})^2$*
$\doteq (3.14)(21.16 \text{ m}^2)$
$\doteq 66.4424 \text{ m}^2$

*Note: $r = \dfrac{9.2 \text{ m}}{2} = 4.6 \text{ m}$

Example 13 Find the area of the semicircles in Figures 13–32 and 13–33. The units are given in inches.

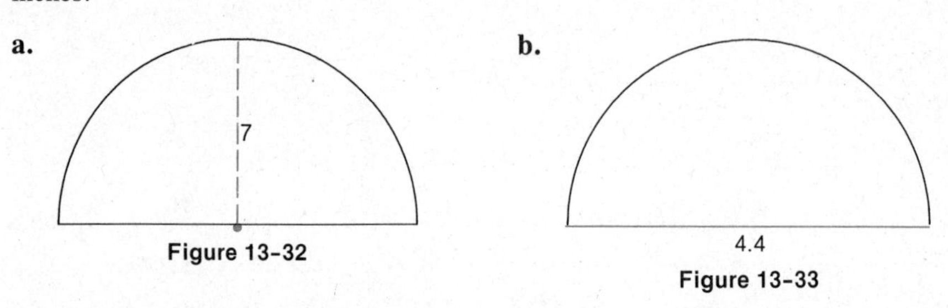

Figure 13-32
Figure 13-33

Solution Since a semicircle is half a circle, its area is found by using the formula $A = \dfrac{\pi r^2}{2}$.

13a. $A = \dfrac{\pi r^2}{2}$
$\doteq \dfrac{(3.14)(7 \text{ in})^2}{2}$
$\doteq \dfrac{(3.14)(49 \text{ in}^2)}{2}$
$\doteq \dfrac{153.86 \text{ in}^2}{2}$
$\doteq 76.93 \text{ in}^2$

13b. $A = \pi r^2$
$\doteq \dfrac{(3.14)(2.2 \text{ in})^2}{2}$
$\doteq \dfrac{(3.14)(4.84 \text{ in}^2)}{2}$
$\doteq \dfrac{15.1976 \text{ in}^2}{2}$
$\doteq 7.5988 \text{ in}^2$

Practice Problem 8 ***Find the area. The units are given in centimeters.***

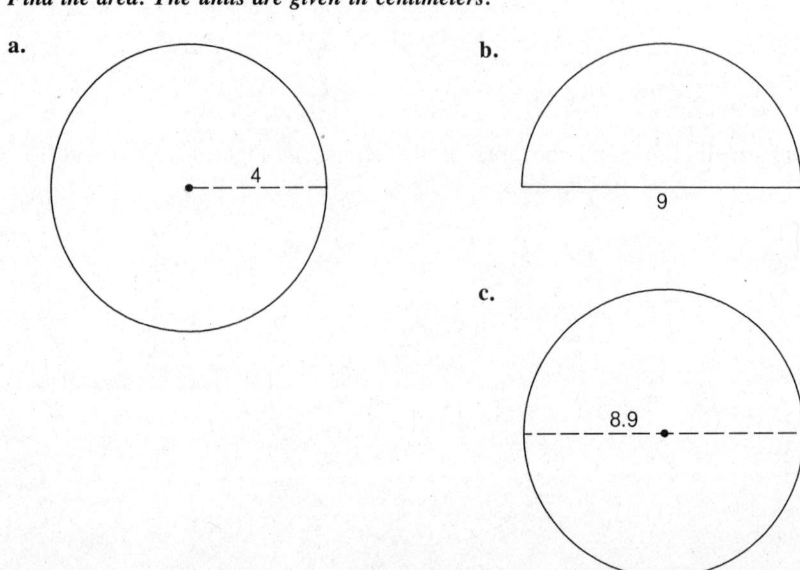

16 Sometimes you may need to find the area of a geometric figure that is a combination of two or more of the figures we have discussed. When this occurs, use the following procedure:

To Find the Area of a Combined Geometric Figure:

1. Divide the figure into rectangles, triangles, circles, line segments, or other familiar figures.
2. Find the area of each figure.
3. Find the sum or difference of these areas.

Example 14 Find the area of Figures 13–34 and 13–35. The units are given in feet.

a.

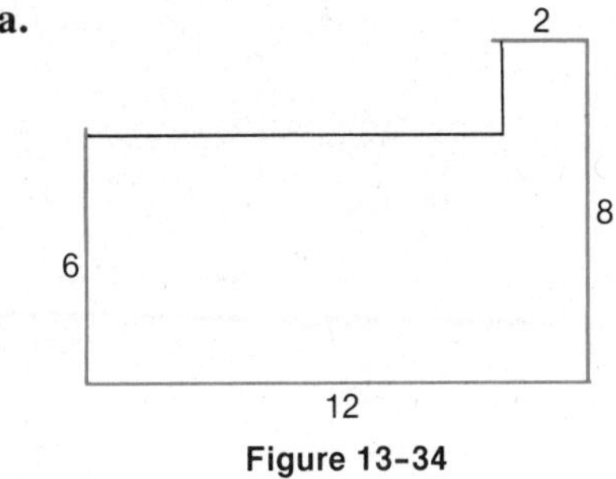

Figure 13-34

b.

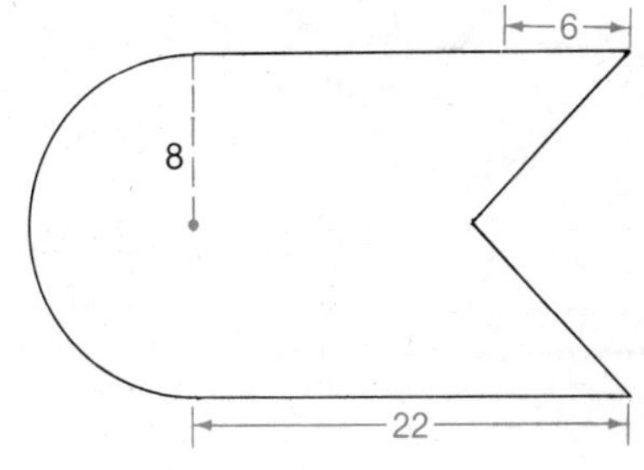

Figure 13-35

Solution To find the area of an unfamiliar figure, first divide the figure into two or more geometric figures. Next, determine the area of these figures. Finally, compute the sum or differences of the areas.

14a.

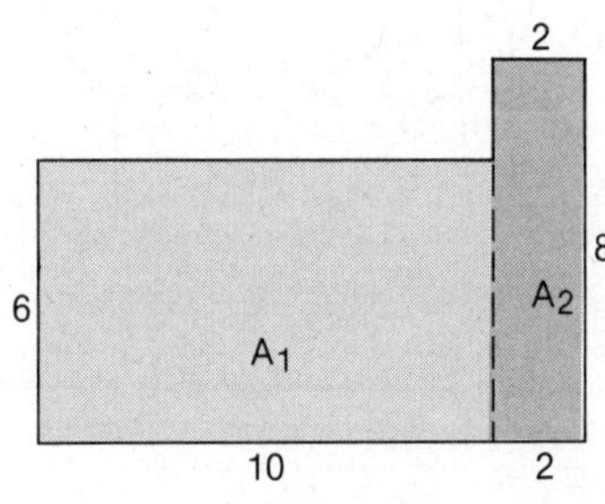

Figure 13-36

Total area is A = Area of rectangle + Area of rectangle

$$\begin{aligned} A &= A_1 + A_2 \\ &= 60 \text{ ft}^2 + 16 \text{ ft}^2 \\ &= 76 \text{ ft}^2 \end{aligned}$$

$$\begin{aligned} A_1 &= l \cdot w = (10 \text{ ft})(6 \text{ ft}) \\ &= 60 \text{ ft}^2 \\ A_2 &= l \cdot w = (2 \text{ ft})(8 \text{ ft}) \\ &= 16 \text{ ft}^2 \end{aligned}$$

14b.

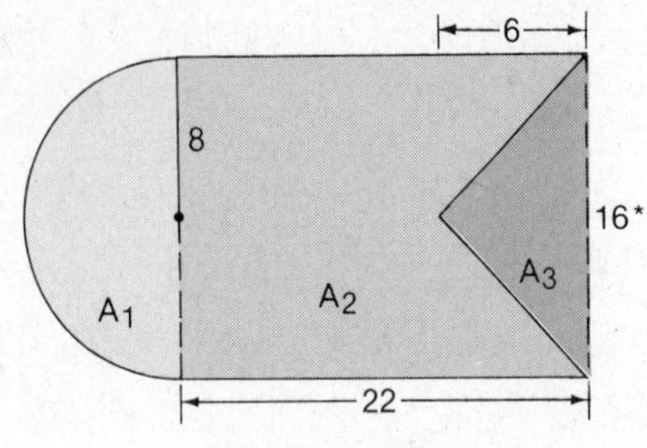

Figure 13-37

*Note: The base of the triangle and the width of the rectangle are the same as the diameter of the semicircle.

Total area is A = Area of semicircle + Area of rectangle − Area of triangle

$$A_1 = \frac{\pi r^2}{2} \doteq \frac{(3.14)(8\text{ ft})^2}{2} \doteq \frac{(3.14)(64\text{ ft}^2)}{2} \doteq 100.48\text{ ft}^2$$

$$A_2 = l \cdot w = (22\text{ ft})(16\text{ ft}) = 352\text{ ft}^2$$

$$A_3 = \frac{1}{2}bh = \frac{1}{2}(16\text{ ft})(6\text{ ft}) = 48\text{ ft}^2$$

Thus,

$$A = A_1 + A_2 - A_3$$
$$\doteq 100.48\text{ ft}^2 + 352\text{ ft}^2 - 48\text{ ft}^2$$
$$\doteq 404.48\text{ ft}^2.$$

Practice Problem 9 ***Find the area. The units are given in meters.***

a.

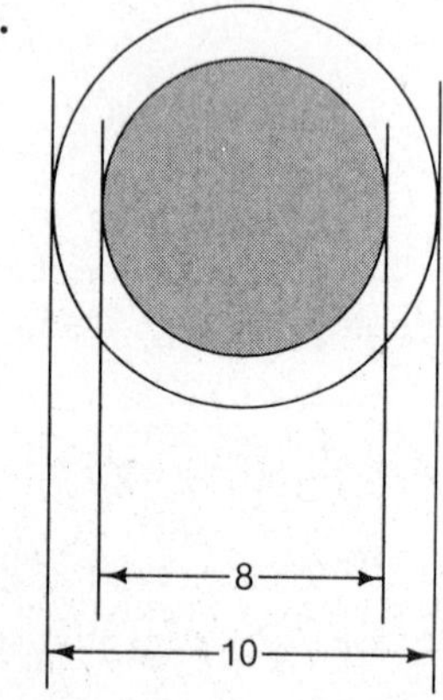

Find area of unshaded region.

b.

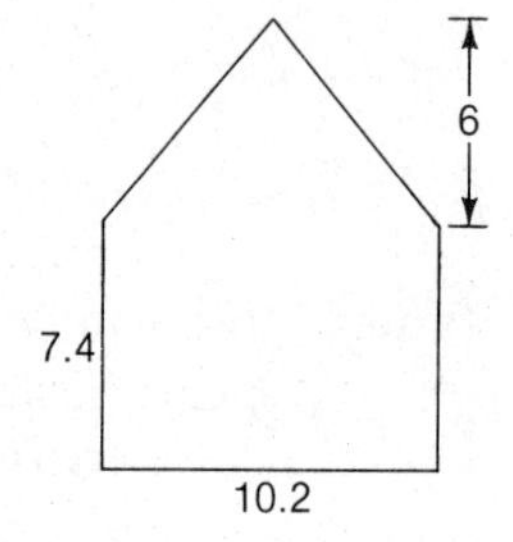

13.2 Exercises

Find the area. The units are given in feet.

1.

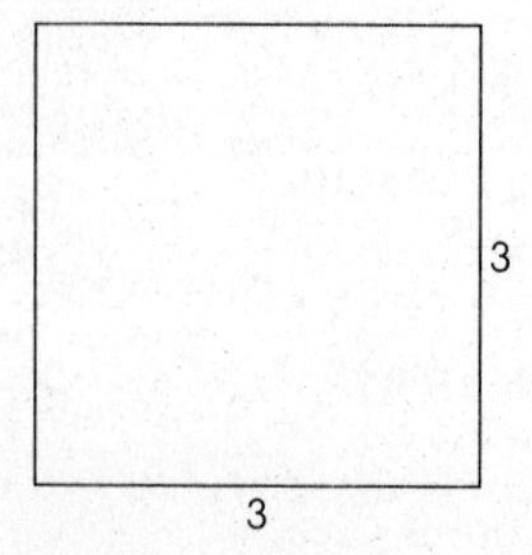

2.

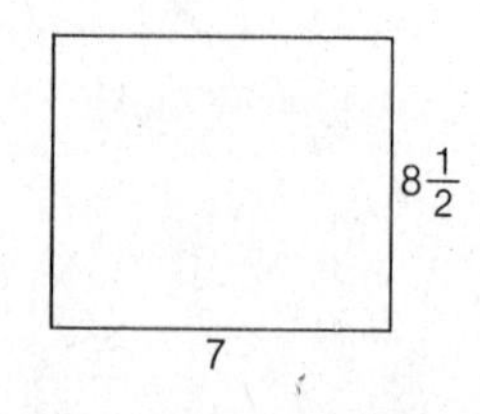

3.

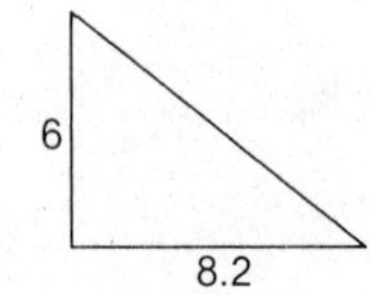

4.

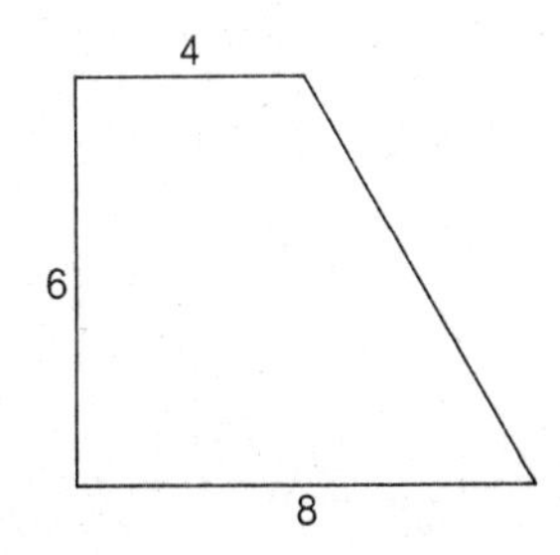

5.

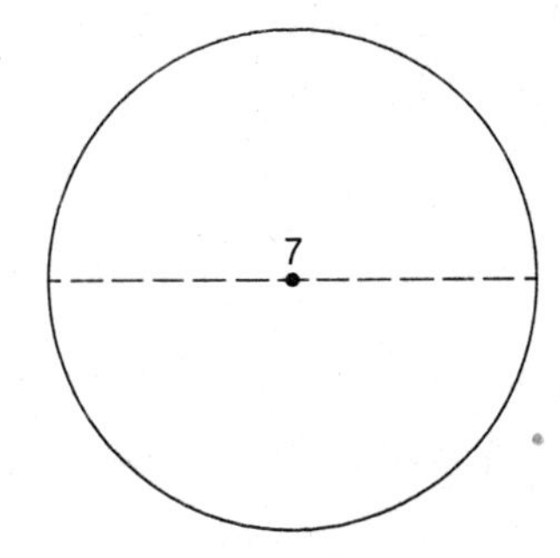

6.

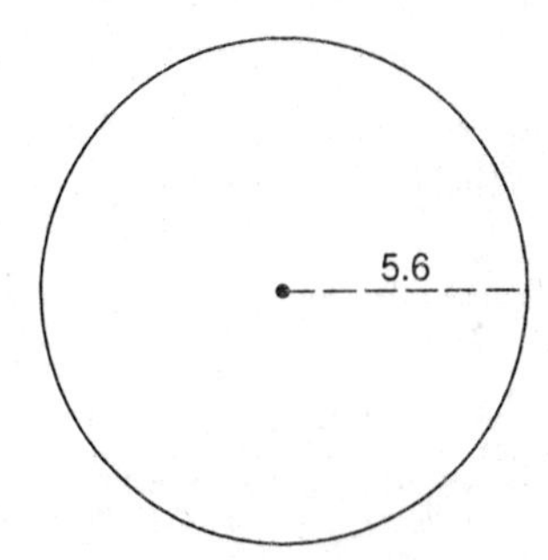

Find the area. The units are given in centimeters.

7.

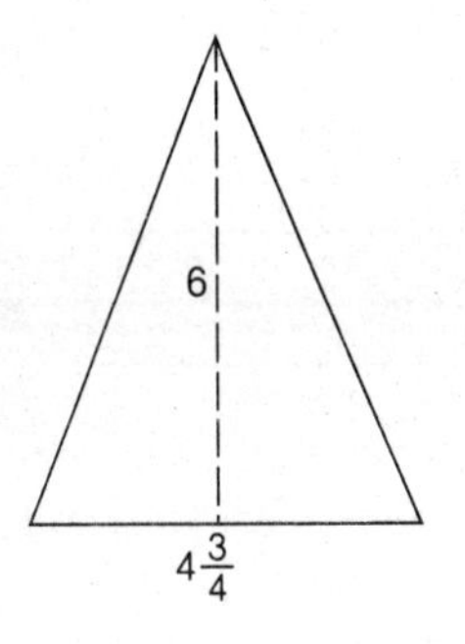

8.

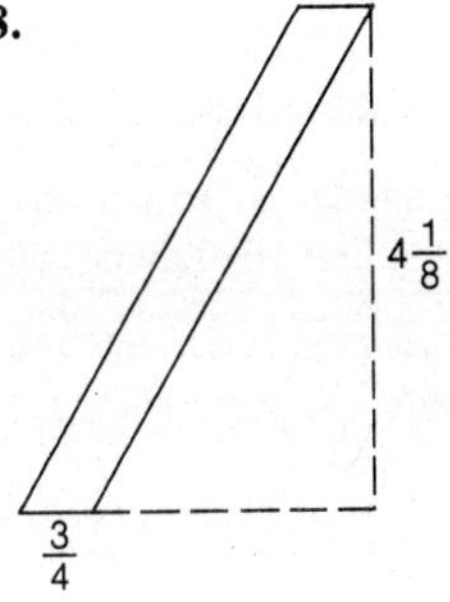

9.

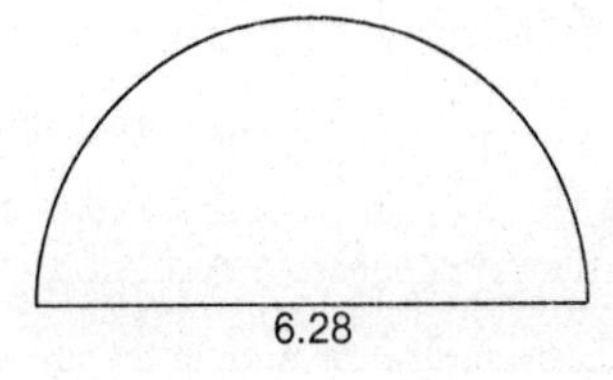

10.

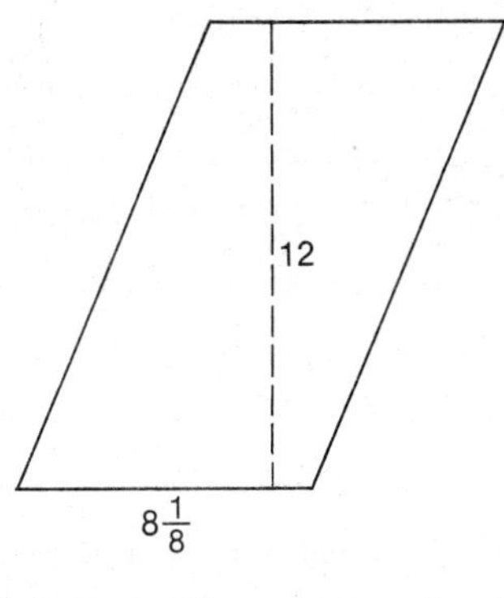

11.

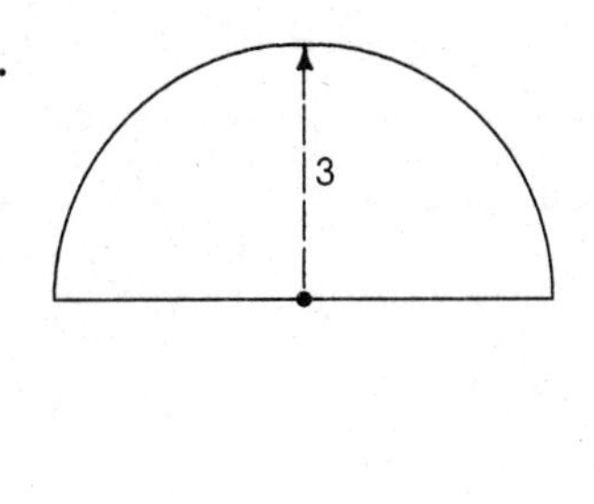

12.

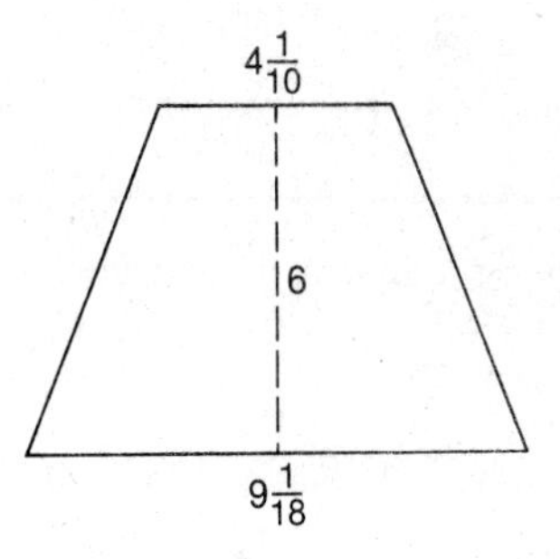

Find the area. The units are given in centimeters.

13.

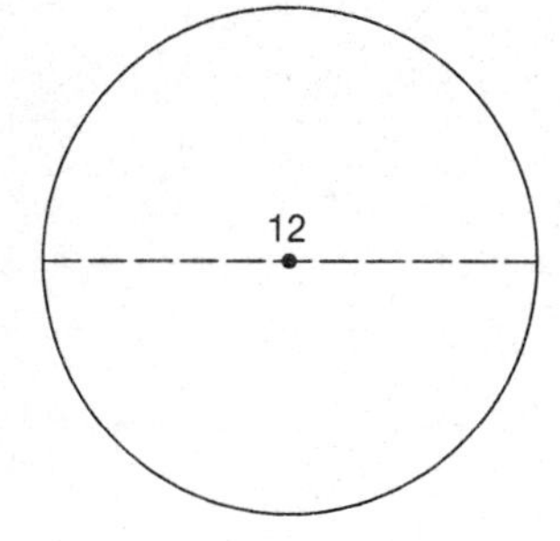

14.

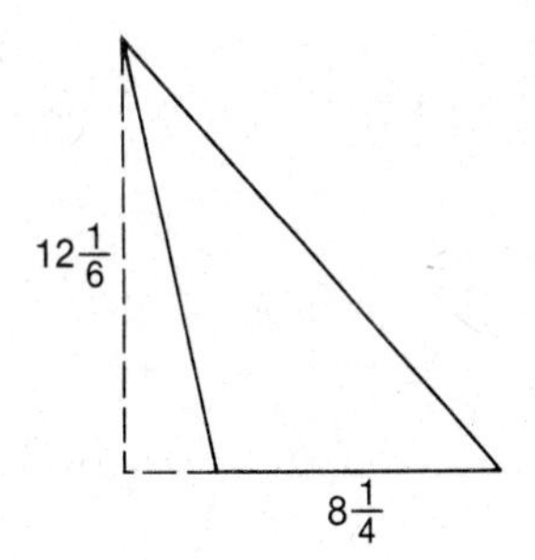

15.

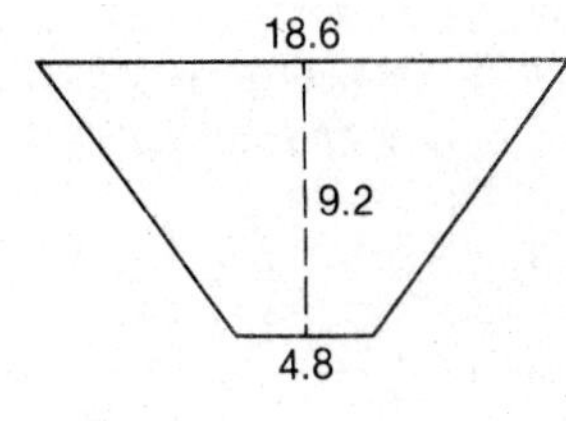

16.

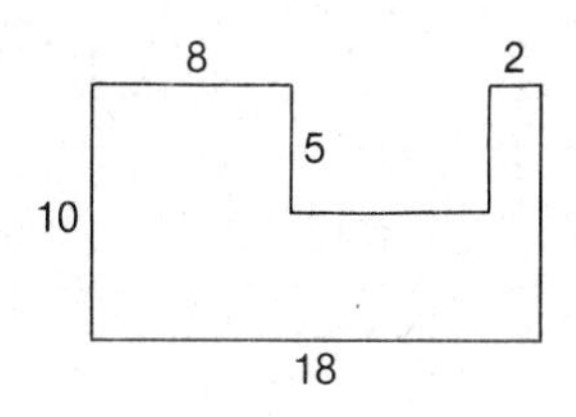

17.

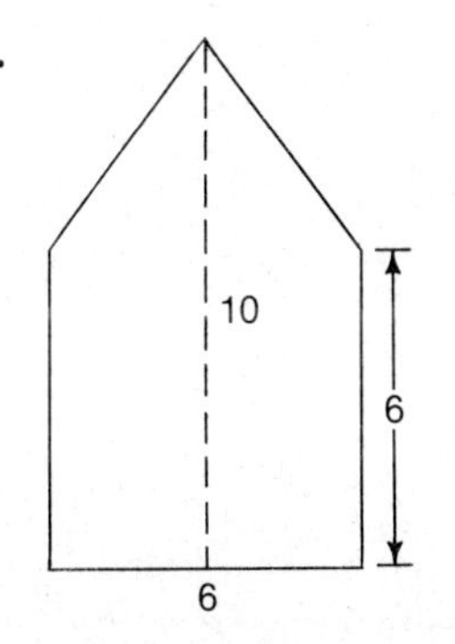

18.

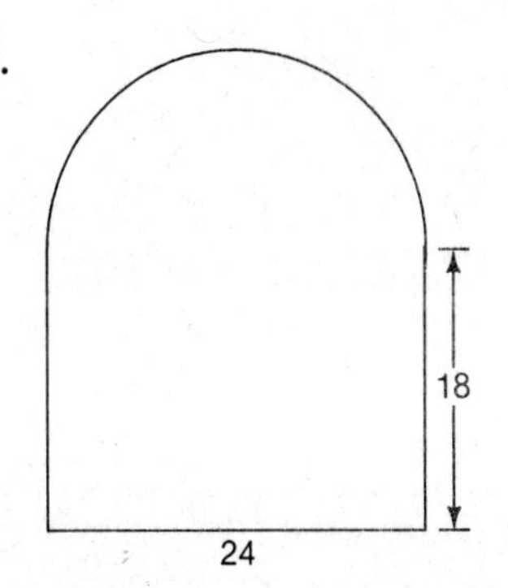

19.

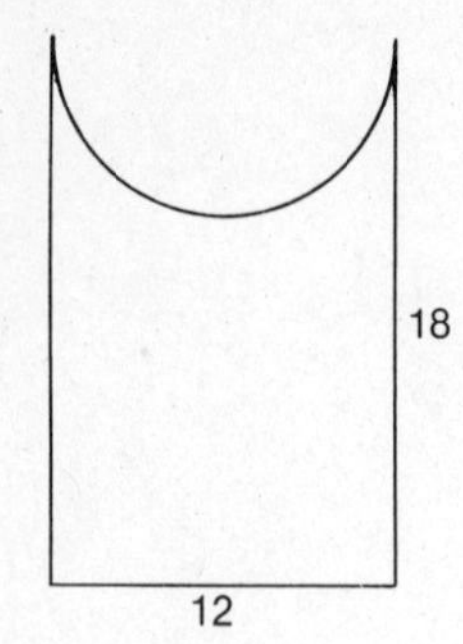

20.

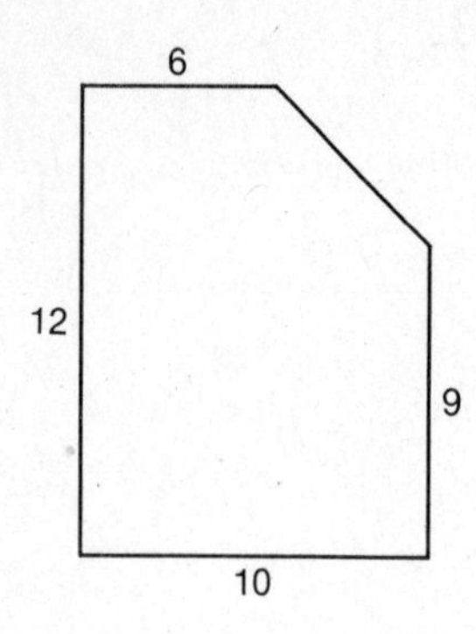

21.

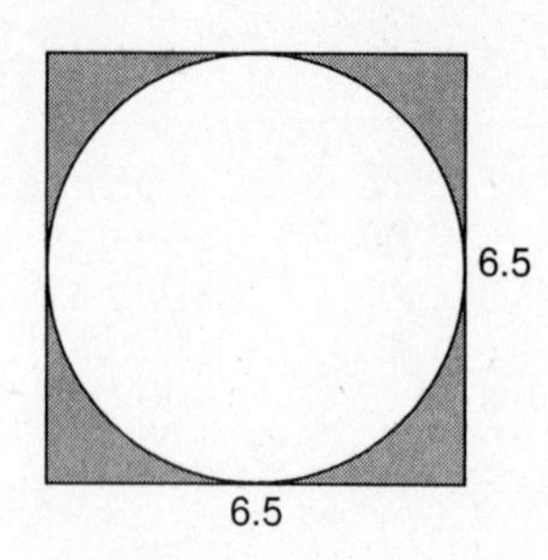

Find the area of the shaded region

Fill in the blanks.

22. ______________ is a measure of a surface, and it is expressed in ______________ units.

23. A ______________ is a four-sided polygon in which the opposite sides are equal and parallel. A ______________ is a three-sided polygon. A ______________ is a four-sided polygon having only two parallel sides.

24. The formulas for finding the areas of rectangles, triangles, parallelograms, trapezoids, and circles are as follows:
a. rectangle: ______________ **b.** triangle: ______________
c. parallelogram: ______________ **d.** trapezoid: ______________
e. circle: ______________

Solve.

25. Ben has a roll of paper that is 34 inches by 55 inches. How many sheets of paper $8\frac{1}{2}$ inches by 11 inches can he cut from this roll of paper?

26. A lot is 40 meters by 20 meters. If a circular swimming pool with a diameter of 12 meters is constructed on the lot, how much area is left over?

27. Tim wants to carpet a rectangular-shaped room that is 80 feet by 90 feet.

a. How many square yards of carpeting will he need? ($1\text{ yd}^2 = 9\text{ ft}^2$)

b. If the carpeting costs $8.50 per square yard and Tim must pay 6% sales tax, what is the total cost of the carpeting?

28. A radio station broadcasts over an area with a 40-mile radius. How much area is this?

29. If one ounce of KEM Weed Killer will effectively treat $1\frac{1}{2}$ square yards of a lawn, how many ounces of KEM Weed Killer will Bill need to treat a triangular-shaped lawn with a height of 15 yards and a base of 10 yards?

30. A square tile (one foot on each side) costs 70¢ per tile.

a. How many square tiles will Vanessa need to cover a rectangular-shaped room that is 19 feet by 23 feet?

b. If Vanessa must pay $5\frac{3}{4}$% sales tax, what is the total cost of the tile? (Round to the nearest cent.)

Answers to Practice Problems **4a.** 18.56 in^2 **b.** 69 in^2 **c.** 71.4025 ft^2 **5a.** 32.5 cm^2 **b.** $39\frac{3}{8}$ cm^2 **6a.** 15 cm^2 **b.** $22\frac{5}{16}$ cm^2 **7a.** $59\frac{1}{2}$ cm^2 **b.** 93 cm^2 **8a.** 50.24 cm^2 **b.** 31.7925 cm^2 **c.** 62.17985 cm^2 **9a.** 28.26 m^2 **b.** 106.08 m^2

13.3 Volume

17 **Volume** is the measure of the capacity within a three-dimensional figure, and it is expressed in cubic units. Two such units—a cubic inch (in^3) and a cubic centimeter (cc or cm^3)—are shown in Figures 13–38 and 13–39.

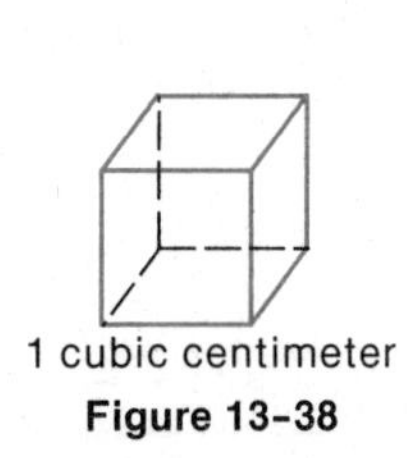

Figure 13–38

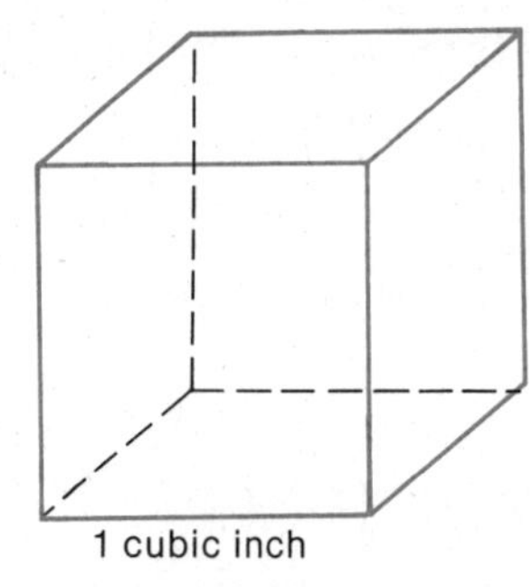

Figure 13–39

18 When you are asked to find the volume of a figure, you will need to determine the number of cubic units contained in that figure.

Example 15 Consider a rectangular box with a length of 4 cm, a width of 3 cm, and a height of 2 cm (see Figure 13–40). You can see that the box contains 12 cubes on the top layer and 12 cubes on the bottom layer. Thus, its volume is 24 cubic centimeters (24 cm^3 or 24 cc).

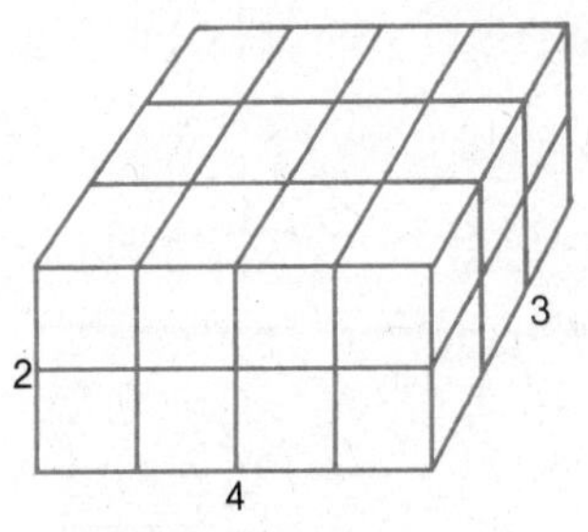

Figure 13–40

19 Obviously, it will be inconvenient to mentally divide a geometric figure into unit cubes and then count these cubes. Therefore, to find the volume of a rectangular box, cylinder, cone, or sphere, use the formulas given in Figure 13–41.

Rectangular box

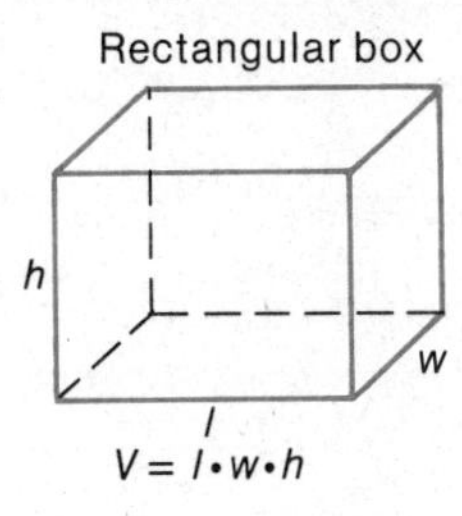

$V = l \cdot w \cdot h$

Cylinder

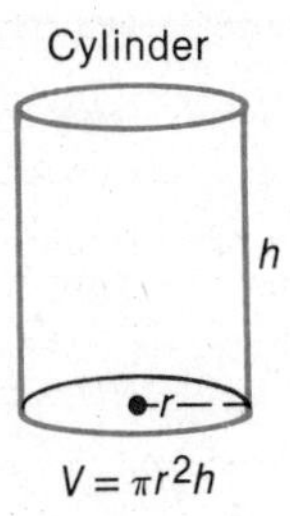

$V = \pi r^2 h$

Cone

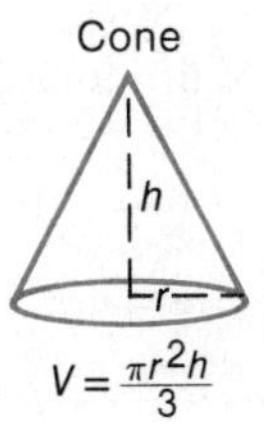

$V = \frac{\pi r^2 h}{3}$

Sphere

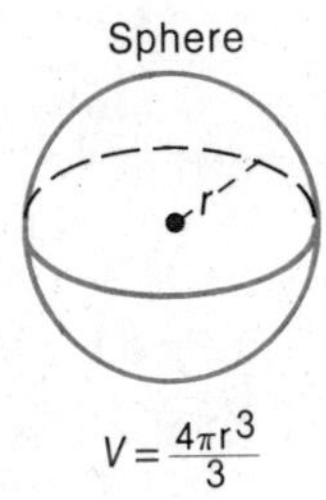

$V = \frac{4\pi r^3}{3}$

Figure 13–41

Example 16 Find the volume of Figures 13–42 through 13–45. The units are given in meters.

a.

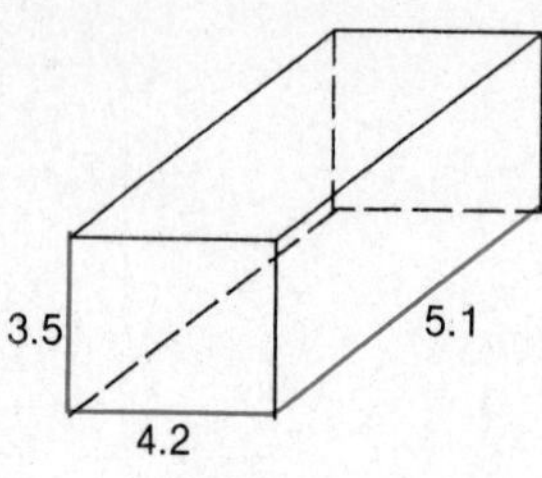

Figure 13-42

b.

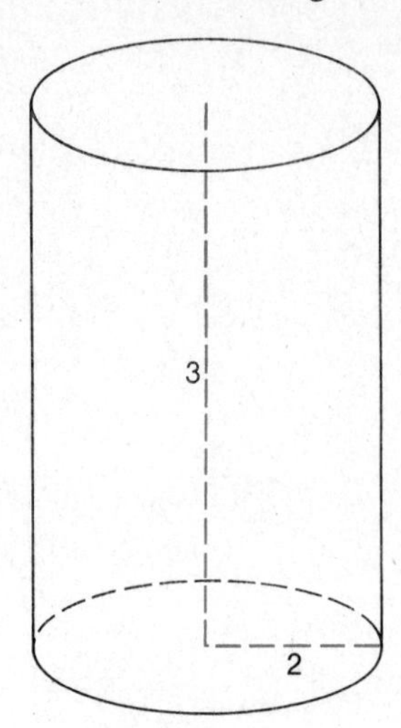

Figure 13-43

c.

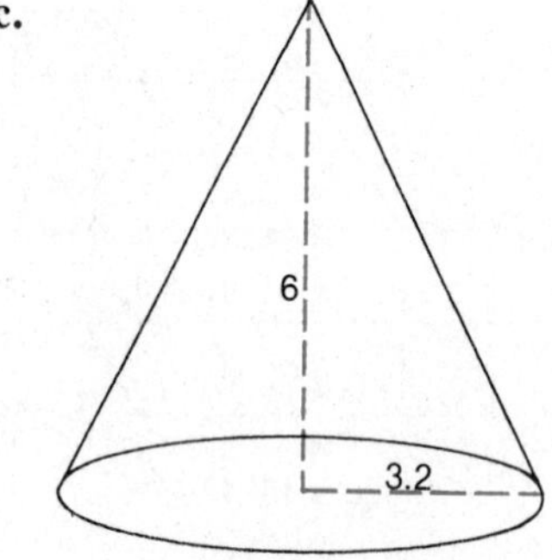

Figure 13-44

d.

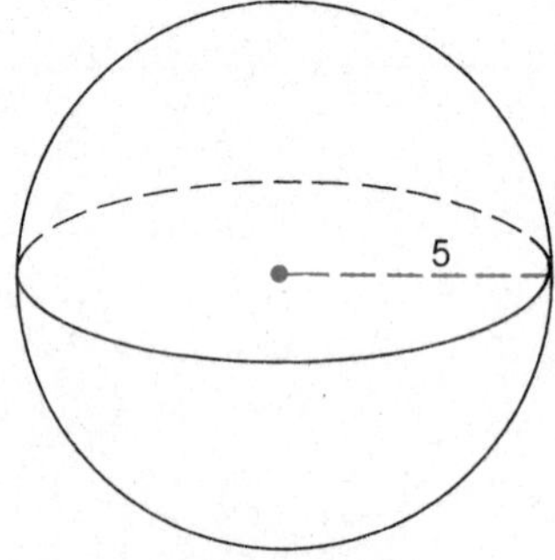

Figure 13-45

Solution **16a.** The volume of the rectangular box (Figure 13–42) is

$$\begin{aligned} V &= lwh \\ &= (4.2\text{ m})(5.1\text{ m})(3.5\text{ m}) \\ &= 74.97\text{ m}^3 \text{ (read as 74.97 cubic meters).} \end{aligned}$$

16b. The volume of the cylinder (Figure 13–43) is

$$\begin{aligned} V &= \pi r^2 h \\ &\doteq (3.14)(2\text{ m})^2(3\text{ m}) \\ &\doteq (3.14)(4\text{ m}^2)(3\text{ m}) \\ &\doteq 37.68\text{ m}^3. \end{aligned}$$

16c. The volume of the cone (Figure 13–44) is

$$\begin{aligned} V &= \frac{\pi r^2 h}{3} \\ &\doteq \frac{(3.14)(3.2\text{ m})^2(6\text{ m})}{3} \\ &\doteq \frac{(3.14)(10.24\text{ m}^2)(6\text{ m})}{3} \\ &\doteq 64.3072\text{ m}^3. \end{aligned}$$

16d. The volume of the sphere (Figure 13–45) is

$$V = \frac{4\pi r^3}{3}$$
$$\doteq \frac{4(3.14)(5\text{ m})^3}{3}$$
$$\doteq \frac{4(3.14)(125\text{ m}^3)}{3}$$
$$\doteq 523.33\text{ m}^3 \text{ (round to nearest hundredth).}$$

Practice Problem 10 ***Find the volume. The units are given in inches.***

a.

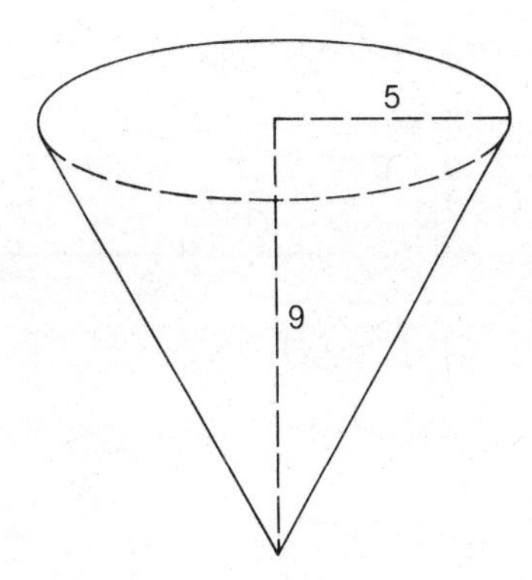

b.

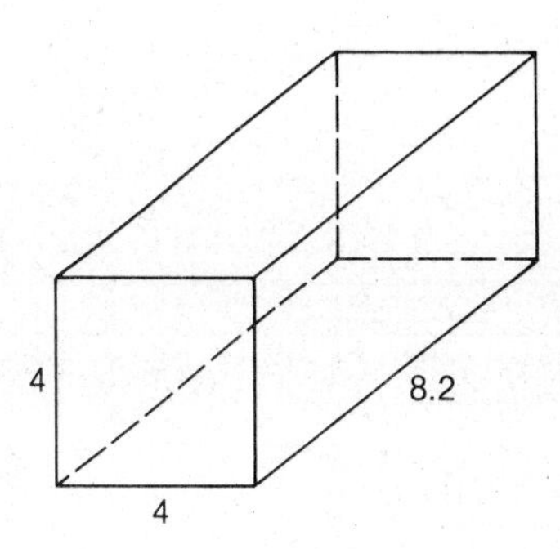

c.

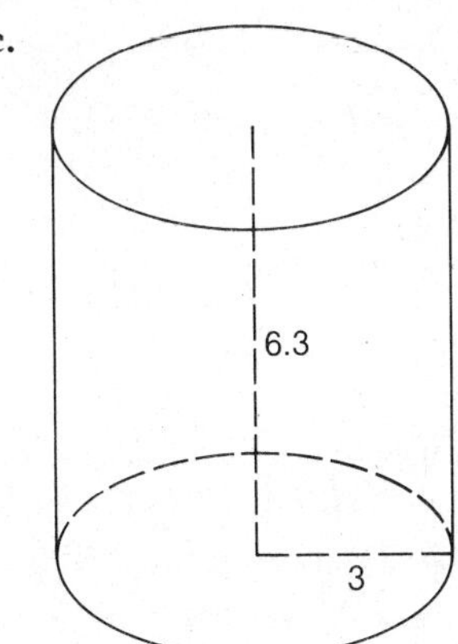

d.

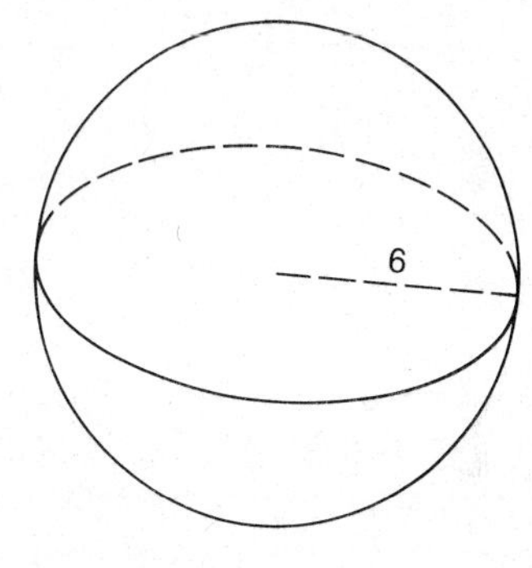

13.3 Exercises

Find the volume. The units are given in inches.

1.

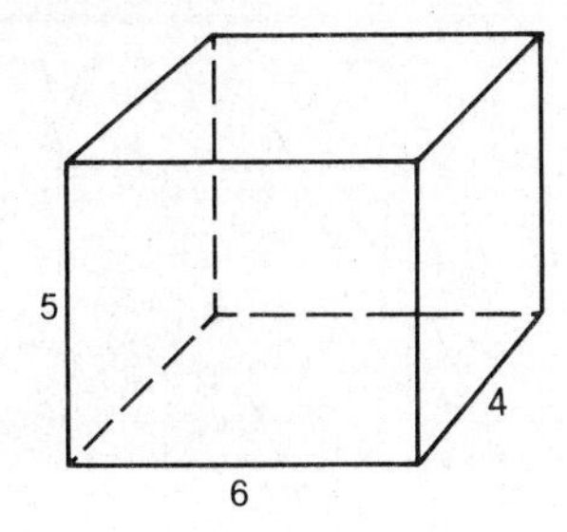

2.

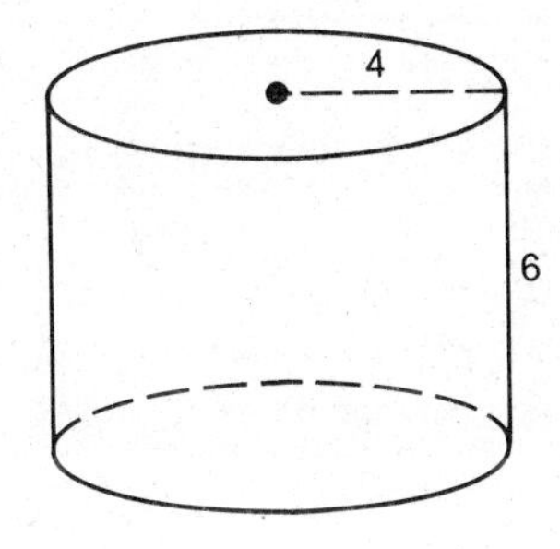

3.

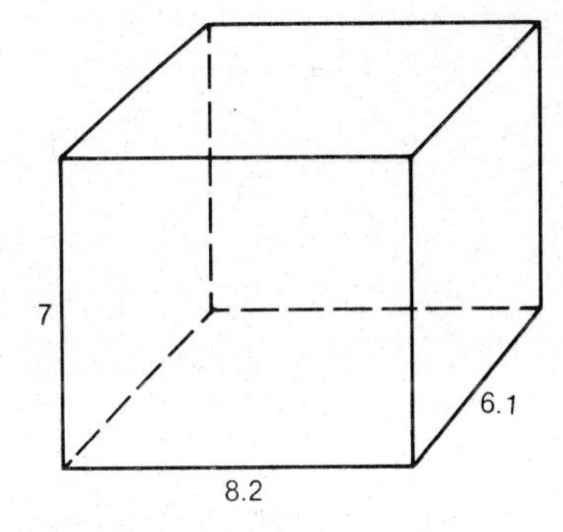

4.

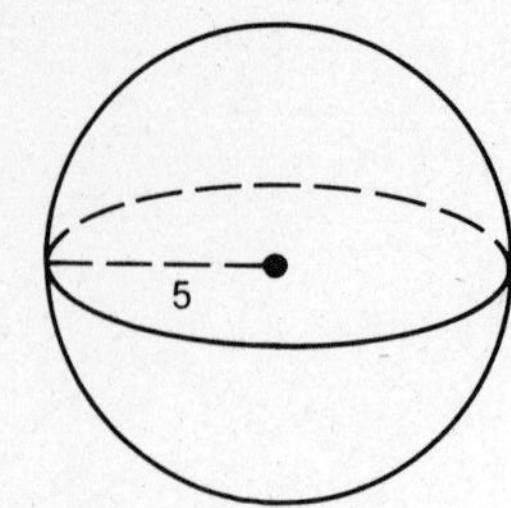

5.

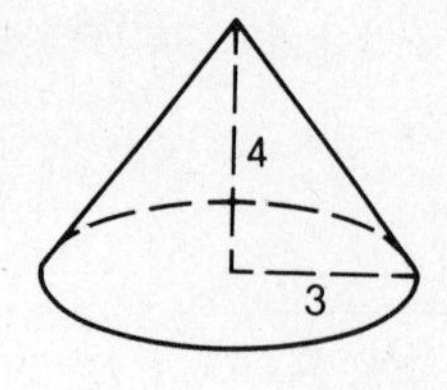

6.

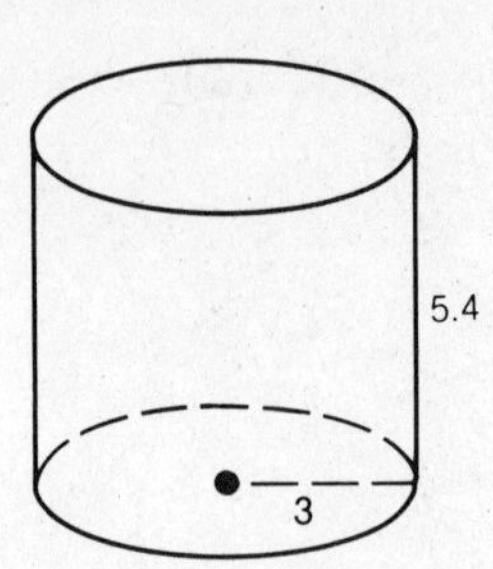

7.

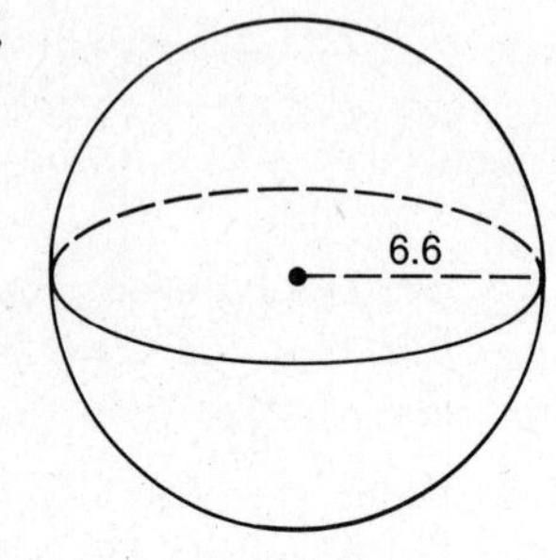

8.

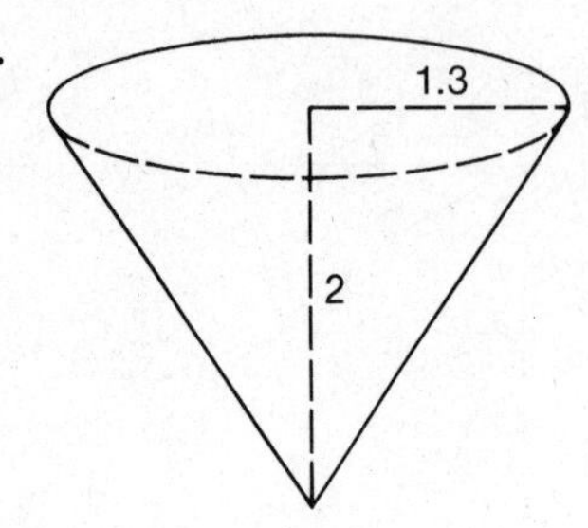

9.

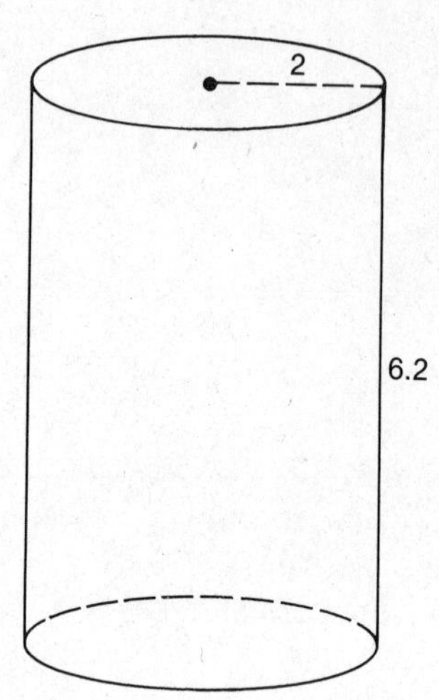

10.

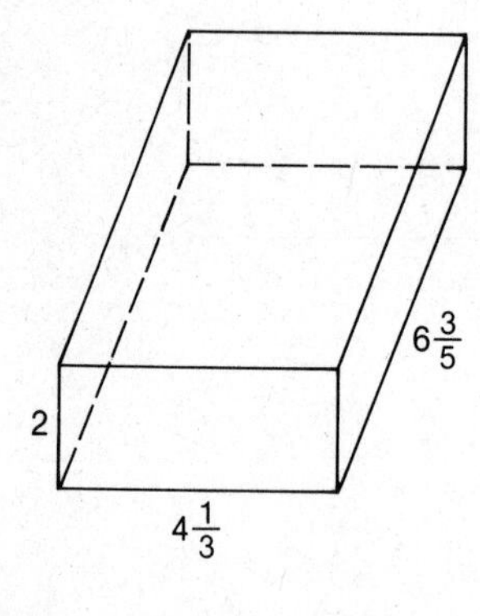

11.

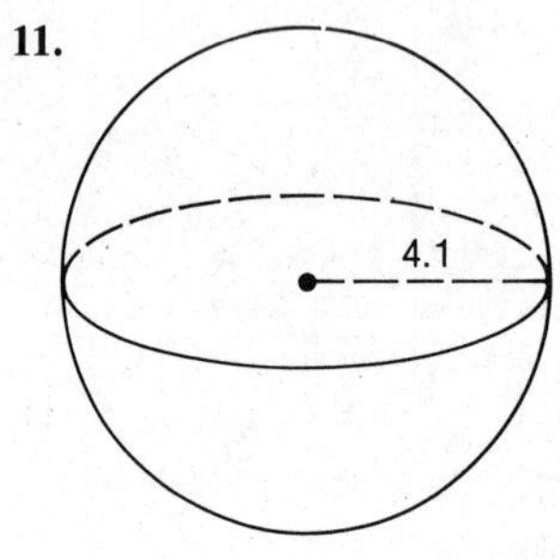

12.

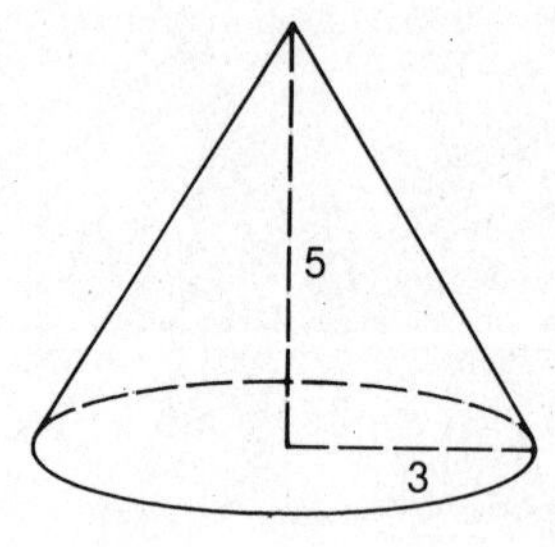

13.

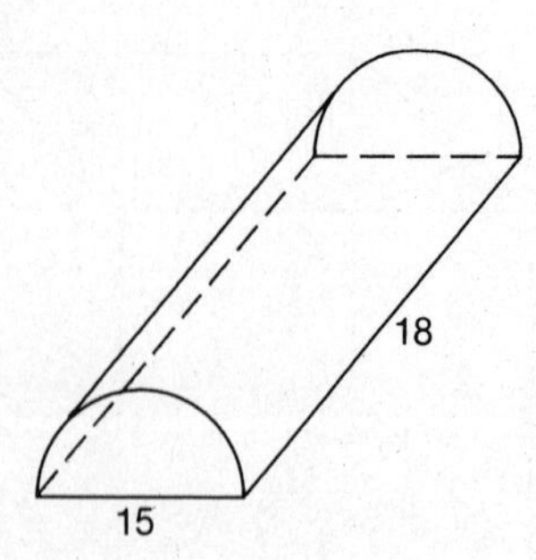

14.

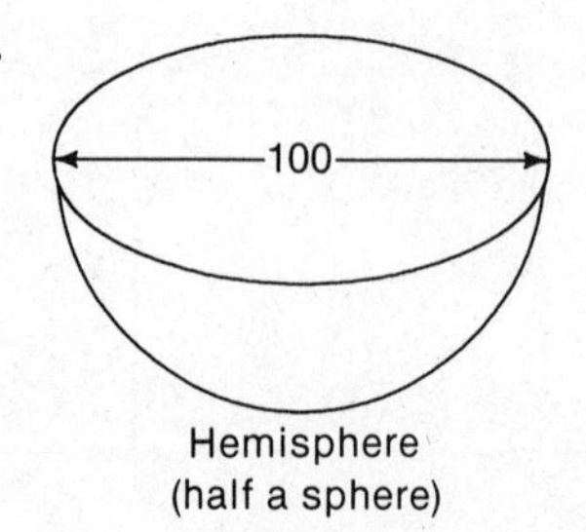

Hemisphere
(half a sphere)

15.

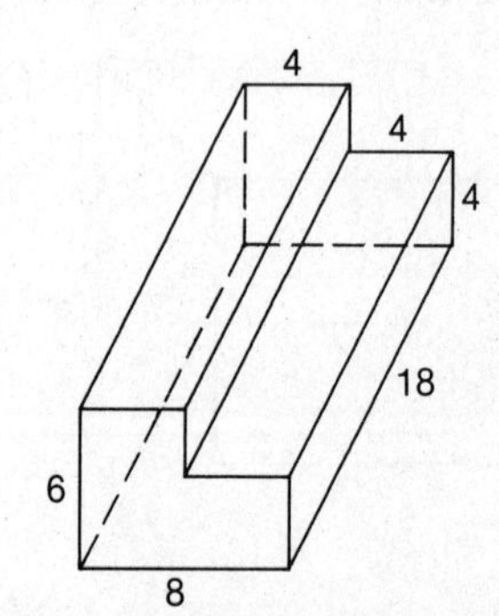

Fill in the blanks.

16. ________________ is the measure of the capacity within a three-dimensional figure, and it is measured in ________________ units.

17. The formulas for finding the volumes of rectangular boxes, cylinders, cones, and spheres are as follows:

a. rectangular box: ________________ **b.** cylinder: ________________

c. cone: ________________ **d.** sphere: ________________

Solve.

18. The figure shown below is a cylinder capped with a hemisphere (half a sphere). Find its volume.

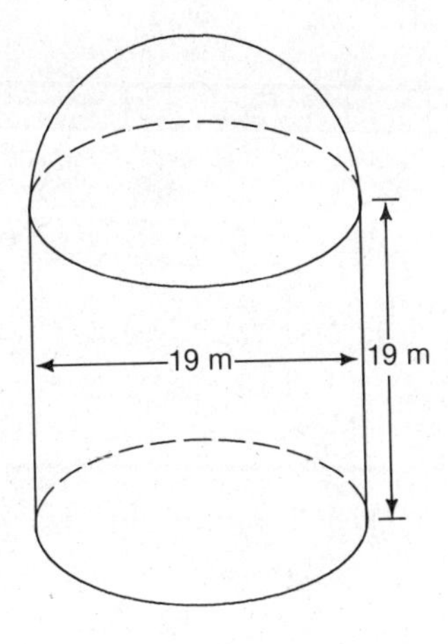

19. Find the volume of air contained in a balloon with a diameter of 8 cm.

20. A hole that is 18 feet wide, 11 feet deep, and six feet long is being made for a storage room. If the bed of a truck holds 11 yd^3, how many truckloads of dirt must be hauled away? (one yard = three feet)

21. A cylindrical-shaped water tank is 44 inches in diameter and 21 inches high. If there are 231 in^3 in one gallon, how many gallons of water can this tank hold?

22. A truck is eight feet wide, six feet high, and 21 feet long. How much dirt can be loaded into the truck?

23. A swimming pool is 40 feet long, 10 feet wide, and six feet deep.

a. How many cubic feet of water will the pool hold?

b. If 1 $ft^3 \doteq 7.5$ gallons, how many gallons of water will the pool hold?

Answers to Practice Problems **10a.** 235.5 in^3 **b.** 131.2 in^3 **c.** 178.038 in^3 **d.** 904.32 in^3

Summary Important Terms

13.1

polygon
rectangle
length
width
square
triangle
perimeter
circle
circumference
pi, π
diameter
radius
semicircle

13.2

area
parallelogram
base
height
trapezoid

13.3

volume
cylinder
cone
sphere

Important Formulas

Figure	Perimeter	Area
Rectangle	$P = 2l + 2w$	$A = lw$
Square	$P = 4s$	$A = s^2$
Triangle	$P = a + b + c$	$A = \frac{1}{2}bh$
Circle	$C = \pi d$	$A = \pi r^2$

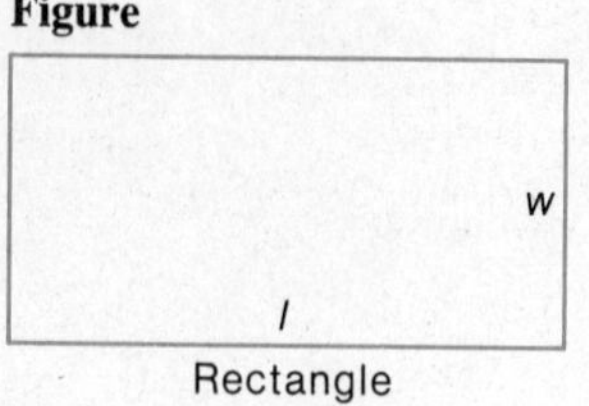

Rectangle

s
s
Square

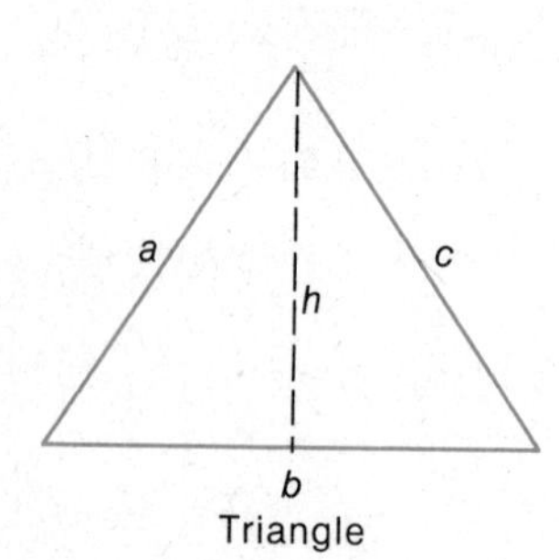

Triangle

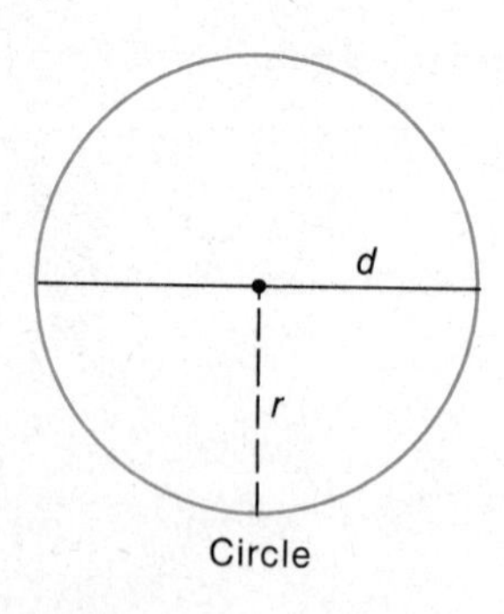

Circle

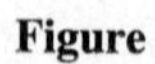

Figure	Area
Parallelogram	$A = bh$
Trapezoid	$A = \frac{1}{2}(a + b) \cdot h$

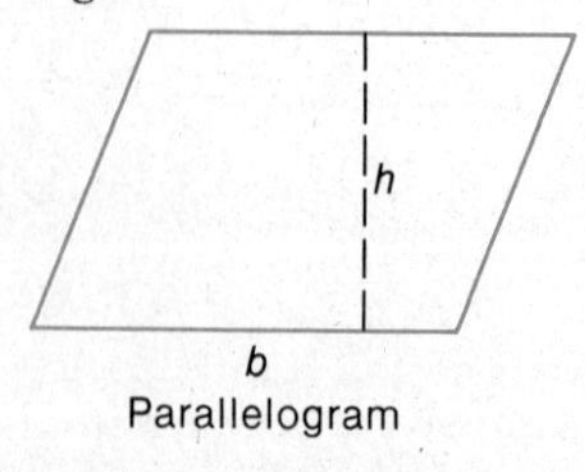

Parallelogram

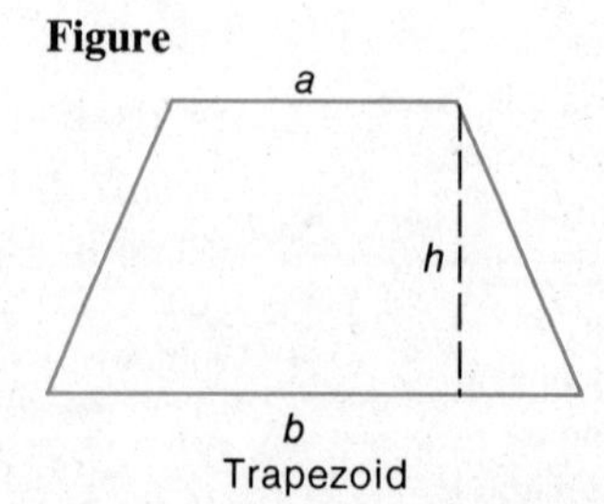

Trapezoid

Figure	Volume	Figure	Volume
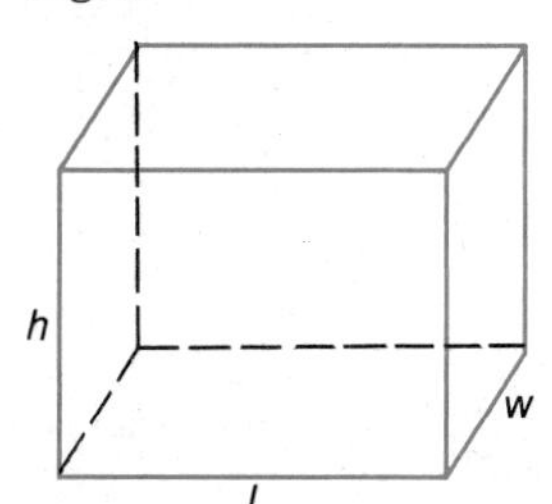 Rectangular box	$V = lwh$	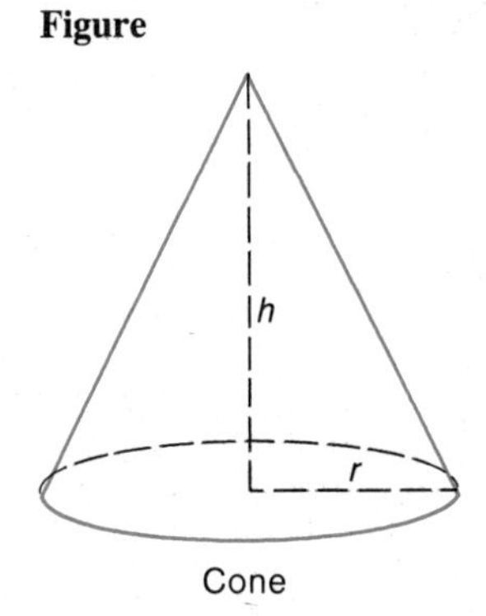 Cone	$V = \dfrac{\pi r^2 h}{3}$
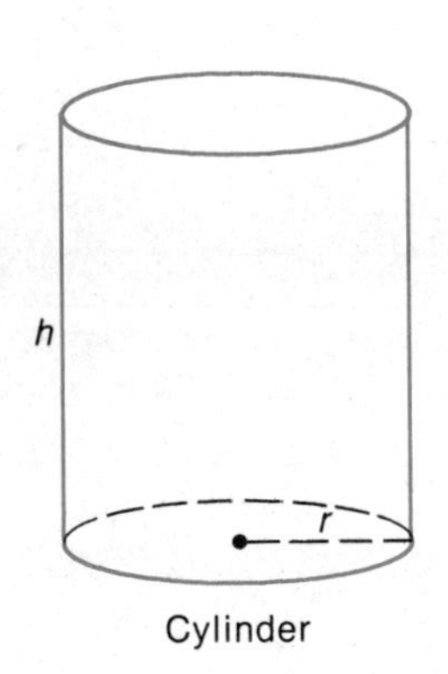 Cylinder	$V = \pi r^2 h$	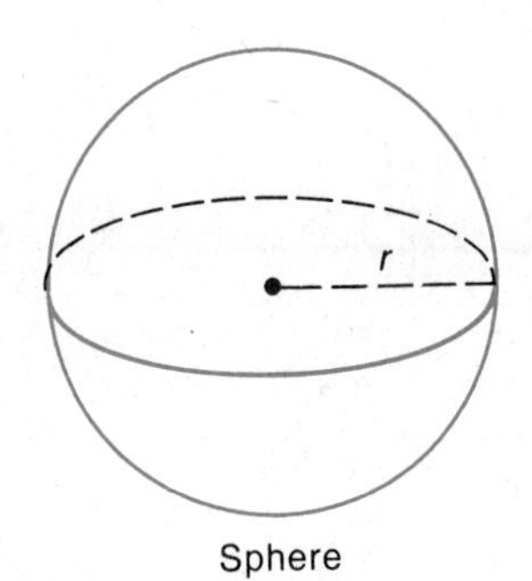 Sphere	$V = \dfrac{4\pi r^3}{3}$

Important Skills

13.1

Finding the perimeter of a polygon
Finding the circumference of a circle
Finding the length of a semicircle
Finding the perimeter of an unfamiliar geometric figure
Solving word problems involving perimeter

13.2

Finding the area of rectangles, parallelograms, triangles, trapezoids, and circles
Finding the area of an unfamiliar geometric figure
Solving word problems involving area

13.3

Finding the volume of boxes, cylinders, cones, and spheres
Finding the volume of unfamiliar geometric figures
Solving word problems involving volume

Review Exercises

Find the perimeter. The units are given in inches.

1.

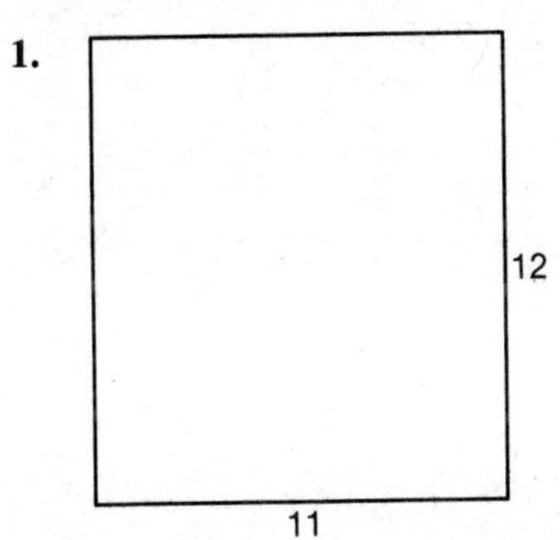

2.

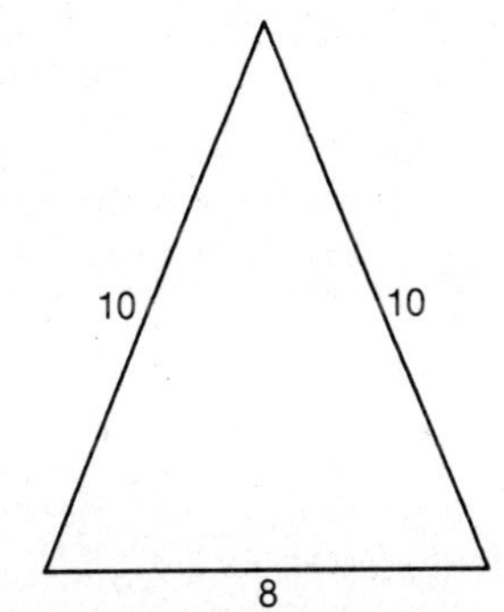

3.

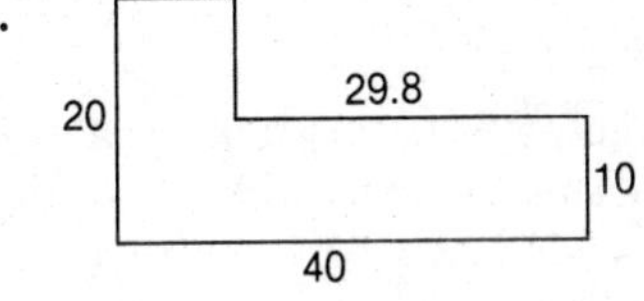

Find the area. The units are given in feet.

4.

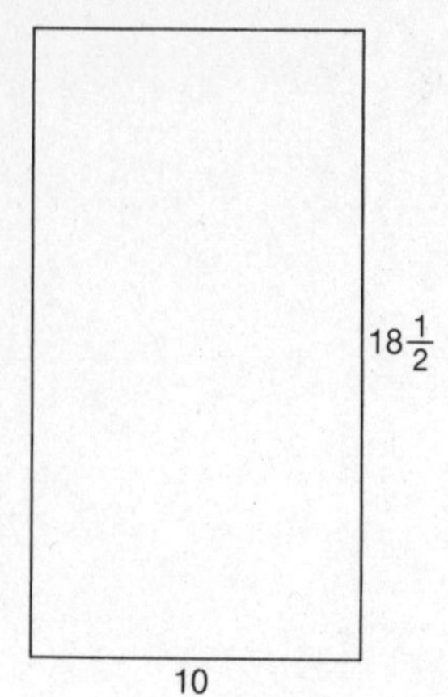

5.

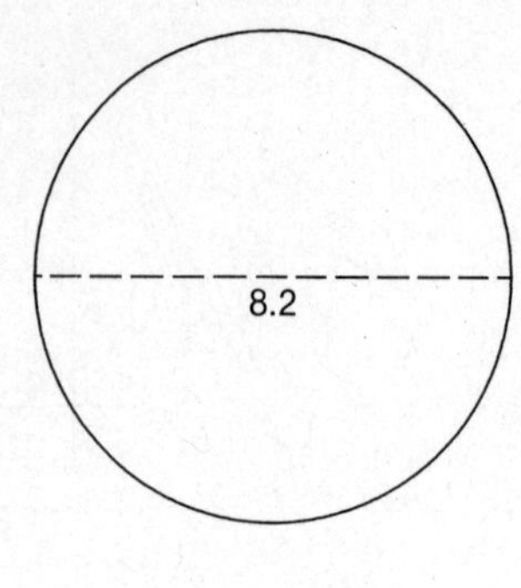

6.

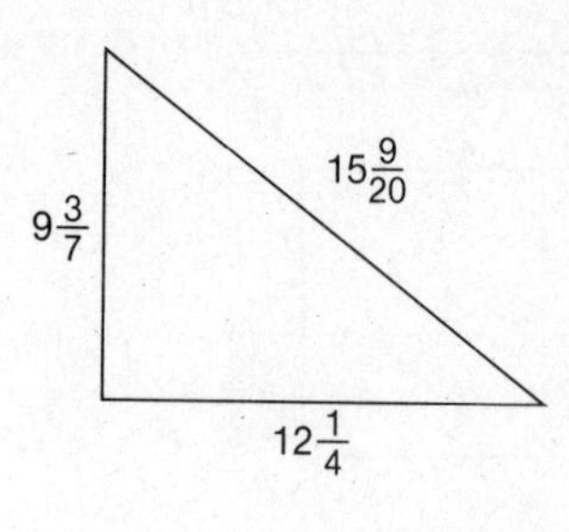

7. Find the circumference of the circle in Problem 5.

8. Find the area of the figure in Problem 3.

Find the volume. The units are given in inches.

9.

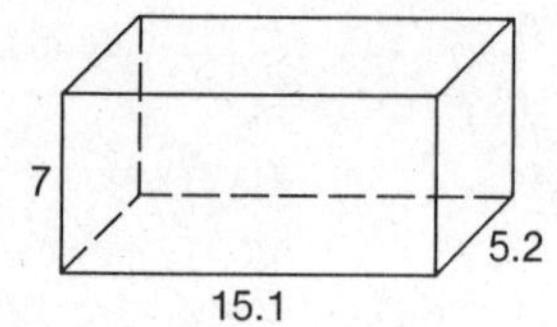

10.

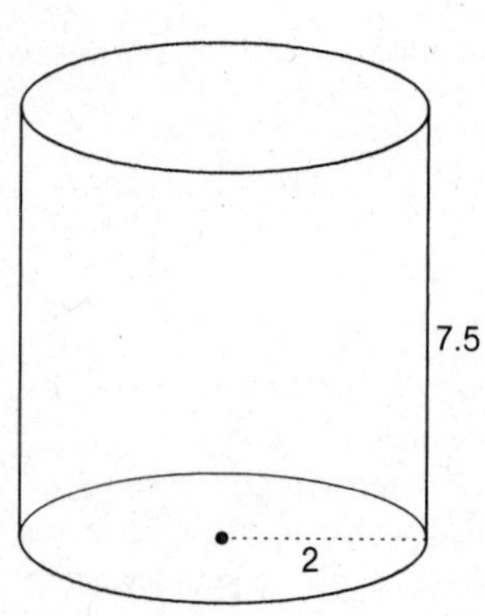

11.

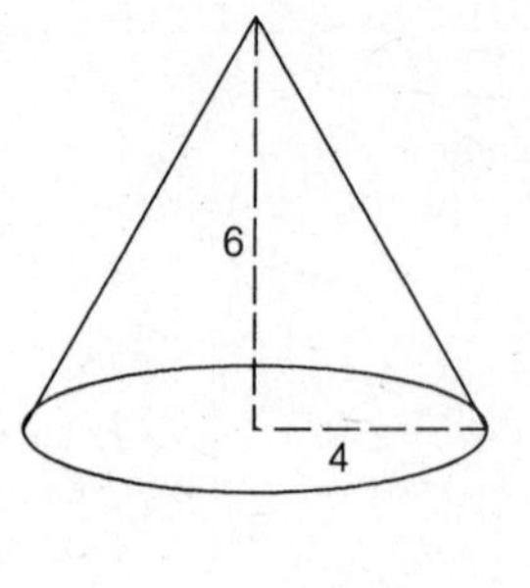

Find the perimeter. The units are given in feet.

12.

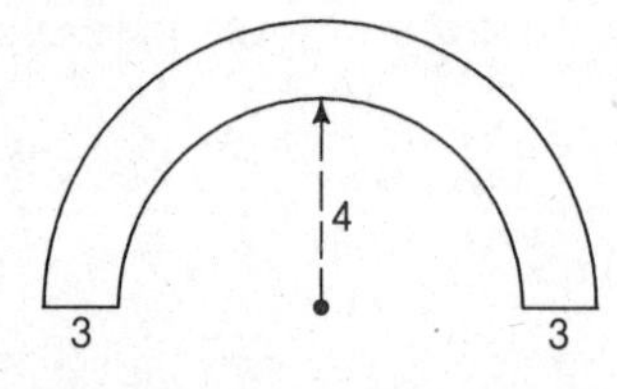

13.

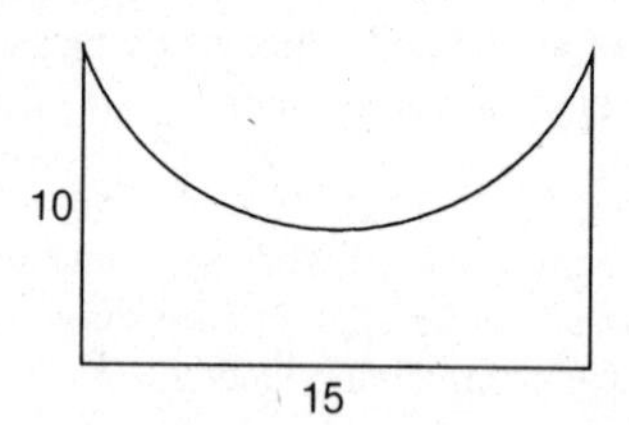

14.

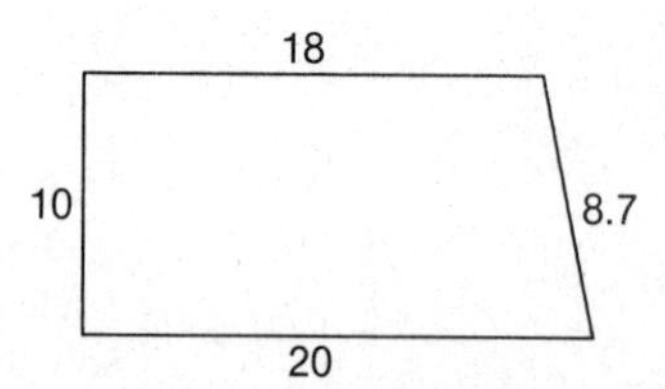

15. Find the area of the figure in Problem 12.

16. Find the area of the figure in Problem 13.

17. Find the area of the figure in Problem 14.

18. A storage tank 8.3 feet long and 4.1 feet wide is filled with water to a depth of 6.9 feet. How many cubic feet of water are in the tank?

19. How many square yards of carpeting will Gilly need to carpet a rectangular-shaped floor that measures 33 feet by 18 feet? (one yard = three feet)

20. A rectangular-shaped playground is $20\frac{1}{4}$ yards by $30\frac{5}{6}$ yards. How many yards of fencing will be needed to build a fence around the playground?

Appendix: Measurement

Table of Squares, Square Roots, and Primes

Answers to Selected Exercises

Index

APPENDIX: MEASUREMENT

Idea **1** The two major systems of measurement used today are the **English system** and the **metric system.** The relationships *within* the two systems are given in Tables A–1 and A–2. Table A–3 summarizes the basic relationships *between* the two systems.

Table A–1 Common English Units of Measure

Volume
1 gallon (gal) = 4 quarts (qt)
1 quart (qt) = 2 pints (pt)

Length
1 foot (ft) = 12 inches (in)
1 yard (yd) = 3 feet (ft)
1 mile (mi) = 5280 feet (ft)

Weight
1 pound (lb) = 16 ounces (oz)
1 ton = 2000 pounds (lb)

Time
1 minute (min) = 60 seconds (sec)
1 hour (hr) = 60 minutes (min)
1 day (da) = 24 hours (hr)

Table A–2 Common Metric Units of Measure

Prefixes
kilo (1000 times)
milli $\left(\frac{1}{1000} \text{ of}\right)$
centi $\left(\frac{1}{100} \text{ of}\right)$

Volume
1 liter (ℓ) = 1000 milliliters (ml)
1 milliliter (ml) = 1 cubic centimeter (cc)

Weight
1 kilogram (kg) = 1000 grams (g)
1 gram (g) = 1000 milligrams (mg)

Length
1 kilometer (km) = 1000 meters (m)
1 meter (m) = 100 centimeters (cm)
1 meter (m) = 1000 millimeters (mm)
1 centimeter (cm) = 10 millimeters (mm)

Table A–3 Approximate English-Metric Conversions

Volume
1.06 quarts (qt) $\doteq$ 1 liter (ℓ)
1 gallon (gal) $\doteq$ 3.8 liters (ℓ)

Weight
2.2 pounds (lb) $\doteq$ 1 kilogram (kg)
1 ounce (oz) $\doteq$ 28.4 grams (g)

Length
1 inch (in) $\doteq$ 2.54 centimeters (cm)
1 foot (ft) $\doteq$ 0.3 meters (m)
1.1 yards (yd) $\doteq$ 1 meter (m)
1 mile (mi) $\doteq$ 1.6 kilometers (km)

2 To make conversions between the English and metric systems (or within the systems), we will use unit fractions. A **unit fraction** is a fraction that is equal to 1. To form unit fractions, use the tables. For example, Table A–1 states that 1 ft = 12 in. Therefore, the ratio of these units $\frac{1 \text{ ft}}{12 \text{ in}}$ or $\frac{12 \text{ in}}{1 \text{ ft}}$ is equal to 1, and $\frac{1 \text{ ft}}{12 \text{ in}}$ and $\frac{12 \text{ in}}{1 \text{ ft}}$ are unit fractions.

To Convert From One Unit of Measure to Another:

1. Express the given unit of measure as a fraction with a denominator of 1.
2. Write a unit fraction that has the same unit of measure in the denominator that appears in the numerator of the preceding fraction.
3. Repeat step 2 until the desired units appear in the numerator of a unit fraction.
4. Multiply and divide out common units of measure.

Example 1 Fill in the missing numbers.

a. 6 gal = ? ℓ **b.** 3.5 m = ? cm **c.** 3 tons = ? oz

Solution

1a. (Table A–3)

$$6 \text{ gal} = \frac{6 \text{ gal}}{1} \cdot \frac{3.8\ \ell}{1 \text{ gal}} = \frac{6(3.8\ \ell)}{1} = 22.8\ \ell$$

1b.

$$3.5 \text{ m} = \frac{3.5 \text{ m}}{1} \cdot \frac{100 \text{ cm}}{1 \text{ m}} = \frac{3.5(100 \text{ cm})}{1} = 350 \text{ cm}$$

1c.

$$3 \text{ tons} = \frac{3 \text{ tons}}{1} \cdot \frac{2000 \text{ lb}}{1 \text{ ton}} \cdot \frac{16 \text{ oz}}{1 \text{ lb}} = \frac{3(2000)(16 \text{ oz})}{1} = 96{,}000 \text{ oz}$$

Exercises

Fill in the missing numbers.

1. 30 ft = ? yd
2. 6 mi = ? ft
3. 40 oz = ? lb
4. 2.5 km = ? m
5. 2500 ml = ? ℓ
6. 13 cc = ? ml
7. 88 km = ? m
8. 12.7 cm = ? in
9. 284 g = ? oz
10. 11 lb = ? kg
11. 19 ℓ = ? gal
12. 11 yd = ? in
13. 15 kg = ? mg
14. 64000 oz = ? ton
15. 11 tons = ? kg
16. 11 pt = ? qt
17. 5 da = ? hr
18. 2640 ft = ? mi
19. 225 cm = ? m
20. 155 mm = ? cm
21. 750 ml = ? ℓ
22. 63.5 cm = ? in
23. 77 lb = ? kg
24. 56 km = ? mi
25. 1 yr = ? sec
26. 1520 ml = ? qt

Table of Squares, Square Roots, and Primes

n	n^2	$\sqrt{n}$	Prime Factorization	n	n^2	$\sqrt{n}$	Prime Factorization	Prime Numbers Through 200
1	1	1.000	—	51	2,601	7.141	3 · 17	2
2	4	1.414	Prime	52	2,704	7.211	2 · 2 · 13	3
3	9	1.732	Prime	53	2,809	7.280	Prime	5
4	16	2.000	2 · 2	54	2,916	7.348	2 · 3 · 3 · 3	7
5	25	2.236	Prime	55	3,025	7.416	5 · 11	11
6	36	2.449	2 · 3	56	3,136	7.483	2 · 2 · 2 · 7	13
7	49	2.646	Prime	57	3,249	7.550	3 · 19	17
8	64	2.828	2 · 2 · 2	58	3,364	7.616	2 · 29	19
9	81	3.000	3 · 3	59	3,481	7.681	Prime	23
10	100	3.162	2 · 5	60	3,600	7.746	2 · 2 · 3 · 5	29
11	121	3.317	Prime	61	3,721	7.810	Prime	31
12	144	3.464	2 · 2 · 3	62	3,844	7.874	2 · 31	37
13	169	3.606	Prime	63	3,969	7.937	3 · 3 · 7	41
14	196	3.742	2 · 7	64	4,096	8.000	2 · 2 · 2 · 2 · 2 · 2	43
15	225	3.873	3 · 5	65	4,225	8.062	5 · 13	47
16	256	4.000	2 · 2 · 2 · 2	66	4,356	8.124	2 · 3 · 11	53
17	289	4.123	Prime	67	4,489	8.185	Prime	59
18	324	4.243	2 · 3 · 3	68	4,624	8.246	2 · 2 · 17	61
19	361	4.359	Prime	69	4,761	8.307	3 · 23	67
20	400	4.472	2 · 2 · 5	70	4,900	8.367	2 · 5 · 7	71
21	441	4.583	3 · 7	71	5,041	8.426	Prime	73
22	484	4.690	2 · 11	72	5,184	8.485	2 · 2 · 2 · 3 · 3	79
23	529	4.796	Prime	73	5,329	8.544	Prime	83
24	576	4.899	2 · 2 · 2 · 3	74	5,476	8.602	2 · 37	89
25	625	5.000	5 · 5	75	5,625	8.660	3 · 5 · 5	97
26	676	5.099	2 · 13	76	5,776	8.718	2 · 2 · 19	101
27	729	5.196	3 · 3 · 3	77	5,929	8.775	7 · 11	103
28	784	5.292	2 · 2 · 7	78	6,084	8.832	2 · 3 · 13	107
29	841	5.385	Prime	79	6,241	8.888	Prime	109
30	900	5.477	2 · 3 · 5	80	6,400	8.944	2 · 2 · 2 · 2 · 5	113
31	961	5.568	Prime	81	6,561	9.000	3 · 3 · 3 · 3	127
32	1,024	5.657	2 · 2 · 2 · 2 · 2	82	6,724	9.055	2 · 41	131
33	1,089	5.745	3 · 11	83	6,889	9.110	Prime	137
34	1,156	5.831	2 · 17	84	7,056	9.165	2 · 2 · 3 · 7	139
35	1,225	5.916	5 · 7	85	7,225	9.220	5 · 17	149
36	1,296	6.000	2 · 2 · 3 · 3	86	7,396	9.274	2 · 43	151
37	1,369	6.083	Prime	87	7,569	9.327	3 · 29	157
38	1,444	6.164	2 · 19	88	7,744	9.381	2 · 2 · 2 · 11	163
39	1,521	6.245	3 · 13	89	7,921	9.434	Prime	167
40	1,600	6.325	2 · 2 · 2 · 5	90	8,100	9.487	2 · 3 · 3 · 5	173
41	1,681	6.403	Prime	91	8,281	9.539	7 · 13	179
42	1,764	6.481	2 · 3 · 7	92	8,464	9.592	2 · 2 · 23	181
43	1,849	6.557	Prime	93	8,649	9.644	3 · 31	191
44	1,936	6.633	2 · 2 · 11	94	8,836	9.695	2 · 47	193
45	2,025	6.708	3 · 3 · 5	95	9,025	9.747	5 · 19	197
46	2,116	6.782	2 · 23	96	9,216	9.798	2 · 2 · 2 · 2 · 2 · 3	199
47	2,209	6.856	Prime	97	9,409	9.849	Prime	
48	2,304	6.928	2 · 2 · 2 · 2 · 3	98	9,604	9.899	2 · 7 · 7	
49	2,401	7.000	7 · 7	99	9,801	9.950	3 · 3 · 11	
50	2,500	7.071	2 · 5 · 5	100	10,000	10.000	2 · 2 · 5 · 5	

Perfect squares are in color

ANSWERS TO SELECTED EXERCISES

CHAPTER 1

CHAPTER 1 Pretest

1a. 10 **b.** 30 **c.** 15 **2a.** −45 **b.** −30 **c.** 9
3a. 10 **b.** −30 **c.** −15 **4a.** −45 **b.** 35
5a. −29 **b.** −10 **6a.** −50 **b.** 30 **7a.** −6 **b.** 6
8a. −2 **b.** −90 **9.** 22

Section 1.1

1. 300 **3.** 100 **5.** −39 **7.** −300 **9.** 35
11. 9 **13.** 95 **15.** 35 **17.** 104 **19.** 724 **21.** 0
23. 5 **25.** 324 **27.** negative

Section 1.2

1. −26 **3.** 6 **5.** −14 **7.** −129 **9.** 70
11. −91 **13.** −105 **15.** 95 **17.** −134
19. −23 **21.** 384 **23.** −500 **25.** 17 **27.** 30
29. 670 **31.** 170 **33.** −421 **35.** 56 **37.** −50
39. 0 **41.** −32 **43.** −10 **45.** −3 **47.** −15
49. −300 **51.** 50 **53.** −6 **55.** −19 **57.** −15
59. 103 **61.** −100 **63.** 35 **65.** negative
67. commutative **69.** −60 **71.** 185 **73.** −8
75. −5

Section 1.3

1. 10 **3.** 4 **5.** −8 **7.** 0 **9.** −100 **11.** −21
13. −3 **15.** −25 **17.** 3 **19.** −205 **21.** −48
23. −18 **25.** −68 **27.** 20 **29.** 185 **31.** −15
33. 189 **35.** 1900 **37.** −284 **39.** −31 **41.** 5
43. −8 **45.** 32 **47.** 0 **49.** −500 **51.** 45
53. −45 **55.** zero **57.** 14,540 feet **59.** −34
61. −5 **63.** 5 **65.** 16 **67.** −2

Section 1.4

1. −27 **3.** −26 **5.** 33 **7.** −40 **9.** −34
11. −160 **13.** −30 **15.** −49 **17.** −33
19. −15 **21.** −18 **23.** −20 **25.** −25 **27.** 22
29. −23 **31.** −41 **33.** −17 **35.** 32 **37.** 1
39. −25 **41.** −110 **43.** subtraction, addition, subtraction, addition **45.** subtraction **47.** account is overdrawn $4 **49.** $8,820

Section 1.5

1. 10 **3.** −15 **5.** −32 **7.** −24 **9.** 20
11. −24 **13.** 0 **15.** 30 **17.** −55 **19.** −28
21. 120 **23.** 24 **25.** −8 **27.** −105 **29.** 28
31. −8 **33.** 16 **35.** −60 **37.** −12 **39.** 120
41. addition **43.** negative **45.** negative
47. absolute values, sign **49.** −30 **51.** 2700
53. −30 **55.** −60

Section 1.6

1. 8 **3.** −3 **5.** −5 **7.** −50 **9.** 15 **11.** 11
13. 1 **15.** −1 **17.** −13 **19.** −10 **21.** 22
23. −4 **25.** 8 **27.** −8 **29.** −100 **31.** 2
33. −13 **35.** 311 **37.** 0 **39.** undefined
41. positive **43.** absolute values, sign **45.** undefined
47. −1,000 **49.** −2 **51.** 2 **53.** −100

Section 1.7

1. 4 **3.** −29 **5.** −65 **7.** 0 **9.** 6 **11.** 46
13. −23 **15.** 4 **17.** −56 **19.** 10 **21.** 7
23. −12 **25.** −19 **27.** 24 **29.** −11 **31.** 16
33. −30 **35.** 213 **37.** −64 **39.** −40
41. multiply, divide, add, subtract **43.** parentheses, multiply, divide, add, subtract **45.** innermost grouping symbol **47.** 280

Section 1.8

1. 10 **3.** −4 **5.** 36 **7.** formula **9.** letters
11. 20 **13.** 130 **15.** 30 **17.** 7 **19.** 3
21. −10 **23.** 8 **25.** 68° **27.** 5° **29.** 86°
31. 14° **33.** −3 **35.** 85

CHAPTER 1 Review Exercises

1. 20 *(Section 1.1)* **2.** 60 *(1.1)* **3.** 95 *(1.1)*
4. 0 *(1.1)* **5.** 20 *(1.3)* **6.** −60 *(1.3)*
7. −95 *(1.3)* **8.** 0 *(1.3)* **9.** −50 *(1.2)*
10. 39 *(1.3)* **11.** 68 *(1.7)* **12.** −40 *(1.5)*
13. −5 *(1.6)* **14.** 6 *(1.6)* **15.** 53 *(1.2)*
16. −120 *(1.5)* **17.** 67 *(1.4)* **18.** −50 *(1.3)*
19. 1 *(1.4)* **20.** undefined *(1.6)* **21.** 220 *(1.7)*
22. 131 *(1.2)* **23.** 45 *(1.3)* **24.** 196 *(1.7)*
25. −14 *(1.6)* **26.** $100 *(1.4)* **27.** 8 *(1.8)*

CHAPTER 2

CHAPTER 2 Pretest

1a. $2 \cdot 2 \cdot 2 \cdot 3$ **b.** $7 \cdot 11$ **c.** $3 \cdot 3 \cdot 3 \cdot 11$
2a. $\frac{7}{8}$ **b.** $\frac{3}{2}$ **3a.** $\frac{10}{3}$ **b.** $\frac{77}{8}$ **4a.** $7\frac{1}{2}$ **b.** $12\frac{1}{8}$
5a. $\frac{21}{10}$ **b.** $91\frac{1}{2}$ **6a.** $\frac{9}{8}$ **b.** $3\frac{279}{427}$ **7a.** $\frac{113}{198}$ **b.** $\frac{125}{132}$
8a. $12\frac{2}{7}$ **b.** $28\frac{23}{24}$ **9a.** $\frac{23}{40}$ **b.** $3\frac{5}{42}$ **10a.** $\frac{8}{15}$ **b.** $-\frac{5}{44}$

Section 2.1

1. numerator = 3; denominator = 14 **3.** numerator = 10; denominator = 9 **5.** $\frac{2}{5}$ **7.** $\frac{3}{4}$ **9.** 3
11. 3 **13.** 3 and 5 **15.** 2, 3, and 5 **17.** $2 \cdot 7$
19. $2 \cdot 2 \cdot 3 \cdot 5$ **21.** $3 \cdot 13$ **23.** $2 \cdot 3 \cdot 7$ **25.** $7 \cdot 11$
27. $2 \cdot 2 \cdot 2 \cdot 11$ **29.** $3 \cdot 3 \cdot 3 \cdot 3$ **31.** $5 \cdot 5$
33. $3 \cdot 5 \cdot 11$ **35.** $3 \cdot 3 \cdot 3 \cdot 5$ **37.** $5 \cdot 23$
39. $2 \cdot 3 \cdot 7 \cdot 11$ **41.** $7 \cdot 11 \cdot 13$ **43.** $2 \cdot 17 \cdot 47$
45. $2 \cdot 2 \cdot 2 \cdot 2 \cdot 3 \cdot 5 \cdot 5$ **47.** $2 \cdot 2 \cdot 2 \cdot 2 \cdot 53$
49. $3 \cdot 3 \cdot 3 \cdot 11$ **51.** $11 \cdot 13 \cdot 17$ **53.** prime
55. prime, quotient, product, prime, quotient **57.** $\frac{17}{30}$

Section 2.2

1. not equal **3.** equal **5.** $\frac{2}{3}$ **7.** $\frac{3}{5}$ **9.** $\frac{3}{4}$ **11.** $\frac{2}{3}$
13. $\frac{1}{9}$ **15.** $\frac{5}{7}$ **17.** $\frac{2}{13}$ **19.** $\frac{7}{11}$ **21.** $\frac{5}{6}$ **23.** $\frac{2}{5}$
25. $\frac{32}{55}$ **27.** $\frac{11}{35}$ **29.** $\frac{5}{6}$ **31.** $\frac{121}{169}$ **33.** $\frac{19}{23}$ **35.** $\frac{26}{37}$
37. $\frac{23}{27}$ **39.** $\frac{2}{3}$ **41.** $\frac{19}{33}$ **43.** $\frac{21}{22}$ **45.** $\frac{6}{35}$ **47.** $\frac{1}{3}$
49. $\frac{25}{66}$ **51.** $\frac{31}{43}$ **53.** cross-products **55.** prime, numerator, denominator, common, numerator, denominator
57. $\frac{1}{8}$ **59.** $\frac{1}{2}$ **61.** $\frac{3}{7}$ **63.** $\frac{1}{3}, \frac{2}{3}$

Section 2.3

1. $\frac{5}{2}$ **3.** $\frac{47}{8}$ **5.** $\frac{92}{9}$ **7.** $\frac{47}{5}$ **9.** $\frac{135}{11}$ **11.** $\frac{227}{16}$
13. $\frac{316}{9}$ **15.** $\frac{662}{9}$ **17.** $\frac{86}{17}$ **19.** $\frac{172}{13}$ **21.** $7\frac{1}{2}$
23. $8\frac{4}{7}$ **25.** $11\frac{2}{3}$ **27.** $6\frac{8}{9}$ **29.** $5\frac{1}{4}$ **31.** $2\frac{1}{8}$
33. $3\frac{3}{5}$ **35.** $5\frac{2}{3}$ **37.** $13\frac{9}{11}$ **39.** $2\frac{3}{5}$ **41.** mixed number **43.** denominator, numerator, quotient, remainder **45.** $\frac{34}{5}$

Section 2.4

1. $\frac{3}{20}$ **3.** $\frac{3}{10}$ **5.** $\frac{5}{42}$ **7.** $\frac{15}{32}$ **9.** $\frac{15}{44}$ **11.** $\frac{270}{7}$
13. $\frac{25}{21}$ **15.** 21 **17.** $\frac{6}{65}$ **19.** $\frac{9}{16}$ **21.** 1 **23.** $\frac{4}{9}$
25. $\frac{12}{7}$ **27.** 1 **29.** 4 **31.** $\frac{891}{200}$ **33.** $\frac{5}{14}$ **35.** $\frac{1}{55}$
37. 600 **39.** $\frac{44}{28175}$ **41.** 2 **43.** 54 **45.** $35\frac{5}{8}$
47. $12\frac{1}{2}$ **49.** $1\frac{8}{21}$ **51.** $6\frac{1}{2}$ **53.** $6\frac{2}{3}$ **55.** $9\frac{1}{6}$
57. 1 **59.** $1\frac{13}{20}$ **61.** $5\frac{5}{6}$ **63.** $1\frac{5}{39}$ **65.** $1\frac{5}{28}$
67. $91\frac{1}{2}$ **69.** $6\frac{2}{3}$ **71.** 8 **73.** $3\frac{3}{4}$ **75.** $\frac{N}{1}$
77. $\$562\frac{1}{2}$ **79.** $\frac{1}{6}$ ton **81.** $195\frac{1}{2}$ miles **83.** 300
85. $\frac{10086}{3773}$ **87.** $\frac{1}{2}$ **89.** 576 **91.** 154 miles
93. $25,125 **95.** $\frac{1}{15}$ ton

Section 2.5

1. 3 **3.** $\frac{5}{4}$ **5.** $\frac{225}{196}$ **7.** $\frac{9}{4}$ **9.** $\frac{150}{7}$ **11.** $\frac{45}{8}$
13. $\frac{5}{3}$ **15.** $\frac{83}{108}$ **17.** $\frac{14}{3}$ **19.** $\frac{1}{81}$ **21.** $\frac{55}{24}$ **23.** 28
25. $\frac{15}{14}$ **27.** 1 **29.** 0 **31.** undefined **33.** 15
35. $\frac{21}{4}$ **37.** 900 **39.** $\frac{205}{224}$ **41.** $\frac{38}{39}$ **43.** $1\frac{8}{125}$
45. $1\frac{3}{7}$ **47.** 1 **49.** $\frac{1}{2}$ **51.** 100 **53.** $\frac{3}{8}$ **55.** 66
57. 4 **59.** undefined **61.** $2\frac{4}{5}$ **63.** $\frac{7}{9}$ **65.** 1
67. $7\frac{5}{7}$ **69.** $14\frac{17}{20}$ **71.** $\frac{27}{35}$ **73.** $2\frac{59}{273}$ **75.** $3\frac{45}{56}$
77. $\frac{8}{15}$ **79.** reciprocal **81.** mixed numbers, improper fractions **83.** $\frac{55}{91}$ **85.** $\frac{2}{3}$ **87.** $\frac{1}{12}$ pound **89.** 6
91. $6\frac{4}{5}$ **93.** 5 **95.** $19\frac{5}{13}$

Section 2.6

1. $\frac{5}{7}$ **3.** $\frac{16}{7}$ **5.** $\frac{2}{3}$ **7.** $\frac{13}{20}$ **9.** $\frac{1}{2}$ **11.** 9 **13.** 3
15. 15 **17.** 7 **19.** 77 **21.** 3 **23.** 15 **25.** 75
27. 35 **29.** 126 **31.** 24 **33.** 105 **35.** 18
37. 40 **39.** 90 **41.** 180 **43.** 600 **45.** 72
47. $\frac{17}{18}$ **49.** $\frac{13}{12}$ **51.** $\frac{47}{40}$ **53.** $\frac{19}{48}$ **55.** $\frac{41}{60}$

57. $\frac{109}{60}$ **59.** $\frac{61}{144}$ **61.** $\frac{1802}{7007}$ **63.** $\frac{29}{150}$ **65.** $\frac{19}{55}$ **67.** $\frac{43}{84}$ **69.** $\frac{3}{16}$ **71.** $\frac{59}{105}$ **73.** $\frac{16}{21}$ **75.** $\frac{49}{80}$ **77.** $\frac{79}{65}$ **79.** $\frac{29}{84}$ **81.** $\frac{71}{72}$ **83.** $\frac{59}{40}$ **85.** $\frac{274}{645}$ **87.** $\frac{589}{588}$ **89.** numerators, common denominator **91.** $\frac{241}{120}$ **93.** $\frac{97}{168}$ **95.** $\frac{3}{4}$ pound **97.** $\frac{31}{60}$ **99.** $\frac{17}{24}$ teaspoon **101.** $\frac{1}{77}$ **103.** $\frac{7}{24}$ **105.** $\frac{19}{40}$ inch

Section 2.7

1. $5\frac{2}{3}$ **3.** $10\frac{3}{8}$ **5.** 9 **7.** $15\frac{3}{5}$ **9.** $5\frac{31}{36}$ **11.** $18\frac{7}{12}$ **13.** $14\frac{9}{20}$ **15.** $15\frac{41}{60}$ **17.** $8\frac{4}{7}$ **19.** $6\frac{2}{15}$ **21.** $7\frac{23}{42}$ **23.** $24\frac{25}{72}$ **25.** $13\frac{31}{48}$ **27.** $24\frac{29}{72}$ **29.** $189\frac{23}{60}$ **31.** $72\frac{1}{20}$ **33.** $11\frac{2}{45}$ **35.** $10\frac{19}{20}$ **37.** $40\frac{31}{72}$ **39.** $135\frac{191}{588}$ **41.** $263\frac{19}{35}$ **43.** $8\frac{3}{4}$ pounds **45.** $346\frac{1}{10}$ miles **47.** $29\frac{1}{24}$ inches **49.** $571\frac{1}{24}$ pounds

Section 2.8

1. $\frac{1}{2}$ **3.** $\frac{3}{5}$ **5.** $\frac{1}{15}$ **7.** $\frac{3}{10}$ **9.** $\frac{1}{12}$ **11.** $\frac{19}{30}$ **13.** $\frac{11}{20}$ **15.** $\frac{13}{24}$ **17.** $\frac{17}{36}$ **19.** $\frac{1}{15}$ **21.** $\frac{1}{4}$ **23.** $\frac{1}{48}$ **25.** $\frac{23}{48}$ **27.** $\frac{13}{15}$ **29.** $\frac{1}{70}$ **31.** $\frac{9}{20}$ **33.** $\frac{1}{84}$ **35.** $\frac{23}{90}$ **37.** $\frac{271}{504}$ **39.** $\frac{77}{200}$ **41.** $\frac{28}{99}$ **43.** $\frac{8}{5}, 6\frac{8}{5}$ **45.** $\frac{16}{9}, 4\frac{16}{9}$ **47.** 13 **49.** 39 **51.** 9 **53.** 110 **55.** $3\frac{2}{7}$ **57.** $3\frac{2}{9}$ **59.** $2\frac{2}{5}$ **61.** $5\frac{1}{3}$ **63.** $3\frac{2}{3}$ **65.** $2\frac{2}{3}$ **67.** $2\frac{3}{4}$ **69.** $3\frac{1}{8}$ **71.** $1\frac{1}{24}$ **73.** $2\frac{23}{36}$ **75.** $2\frac{11}{14}$ **77.** $27\frac{13}{48}$ **79.** $1\frac{53}{90}$ **81.** $58\frac{65}{96}$ **83.** $2\frac{1}{4}$ **85.** $4\frac{1}{3}$ **87.** $8\frac{5}{8}$ **89.** $5\frac{31}{63}$ **91.** $12\frac{7}{90}$ **93.** numerators, denominator **95.** whole, fractional, whole, minuend, fractional **97.** $3\frac{7}{12}$ **99.** $\frac{17}{24}$ **101.** $1\frac{1}{2}$ **103.** $7\frac{19}{24}$ **105.** $35\frac{3}{4}$ **107.** $\frac{67}{396}$ **109.** $49\frac{1}{2}$ yards

Section 2.9

1. $\frac{3}{5}$ **3.** $\frac{13}{8}$ **5.** $\frac{5}{8}$ **7.** $-\frac{5}{9}$ **9.** $-\frac{3}{40}$ **11.** $-\frac{5}{24}$ **13.** $\frac{2}{7}$ **15.** $\frac{13}{48}$ **17.** $-\frac{9}{70}$ **19.** $\frac{1}{2}$ **21.** $-\frac{32}{5}$ **23.** $-\frac{2}{25}$ **25.** $-\frac{5}{3}$ **27.** -15 **29.** $\frac{6}{5}$ **31.** $-\frac{1}{98}$ **33.** $-\frac{24}{5}$ **35.** $\frac{3}{40}$ **37.** $-\frac{7}{8}$ **39.** -2 **41.** $-\frac{13}{42}$ **43.** $\frac{2}{9}$ **45.** $-\frac{71}{60}$ **47.** $-\frac{17}{36}$ **49.** $-\frac{13}{12}$ **51.** $-\frac{25}{3}$ **53.** $-\frac{10}{21}$ **55.** $\frac{13}{72}$ **57.** -72

CHAPTER 2 Review Exercises

1. $2 \cdot 2 \cdot 3 \cdot 3$ *(Section 2.1)* **2.** $11 \cdot 11$ *(2.1)* **3.** $2 \cdot 3 \cdot 3 \cdot 29$ *(2.1)* **4.** $7 \cdot 11 \cdot 13$ *(2.1)* **5.** $\frac{3}{5}$ *(2.2)* **6.** $\frac{5}{13}$ *(2.2)* **7.** $\frac{3}{7}$ *(2.2)* **8.** $\frac{7}{9}$ *(2.2)* **9.** $3\frac{2}{5}$ *(2.3)* **10.** $12\frac{5}{6}$ *(2.3)* **11.** $13\frac{4}{9}$ *(2.3)* **12.** $\frac{15}{4}$ *(2.3)* **13.** $\frac{61}{8}$ *(2.3)* **14.** $\frac{331}{4}$ *(2.3)* **15.** $\frac{14}{15}$ *(2.4)* **16.** $-\frac{41}{60}$ *(2.9)* **17.** $\frac{29}{18}$ *(2.6)* **18.** $-\frac{15}{14}$ *(2.9)* **19.** $5\frac{1}{4}$ *(2.4)* **20.** $\frac{23}{45}$ *(2.8)* **21.** $3\frac{11}{36}$ *(2.8)* **22.** $\frac{4}{27}$ *(2.5)* **23.** $12\frac{11}{15}$ *(2.7)* **24.** $5\frac{5}{6}$ *(2.5)* **25.** $8\frac{1}{3}$ *(2.8)* **26.** -210 *(2.9)* **27.** $\frac{1}{4}$ *(2.8)* **28.** $\frac{2}{9}$ *(2.5)* **29.** \$750 *(2.4)* **30.** $\$3\frac{5}{8}$ *(2.8)* **31.** $9\frac{5}{24}$ cups *(2.7)*

CHAPTER 3

CHAPTER 3 Pretest

1a. three hundredths **b.** two and four hundred fifteen thousandths **2a.** 3.41 **b.** 52.668 **3a.** 1.771 **b.** 10.672 **4a.** 0.16 **b.** 28.397 **5a.** 0.32 **b.** 8.0 **6a.** 33.1 **b.** 2.03 **7a.** 0.6 **b.** 2.67 **8a.** $\frac{7}{20}$ **b.** $\frac{25}{8}$ **9a.** 0.45 **b.** -9.565 **c.** 10.56 **d.** -32.3

Section 3.1

1. ones **3.** tenths **5.** thousandths **7.** hundred-thousandths **9.** thousands **11.** eight hundred thirteen thousandths **13.** one thousand three hundred fifty seven ten-thousandths **15.** thirty-four millionths **17.** six and thirty-four thousandths **19.** thirty-one thousandths **21.** six and seven hundred three thousandths **23.** seven and seven tenths **25.** 8.17 **27.** 0.0046 **29.** 10.000005 **31.** decimal **33.** whole, and, whole, place value

Section 3.2

1. 0.95 **3.** 1.3 **5.** 17.72 **7.** 35.34 **9.** 4.288 **11.** 7.689 **13.** 25.234 **15.** 110.6628 **17.** 14.9691 **19.** 128.171 **21.** 7.2 **23.** 2.12 **25.** 0.1 **27.** 4.077 **29.** 0.1 **31.** 62.36 **33.** 73.92 **35.** 16.88 **37.** 80.315 **39.** 1.31 **41.** 20.851 **43.** 0.19 **45.** 4.36 **47.** 8.605 **49.** 24.66 **51.** 7726 **53.** 1984 **55.** 339.49 **57.** 1195.05 **59.** 0.0022 **61.** 22.79 **63.** 13.398 **65.** 5.1264 **67.** 20 **69.** 151.05 **71.** 2099.88 **73.** 23.81 **75.** 216.7 **77.** 4206.09 **79.** 14.7461 **81.** 35.82 **83.** decimal points, whole, decimal points **85.** place value **87.** 114.516 **89.** 28.725 inches **91.** $18.91 **93.** 142.6 **95.** $25.92 **97.** 27.08 **99.** $318.43 **101.** $111.58 **103.** $702.52 **105.** $133.50

Section 3.3

1. 0.5742 **3.** 5.742 **5.** .05742 **7.** 0.96 **9.** 1.088 **11.** 0.0016 **13.** 79.56 **15.** 0.10104 **17.** 23.63 **19.** 0.0001639 **21.** 1676.47 **23.** 133.375 **25.** 670.5 **27.** 0.6272 **29.** 1.0798137 **31.** 5012.596 **33.** 13.86 **35.** 3.75 **37.** 21.3444 **39.** 0.2606225 **41.** digits, right **43.** zeros **45.** $42.75 **47.** $28.12 **49.** 250.6 **51.** 0.621 **53.** 0.28608 **55.** $1671 **57.** 334,800 **59.** $178.50 **61.** 107.1

Section 3.4

1. 0.2 **3.** 0.3 **5.** 18.1 **7.** 0.9 **9.** 15.8 **11.** 0.85 **13.** 0.17 **15.** 1.38 **17.** 2.75 **19.** 0.146 **21.** 0.679 **23.** 5.678 **25.** 95.333 **27.** 10.000 **29.** 1335.375 **31.** 1300 **33.** 1335.374590 **35.** 1335.3746 **37.** approximately equal to **39.** hundredth **41.** 75.78 **43.** 0.739 **45.** 5.0

Section 3.5

1. 33.2 **3.** 6.81 **5.** 3.5 **7.** 5.3 **9.** 3.29 **11.** 0.0042 **13.** 0.0005 **15.** 0.2 **17.** 5.46 **19.** 1.8 **21.** 0.04 **23.** 65.8 **25.** 5780 **27.** 125,000 **29.** 0.178 **31.** 35.58 **33.** 0.04 **35.** 0.49 **37.** 0.27 **39.** 45.1 **41.** dividend, quotient **43.** $3.42 **45.** 21.2 **47.** 8.5 **49.** 5.1 **51.** 8.5 **53.** 3.68 **55.** $2.15 **57.** 20.4 **59.** 24 months

Section 3.6

1. 0.6 **3.** 0.25 **5.** 1.125 **7.** 0.28 **9.** 0.33 **11.** 0.71 **13.** 0.425 **15.** 0.62 **17.** 0.9 **19.** 0.256 **21.** 0.75 **23.** 3.5 **25.** 0.125 **27.** 0.06 **29.** 0.01 **31.** 0.02 **33.** 6.2 **35.** 0.19 **37.** 1.17 **39.** 0.014 **41.** $\frac{3}{10}$ **43.** $\frac{31}{1000}$ **45.** $\frac{1}{20}$ **47.** $\frac{5}{2}$ **49.** $\frac{8}{1}$ **51.** $\frac{128}{25}$ **53.** $\frac{61103}{10,000}$ **55.** $\frac{165}{1}$ **57.** $\frac{1}{8}$ **59.** $\frac{5}{16}$ **61.** $\frac{9}{20}$ **63.** $\frac{111}{1000}$ **65.** $\frac{13}{1}$ **67.** $\frac{3}{10}$ **69.** $\frac{7}{2}$ **71.** $\frac{9}{40}$ **73.** $\frac{701}{100}$ **75.** $\frac{9}{25}$ **77.** $\frac{3}{20}$ **79.** $\frac{1234}{5}$ **81.** denominator, numerator, round **83.** decimal point, place value, reduce **85.** $\frac{33}{8}$, 4.125 **87.** $\frac{25}{2}$, 12.5 **89.** $\frac{1}{1}$, 1 **91.** $0.43 **93.** $\frac{9}{16}$ **95.** 3.25 **97.** 7.5 **99.** 0.0017 **101.** $\frac{1}{8}$

Section 3.7

1. 3.21 **3.** 8.231 **5.** 0.85 **7.** 0.96 **9.** 2.06 **11.** 8.58 **13.** 0.521 **15.** 5.7 **17.** −0.91 **19.** 7.09 **21.** 0.069 **23.** −0.93 **25.** 13.54 **27.** −0.219 **29.** −0.15 **31.** −0.024 **33.** −1.83 **35.** −22.2 **37.** −100 **39.** −4.34 **41.** 0.271 **43.** 11.31 **45.** −3.03 **47.** −1.82 **49.** 20.2 **51.** −0.09 **53.** −9.44 **55.** 5.55 **57.** 0.009 **59.** −1.08 **61.** 0.05 **63.** 4 **65.** −8.85 **67.** 0.21 **69.** −14.4 **71.** 0.012 **73.** −0.024 **75.** −6 **77.** 44.4 **79.** −20 **81.** −4.7 **83.** −3.9752 **85.** −22.11 **87.** 0.0012 **89.** −52.1 **91.** 22.11 **93.** −0.034

CHAPTER 3 Review Exercises

1. five and nine thousandths *(Section 3.1)* **2.** two thousand three hundred eighteen ten-thousandths *(3.1)* **3.** thirty-five hundredths *(3.1)* **4.** 3.05 *(3.6)* **5.** $\frac{31}{100}$ *(3.6)* **6.** 0.33 *(3.6)* **7.** 0.58 *(3.6)* **8.** $\frac{13}{20}$ *(3.6)* **9.** $\frac{1}{8}$ *(3.6)* **10.** 1751.859 *(3.4)*

11. 1800 *(3.4)* **12.** 14.84 *(3.2)* **13.** 0.915 *(3.2)*
14. 0.003945 *(3.3)* **15.** 453.47 *(3.5)*
16. −11.915 *(3.7)* **17.** −110 *(3.7)*
18. −8.466 *(3.7)* **19.** 8.925 *(3.3)* **20.** 0.6625 *(3.2)*
21. 0.568 *(3.2)* **22.** −0.05 *(3.7)* **23.** 17.881 *(3.2)*
24. $15.08 *(3.3)* **25.** $30.75 *(3.5)* **26.** $6.52 *(3.2)*
27. $314.60 *(3.6)*

CHAPTER 4

CHAPTER 4 Pretest

1a. $\frac{13}{20}$ **b.** $\frac{1}{1250}$ **c.** $\frac{3}{800}$ **2a.** 0.65 **b.** 0.0008
c. 0.00375 **3a.** 45% **b.** 0.95% **c.** 320%
4a. 75% **b.** 12.5% **c.** $66\frac{2}{3}$% **5a.** $\frac{3}{8}$
b. 0.03222 **c.** 19 **d.** 33

Section 4.1

1. $\frac{13}{100}$ **3.** $\frac{31}{100}$ **5.** $\frac{11}{20}$ **7.** $\frac{1}{50}$ **9.** $\frac{19}{300}$ **11.** 1
13. $\frac{3}{8}$ **15.** $\frac{3}{200}$ **17.** $\frac{3}{2}$ **19.** $\frac{1}{25{,}000}$ **21.** $\frac{17}{100}$
23. $\frac{9}{50}$ **25.** $\frac{23}{20}$ **27.** $\frac{13}{1250}$ **29.** $\frac{1}{12}$ **31.** $\frac{3}{400}$
33. 5 **35.** $\frac{441}{800}$ **37.** $\frac{3}{50{,}000}$ **39.** $\frac{13}{20}$ **41.** 0.13
43. 0.31 **45.** 0.55 **47.** 0.02 **49.** 0.0625 **51.** 1
53. 0.00125 **55.** 2.5 **57.** 0.1075 **59.** 0.034
61. 0.17 **63.** 0.16 **65.** 1.15 **67.** 0.0104
69. 0.0833 **71.** 0.0075 **73.** 5 **75.** 0.55125
77. 0.00006 **79.** 0.65 **81.** percent **83.** 0.01,
percent symbol **85.** 23, 100 **87.** $\frac{9}{40}$ **89.** 0.125
91. $\frac{131}{100}$ **93.** 0.00375 **95.** $\frac{297}{800}$

Section 4.2

1. 65% **3.** 230% **5.** 0.6% **7.** 81% **9.** 75%
11. 60% **13.** 40% **15.** 5120% **17.** 0.7%
19. $33\frac{1}{3}$% **21.** 13% **23.** 800% **25.** 0.8%
27. 50% **29.** 11.1% **31.** 810% **33.** 0.45%
35. 60% **37.** $55\frac{1}{2}$% **39.** $80\frac{1}{2}$% **41.** 60%
43. 25% **45.** 112.5% **47.** 28% **49.** $33\frac{1}{3}$%
51. $71\frac{3}{7}$% **53.** 42.5% **55.** 62% **57.** 90%
59. 75% **61.** 10% **63.** $66\frac{2}{3}$% **65.** 27.5%
67. 70% **69.** 150% **71.** 0.7% **73.** 500%
75. 2.5% **77.** $31\frac{1}{4}$% **79.** 110% **81.** decimal point,
right **83.** fraction, decimal, decimal, percent **85.** 60%
87. 14.55% **89.** 15.5% **91.** $114\frac{2}{7}$% **93.** $52\frac{1}{8}$%
95. 18.7%

Section 4.3

1. 15 **3.** 14.4 **5.** 44 **7.** $\frac{7}{8}$ **9.** $\frac{42}{25}$ **11.** $\frac{10}{3}$
13. 49.2 **15.** $\frac{200}{3}$ **17.** $\frac{8}{9}$ **19.** 16.65 **21.** percent,
multiply, number **23.** 65% **25.** $216 **27.** $292.50
29. $350 **31.** $448 **33.** $179.78 **35.** $7.50
37. 12 **39.** 45 **41.** 15 **43.** $46 **45.** $307.50
47. $\frac{873}{16}$ **49.** $385 **51.** 24 **53.** $\frac{7}{18}$

CHAPTER 4 Review Exercises

1. $\frac{1}{2}$, 50% *(Section 4.2)* **2.** 0.2, 20% *(4.2)* **3.** 0.25,
$\frac{1}{4}$ *(4.1)* **4.** $\frac{61}{100}$, 61% *(4.2)* **5.** 0.25, 25% *(4.2)*
6. 0.8, $\frac{4}{5}$ *(4.1)* **7.** $\frac{1}{8}$, $12\frac{1}{2}$% *(4.2)* **8.** $0.33\frac{1}{3}$,
$33\frac{1}{3}$% *(4.2)* **9.** $0.66\frac{2}{3}$, $\frac{2}{3}$ *(4.1)* **10.** $\frac{3}{5}$, 60% *(4.2)*
11. 0.75, 75% *(4.2)* **12.** 0.375, $\frac{3}{8}$ *(4.1)* **13.** $\frac{5}{1}$,
500% *(4.2)* **14.** 0.006, $\frac{3}{500}$ *(4.1)* **15.** 0.014,
1.4% *(4.2)* **16.** $\frac{11}{25}$ *(4.3)* **17.** 175 *(4.3)*
18. $500.25 *(4.3)* **19.** $139,500 *(4.3)*
20. $8100 *(4.3)* **21.** $\frac{1}{8}$ *(4.1)*

CHAPTER 5

CHAPTER 5 Pretest

1a. 8 **b.** $\frac{1}{8}$ **c.** −8 **2a.** 7^{50} **b.** x^{12} **3a.** 7^{13}
b. $\frac{1}{x^{18}}$ **4a.** x^8 **b.** $\frac{1}{7^{15}}$ **5a.** $3^{11} \cdot 6^{11}$ **b.** $81x^4$
6a. $\frac{8}{27}$ **b.** $-\frac{x^3}{125}$ **7a.** 6.8×10^4 **b.** 6.8×10^{-5}

Section 5.1

1. $3 \cdot 3 \cdot 3 \cdot 3 \cdot 3$ **3.** $5 \cdot 5 \cdot 5 \cdot 5 \cdot 5 \cdot 5 \cdot 5$ **5.** $7 \cdot 7 \cdot 7 \cdot 7 \cdot 7 \cdot 7 \cdot 7$ **7.** 9^2 **9.** 6^5 **11.** $(.5)^2$ **13.** 4^6 **15.** 9 **17.** 36 **19.** $\frac{1}{49}$ **21.** $\frac{1}{1000}$ **23.** $\frac{1}{16}$ **25.** 1 **27.** -1 **29.** 0.09 **31.** 1.331 **33.** 81 **35.** -81 **37.** $\frac{1}{36}$ **39.** $-\frac{1}{343}$ **41.** $-\frac{1}{32}$ **43.** 0.0000000001 **45.** base **47.** fraction, one, positive

Section 5.2

1. 3^{11} **3.** 8^{15} **5.** $\frac{1}{4^7}$ **7.** $\frac{1}{2^{11}}$ **9.** 1 **11.** m^7 **13.** $\frac{1}{6^7}$ **15.** x^{19} **17.** $\frac{1}{5^9}$ **19.** $\frac{1}{3^6}$ **21.** $\frac{1}{3^8}$ **23.** a^{21} **25.** $\frac{1}{m^{41}}$ **27.** $\frac{1}{x^{16}}$ **29.** $\frac{1}{7^6}$ **31.** n^2 **33.** 1 **35.** $\frac{1}{8}$ **37.** 32 **39.** $\frac{1}{x^7}$ **41.** $\frac{1}{n^{20}}$ **43.** 7^{15} **45.** x^4 **47.** x^4 **49.** $\frac{1}{x^{55}}$ **51.** base, add **53.** exponential expression, operation **55.** 49 **57.** $\frac{1}{4}$ **59.** 63 **61.** 4

Section 5.3

1. 2^{27} **3.** y^8 **5.** $\frac{1}{3^{18}}$ **7.** $\frac{1}{5^{16}}$ **9.** $\frac{1}{x^{27}}$ **11.** x^{ab} **13.** $\frac{1}{x^{12}}$ **15.** x^6 **17.** 2^{80} **19.** 16 **21.** $3^7 \cdot 5^7$ **23.** $5^{10}n^{10}$ **25.** $-27p^3$ **27.** $16x^2y^2$ **29.** $a^8b^8c^8d^8$ **31.** $49x^2$ **33.** $\frac{16}{81}$ **35.** $\frac{81}{n^4}$ **37.** $-\frac{x^3}{8}$ **39.** $\frac{b^8}{5^8}$ **41.** $\frac{x^n}{y^n}$ **43.** $-\frac{n^3}{125}$ **45.** $32x^5$ **47.** $-8x^3$ **49.** $x^9y^9z^9$ **51.** $49x^2$ **53.** $\frac{1}{36}$ **55.** $\frac{n^4}{81}$ **57.** $-\frac{x}{343}$ **59.** 25 **61.** factor, power **63.** numerator, denominator, power **65.** $16x^8y^{12}$ **67.** $\frac{1}{3^{ab}}$ **69.** $\frac{1}{m^7}$ **71.** $\frac{a^{2n}}{b^{4n}}$

Section 5.4

1. 372 **3.** 0.0000000857 **5.** 0.875 **7.** 300,000 **9.** 0.00000007 **11.** -0.000677 **13.** 81,000 **15.** 0.000535 **17.** $-3,120,000$ **19.** 810,000,000,000 **21.** 2.75×10^2 **23.** 2.75×10^{-4} **25.** 3.18×10^1 **27.** 1.1×10^4 **29.** 7.5×10^8 **31.** 5×10^{-4} **33.** 6.19×10^{-6} **35.** 3.78×10^{14} **37.** 6.25×10^2 **39.** 8.75×10^{-2} **41.** 3.5×10^{10} **43.** 5.1×10^{-8} **45.** n, right **47.** one, ten, decimal point, one, ten **49.** 30.8 **51.** 32,000,000 **53.** 0.0000000004 **55.** 1,920,000,000,000 **57.** 30,000,000,000 **59.** 0.000007 **61.** -0.000000000155 **63.** 9.3×10^7 **65.** 5.19×10^{-5} **67.** 2.25×10^8

CHAPTER 5 Review Exercises

1. 81 *(Section 5.1)* **2.** $\frac{1}{81}$ *(5.1)* **3.** 81 *(5.1)* **4.** -81 *(5.1)* **5.** $\frac{1}{x^{20}}$ *(5.2)* **6.** $\frac{121}{16}$ *(5.3)* **7.** x^3 *(5.2)* **8.** x^{30} *(5.3)* **9.** $-8p^3$ *(5.3)* **10.** 5^{53} *(5.2)* **11.** $16\frac{1}{8}$ *(5.1)* **12.** $\frac{1}{9^{24}}$ *(5.2)* **13.** $\frac{9}{16}$ *(5.3)* **14.** 2^{15} *(5.3)* **15.** $2^{15}x^{15}$ *(5.3)* **16.** $\frac{8}{9}$ *(5.1)* **17.** $81x^8$ *(5.3)* **18.** x^{20} *(5.2)* **19.** $\frac{1}{5^{29}}$ *(5.3)* **20.** 347,000,000 *(5.4)* **21.** 0.000000406 *(5.4)* **22.** $-2,470$ *(5.4)* **23.** 1.5×10^7 *(5.4)* **24.** 3.12×10^{-7} *(5.4)* **25.** 3.25×10^{-4} *(5.4)* **26.** 1.1552×10^{12} *(5.4)* **27.** 18,600 *(5.4)* **28.** 1,230,000 *(5.4)*

CHAPTER 6

CHAPTER 6 Pretest

1. $-7x$ and $-8x$ **2a.** $-17x$ **b.** $-7x + 7$ **3a.** 69 **b.** 2.088 **4a.** $6x - 13$ **b.** x **5a.** $x + 1$ **b.** $-3.6x^2 - 2.2x - 1.3$ **6.** $21x^9y$ **7.** $15x^3y^3 + 9x^2y^3 - 6xy^2$ **8.** $4x^2 - 9$ **9.** $-\frac{7x^3y}{z}$ **10.** $5x^3 - 6x + \frac{4}{x^2}$ **11.** $3x^4 - 2$

Section 6.1

1. polynomial **3.** not a polynomial **5.** $7x^3, -7x^2, 8$; coefficients: 7, -7, 8 **7.** $-3x^2, -7x, 5$; coefficients: $-3, -7, 5$ **9.** trinomial **11.** monomial **13.** monomial **15.** $7x^3$ and $3x^3$ **17.** $3x^3$ and $5x^3$; $-7x^2$ and $-x^2$ **19.** $3xy^3$, $-6xy^3$, and xy^3 **21.** $7x^2$ and $-6x^2$, 9 and -2 **23.** $5x^3$ and $-x^3$, $3x^2$ and x^2, 5 and -9 **25.** $3y^2$ **27.** $-2x^5$ **29.** $-6x$ **31.** $6x^3y + 6xy^3$ **33.** $9x^3 + 7x$ **35.** $2xy + 8y$ **37.** $18x^2$ **39.** $-2xy$ **41.** $0.81x + x^2$ **43.** $-0.5x^3 + 3x^2$ **45.** $\frac{11}{72}x^2 + \frac{1}{3}x$ **47.** $-3xy$ **49.** $-0.21x$ **51.** $\frac{1}{24}x^2y$ **53.** $-9y$ **55.** $\frac{19}{48}x^2 - \frac{1}{8}y^2$ **57.** $-0.08a$ **59.** $4.2b$ **61.** $\frac{16}{45}x^3$ **63.** $0.5x$ **65.** $\frac{1}{4}x$ **67.** $-6.1y$ **69.** degree of terms:

3, 2, 1, 0; degree of polynomial = 3 **71.** monomial
73. terms **75.** like, similar

Section 6.2

1. 2 **3.** -15 **5.** 76 **7.** -71 **9.** -27
11. 1.59 **13.** -3.73 **15.** 3.93 **17.** $-\frac{4}{27}$ **19.** $\frac{49}{8}$
21. 23 **23.** 8.14 **25.** $\frac{2}{7}$ **27.** evaluating
29. \$9500 **31.** \$10,750 **33.** 10 **35.** \$104,950
37. 4.35 **39.** $-\frac{133}{90}$

Section 6.3

1. $5x^2 + 6x - 7$ **3.** $6n^3 - 7n^2 + n - 5$
5. $-8x^6 - 9x^5 + 7x^4 + 7x^3 + 6x + 5$ **7.** $5n - 2$
9. $-3x - 5$ **11.** $-x - 4y$ **13.** $-3n^3 - 10n^2 - 16$
15. $-2x - 15y + 10z$ **17.** $4s^3 + 2s^2 + 7s$
19. $7a^2b + 2ab - b^2$ **21.** $8x^4 - 3x^3 - 11x^2 - 16x - 3$
23. $-3x^3 + 4x^2 + 3x$ **25.** $-3a^2 - 12ac + 9c^2$
27. $5.3x - 1.5$ **29.** $3.3x^2 + 4.3xy - 3.81y^2$
31. $-0.81x^2 + 5x - 2.9$ **33.** $\frac{9}{7}x - \frac{69}{8}$
35. $\frac{97}{396}x^2 + \frac{19}{40}x + \frac{28}{11}$ **37.** $\frac{2}{5}x^2 - \frac{7}{30}x + \frac{1}{6}$
39. alphabetical, first **41.** $5x^2 - 3x + 1$
43. $\frac{19}{12}x^2 - \frac{17}{40}$ **45.** $1.45x + 3.18y$ **47.** $-2.1x + 5z$

Section 6.4

1. $-2x + 6$ **3.** $4x + 5$ **5.** $-x + 3y - 4z$
7. $6x^2 - 2$ **9.** $4a + 12$ **11.** $-3x + 2$
13. $-2x^2 + x - 4$ **15.** $2x^2 + 2y^2$
17. $10n^2 - 2n - 3$ **19.** $-2y^2 - 2y + 2$
21. $3x^3 - 12x^2 + 10$ **23.** $2xy^2 - 2xy + 10$
25. $-2a - 10b - c$ **27.** $0.9x - 2.3$
29. $-3.6x^2 - 0.8x + 0.85$ **31.** $\frac{2}{5}x - \frac{2}{3}$
33. $\frac{5}{24}a^2 + \frac{2}{5}a - \frac{19}{10}$ **35.** $\frac{9}{35}x^2 + \frac{13}{60}x + \frac{1}{18}$
37. additive inverse, subtrahend, minuend **39.** $2x + 1$
41. $-x - 6$ **43.** $-3.8x^2 - 7y + 4.79$
45. $7x^3 - 10x^2 - 2x + 11$ **47.** $-x^2 - \frac{11}{40}x + \frac{19}{5}$

Section 6.5

1. $-15x^9$ **3.** $-15n^9$ **5.** $-45x^4y^2$ **7.** $56x^5$
9. $9xy$ **11.** $30x^6$ **13.** $0.93x^8$ **15.** $-12n^{12}$
17. $-0.407n^7$ **19.** $-\frac{3}{20}x^7$ **21.** $-21x^7$
23. $2x^2 + 10x$ **25.** $7x^5 - 49x^4 + 14x^2$
27. $-15x^4 + 12x^3 - 6x^2$ **29.** $12n^7 - 28n^6 - 32n^5$
31. $0.06x^6 - 0.62x^5$ **33.** $-1.55x^3 + 0.1x^2 - 1.05x$
35. $\frac{20}{81}x^3 - \frac{10}{21}x^2$ **37.** $x^2 + 8x + 15$
39. $21x^2 - 41xy + 10y^2$ **41.** $x^3 + 3x^2 - 8x - 4$
43. $24n^3 - 46n^2 + 37n - 12$
45. $7n^4 + 23n^3 - 68n^2 + 39n - 7$
47. $x^4 - 2x^3 - 12x^2 + x + 2$ **49.** numerical coefficients, exponential expressions, numerical coefficient
51. multiplied, sum, like **53.** $-32x^4y^2$ **55.** $72xy$
57. $-0.06x^8$ **59.** $-\frac{12}{55}x^6y^2$ **61.** $24y^5 - 32y^4 + 20y^3$
63. $-16x^3 + 48x^2y - 8xy^2$
65. $30x^3y^2 + 24x^2y^2 - 36xy$
67. $10x^2y^3 + 12x^2y^4 + 6xy^4$ **69.** $0.06x^5 - 0.62x^3$
71. $24.6c^4 - 1.23c^3 + 0.246c^2$ **73.** $\frac{3}{10}x^6 - 8x^4$
75. $\frac{3}{28}a^3b^3 - \frac{3}{8}a^3b^4$ **77.** $x^2 - 16$
79. $x^2 + 8x + 16$ **81.** $x^2 + 10x + 25$
83. $12x^3 - 22x^2 - 3x + 3$ **85.** $8x^2 - 1.4x - 0.15$
87. $0.6x^3 - 2.34x^2 - 8.02x + 15.5$
89. $x^8 + 2x^4 - x^2 + 2$
91. $2x^2 - 3xy + 9xz - 9y^2 + 18yz - 5z^2$
93. $x^3 + y^3$ **95.** $3x^3 - 4x^2 - 11x + 14$
97. $-12x^3 + 15x^2 - 3x$ **99.** $\frac{9}{25}x^4$ or $0.36x^4$

Section 6.6

1. $4x^3$ **3.** $-\frac{2}{3}x^3y^3$ **5.** $\frac{7x}{y}$ **7.** $\frac{mn^4}{5}$ **9.** $-\frac{3y^2}{x^2}$
11. $70x^3$ **13.** $\frac{xy^6}{5}$ **15.** $\frac{11}{13}x^2y$ **17.** $\frac{3}{4}x^2$
19. $3x^3 + 4x$ **21.** $4x^5 + 3x^3 - 2x$
23. $-5x^5 + 3x^4$ **25.** $\frac{3a}{b^2} - 2b^2$ **27.** $4x - 3 + \frac{5}{3x}$
29. $\frac{3}{5}n - \frac{6}{5m}$ **31.** $-\frac{15a^3}{b} - 5a^2b + \frac{32b^3}{a}$
33. $\frac{3}{2}x^4y^2 - 2x^2 + \frac{1}{y^2} + \frac{3}{x^2y^3} + \frac{9}{2x^3y^4}$
35. polynomial, monomial, sum **37.** $5x^6$ **39.** $-\frac{111}{x^4}$
41. $-3x$ **43.** $\frac{2}{3}y^2$ **45.** $-130x^6$ **47.** $\frac{3}{x^4}$ **49.** $\frac{5a}{7b}$
51. $-\frac{11}{13}a^2b^2$ **53.** $\frac{5b}{4a}$ **55.** $0.13x^7$
57. $3x^4 - 5x^2 + 7x$ **59.** $5x^4 - 4x^2$
61. $-5x^5 + 6x^3 - 7$ **63.** $5a - \frac{2}{a} + \frac{1}{a^2}$
65. $\frac{2x}{y^3} - 3y^2$ **67.** $5x - 4 + \frac{3}{2x^2}$ **69.** $\frac{6}{5}x - \frac{8}{5y}$
71. $-\frac{10x^4}{y} - 4x^3y^2 + \frac{11x^3}{y}$ **73.** $-x^{12} - x^{10} - x^8 + x^6$
75. $2x + 3y - 5z$ **77.** $-\frac{6}{x} + \frac{8}{x^3} - \frac{4}{x^4}$ **79.** $-\frac{4}{5x^7}$
81. $3x^2 - 4yz - 5xy$

Section 6.7

1. $x + 3$ **3.** $3x + 2 - \frac{1}{3x - 5}$ **5.** $3x + 1$ **7.** $5x - 3y$ **9.** $5x^2 - 2xy - 2y^2$ **11.** $4x^2 + 3x + 6$ **13.** $2x - 3y$ **15.** $x + 6 + \frac{13}{x - 3}$ **17.** $3x - 2 - \frac{12}{2x - 3}$ **19.** $3a^2 + 6a - 5$ **21.** $x + y$ **23.** $x^2 + xy + y^2$ **25.** $a^3 - a^2 + 2 + \frac{-3a - 3}{2a^2 - a - 1}$ **27.** $3x^2 - 1$ **29.** $x + 2$ **31.** $a^2 + 2a + 4$ **33.** $3x^2 - 8x + 7 - \frac{12}{2x + 2}$ **35.** $5x - 7 + \frac{14}{2x + 1}$ **37.** $3x^4 - 2$ **39.** $3t + 2 + \frac{1}{3t + 1}$ **41.** descending, long division, whole **43.** $x - 2$ **45.** $a - b$

CHAPTER 6 Review Exercises

1. $6x^2$ and $-16x^2$, $-7x$ and $3x$ *(Section 6.1)* **2.** $-5x^2y$ and x^2y *(6.1)* **3.** $-3x^2 + 3x$ *(6.1)* **4.** $5x^2 + 0.91x$ *(6.1)* **5.** $\frac{29}{24}x^3y$ *(6.1)* **6.** $-\frac{61}{396} + \frac{1}{36}y^2$ *(6.1)* **7.** 35 *(6.2)* **8.** $\frac{68}{9}$ *(6.2)* **9.** $\frac{17}{30}x - \frac{49}{60}$ *(6.3)* **10.** $-\frac{5x^2}{y^2}$ *(6.6)* **11.** $-36.4x^6y^5$ *(6.5)* **12.** $-2x^3 + 7x^2 + y^2$ *(6.4)* **13.** $2x^3 - 2x^2 - 4x - 2$ *(6.3)* **14.** $x^2 + 2x + 4$ *(6.7)* **15.** $-1.5x^4 + 0.05x^3 - 2x^2$ *(6.5)* **16.** $-3a^2 + 4a - \frac{5}{2a}$ *(6.6)* **17.** $3x^3 - 26x^2 + 17x - 6$ *(6.5)* **18.** $-\frac{3x^5}{4z}$ *(6.6)* **19.** $8x^2 - 10xy - 11y^2$ *(6.4)* **20.** 39 *(6.2)*

CHAPTER 7

CHAPTER 7 Pretest

1a. yes **b.** no **2a.** 4 **b.** -2.25 **3a.** -18 **b.** $\frac{5}{3}$ **4a.** 18 **b.** $\frac{1}{2}$ **5a.** 15.1 **b.** 8 **6a.** $\frac{77}{100}$ **b.** 3 **7a.** $\frac{c}{a + b}$ **b.** $\frac{P - ad}{a}$ **8a.** $x < -19$ **b.** $x \le -9$ **9a.** $x < -30$ **b.** $x \ge 15$ **10a.** $x \ge -\frac{5}{3}$ **b.** $x \le \frac{9}{2}$ **c.** $x \ge \frac{4}{23}$

Section 7.1

1. not a linear equation in one variable **3.** not a linear equation in one variable **5.** not a linear equation in one variable **7.** 4 is not the solution **9.** -2 is not the solution **11.** 1.9 is the solution **13.** $-\frac{3}{8}$ is not the solution **15.** 4 is not the solution **17.** 11.9 is not the solution **19.** 3 is not the solution **21.** -3 is not the solution **23.** -2 is not the solution **25.** 5 is not the solution **27.** 5 is the solution **29.** solution **31.** -2.8 is the solution **33.** 4 is not the solution **35.** -1.64 is the solution **37.** $-4\frac{1}{3}$ is the solution **39.** -10 is not the solution **41.** $-\frac{5}{2}$ is not the solution **43.** -5 is not the solution **45.** -3 is not the solution

Section 7.2

1. 5 **3.** -6 **5.** -5 **7.** 5 **9.** -18 **11.** -1.1 **13.** -2.71 **15.** -1 **17.** $-\frac{37}{40}$ **19.** -44 **21.** $\frac{1}{42}$ **23.** 2 **25.** -7 **27.** $-\frac{3}{2}$ **29.** 28 **31.** -0.09 **33.** -24 **35.** $-\frac{2}{5}$ **37.** -3 **39.** 2.5 **41.** -40 **43.** additive inverse **45.** 5.5 **47.** -24 **49.** 6 **51.** -2 **53.** -15 **55.** 15 **57.** 21 **59.** 1 **61.** $\frac{5}{6}$ **63.** $-\frac{3}{2}$ **65.** 35 **67.** 6 **69.** $-\frac{21}{2}$ **71.** 21 **73.** -0.015 **75.** -41 **77.** -110 **79.** -3.7 **81.** -0.7 or $-\frac{7}{10}$ **83.** $-\frac{2}{5}$ or -0.4 **85.** $\frac{7}{6}$ **87.** $\frac{15}{2}$

Section 7.3

1. 90 **3.** 5.85 **5.** -6 **7.** $\frac{35}{3}$ **9.** 4 **11.** 2 **13.** -6 **15.** 16.5 **17.** $\frac{1}{2}$ **19.** $\frac{15}{2}$ **21.** 6 **23.** $\frac{2}{7}$ **25.** $-\frac{5}{3}$ **27.** -3.8 **29.** $-\frac{7}{4}$ **31.** 0.4 **33.** -1.2 **35.** 6 **37.** -1 **39.** 3.2 **41.** -9 **43.** 4 **45.** -20 **47.** -4 **49.** 2 **51.** 3 **53.** $\frac{5}{3}$ **55.** 2 **57.** $\frac{11}{7}$ **59.** 5 **61.** $-\frac{9}{2}$ **63.** 1 **65.** -50 **67.** 0.09 **69.** -3.2 **71.** -0.7 **73.** -4 **75.** $-\frac{4}{3}$ **77.** 20 **79.** -4.2 **81.** 0.1 **83.** combine, addition

Section 7.4

1. 1 **3.** 0 **5.** -5 **7.** 1 **9.** 3 **11.** $\frac{7}{3}$ **13.** 7 **15.** $\frac{32}{3}$ **17.** 4 **19.** 1 **21.** 17 **23.** $-\frac{5}{3}$ **25.** $\frac{1}{2}$ **27.** -3 **29.** $\frac{1}{4}$ **31.** $\frac{2}{3}$ **33.** -18 **35.** 12 **37.** 20 **39.** 36 **41.** 4 **43.** 0 **45.** 30 **47.** -3 **49.** -17 **51.** -4 **53.** 14 **55.** -4 **57.** $-\frac{60}{11}$ **59.** -12 **61.** $-\frac{8}{5}$ **63.** perform the indicated operation **65.** -6 **67.** 6 **69.** $-\frac{3}{2}$ **71.** $-\frac{21}{16}$ **73.** $-\frac{1}{2}$ **75.** -8 **77.** $\frac{9}{8}$ **79.** 43 **81.** $\frac{17}{8}$ **83.** 3 **85.** -72 **87.** 2

Section 7.5

1. $\frac{d}{r}$ **3.** $\frac{2A}{h}$ **5.** $\frac{2h - 440}{11}$ **7.** $\frac{2S - an}{n}$ **9.** $\frac{2}{a - b}$ **11.** $\frac{E - IR}{I}$ **13.** $\frac{A - P}{pt}$ **15.** $\frac{d + 4a}{4}$ **17.** $\frac{V}{wh}$ **19.** $(a - c)x - cd$ **21.** $\frac{1 + a}{ac - a}$ **23.** $5 - y$ **25.** $\frac{12 - 3y}{4}$ **27.** $\frac{c + 5a}{3}$ **29.** $\frac{5a + 10}{2}$ **31.** $\frac{ac}{a - c}$ **33.** $\frac{I}{Pt}$ **35.** $\frac{6}{a + b}$ **37.** $\frac{3 - ab}{a}$ **39.** $\frac{2A - bh}{h}$ **41.** literal, constant

Section 7.6

1. $x < -9$ (number line: −9 −6 −3 0 3)

3. $x \leq -6$ (number line: −9 −6 −3 0 3)

5. $x \leq -0.59$ (number line: −2 −1 0)

7. $x < 4.7$ (number line: −1 0 1 2 3 4 5)

9. $x \geq -10$ (number line: −10 −5 0 5)

11. $x \geq \frac{14}{15}$ (number line: −1 0 1 2)

13. $x < 2$ (number line: −1 0 1 2 3)

15. $x \leq 2$ (number line: −1 0 1 2)

17. $x \geq -\frac{11}{2}$ (number line: −6 −4 −2 0 2)

19. $x \leq 10$ (number line: −10 0 10 20)

21. $x < 1.2$ (number line: −1 0 1 2)

23. $x \geq -30$ (number line: −45 −30 −15 0 15)

25. $x < -15$ (number line: −30 −15 0 15)

27. $x < -\frac{8}{3}$ (number line: −3 −2 −1 0 1)

29. $x < -3$ (number line: −6 −3 0 3)

31. $x > 0.02$ (number line: −1 0 1)

33. $x \geq -1.7$ (number line: −2 −1 0 1)

35. $x \geq -20$ (number line: −40 −20 0 20)

37. $x > 12$ (number line: −12 0 12 24)

39. $x > -10$ (number line: −20 −10 0 10)

41. $x \geq 15$ (number line: −15 0 15 30)

43. $x < \frac{13}{24}$ (number line: −1 0 1 2)

45. $x \leq 1.7$ (number line: −1 0 1 2)

47. $x < -3$ (number line: −6 −3 0 3)

49. $x > -1$ (number line: −2 −1 0 1)

51. $x \leq 0.2$ (number line: −1 0 1)

53. $x \leq -5$ (number line: −10 5 0 5 10)

55. $x < \frac{10}{7}$ (number line: −1 0 1 2)

57. $x \geq -1.2$ (number line: −2 −1 0 1)

59. $x > -10.8$ (number line: −15 −10 −5 0 5)

61. $x \geq -10$ (number line: −20 −10 0 10)

63. $x > 0.636$ (number line: −1 0 1 2)

65. $x > 12$ (number line: −12 0 12 24)

67. $x \leq -8.6$ (number line: −16 −8 0 8)

69. $x < -0.12$ (number line: −2 −1 0 1)

71. $x < 14$ (number line: −28 −14 0 14)

73. $x > 8$ (number line: −8 0 8 16)

75. $x \leq -8$ (number line: −16 −8 0 8)

77. $x \leq \frac{15}{14}$ (number line: −1 0 1 2)

79. $x \geq -\frac{2}{7}$ (number line: −2 −1 0 1)

81. $x \leq 1$ (number line: −2 −1 0 1)

83. $x \geq -1$ (number line: −2 −1 0 1 2)

85. inequality, greater

87. not, one, solutions

Section 7.7

1. $x \leq 2$ **3.** $x > 2$

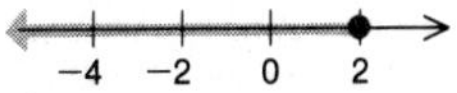

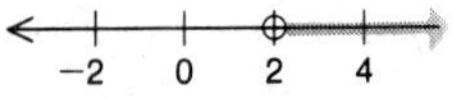

5. $x < 6$ **7.** $x \leq 8$

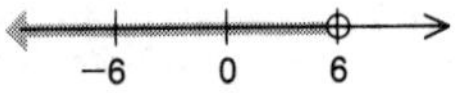

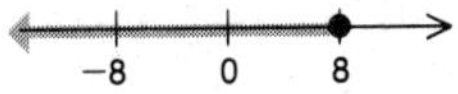

9. $x < 10$ **11.** $x \leq -5$

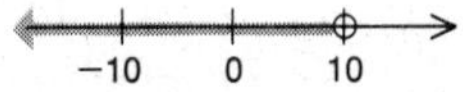

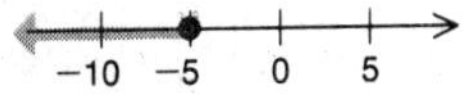

13. $x \leq -2$ 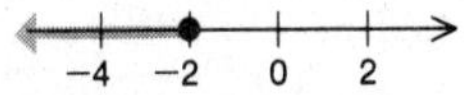 **15.** $x < 2$

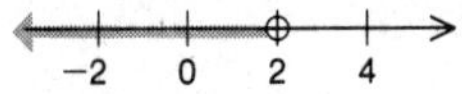

17. $x \geq 4$ 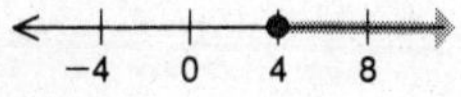 **19.** $x > -2.5$

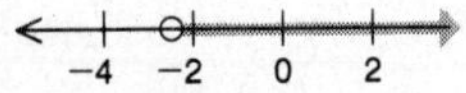

21. $x \geq 2$ 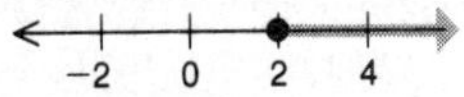 **23.** $x < 2$

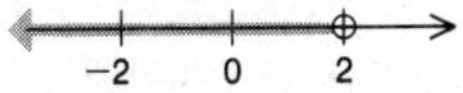

25. $x \leq -4$ 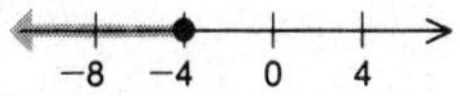 **27.** $x > -2$

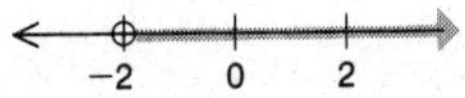

29. $x \leq -3$ 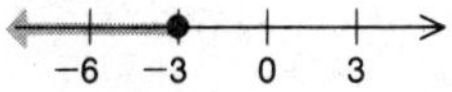 **31.** $x < -\frac{1}{4}$

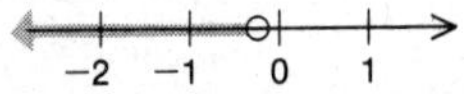

33. $x \leq \frac{42}{5}$ 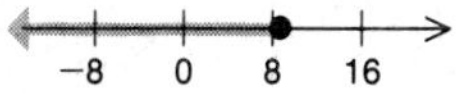 **35.** no solution

37. $x \geq \frac{9}{8}$ **39.** $x < \frac{14}{17}$

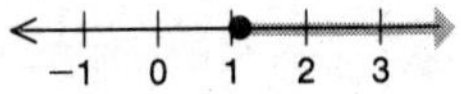

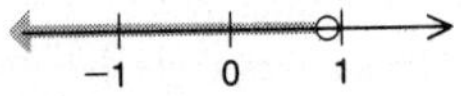

41. $x < -72$ **43.** $x \leq -\frac{39}{10}$

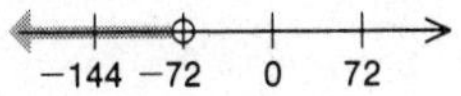

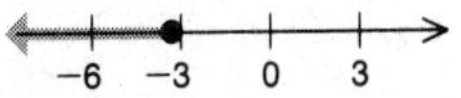

45. $x < -\frac{12}{5}$ **47.** $x > -\frac{3}{5}$

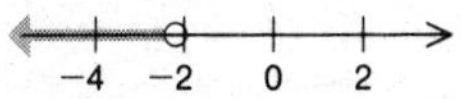

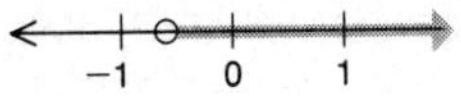

49. $x \leq -2$ 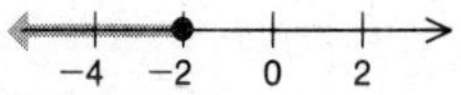 **51.** $x \leq 100$

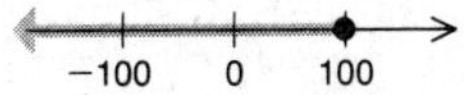

53. $x \leq 4$ 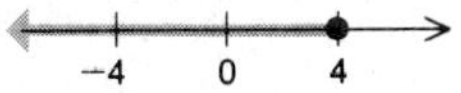 **55.** $x > 2$

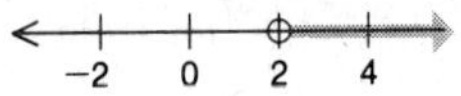

57. $x < 3$ **59.** $x > -2$

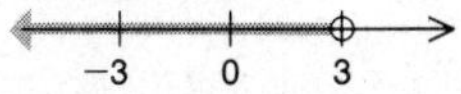

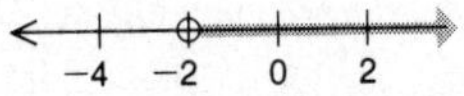

61. $x \geq 10$ 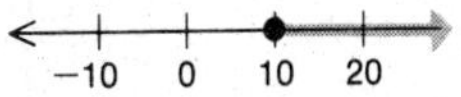 **63.** $x < -6$

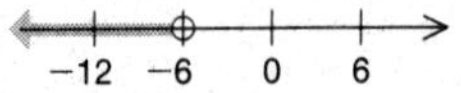

65. infinite number of solutions

CHAPTER 7 Review Exercises

1. yes *(Section 7.1)* **2.** no *(7.1)* **3.** yes *(7.1)*
4. -7 *(7.2)*
5. $x > -2$ *(7.6)*

6. -27 *(7.2)* **7.** -3 *(7.2)* **8.** 33 *(7.3)*
9. 15 *(7.4)* **10.** 0.39 *(7.2)*
11. $x > 9$ *(7.7)*

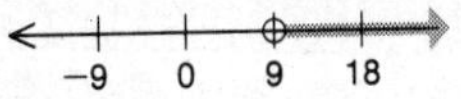

12. $\frac{c-b}{a}$ *(7.5)* **13.** -20 *(7.3)* **14.** $-\frac{5}{2}$ *(7.4)*

15. $\frac{9}{128}$ *(7.4)* **16.** $\frac{yz}{y+z}$ *(7.5)*

17. $x < -6$ *(7.6)*

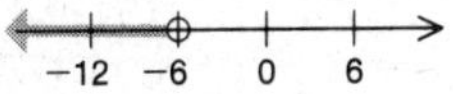

18. $x \geq -25$ *(7.6)*

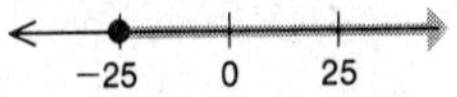

19. $x \geq 3.62$ *(7.6)*

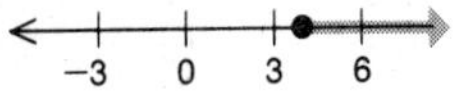

20. $\frac{7}{3}$ *(7.4)*

21. $x \leq \frac{11}{2}$ *(7.7)*

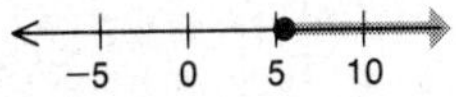

22. $x < 24$ *(7.7)*

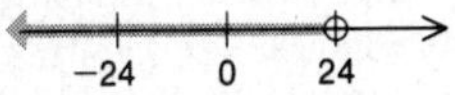

23. $\frac{b}{a-c}$ *(7.5)* **24.** $-\frac{5}{29}$ *(7.4)*

CHAPTER 8

CHAPTER 8 Pretest

1. x, $12 - x$ **2.** 30 **3.** $\frac{1}{12}$ **4.** \$25 **5.** 5

6. \$3500 **7.** 25 liters **8.** $3\frac{1}{2}$

Section 8.1

1. $x + 10$ **3.** $\frac{2}{3}x$ **5.** $2x$ **7.** $x - 6$ **9.** $6x$
11. $x - 8$ **13.** $x - 50$ **15.** $x - 7$ **17.** $x - 80$
19. $\frac{x}{12}$ **21.** $\frac{x}{10}$ **23.** $x - 2$ **25.** $x + 90$ **27.** $9x$
29. $3 + \frac{x}{2}$ **31.** $4x + 7$ **33.** $\frac{2}{4} - 2x$ **35.** $2x + 8$
37. $\frac{x}{2} - 10$ **39.** $9(x + 7)$ **41.** $\frac{1}{4}x + 5$ **43.** $x + \frac{2}{6}$
45. $x + 130$ **47.** $l - 13$ **49.** $n + 2$ **51.** $d - 35$
53. $s + 75$ **55.** $10 - x$ **57.** $40 - x$ **59.** $\frac{x}{2}$
61. x, $2x$ **63.** x = John's height, $x + 8$ = Joe's height
65. x = \$ in Dan's account, $x - 1025$ = \$ in Tom's account **67.** x = length of rectangle, $x + 3$ = width of rectangle **69.** x, $60 - x$ **71.** x, $1000 - x$
73. $2b - 5$ **75.** x, $4x$ **77.** x, $50 - x$
79. $300 + x$ **81.** x, $5000 - x$ **83.** $10 - x$
85. x, $x + 1$, $x + 2$ **87.** 25, 35 **89.** 17, 18
91. -5 **93.** 14 **95.** 10 **97.** \$425 **99.** 39, 40
101. 93, 95 **103.** 10, 12, 14 **105.** 23, 33
107. Gene received 122 votes; Howard received 262 votes
109. 30 **11.** 20°, 100°, 60° **113.** 171 **115.** 625
117. 37, 39, 41 **119.** 10 teachers, 11 tutors

Section 8.2

1. $\frac{2}{3}$ **3.** $\frac{73}{110}$ **5.** $\frac{15}{7}$ **7.** $\frac{1}{15}$ **9.** $7\ \frac{\text{feet}}{\text{minute}}$
11. $15\ \frac{\text{dollars}}{\text{hour}}$ **13.** 5:9 **15.** 3:7 **17.** 2.5¢
19. 80¢ **21.** extremes: 5, 30; means: 15, 10
23. extremes: 12, 65; means: 60, 13 **25.** equal
27. not equal **29.** 9 **31.** 27 **33.** 60 **35.** 4
37. $\frac{27}{5}$ **39.** 0.5 **41.** 3.5 **43.** 5 **45.** $-\frac{10}{3}$
47. 4 **49.** 78 **51.** same, units **53.** proportions, product, product **55.** 20, 25 **57.** \$63.75 **59.** 105 women, 60 men **61.** \$3750 in stocks, \$5000 in bonds, \$6250 in real estate **63.** Phyllis received \$8000, Pam received \$14,000 **65.** \$24,000; \$40,000; \$56,000
67. 20 **69.** 595 **71.** 230.4 **73.** 15 **75.** 42
77. 52 **79.** 966

Section 8.3

1. 60 **3.** 25% **5.** 25 **7.** 226 **9.** 70%
11. 1300 **13.** $66\frac{2}{3}$% **15.** 1400 **17.** 1.44
19. 73.77% **21.** 20 **23.** \$30 **25.** 25 **27.** \$2,800
29. 8% **31.** 20 **33.** markup is \$90, selling price is \$240 **35.** \$16,000 **37.** 10% **39.** original cost is \$50,000; new cost is \$70,000 **41.** \$11,200 **43.** \$10
45. \$80,000

Section 8.4

1. \$80 **3.** \$240 **5.** \$30 **7.** 6.25% **9.** 15%
11. \$15,000 **13.** 12 **15.** 4 **17.** \$16.20
19. \$20,000 **21.** 9 **23.** \$2,000 at 15%; \$3,000 at 16% **25.** \$850 **27.** \$3,000 **29.** \$2,000 at 10%; \$4,000 at 12%

Section 8.5

1. 9 gallons of 15% solution; 3 gallons of 75% solution
3. 3.6 ounces of 12% solution; 2.4 ounces of 7% solution
5. 17 **7.** $8\frac{1}{3}$ ounces **9.** 400 gallons **11.** $\frac{2}{3}$ quarts
13. 5 **15.** 17 pints **17.** 12 ounces **19.** 20 barrels at \$80 per barrel; 30 barrels at \$100

Section 8.6

1. 110 **3.** 50 mph **5.** 3 **7.** 110 mph **9.** 12.54 mph **11.** $\frac{1}{2}$ hour **13.** 10:24 A.M. **15.** 3
17. Tim's rate is 34 mph; Sam's rate is 42 mph
19. 10:40 A.M. **21.** faster boat travels 30 miles; slower boat travels 20 miles

CHAPTER 8 Review Exercises

1. $x - 5$ *(Section 8.1)* **2.** x, $50 - x$ *(8.1)*
3. 131,132 *(8.1)* **4.** 60% *(8.3)* **5.** $\frac{2}{9}$ *(8.2)*
6. \$600 *(8.4)* **7.** 20 *(8.5)* **8.** \$80,000 *(8.3)*
9. \$12,480 *(8.2)* **10.** 34 mph, 42 mph *(8.6)*
11. \$5000 at 11%, \$3000 at 10% *(8.4)* **12.** 40 *(8.2)*
13. 80 mph *(8.6)* **14.** 50 ounces *(8.5)*
15. 80.5 *(8.1)* **16.** \$5600; \$7840; \$8960 *(8.2)*

CHAPTER 9

CHAPTER 9 Pretest

1.

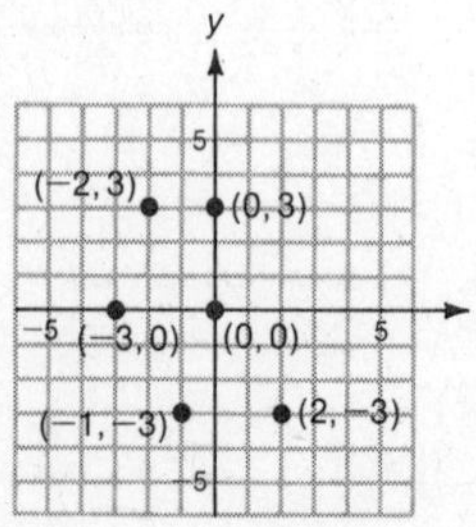

y
5
(−2, 3)
(0, 3)
−5
(−3, 0)
(0, 0)
5
x
(−1, −3)
(2, −3)
−5

2. $A = (2, 2)$, $B = (5, -3)$, $C = (-5, 0)$, $D = (-5, -4)$, $E = (0, -5)$

3a.

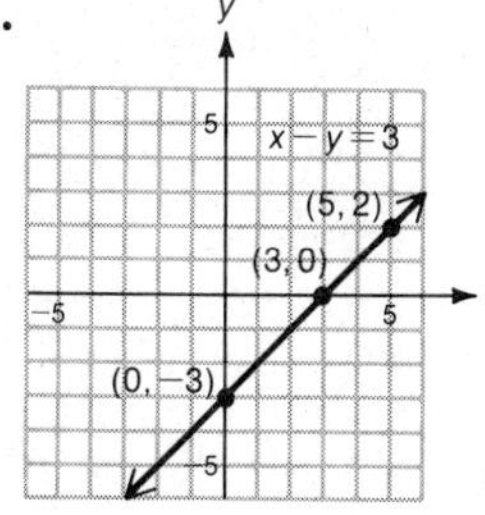

3b.

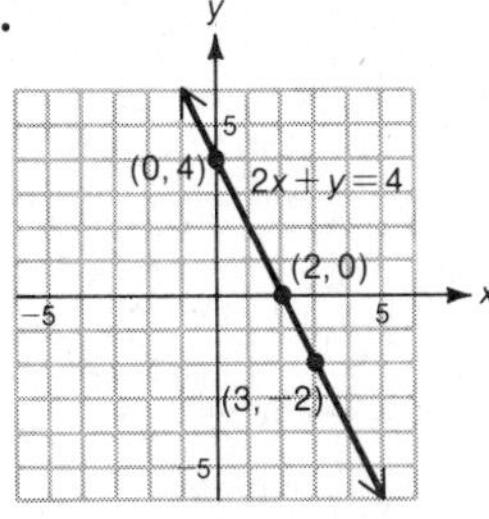

3c.

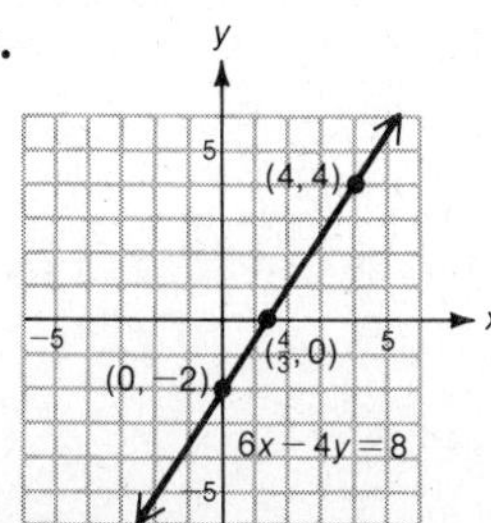

3d.

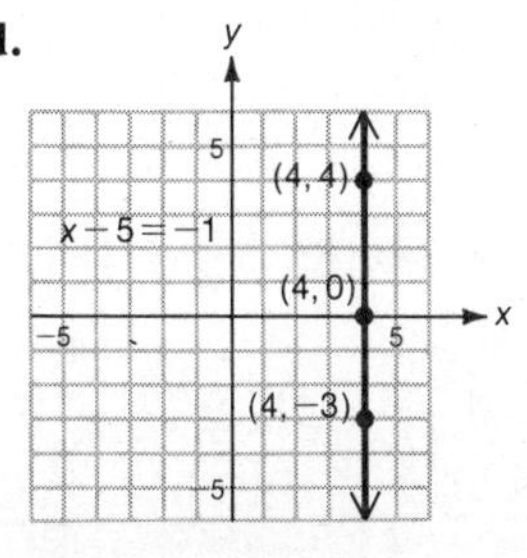

4a.

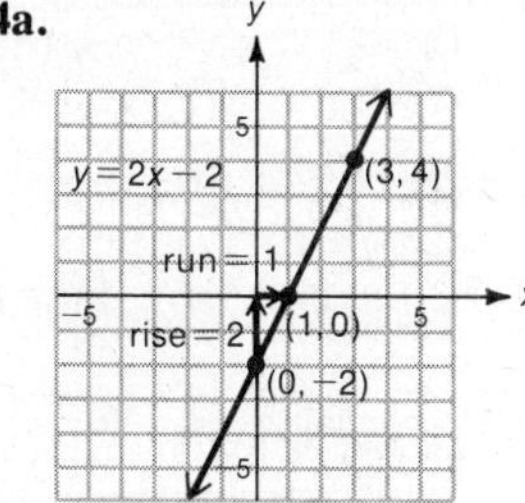

4b.

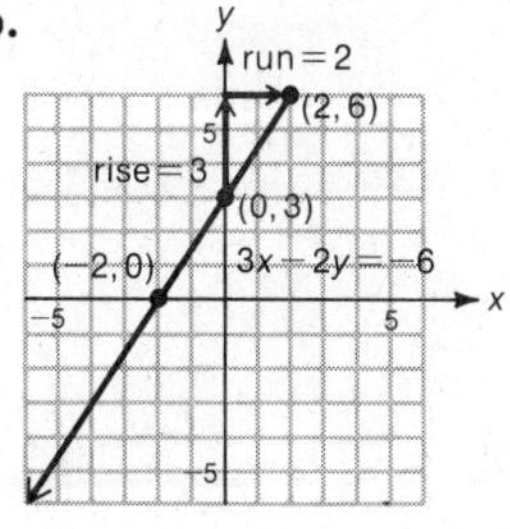

5a.

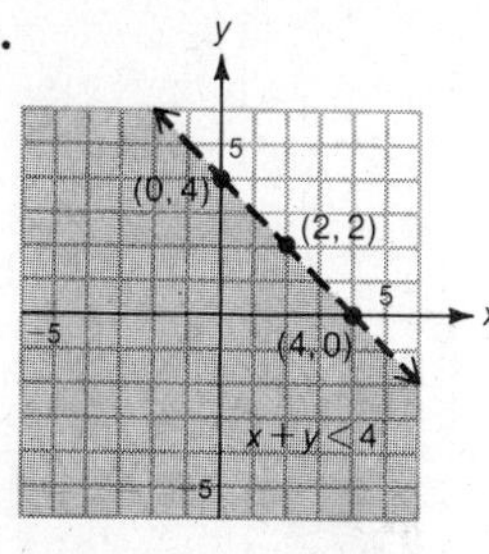

5b.

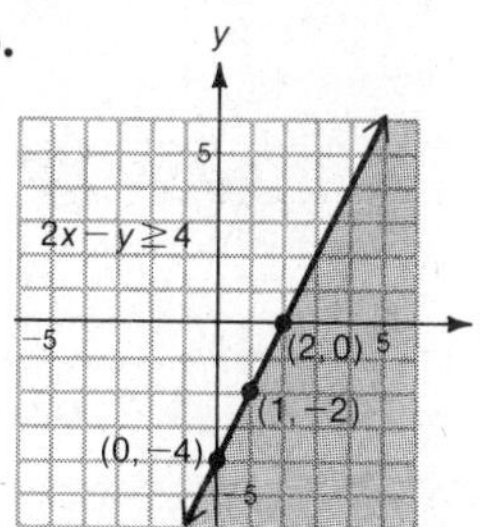

5c.

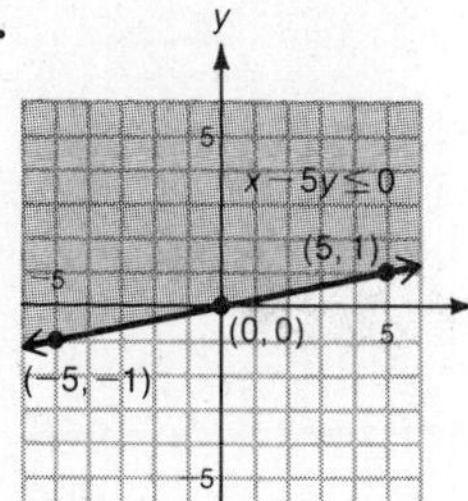

5d.

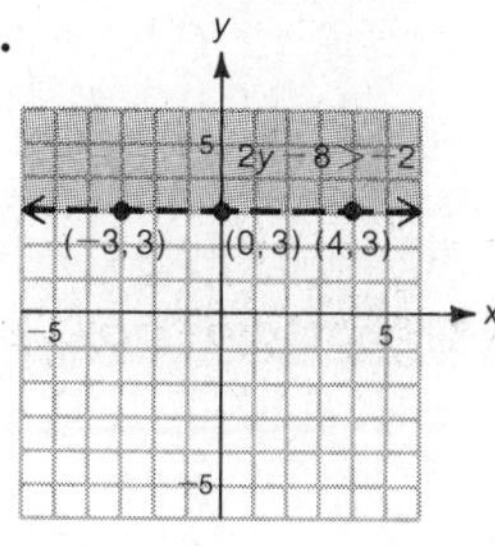

Section 9.1

1.

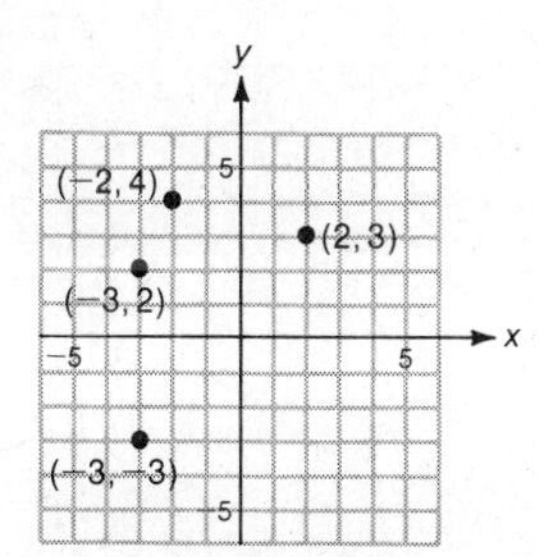

3.

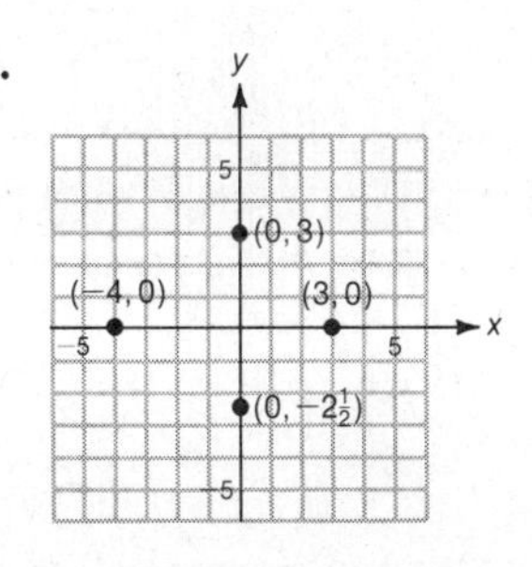

5.

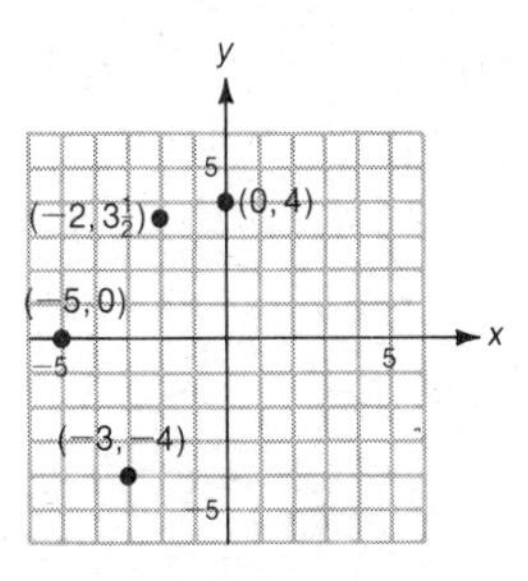

7. $A = (2, 3), B = (-5, -2), C = (0, -5),$
$D = (-6, 5), E = (5, -4), F = (5, 6)$
9. $A = (-6, -2), B = (-4, -6), C = (0, 0), D = (5, 0),$
$E = (-4, 3), F = (3, 7)$ **11.** quadrants
13. ordered pairs **15.** x-coordinate, abscissa, y-coordinate, ordinate **17.** vertical, x, coordinate, horizontal, y, second, coordinates, ordered pair

Section 9.2

1. $(0, 1), (-2, -5), (1, 4)$ **3.** $(0, -2), (7, 1), \left(5, \frac{1}{7}\right)$

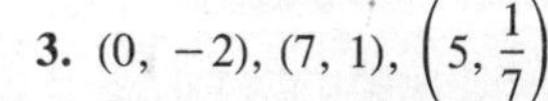

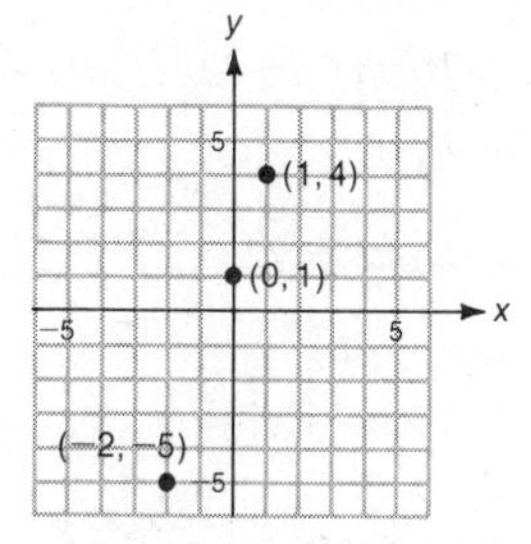

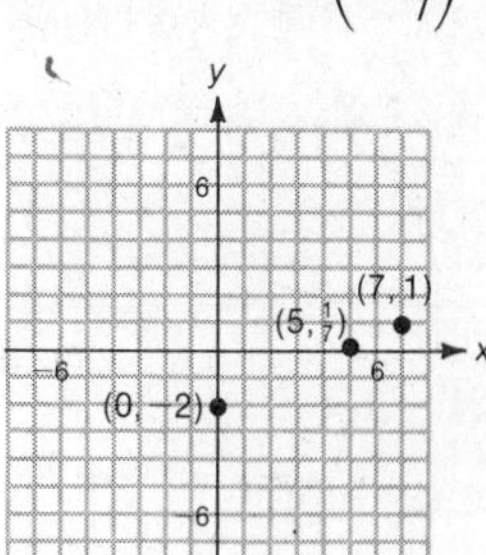

5. $(0, -2), (1, -2), (-3, -2)$

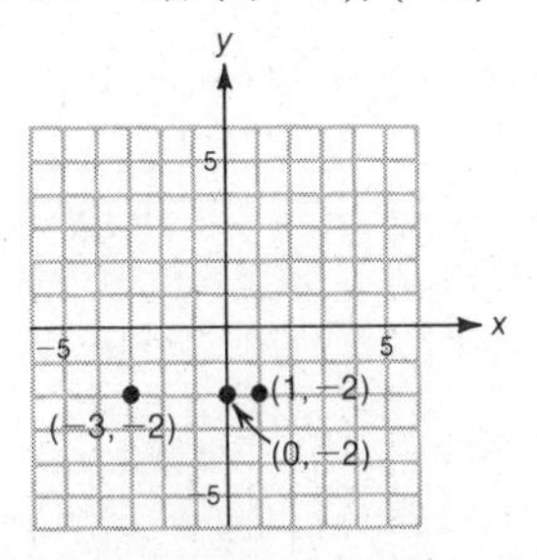

7. x-intercept $= (2, 0)$, y-intercept $= (0, -6)$
9. x-intercept $= (-2, 0)$, y-intercept $= (0, 6)$
11. x-intercept $= (3, 0)$, y-intercept $= (0, 2)$
13. x-intercept $= (12, 0)$, y-intercept $= (0, -6)$

15.

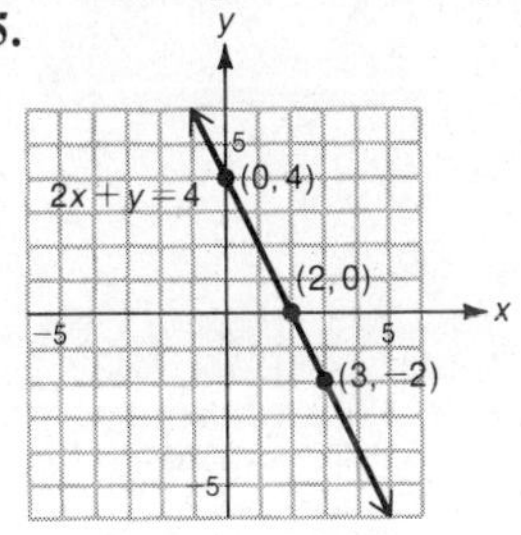

17.

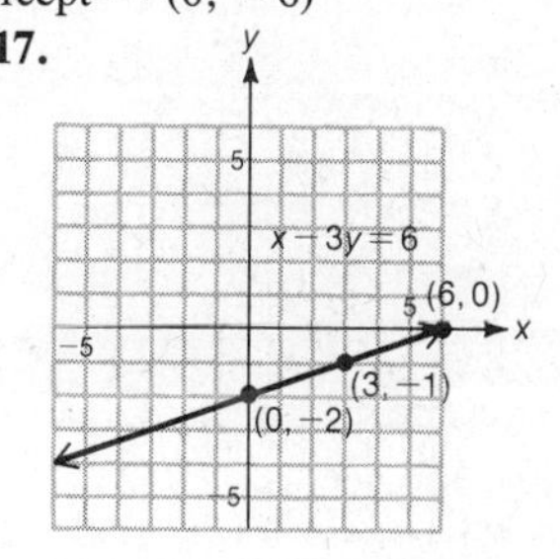

19.

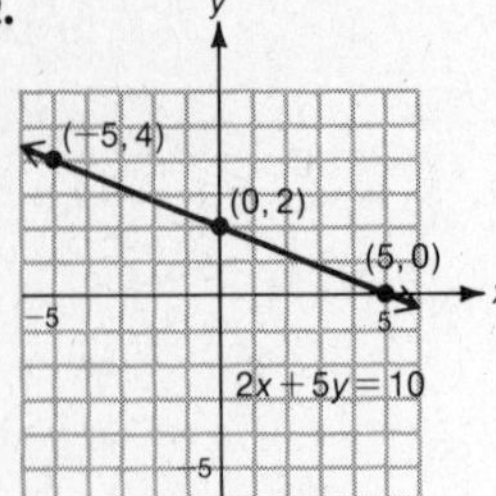

21.

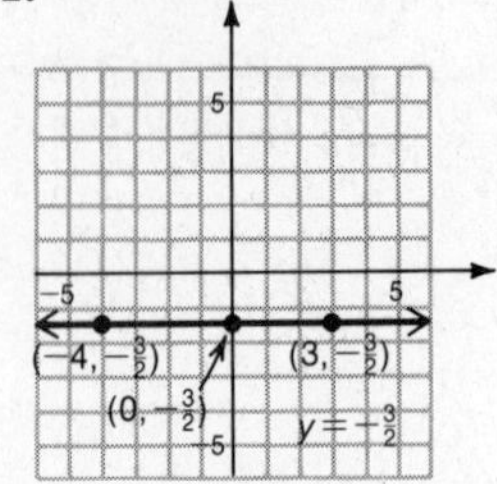

23.

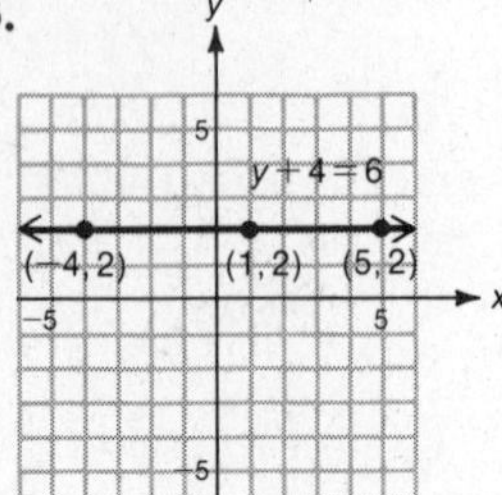

25.

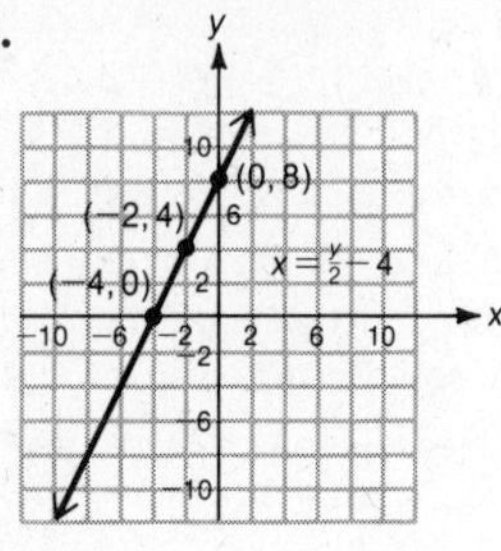

27.

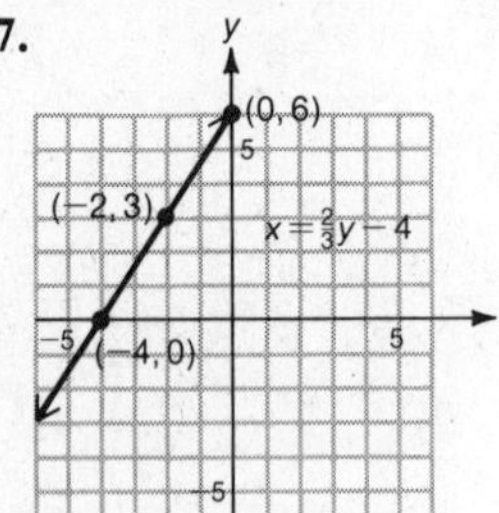

29.

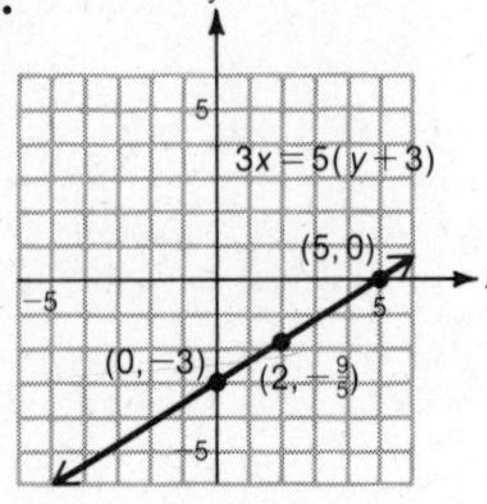

31.

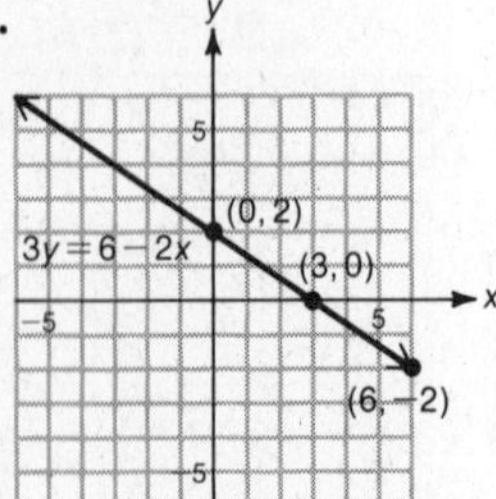

33.

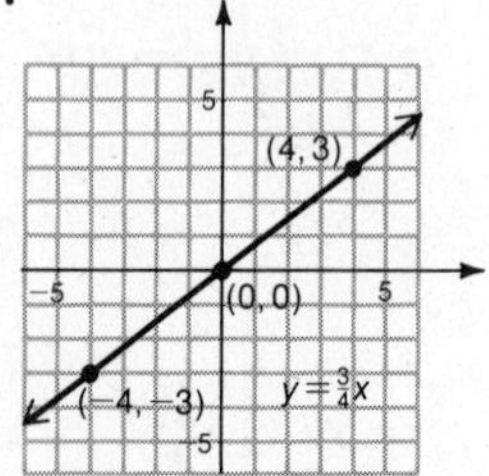

35.

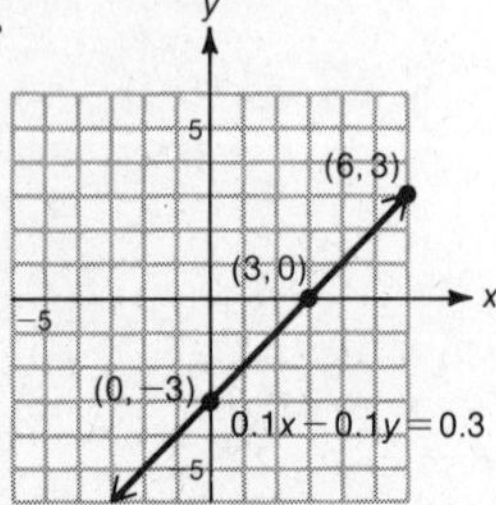

37.

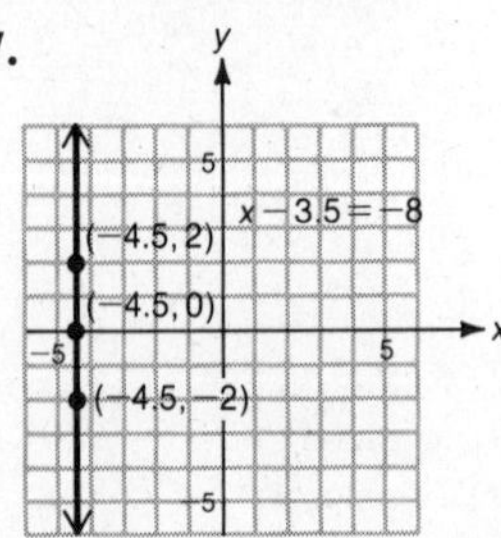

39.

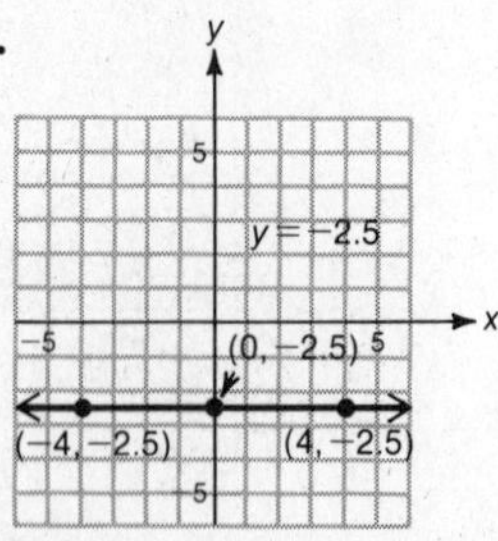

41. straight line **43.** ordered pairs, satisfy, ordered pairs, straight line **45a.** $y = 10x + 10$
45b. (2, 30), (4, 50), (6, 70)
c.

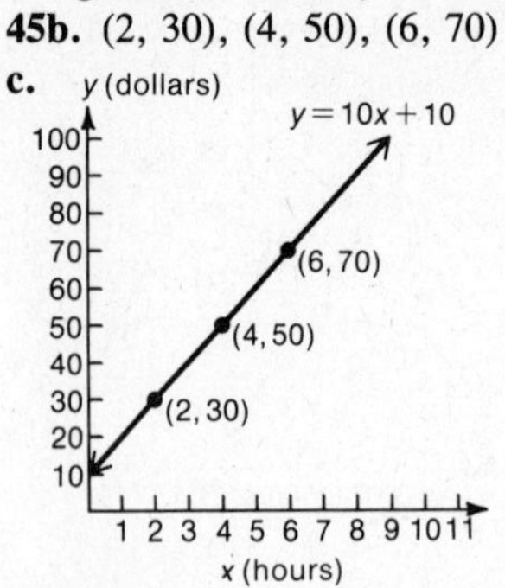

d. 7.5 hours

Section 9.3

1. $y = -2x + 6$, slope $= -2$, y-intercept $= (0, 6)$

3. $y = \frac{2}{5}x - 6$, slope $= \frac{2}{5}$, y-intercept $= (0, -6)$

5. $y = \frac{1}{4}x - 2$, slope $= \frac{1}{4}$, y-intercept $= (0, -2)$

7. $y = -\frac{3}{2}x + 3$, slope $= -\frac{3}{2}$, y-intercept $= (0, 3)$

9. $y = -3x + 6$, slope $= -3$, y-intercept $= (0, 6)$

11. $y = 5x - 8$, slope $= 5$, y-intercept $= (0, -8)$

13. $y = -\frac{3}{2}x + \frac{5}{2}$, slope $= -\frac{3}{2}$, y-intercept $= \left(0, \frac{5}{2}\right)$

15. $y = \frac{3}{5}x - \frac{11}{5}$, slope $= \frac{3}{5}$, y-intercept $= \left(0, -\frac{11}{5}\right)$

17.

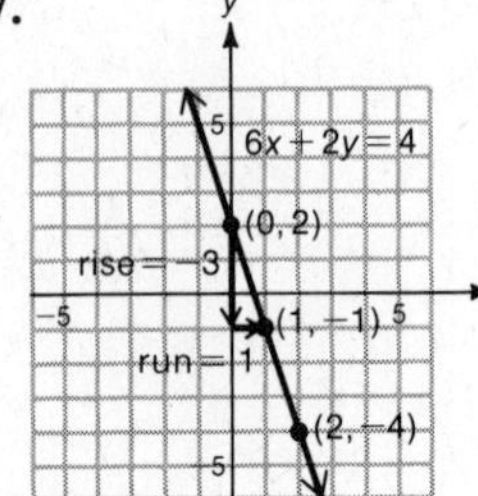

19.

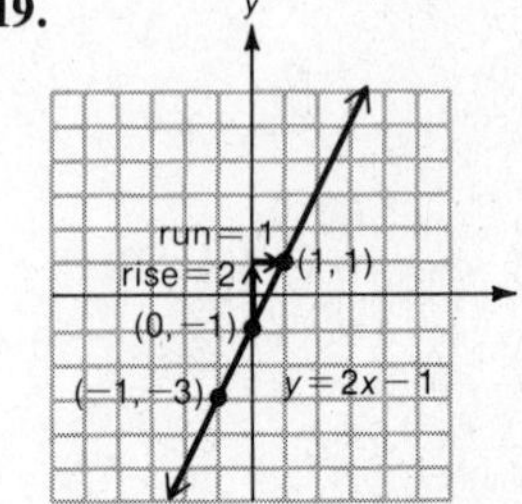

21.

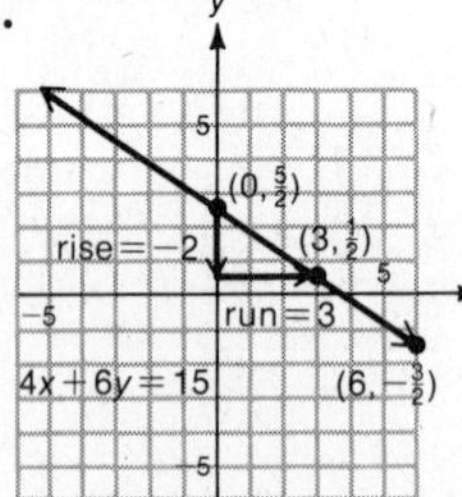

23.

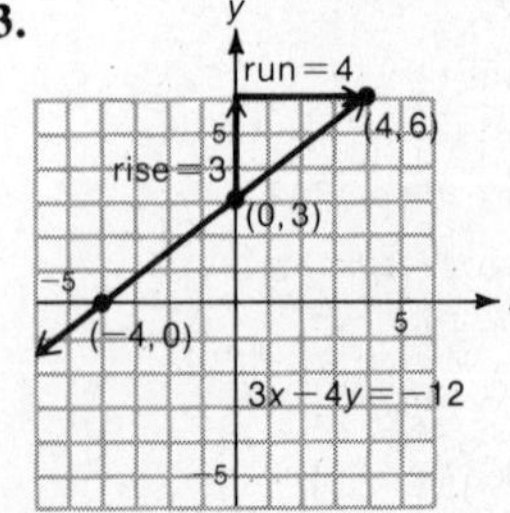

25.

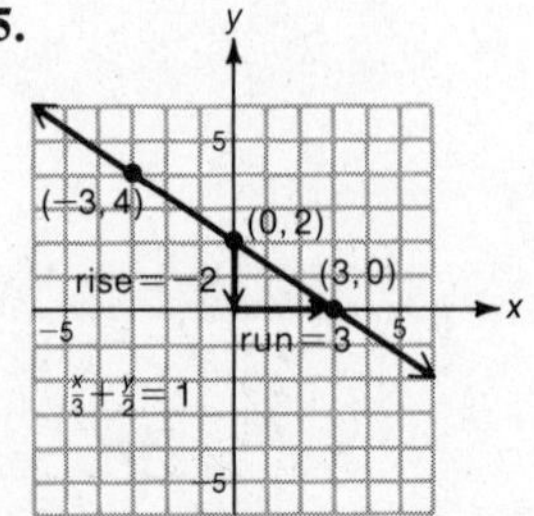

27.

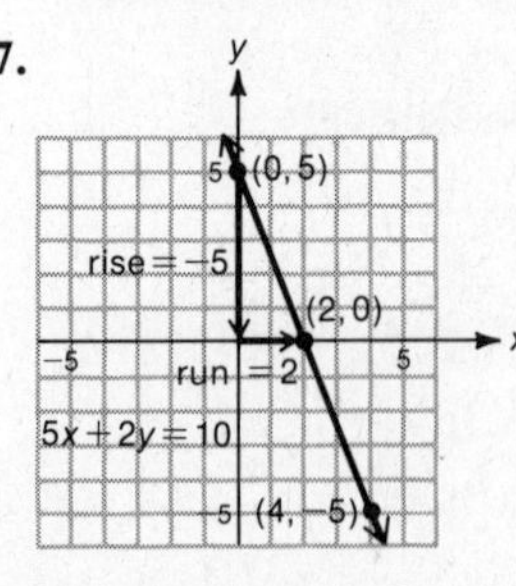

29.

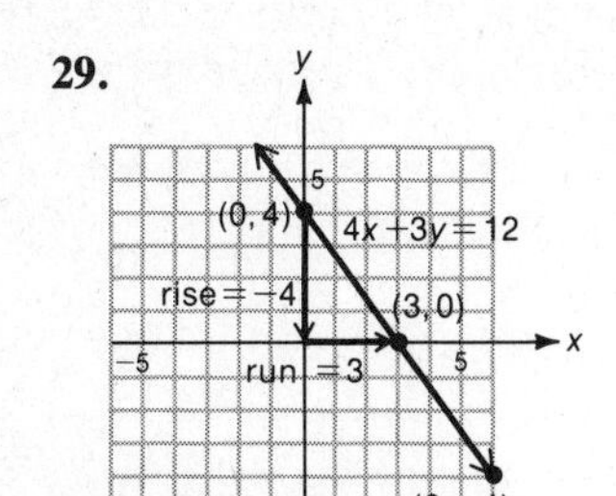

31.

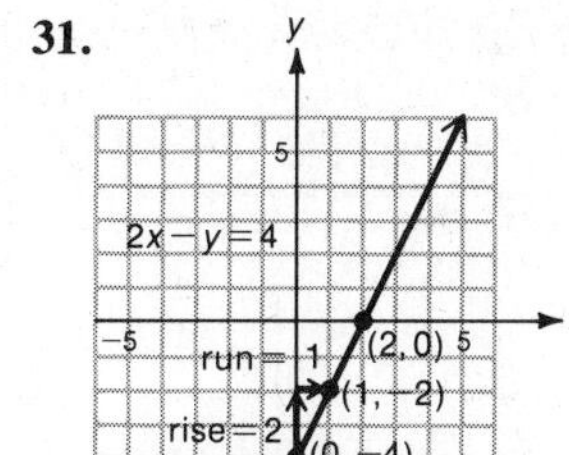

33.

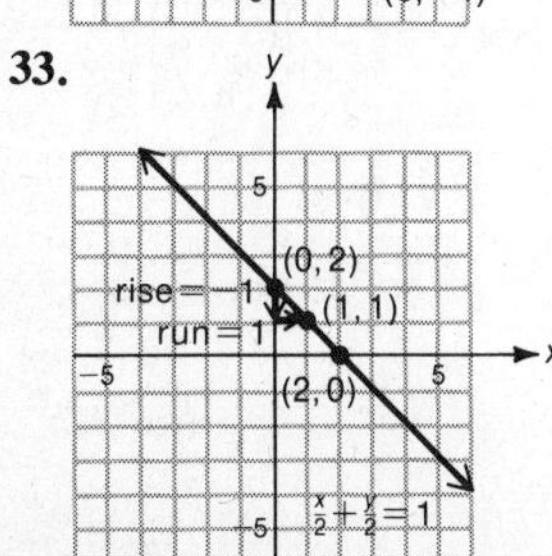

35.

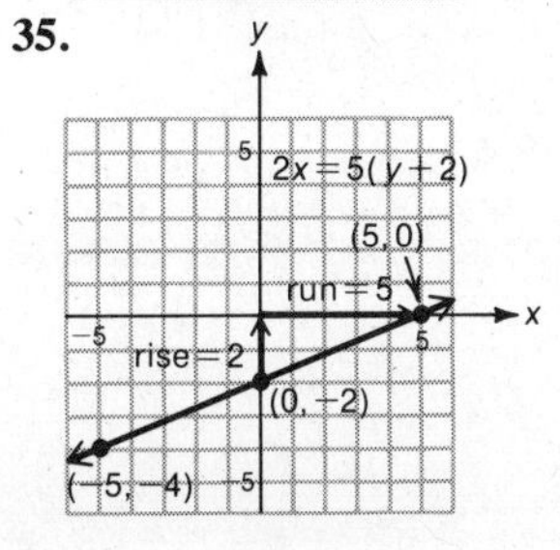

37.

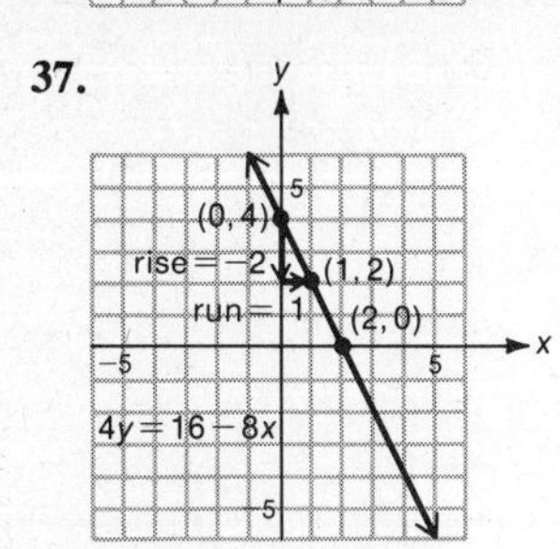

39.

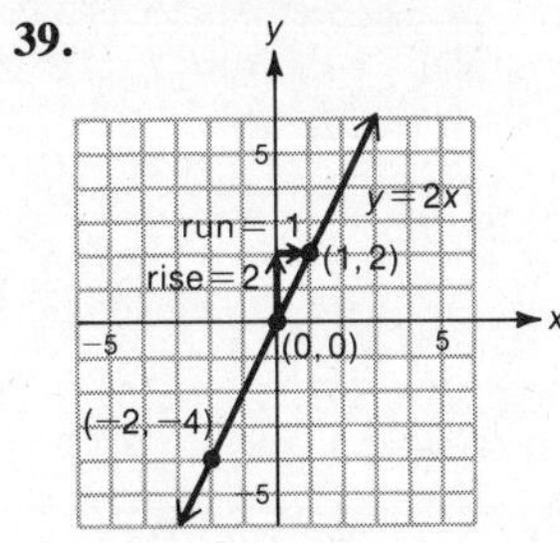

41.

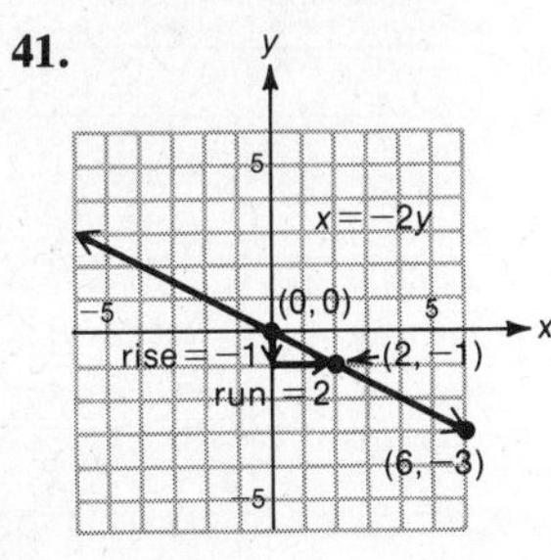

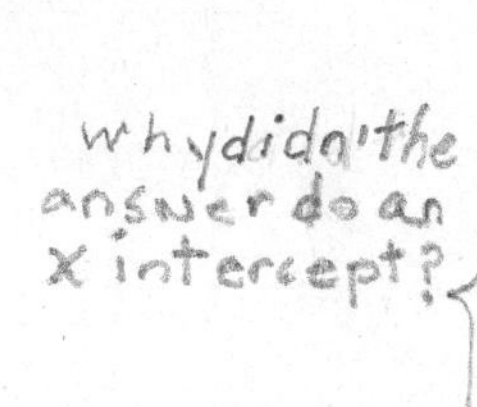

43. $y = mx + b$ **45.** right **47.** rise, run

Section 9.4

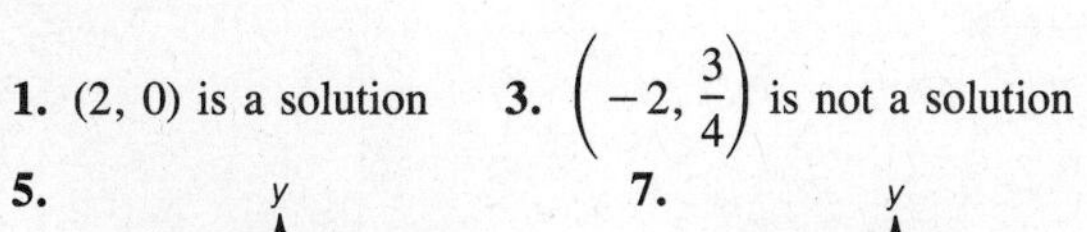

1. (2, 0) is a solution **3.** $\left(-2, \frac{3}{4}\right)$ is not a solution

5.

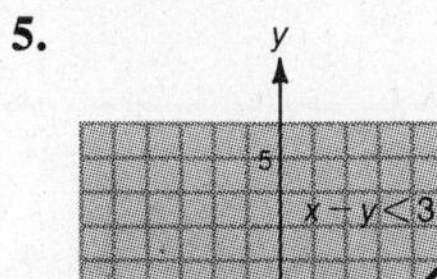

7.

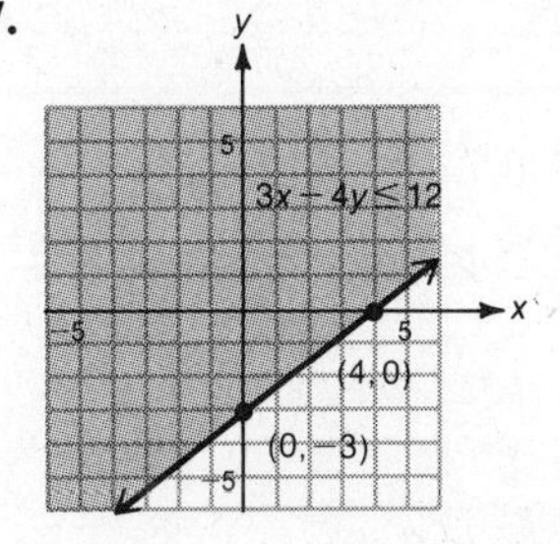

9.

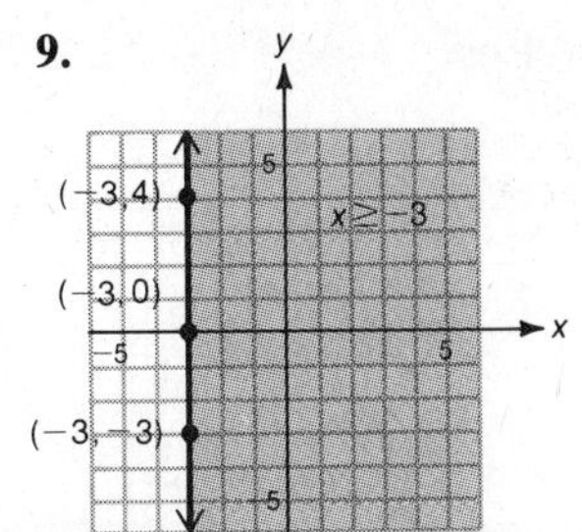

11.

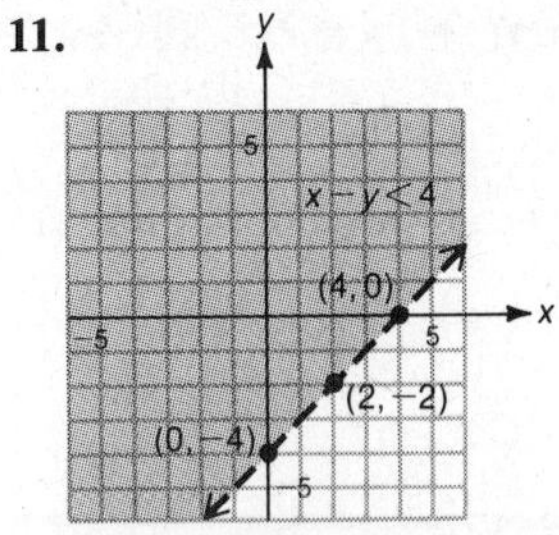

13.

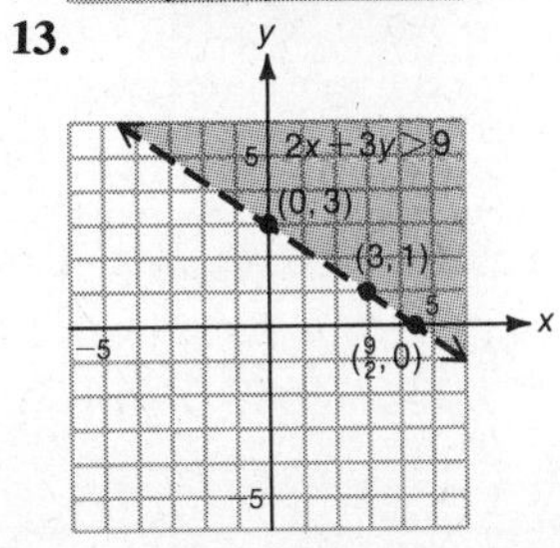

15.

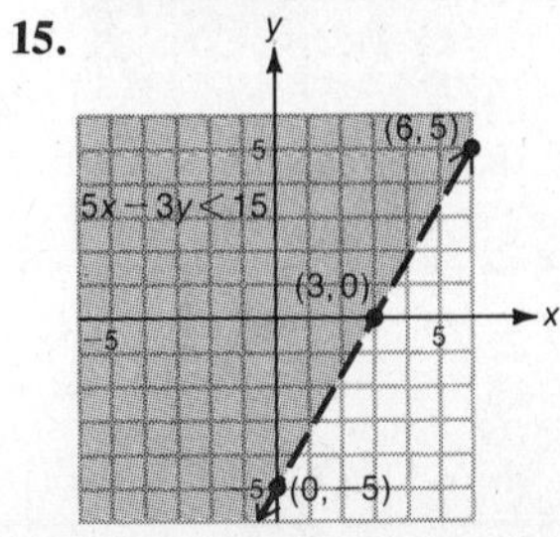

17.

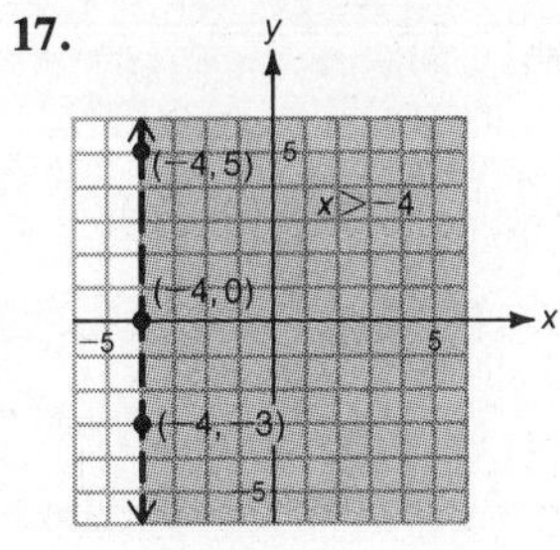

19.

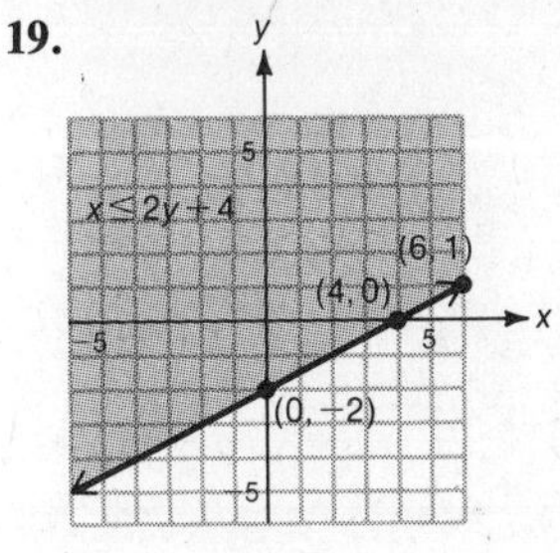

21.

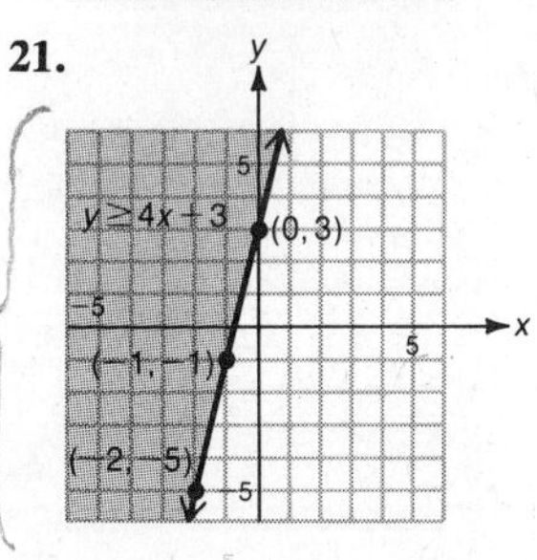

23.

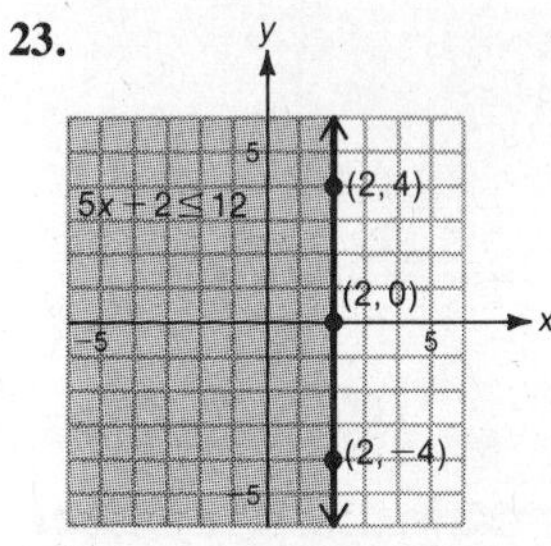

25.

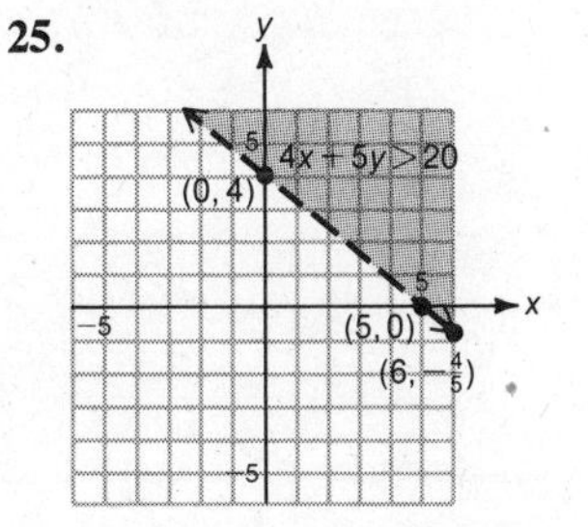

27.

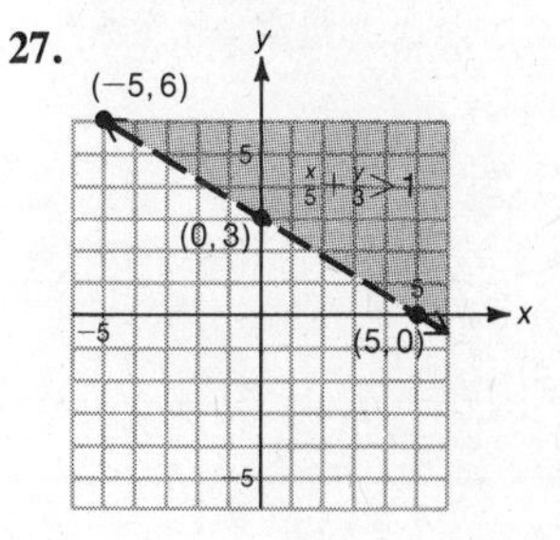

29.

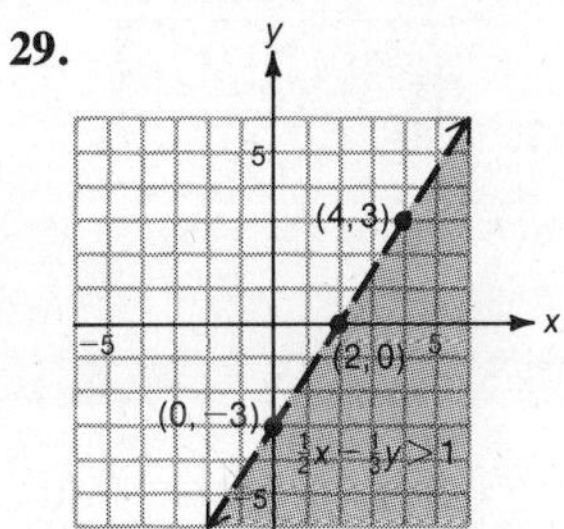

31. equation, equal **33.** dotted, not part **35.** boundary line, line, solution, shade

CHAPTER 9 Review Exercises

1. *(Section 9.1)*

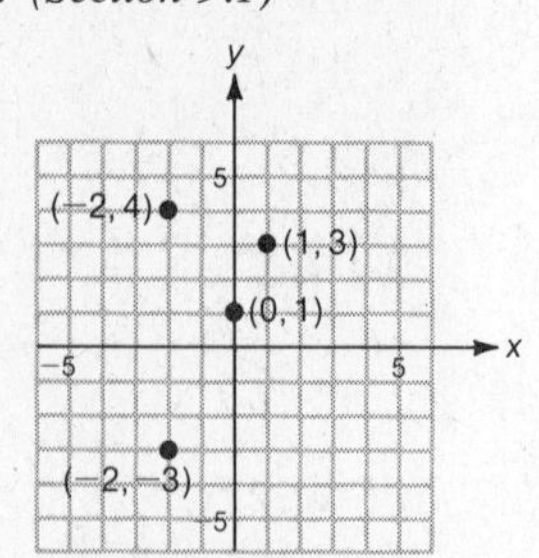

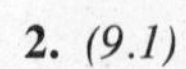
2. *(9.1)*

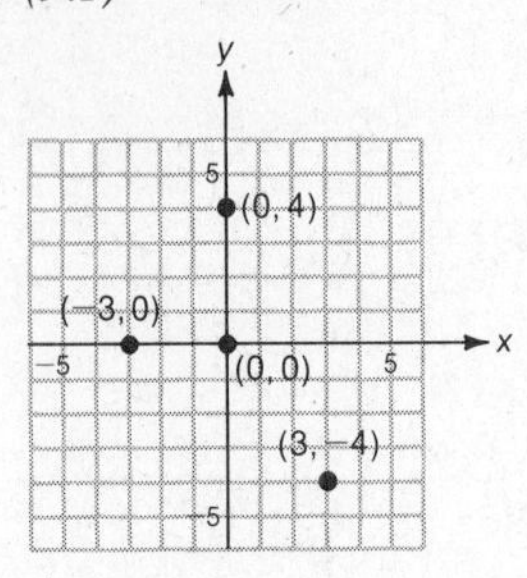

3. $A = (0, 4)$, $B = (-5, 0)$, $C = (3, -2)$, $D = (0, -5)$, $E = (-4, -5)$ *(9.1)* **4.** $A = (4, 1)$, $B = (-3, 1)$, $C = (0, -2)$, $D = (-6, -2)$, $E = (3, -5)$ *(9.1)*

5. *(9.2)*

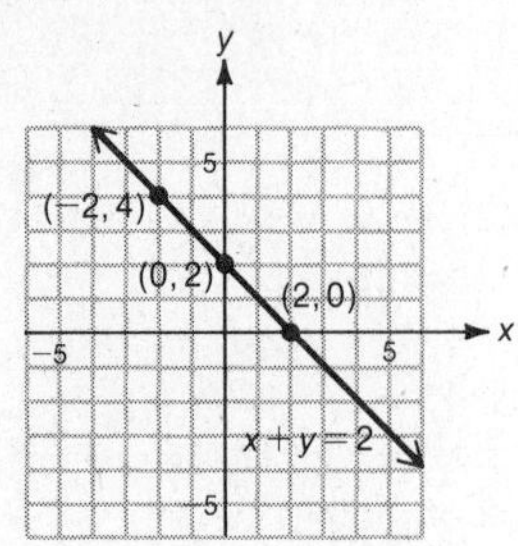

6. *(9.2)*

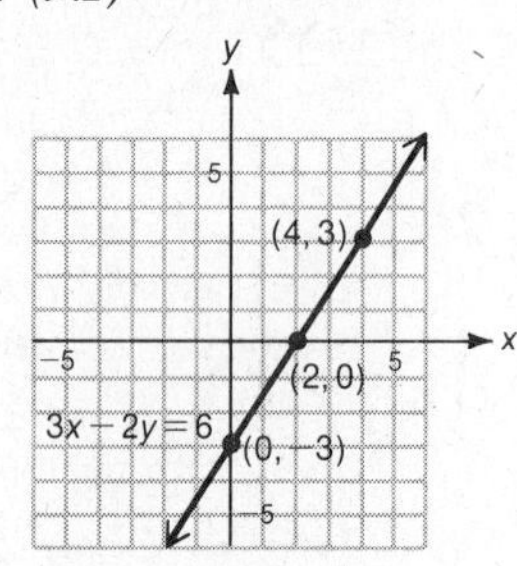

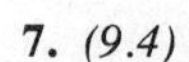
7. *(9.4)*

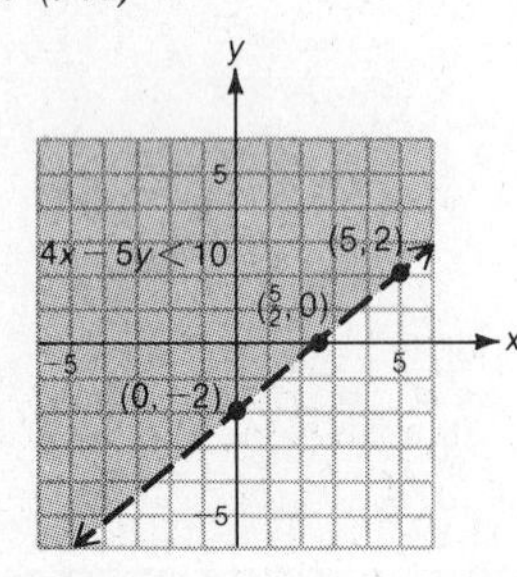

8. *(9.4)*

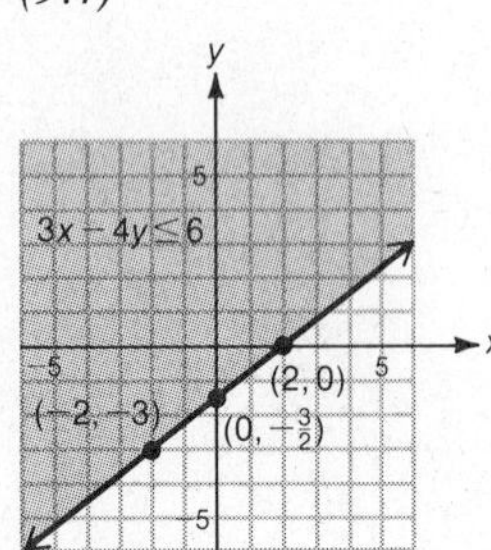

9. *(9.2)*

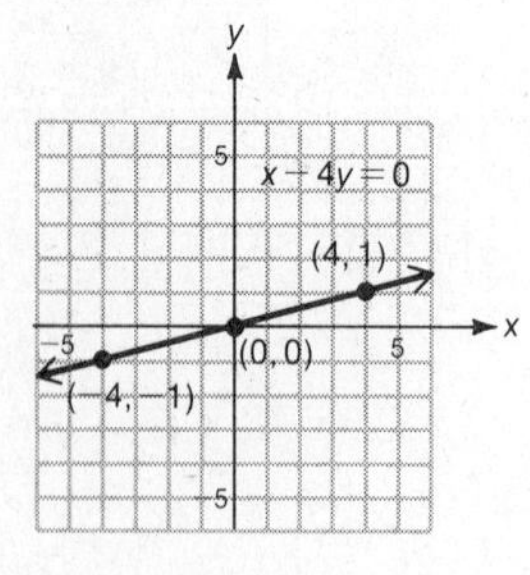

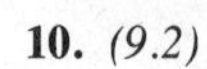
10. *(9.2)*

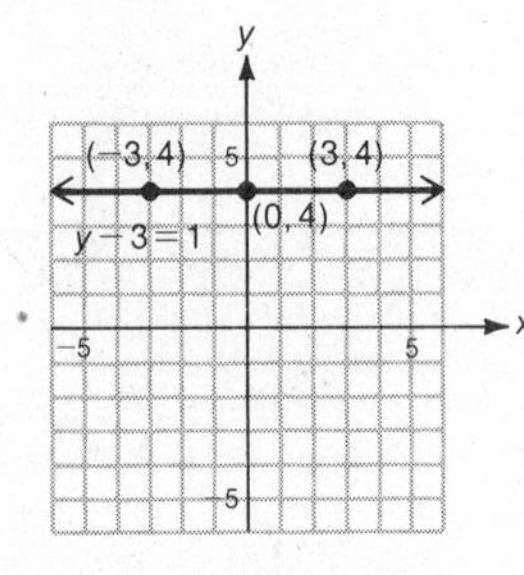

11. *(9.4)*

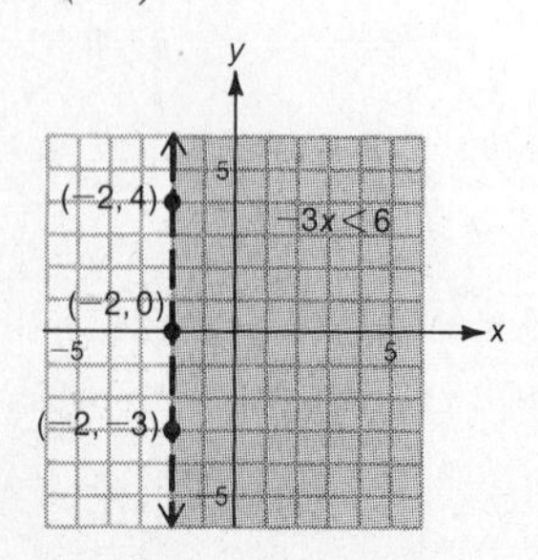

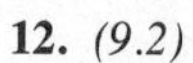
12. *(9.2)*

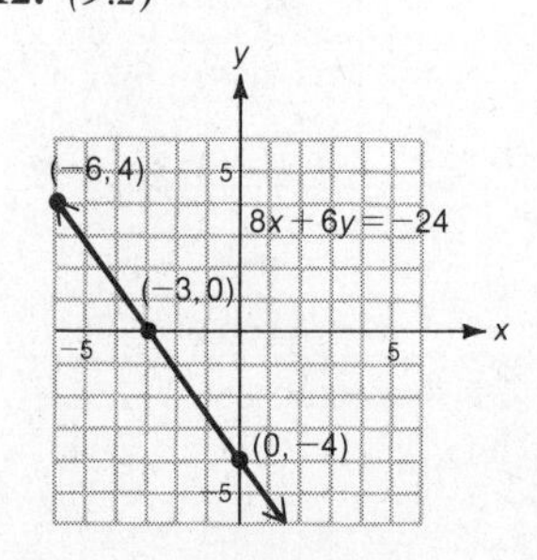

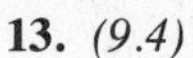
13. *(9.4)*

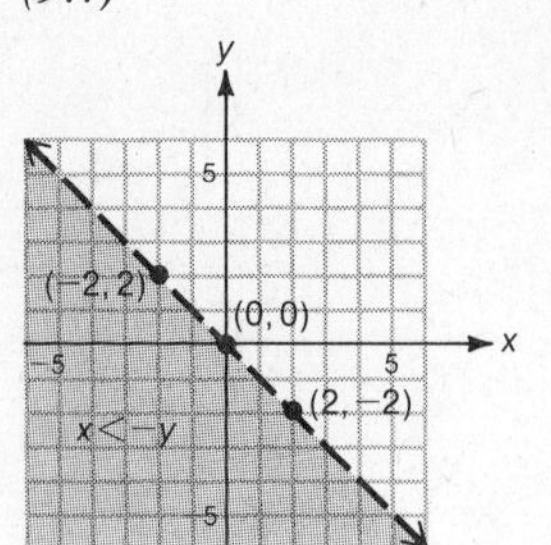

14. *(9.2)*

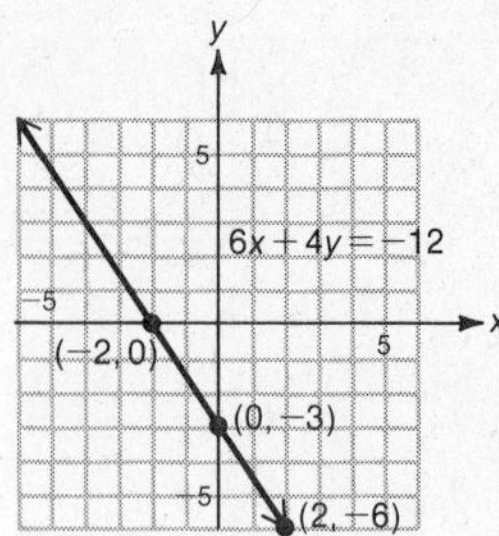

15. *(9.3)*

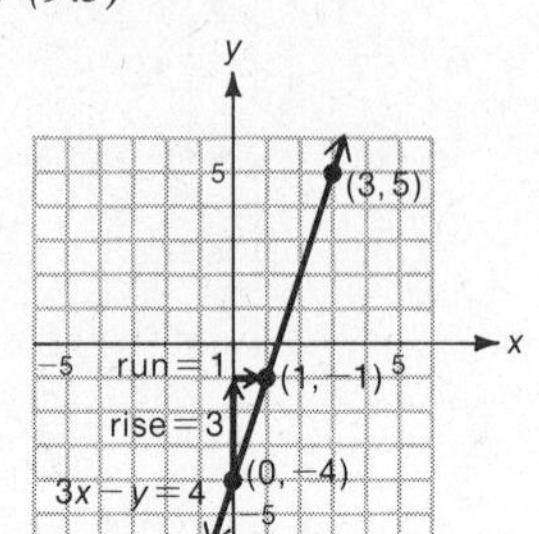

16. *(9.3)*

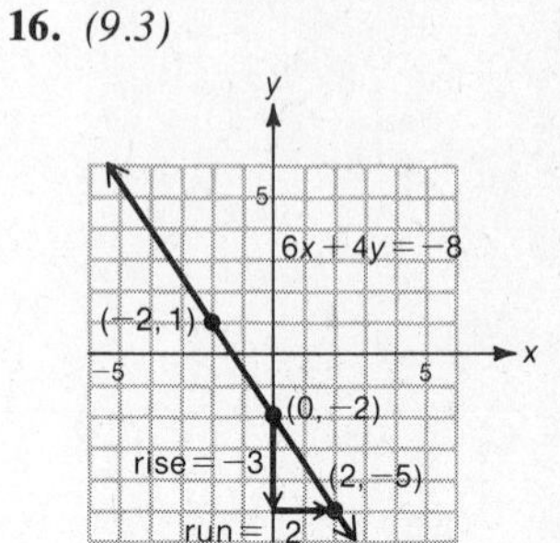

17. *(9.3)*

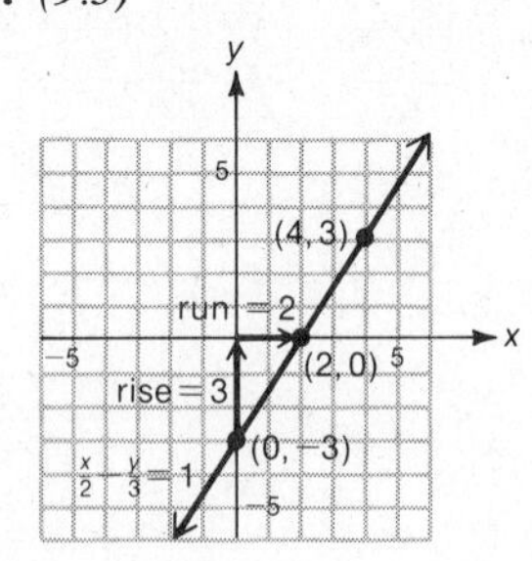

18. *(9.3)*

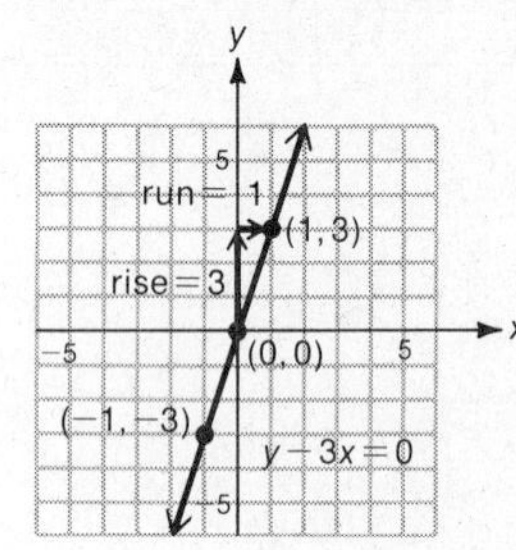

19a. when $n = 2$, $b = 3200$; when $n = 3$, $b = 2800$; when $n = 5$, $b = 2000$ *(9.2)*
b. *(9.2)*

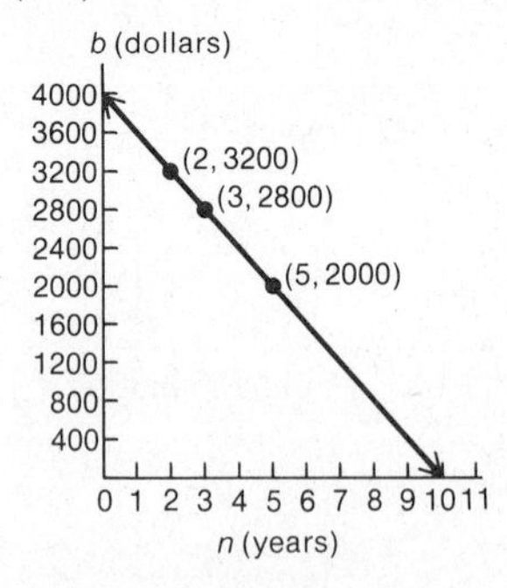

c. After 1 year the book value is $3600 *(9.1)* **d.** when $n = 8$ or in 8 years *(9.1)*

CHAPTER 10

CHAPTER 10 Pretest

1a.

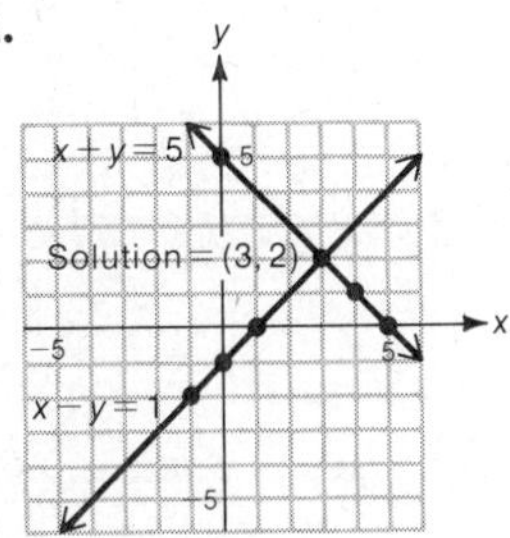

1b.

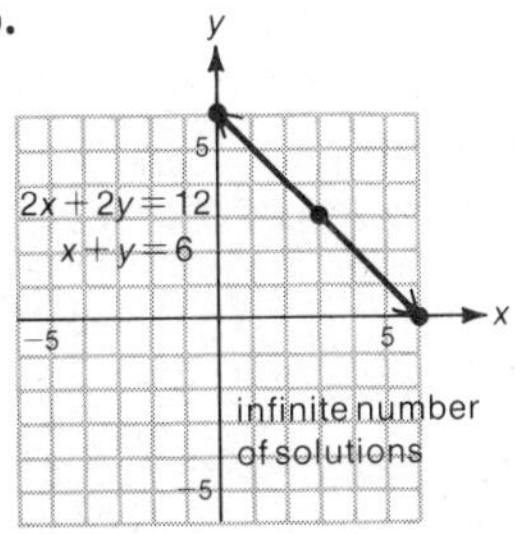

1c.

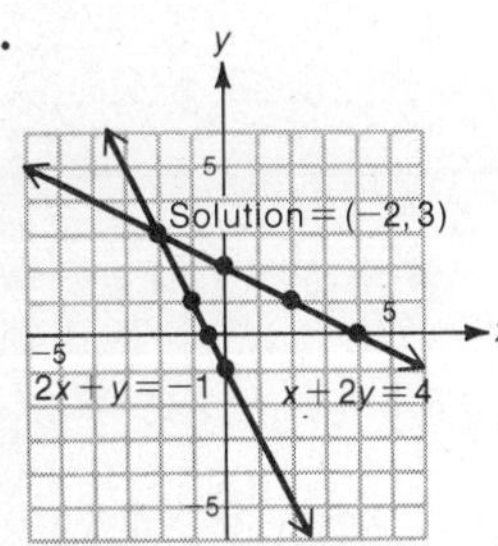

1d.

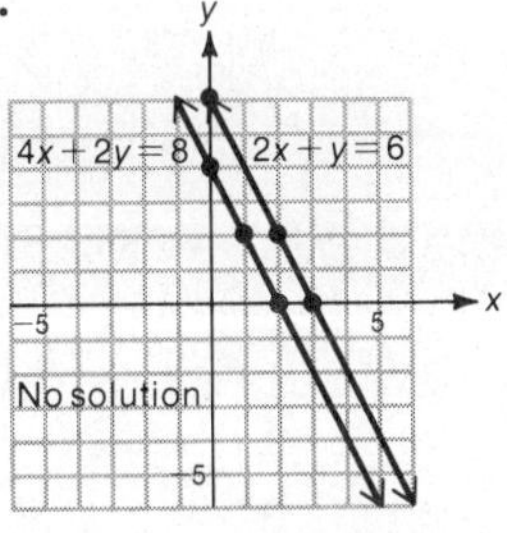

2a. no solution **b.** (1, 3) **c.** $\left(\frac{3}{2}, -2\right)$ **d.** $\left(\frac{40}{7}, \frac{2}{7}\right)$

3a. no solution **b.** infinite number of solutions **c.** $\left(\frac{15}{7}, -\frac{3}{7}\right)$ **d.** (−4, 0) **e.** Guerin earns \$205, Cazzie earns \$410

Section 10.1

1. solution **3.** not the solution **5.** not the solution **7.** not the solution

9.

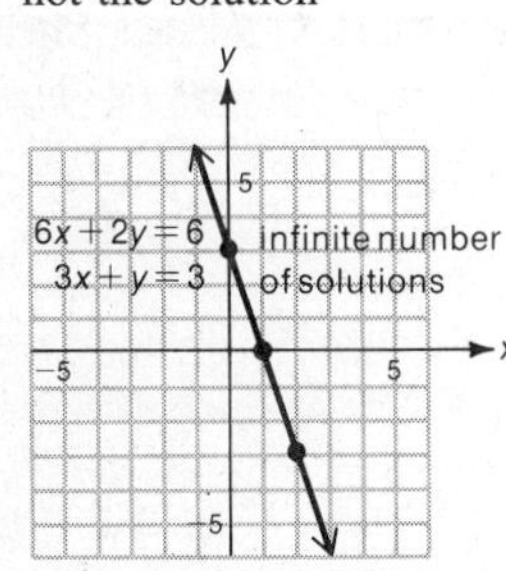

11.

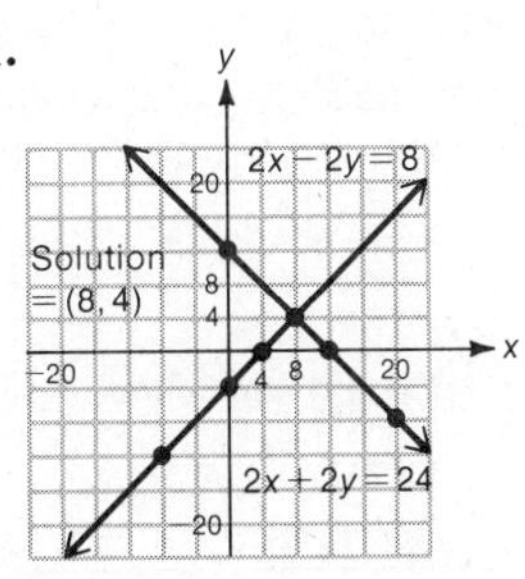

13.

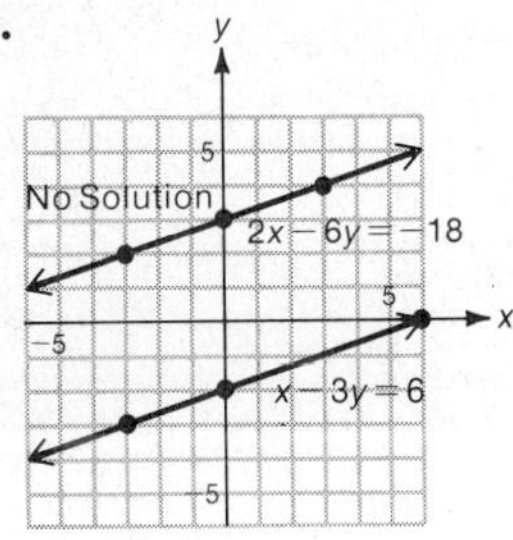

15.

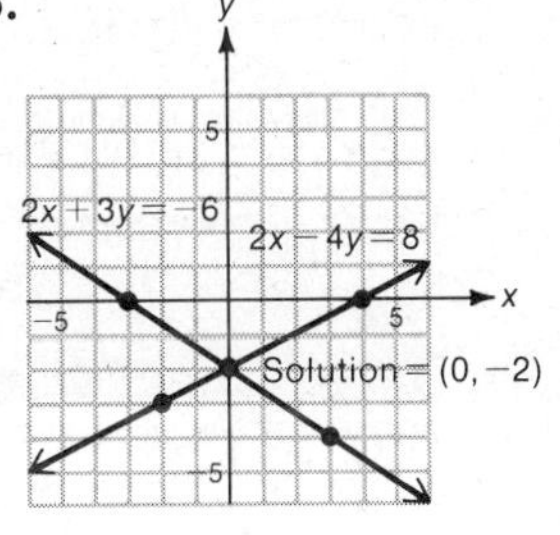

17.

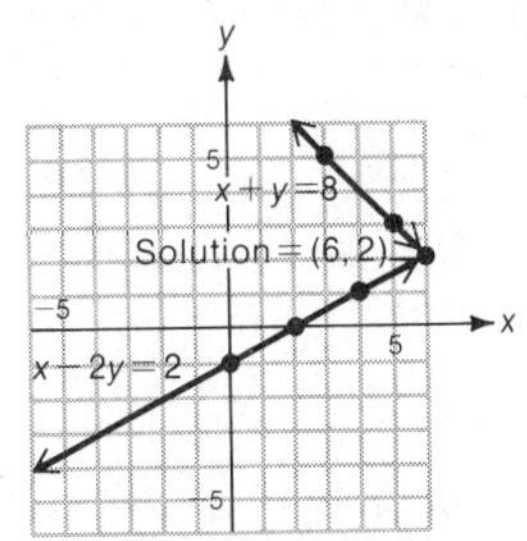

19.

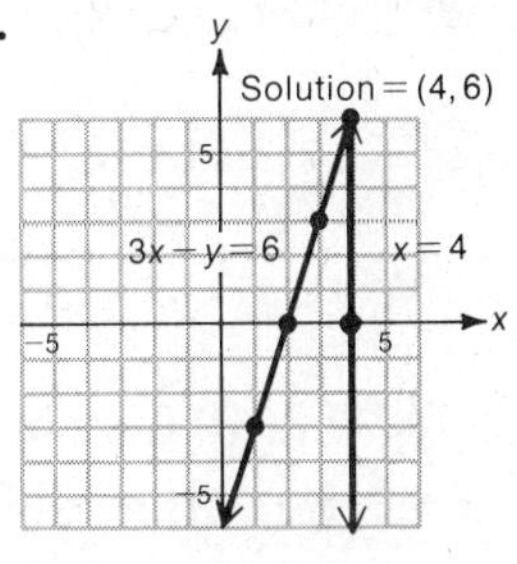

21.

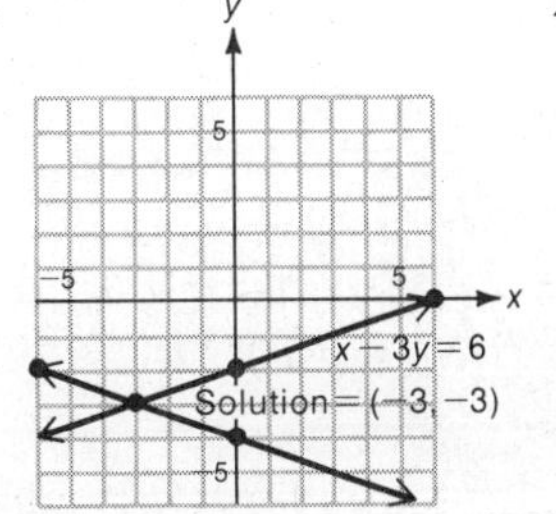

23.

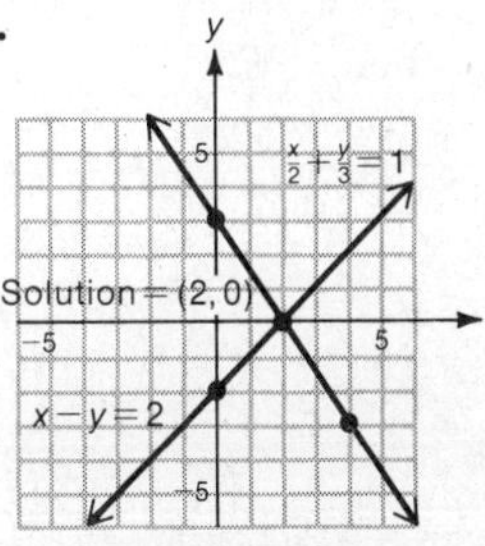

25. consistent, one solution **27.** inconsistent, no solution **29.** dependent, infinite number of solutions **31.** consistent, one solution **33.** dependent, infinite number of solutions **35.** consistent, one solution **37.** consistent, one solution **39.** dependent, an infinite number of solutions **41.** equation, interpret

43.

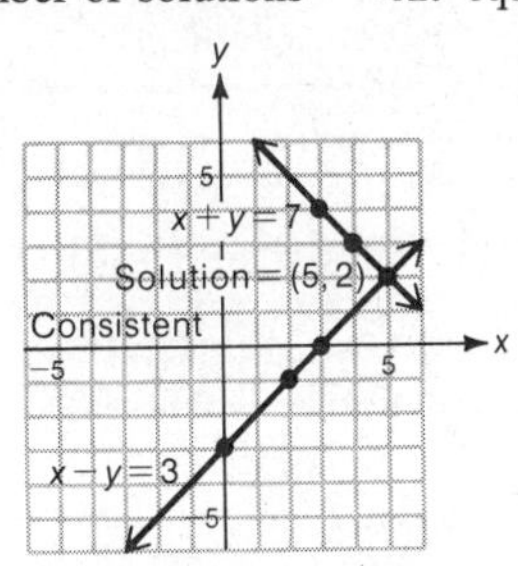

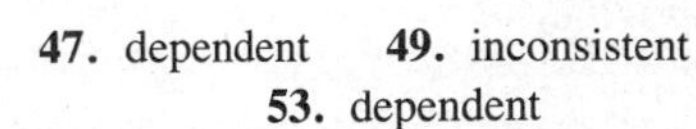

45. inconsistent **47.** dependent **49.** inconsistent

51.

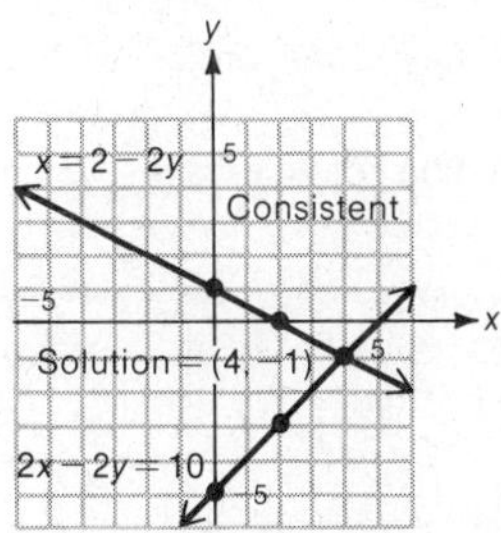

53. dependent

55.

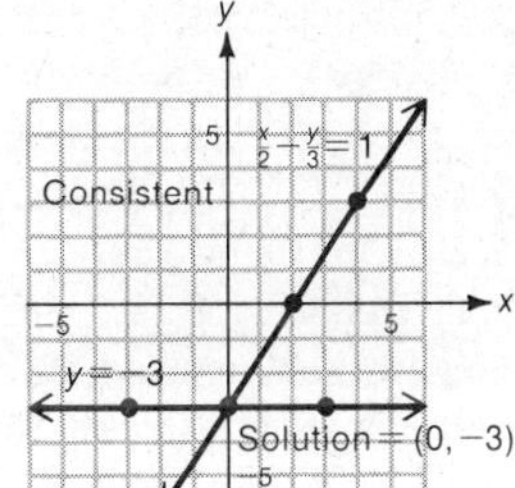

57. dependent

59.

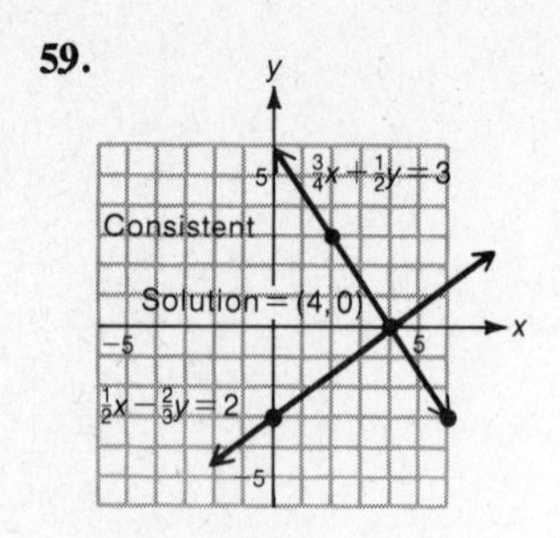

Section 10.2

1. (1, 6) **3.** no solution **5.** $\left(\frac{1}{2}, 1\right)$ **7.** (3, 0) **9.** $\left(-\frac{45}{34}, -\frac{38}{17}\right)$ **11.** (4, 2) **13.** infinite number of solutions **15.** (2, −1) **17.** $\left(-\frac{2}{5}, \frac{12}{5}\right)$ **19.** $\left(\frac{122}{39}, -\frac{110}{39}\right)$ **21.** (6, 2) **23.** $\left(\frac{69}{34}, \frac{31}{34}\right)$ **25.** (3, 1) **27.** $\left(\frac{2}{3}, 4\right)$ **29.** infinite number of solutions **31.** $ax + by = c$, no solution, infinite number of solutions **33.** inconsistent, no **35.** 2.4 gallons of the 7% solution, 3.6 gallons of the 12% solution **37.** \$1000 for the Ford, \$3000 for the BMW **39.** $-\frac{82}{7}$ is the first number, $\frac{146}{7}$ is the second number **41.** \$10 for typesetting, 50¢ per brochure **43.** 6.4 gallons of the 15% solution, 9.6 gallons of the 40% solution **45.** width is 38.7 feet, length is 32.3 feet

Section 10.3

1. (6, 0) **3.** (0, 6) **5.** (3, 5) **7.** (3, 4) **9.** $\left(\frac{17}{6}, \frac{25}{6}\right)$ **11.** infinite number of solutions **13.** $\left(-\frac{44}{7}, \frac{20}{7}\right)$ **15.** no solution **17.** $\left(\frac{8}{15}, -\frac{18}{5}\right)$ **19.** $\left(\frac{1}{5}, -\frac{4}{3}\right)$ **21.** $\left(-\frac{11}{3}, -\frac{4}{3}\right)$ **23.** (2, −5) **25.** (−4, 0) **27.** (−2, 0) **29.** (3, −10) **31.** (1.8, −2.2) **33.** eliminate, eliminate, substitution, addition **35a.** one, one **b.** substitute, other equation **c.** one, one **d.** substituting, equation **e.** ordered pair **f.** check **37.** Gene received 122 votes, Howard received 262 votes **39.** the lot costs \$20,000; the house costs \$35,000 **41.** \$400 at 6.25%, \$1100 at 13% **43.** 9 gallons of the 15% solution; 3 gallons of the 75% solution **45.** 50 mph on the highway, 30 mph on the unpaved road **47.** the faster boat traveled 30 miles, the slower boat traveled 20 miles

CHAPTER 10 Review Exercises

1. *(Section 10.1)*

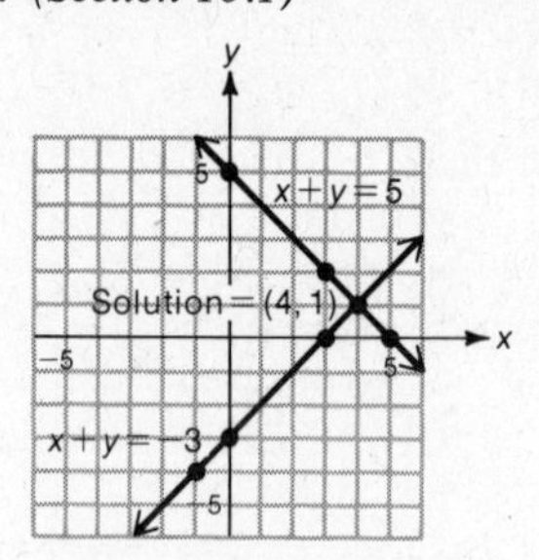

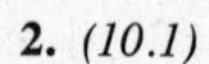

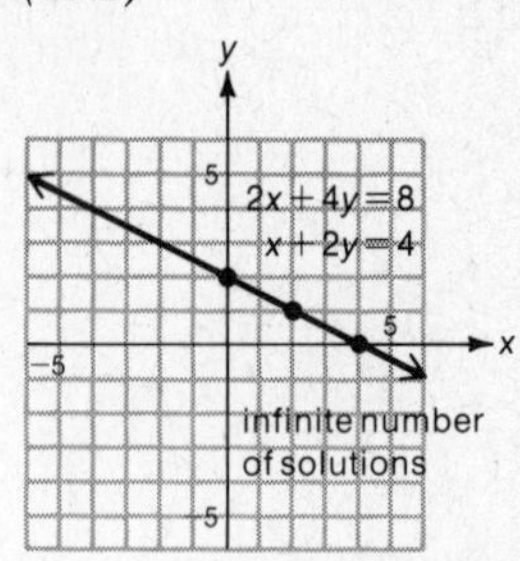

3. *(10.1)*

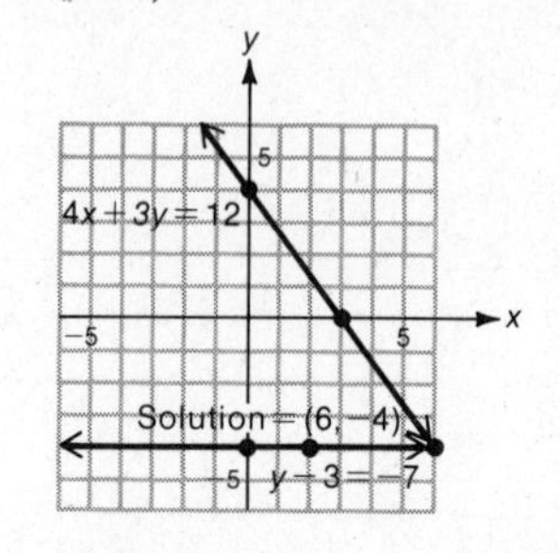

4. $\left(\frac{1}{2}, 4\right)$ *(10.2)* **5.** no solution *(10.2)* **6.** (0, 2) *(10.2)* **7.** (1, 4) *(10.3)* **8.** (1, 1) *(10.3)* **9.** $\left(\frac{1}{2}, -\frac{11}{6}\right)$ *(10.3)* **10.** no solution *(10.1)* **11.** $\left(1, \frac{5}{2}\right)$ *(10.2)* **12.** (−2, 1) *(10.2)* **13.** (5, −1) *(10.3)* **14.** (1.5, −2) *(10.2)* **15.** $\left(\frac{51}{4}, -\frac{13}{2}\right)$ *(10.2)* **16.** 38, 32 *(10.2)* **17.** 10 pounds of pumpkin seeds, 20 pounds of sunflower seeds *(10.2)* **18.** \$15,000 at 15%; \$10,000 at 25% *(10.2)* **19.** Ed's rate was 27 feet per second; Mike's rate was 24 feet per second *(10.3)* **20.** interest rate for the real estate was 12%, interest rate for stocks was 15% *(10.2)*

CHAPTER 11

CHAPTER 11 Pretest

1a. $x^2 - 2x - 15$ **b.** $3x^2 - 13x - 10$ **2a.** $x^2 - 6x + 9$ **b.** $9x^2 + 30x + 25$ **3a.** $x^2 - 16$ **b.** $9x^2 - 1$ **4a.** $2x^5(8x^8 - 9)$ **b.** $x^2(15x^4 - 18x^3 + 10)$ **5a.** $(x - 4)(x - 2)$ **b.** $(x + 8)(x - 3)$ **6a.** $(6x - 7)(x + 1)$ **b.** $2x(3x - 2)(x + 5)$ **7a.** $(x - 4)(x + 4)$ **b.** $(4x - 5)(4x + 5)$ **8a.** $(x + 3)^2$ **b.** $(3x - 5)^2$ **9a.** $(x - y)(a + b)$ **b.** $(x - 1 - 9y)(x - 1 + 9y)$

Section 11.1

1. $6x^2 + 13x + 6$ **3.** $4x^2 + 19x - 5$
5. $5x^2 - 44x - 9$ **7.** $x^2 - 0.1x - 0.06$
9. $2x^2z^2 - 8xz - 10$ **11.** $x^2 + xy - 6y^2$
13. $6x^2 - 11x + 3$ **15.** $x^2 + 8x + 16$
17. $6x^2 - 25x + 25$ **19.** $x^2 - 25$
21. $x^2 + 2xy - 3y^2$ **23.** $9x^2 - 16$
25. $6x^2 - 17xy + 5y^2$ **27.** $5x^2 + 14xy - 3y^2$
29. $18x^2 - 9x - 2$ **31.** $x^2 - 0.4x + 0.04$
33. $x^2 - 0.25y^2$ **35.** $x^2 - \frac{11}{6}x - \frac{5}{3}$
37. $x^2 + 4x + 4$ **39.** $x^2 - 14x + 49$
41. $4x^2 + 4xy + y^2$ **43.** $4x^2 + x + \frac{1}{16}$
45. $25x^2 + 20xy + 4y^2$ **47.** $x^2 + 18x + 81$
49. $x^2 - 6x + 9$ **51.** $x^2 + 14x + 49$
53. $x^2 - 12x + 36$ **55.** $x^2 + 6xy + 9y^2$
57. $x^2 - 10xy + 25y^2$ **59.** $4x^2 - 12xy + 9y^2$
61. $x^2 + 0.4x + 0.04$ **63.** $4x^2 + 2x + \frac{1}{4}$
65. $9x^2 + 0.6x + 0.01$ **67.** $4x^2 + 28xy + 49y^2$
69. $x^4 + 26x^2 + 169$ **71.** $x^2 - 9$ **73.** $25x^2 - 1$
75. $9x^4 - 100$ **77.** $a^2b^2 - 1$ **79.** $x^2 - \frac{9}{16}$
81. $\frac{x^2}{4} - 25$ **83.** square, twice, square
85. $20x^2 - 19x + 3$ **87.** $x^2 + 18x + 81$
89. $x^2 + 6x + 8$ **91.** $9x^2 - 12x + 4$
93. $x^2 - 4$ **95.** $x^2 - \frac{3}{2}x + \frac{9}{16}$ **97.** $x^2 - \frac{1}{4}$
99. $9x^2 - 25y^2$ **101.** $2x^2 + 3x - 35$
103. $x^2 - 16x + 64$ **105.** $6x^6 + x^3y^2 - y^4$

Section 11.2

1. $2(x + 4)$ **3.** $3xy(3x + 5)$ **5.** $5y(5x - 7y)$
7. $5(2x^3 - 5x^2 + 4)$ **9.** $50x^2(x^3 + 2x - 3)$
11. $8xy(2x^2 + x + 3)$ **13.** $19x^2(x + 2y^2)$
15. $6a^2d^2(2a - 3d)$ **17.** $-9b(3b^3 + 2b^2 - 4)$
19. $14xy^6(x^4 - 3x^2y + 2y^2)$ **21.** $6a^2b^2(2a - 3b)$
23. $11a^{10}b^5(4a^4b^2 - 3)$ **25.** $2xy^3(6x^7y^6 + 9x^4y - 10)$
27. $ab(15a^{11} - 8b^{11} + 9)$
29. $-15a^2b^2(2ab^2 + 3a^6b^5 + 1)$
31. $6(3x^3 - 2y^2 - 8x^4)$ **33.** $8xy^2(2x^2 + 3y - 5x^3)$
35. $-5x^2(4x^3 + 2x - 1)$ **37.** $11x^2y^2z(4xy^2 - 5)$
39. $40a^2b^2c^3(bc - 2a)$ **41.** 12 **43.** 13 **45.** 21
47. 3 **49.** 15 **51.** 26 **53.** 77
55. $13x^2(5x - 7)$ **57.** $35xy(2x^2 - 3xy + 4y^2)$
59. $11(11x^4 - 13x^3 + 17)$ **61.** $-12x(4x^2y - 5x + 7y)$
63. greatest common factor **65.** greatest common monomial factor, term, greatest common monomial factor

Section 11.3

1. $(x + 7)(x + 5)$ **3.** $(x - 5)(x - 1)$
5. $(x - 6)(x + 5)$ **7.** $(x - 8)(x - 3)$
9. $(x - 12)(x + 3)$ **11.** $(x + 7)(x - 5)$
13. $(x - 7)(x - 4)$ **15.** not factorable
17. $(5x + 2)(x - 3)$ **19.** $(6x + 5)(x - 1)$
21. $(2x - 5)(x - 4)$ **23.** $(3x + 4)(x - 2)$
25. $(5x - 1)(4x - 3)$ **27.** $(6x + 7)(2x - 5)$
29. $(9x - 1)(4x - 1)$ **31.** $(y + 5)(y - 3)$
33. $(x + 3)(x - 2)$ **35.** $(a - 6)(a + 3)$
37. $(7x - 5)(3x - 2)$ **39.** $3(x - 18)(x + 2)$
41. $3(3x + 4)(3x + 2)$ **43.** $(7x^2 - 1)(x^2 + 10)$
45. $(x - 9y)(x + 2y)$ **47.** not factorable
49. $2(3x + 2)(2x - 5)$ **51.** $x^5(x - 7)(x + 2)$
53. $(x + 7)(x - 6)$ **55.** $(4x^2 + 3)(2x^2 - 3)$
57. $(8x + 5)(x + 1)$ **59.** $(3x + 4)(2x - 5)$
61. $(5x + 3)(3x - 2)$ **63.** $(5x - 8)(2x + 3)$
65. $(5x^2 - 4)(3x^2 - 2)$ **67a.** same, first term
b. product, sum, coefficient **c.** last terms
d. product, same **69.** common monomial
71. $(5x - 2y)(5x + y)$ **73.** $(4s + 3)(s - 2)$
75. $(8x - 3)(2x - 3)$ **77.** $2(a + 5b)(a - 3b)$
79. $(4a^2 - 5b^2)(3a^2 + 2b^2)$ **81.** $(9c + 2)(7c - 5)$

Section 11.4

1. $(x - 3)(x + 3)$ **3.** $(3x - 2)(3x + 2)$
5. $(x - 8)(x + 8)$ **7.** $(3s + 1)(3s - 1)$
9. $(x^2 + 4)(x - 2)(x + 2)$ **11.** not factorable
13. $(9x + 1)(9x - 1)$ **15.** $(5y + 2)(5y - 2)$
17. $(x + 2)^2$ **19.** $(a - 5)^2$ **21.** $(x - 4)^2$
23. not factorable **25.** not factorable
27. not factorable **29.** $(x + 8)^2$ **31.** not factorable
33. $(x + y)(c + d)$ **35.** $(x - 1)(b + a)$
37. $(y + 3)(y^2 + 4)$ **39.** $(y - 3)(y - 3)(y + 3)$
41. $(z - x + 2y)(z + x - 2y)$ **43.** $(2b - c)(3b - 7)$
45. $(x + 4 - 5y)(x + 4 + 5y)$ **47.** $(x + y)(b + c)$
49. $(c - d)(w - z)$ **51.** $(b + 1)(y - 1)$
53. $(x - y)(x + y + 5)$ **55.** $(s - 3t)(4s + 7)$
57. $(x - 3 - 6y)(x - 3 + 6y)$
59. $(y - x - 6)(y + x + 6)$
61. $(x + y)(x - y + 1)$ **63.** perfect squares, twice, perfect squares, additive inverse **65.** common, factor
67. $(3x + 8)^2$ **69.** $(x - a)(y + p)$
71. $2(2x - 3)^2$ **73.** $2(4x + 1)^2$ **75.** $3(x - 5)(x + 5)$

CHAPTER 11 Review Exercises

1. $(2x + 3)^2$ *(Section 11.4)* **2.** $(x - 5)(x + 2)$ *(11.3)*
3. $(x - 6)(x - 5)$ *(11.3)*
4. $(x^2 + 9)(x - 3)(x + 3)$ *(11.4)*
5. $2(x + 7)(x - 1)$ *(11.3)* **6.** $(5x - 1)^2$ *(11.4)*
7. not factorable *(11.4)* **8.** $(b - 2)(b + 2)(b - 2)$ *(11.4)*
9. not factorable *(11.4)* **10.** $2(4x - 1)(4x + 1)$ *(11.4)*
11. $(2x - 1)(2x + 3)$ *(11.3)*
12. $6x^3(3x^2 - 4x + 6)$ *(11.2)*
13. $12x^4(5x^5 + 7x^4 - 8x^3 + 2)$ *(11.2)*
14. $(2x - 3 - 5y)(2x - 3 + 5y)$ *(11.4)*
15. $(2x - 1)(x + 3)$ *(11.3)* **16.** $(4x + 3)(3x - 4)$ *(11.3)*
17. $(6x + 5)(x - 2)$ *(11.3)*
18. $2(x^2 + 11x + 14)$ *(11.2)*
19. $(8x^2 + 1)(8x^2 - 1)$ *(11.4)*
20. $(6x - 7y)(4x + 3y)$ *(11.3)*
21. $3(3x^2 - 14x + 7)$ *(11.2)*
22. $3(5x^4 - 9x^3 - 3)$ *(11.2)*
23. not factorable *(11.3)* **24.** $(x^3 + 2)(x^3 - 2)$ *(11.4)*

25. $(6x - 5)(3x + 2)$ *(11.3)* **26.** $9x^2 - 21x + 10$ *(11.1)* **27.** $4x^2 + 20x + 25$ *(11.1)* **28.** $x^2 - 81$ *(11.1)* **29.** $x^2 - 0.4x + 0.04$ *(11.1)* **30.** $x^2 - x - 72$ *(11.1)* **31.** $\frac{1}{16}x^2 - 4$ *(11.1)* **32.** $x^2 - 0.2x - 0.15$ *(11.1)* **33.** $x^2 - \frac{5}{12}x - \frac{1}{4}$ *(11.1)*

CHAPTER 12

CHAPTER 12 Pretest

1a. 4 **b.** $-y^4$ **2a.** 3.606 **b.** 9.849 **3a.** $\sqrt{22}$ **b.** 9 **4a.** $2x\sqrt{5x}$ **b.** $7\sqrt{11}$ **5a.** $4\sqrt{x}$ **b.** $8x^3$ **6a.** $\frac{\sqrt{3}}{3}$ **b.** $\frac{\sqrt{15}}{10}$ **7a.** $-3\sqrt{3}$ **b.** $19\sqrt{2}$ **8a.** 1, -7 **b.** $\frac{\sqrt{6}}{2}, -\frac{\sqrt{6}}{2}$ **9a.** 0, 5 **b.** $-1, \frac{1}{2}$ **10a.** -3 **b.** $\frac{-2 + \sqrt{6}}{2}, \frac{-2 - \sqrt{6}}{2}$

Section 12.1

1. 2, -2 **3.** 4, -4 **5.** 9, -9 **7.** 1, -1 **9.** 7, -7 **11.** -5 **13.** -1 **15.** x^8 **17.** -2^5 **19.** $-x^{50}$ **21.** -7^{35} **23.** $-a^{10}$ **25.** 5^{35} **27.** 5.657 **29.** 6.856 **31.** 3.742 **33.** 8.944 **35.** -3.606 **37.** -2.646 **39.** positive, negative **41.** radicand, radical **43.** square, given number **45.** number, $\sqrt{n}$

Section 12.2

1. $2\sqrt{2}$ **3.** $2\sqrt{10}$ **5.** $-5\sqrt{5}$ **7.** $2\sqrt{7}$ **9.** $x^2\sqrt{x}$ **11.** $x^{16}\sqrt{x}$ **13.** $3x^2\sqrt{6}$ **15.** $4\sqrt{3}$ **17.** $5\sqrt{2}$ **19.** $6x^2\sqrt{2}$ **21.** $7\sqrt{2}$ **23.** $x^9\sqrt{x}$ **25.** $-x^3\sqrt{x}$ **27.** $-x^{15}\sqrt{x}$ **29.** $5x^2\sqrt{x}$ **31.** $4\sqrt{x}$ **33.** $c^3\sqrt{c}$ **35.** $-3\sqrt{10}$ **37.** 10 **39.** $-7\sqrt{10}$ **41.** $14\sqrt{3x}$ **43.** $-30x^4\sqrt{3}$ **45.** $\sqrt{15}$ **47.** $\sqrt{13x}$ **49.** 19 **51.** 28 **53.** 10 **55.** $3x\sqrt{6x}$ **57.** $10x^6\sqrt{2y}$ **59.** 2 **61.** 8 **63.** $5x^2\sqrt{x}$ **65.** $5x^4\sqrt{3x}$ **67.** $5x^5\sqrt{2}$ **69.** $\frac{2}{3}$ **71.** $\frac{6}{7}$ **73.** $\frac{\sqrt{5}}{3}$ **75.** $\frac{\sqrt{11}}{5}$ **77.** $\frac{5\sqrt{2}}{2}$ **79.** $\frac{\sqrt{6}}{6}$ **81.** $\frac{\sqrt{6}}{3}$ **83.** $\frac{\sqrt{15}}{5}$ **85.** $\frac{\sqrt{2}}{2}$ **87.** $\frac{1}{2}$ **89.** $\frac{\sqrt{x}}{x}$ **91.** $\frac{4x\sqrt{6}}{3}$ **93.** $\frac{\sqrt{10x}}{4x}$ **95.** $\frac{2x^2\sqrt{6}}{3}$ **97.** square root, product **99.** radicand, perfect square factor **101.** numerator, denominator **103.** $\frac{2}{5}$ **105.** $\frac{\sqrt{3}}{2}$ **107.** $\frac{\sqrt{3}}{2}$ **109.** $0.4\sqrt{3}$ **111.** $\frac{\sqrt{2}}{4}$ **113.** $\frac{1}{2}$ **115.** $\frac{\sqrt{7}}{2}$ **117.** $x^7y^{10}z\sqrt{xz}$

Section 12.3

1. $9\sqrt{2}$ **3.** $-2\sqrt{7}$ **5.** $12\sqrt{x}$ **7.** $11\sqrt{13}$ **9.** $-4\sqrt{2}$ **11.** $2\sqrt{3}$ **13.** $-\sqrt{2}$ **15.** $\sqrt{3}$ **17.** $9\sqrt{5}$ **19.** $3\sqrt{2} + 12\sqrt{3} - 14\sqrt{7}$ **21.** radicand **23.** like square roots **25.** $\sqrt{2} + \sqrt{3} + \sqrt{5}$ **27.** $9\sqrt{3}$ **29.** $-\frac{1}{2}\sqrt{2}$ **31.** $\frac{10\sqrt{21}}{21}$ **33.** $3\sqrt{6}$ **35.** $\sqrt{2} + 2\sqrt{3}$ **37.** $7\sqrt{a}$ **39.** $5x^2\sqrt{10}$ **41.** $-6\sqrt{11}$

Section 12.4

1. $4x^2 - 7x - 9 = 0$ **3.** $3x^2 + 3x - 8 = 0$ **5.** $3x^2 - 17x + 2 = 0$ **7.** 3, -3 **9.** 6, -6 **11.** $2\sqrt{3}, -2\sqrt{3}$ **13.** $\sqrt{7}, -\sqrt{7}$ **15.** no real solution **17.** $\frac{1}{2}, -\frac{1}{2}$ **19.** $\frac{2}{3}, -\frac{2}{3}$ **21.** $\frac{\sqrt{10}}{3}, -\frac{\sqrt{10}}{3}$ **23.** $\frac{\sqrt{2}}{4}, -\frac{\sqrt{2}}{4}$ **25.** $\frac{5}{2}, -\frac{5}{2}$ **27.** $\frac{2\sqrt{10}}{5}, -\frac{2\sqrt{10}}{5}$ **29.** no real solutions **31.** $\frac{5\sqrt{3}}{2}, -\frac{5\sqrt{3}}{2}$ **33.** $\frac{4\sqrt{2}}{3}, -\frac{4\sqrt{2}}{3}$ **35.** $\frac{4}{3}, -\frac{8}{3}$ **37.** 1, 5 **39.** $\frac{5}{3}$, 3 **41.** $-\frac{3}{2}, \frac{13}{6}$ **43.** $\frac{2 - 3\sqrt{3}}{3}, \frac{2 + 3\sqrt{3}}{3}$ **45.** $\frac{5 + 7\sqrt{2}}{2}, \frac{5 - 7\sqrt{2}}{2}$ **47.** 12 **49.** $\sqrt{97}$ **51.** 2 **53.** 20 **55.** quadratic **57.** hypotenuse, legs **59.** 4.5 **61.** 5 **63.** 84.84 feet **65.** 8

Section 12.5

1. -4, 2 **3.** -7, -5 **5.** 2, 3 **7.** -5, 2 **9.** 4, -4 **11.** 7 **13.** $\frac{3}{2}, \frac{7}{5}$ **15.** 2, 4 **17.** $-\frac{1}{4}$, 0 **19.** 0, 9 **21.** 0, $\frac{2}{3}$ **23.** -1, 0 **25.** -8, -3 **27.** 2, 6 **29.** 3 **31.** $-\frac{5}{4}, \frac{4}{3}$ **33.** $\frac{6}{5}, -\frac{6}{5}$ **35.** $-\frac{3}{4}$, 2 **37.** product, zero **39.** double root **41.** -2 **43.** $-2, \frac{5}{2}$ **45.** $-\frac{9}{2}$, 0 **47.** 8, 9 **49.** -8, -3 or 3, 8 **51.** length is 8 meters, width is 4 meters **53.** base is 4 inches, height is 10 inches **55.** width is 5, length is 10

Section 12.6

1. 2, 4 **3.** no real solutions **5.** $\frac{5}{2}$, 4 **7.** 1 **9.** no real solutions **11.** $-\frac{4}{3}$, 2 **13.** $-\frac{4}{3}, \frac{4}{3}$ **15.** $\frac{5}{3}$, 0 **17.** $\frac{7 - \sqrt{61}}{6}, \frac{7 + \sqrt{61}}{6}$ **19.** no real solutions

21. $3 + 2\sqrt{3}, 3 - 2\sqrt{3}$ **23.** $3 + 2\sqrt{2}, 3 - 2\sqrt{2}$ **25.** $2 + 2\sqrt{2}, 2 - 2\sqrt{2}$ **27.** $-1 + \sqrt{5}, -1 - \sqrt{5}$ **29.** $-2 - 2\sqrt{2}, -2 + 2\sqrt{2}$ **31.** coefficient, coefficient, constant **33.** standard, *a*, *b*, *c*, **35.** $-2.3, 0.3$ **37.** $-0.3, 1.6$ **39.** no real solutions **41.** 4.9155 meters, 0.9155 meters **43.** width is 5.6535 meters, length is 2.6535 meters **45.** 1.5615 feet, 2.5615 feet

CHAPTER 12 Review Exercises

1. 6 *(Section 12.1)* **2.** -9 *(12.1)* **3.** $2\sqrt{7}$ *(12.2)* **4.** $2x\sqrt{10}$ *(12.2)* **5.** 4 *(12.2)* **6.** $\frac{2\sqrt{3}}{7}$ *(12.2)* **7.** $5\sqrt{3}$ *(12.3)* **8.** $\sqrt{5}$ *(12.3)* **9.** $\frac{\sqrt{2}}{4}$ *(12.2)* **10.** $\frac{\sqrt{15}}{5}$ *(12.2)* **11.** $9\sqrt{11}$ *(12.2)* **12.** $\frac{\sqrt{2}}{5}$ *(12.2)* **13.** $\frac{6}{5}\sqrt{5}$ *(12.3)* **14.** $5x^3\sqrt{3}$ *(12.2)* **15.** $16\sqrt{2}$ *(12.3)* **16.** 6.557 *(12.1)* **17.** 15.876 *(12.2)* **18.** 1.118 *(12.2)* **19.** $-\frac{2}{3}, \frac{5}{2}$ *(12.5)* **20.** $2 - \sqrt{5}, 2 + \sqrt{5}$ *(12.6)* **21.** $\frac{\sqrt{66}}{2}, -\frac{\sqrt{66}}{2}$ *(12.4)* **22.** $-4, 2$ *(12.5)* **23.** 7, -7 *(12.4)* **24.** $0, \frac{5}{4}$ *(12.5)* **25.** no real solutions *(12.6)* **26.** $-\frac{7}{2}, \frac{1}{2}$ *(12.5)* **27.** $-\frac{2}{3}, 1$ *(12.5)* **28.** $\frac{4}{3}, 4$ *(12.4)* **29.** $-8 + \sqrt{70}$, $-8 - \sqrt{70}$ *(12.6)* **30.** 14 yards *(12.4)* **31.** 9, 10 *(12.5)* **32.** height is 4.7015 yards; base is 1.7015 yards *(12.6)* **33.** 10 feet, 14 feet *(12.5)*

CHAPTER 13

CHAPTER 13 Pretest

1a. 18.65 in **b.** $11\frac{5}{6}$ in **2a.** 25.12 ft **b.** 14.444 ft **3a.** 22.5 m^2 **b.** $13\frac{17}{96}$ m^2 **c.** $4\frac{1}{6}$ m^2 **d.** 17.5053 m^2 **4a.** 267.95 m^3 **b.** 64 yd^3 **c.** 16.75 m^3 **d.** 169.56 ft^3

Section 13.1

1. 40 inches **3.** 12 inches **5.** $9\frac{2}{3}$ inches **7.** 22.8 feet **9.** 35.22 feet **11.** 27.004 feet **13.** 21.98 feet **15.** 10.8016 feet **17.** 49.39 yards **19.** 138 yards **21.** 53.1 yards **23.** perimeter **25.** sides, sum, sides **27.** $C = \pi d$ **29.** 84 minutes **31.** 103.62 inches **33.** \$144

Section 13.2

1. 9 ft^2 **3.** 24.6 ft^2 **5.** 38.465 ft^2 **7.** $14\frac{1}{4}$ cm^2 **9.** 15.479572 cm^2 **11.** 14.13 cm^2 **13.** 113.04 cm^2 **15.** 107.64 cm^2 **17.** 48 cm^2 **19.** 159.48 cm^2 **21.** 9.08375 cm^2 **23.** parallelogram, triangle, trapezoid **25.** 20 **27a.** 800 yd^2 **b.** \$7,208 **29.** 50

Section 13.3

1. 120 in^3 **3.** 350.14 in^3 **5.** 523.33 in^3 **7.** 1203.6499 in^3 **9.** 77.872 in^3 **11.** 288.55 in^3 **13.** 1589.625 in^3 **15.** 720 in^3 **17a.** $V = lwh$ **b.** $V = \pi r^2 h$ **c.** $V = \frac{\pi r^2 h}{3}$ **d.** $V = \frac{4\pi r^3}{3}$ **19.** 267.95 cm^3 **21.** 138.16 **23a.** 2400 **b.** 18,000

CHAPTER 13 Review Exercises

1. 46 in *(Section 13.1)* **2.** 28 in *(13.1)* **3.** 120 in *(13.1)* **4.** 185 ft^2 *(13.2)* **5.** 52.7834 ft^2 *(13.2)* **6.** $57\frac{3}{4}$ ft^2 *(13.2)* **7.** 25.748 ft *(13.1)* **8.** 502 in^2 *(13.2)* **9.** 549.64 in^3 *(13.3)* **10.** 94.2 in^3 *(13.3)* **11.** 100.48 in^3 *(13.3)* **12.** 40.54 ft *(13.1)* **13.** 58.55 ft *(13.1)* **14.** 56.7 ft *(13.1)* **15.** 51.81 ft^2 *(13.2)* **16.** 61.6875 ft^2 *(13.2)* **17.** 190 ft^2 *(13.2)* **18.** 234.807 *(13.3)* **19.** 66 *(13.2)* **20.** $102\frac{1}{6}$ *(13.1)*

APPENDIX

1. 10 **3.** 2.5 **5.** 2.5 **7.** 88000 **9.** 10 **11.** 5 **13.** 15,000,000 **15.** 10,000 **17.** 120 **19.** 2.25 **21.** 0.75 **23.** 35 **25.** 3600

INDEX

A

B

C

D

E

F

G

I

L

M

N

O

P

Q

R

S

T

W